ATLAS
OF
Renal and Urinary Tract Cytology and Its Histopathologic Bases

G. BERRY SCHUMANN, M.D.
ASSISTANT PROFESSOR OF PATHOLOGY
DIRECTOR, CYTOLOGY DIVISION
DEPARTMENT OF PATHOLOGY
UNIVERSITY OF UTAH MEDICAL CENTER
SALT LAKE CITY, UTAH

MARK A. WEISS, M.D.
ASSISTANT PROFESSOR OF PATHOLOGY
DEPARTMENT OF PATHOLOGY
UNIVERSITY OF CINCINNATI MEDICAL CENTER
CINCINNATI, OHIO

OF

Renal and Urinary Tract Cytology and Its Histopathologic Bases

PHILADELPHIA J.B. LIPPINCOTT COMPANY TORONTO

The authors and publisher have exerted every effort to ensure that drug selection and dosage set forth in this text are in accord with current recommendations and practice at the time of publication. However, in view of ongoing research, changes in government regulations, and the constant flow of information relating to drug therapy and drug reactions, the reader is urged to check the package insert for each drug for any change in indications and dosage and for added warnings and precautions. This is particularly important when the recommended agent is a new or infrequently employed drug.

 Printed in the United States of America. For information address, J. B. Lippincott Company, East Washington Square, Philadelphia, Penna. 19105.

6 5 4 3 2 1

Library of Congress Cataloging in Publication Data

Schumann, Gilbert Berry,
Atlas of renal and urinary tract cytology and its histopathologic bases.

Bibliography
Includes index.
1. Urine—Examination. 2. Diagnosis, Cytologic. 3. Histology, Pathological. I. Weiss, Mark A., joint author. II. Title. [DNLM: 1. Cytodiagnosis—Atlases. 2. Kidney diseases—Diagnosis—Atlases. 3. Urologic diseases—Atlases. 4. Urine—Cytology—Atlases. WJ17 S392a]
RB53.S42 616.6'0782'0222 80-15820
ISBN 0-397-50443-8

To our mentors, parents and family

CONTENTS

INTRODUCTION

The examination of urine (urinalysis) represents one of the oldest medical laboratory procedures for the evaluation of renal parenchymal and urinary tract disorders. Early popularity was based upon the ease of specimen collection. Although crude procedures, such as smelling, tasting, boiling and adding reagents (acids) were initially described in medical textbooks, it was not until the microscope became available that urinalysis became a valuable method of clinical diagnosis. Fascination with the capabilities of this simple technique frequently led physicians into the habit of examining the urine before the patient.

Historically, the laboratory urinalysis procedure has been divided into *macroscopic* and *microscopic* components. The *macroscopic* urinalysis, as a measure of functional change, involved physicochemical testing for the assessment of color, appearance (turbidity), specific gravity, pH and chemical constituents (glucose, protein, ketones, etc.). The *microscopic* urinalysis involved an interpretation of urine sediment to provide structural (morphologic) evidence of infections, hematuria and inflammation.

In the mid 1960's, the development of reagent-strip physicochemical testing led to a major modernization of the urinalysis laboratory. When properly performed, this provided a cost effective semi-quantitative method for screening and functional monitoring of patients with renal disease. Interest in the microscopic urinalysis declined, because of crude microscopic methods, poor clinico-pathologic correlations and the inability to adapt to mass screening. Requirements for quality control and continuing education widened the gap between the microscopic and macroscopic urinalysis and resulted in routine urinalyses being performed in the clinical chemistry laboratory.

Attempts to improve the urine sediment examination, especially for renal disease, primarily occurred in microscope modifications, urine staining and counting chamber methods. The wet mount (unstained brightfield microscopy) was criticized, and phase-contrast and interference-contrast microscopy were suggested. Supra-vital staining techniques, such as the Sternheimer-Malbin, never gained popularity.

Poor visualization, standardization and quality control continued to be major deficiencies.

While interest in utilizing sediment examinations for renal disease declined, the detection of cancer cells in the sediment using cytologic techniques was rapidly developing. In the early 1950's, the field of diagnostic cytology was detecting and modifying the clinical course of cervical cancer. By using the Papanicolaou staining method, impressive cellular detail could be achieved that correlated with histologic material. Other microscopists, especially in Europe, found the permanent Wright-Giemsa staining method valuable. Unfortunately, cytologists concentrated primarily on neoplastic disease of the lower urinary tract.

With the advent of renal transplantation, the common use of immunosuppressive and chemotherapeutic agents and industrial exposure to nephrotoxic agents and carcinogens, a new approach to the sediment examination was required. Semi-quantitative methods were needed to document chronologic sediment changes.

This atlas on renal and urinary tract cytology provides the histologic bases for urine sediment findings. In addition, it provides a practical diagnostic approach, termed *cytodiagnostic urinalysis,* for the evaluation of urine sediment. *Atlas of Renal and Urinary Tract Cytology and Its Histopathologic Bases* is divided into four parts. Part I briefly discusses the cytologic and histologic techniques used. Part II defines common morphologic entities and criteria for accurate identification. Parts III and IV utilize case material for the demonstration of various types of urine sediment patterns and their differential diagnoses.

Data derived from an accurate urine sediment examination will provide important information for the diagnosis and management of patients with urinary system disease. In the future, with incorporation into the cytodiagnostic urinalysis of more sophisticated techniques, such as transmission and scanning electron microscopy, immunofluorescence, cytochemistry, and chemical analyses of heavy metals, the urine sediment evaluation is sure to regain medical popularity.

PREFACE

Accurate interpretation of morphologic urinary sediment findings is essential for defining and documenting the dynamic changes involving the urinary system during disease. In the past, little information was derived from the urine sediment examination because imprecise techniques were utilized. The authors have evaluated several thousand urine sediments using the Papanicolaou staining technique and found this method far superior in demonstrating the cellular detail required for accurate interpretation and precise diagnosis.

Over the last two years, numerous urine sediments have been examined by a cytologist and nephropathologist to establish a histologic basis. Unlike most traditional textbooks on microscopic urinalysis or urine cytology, this atlas provides correlates of renal parenchymal and lower urinary tract conditions. Numerous photographs demonstrate the remarkable morphologic similarity between exfoliated cells or structures and their appearance in histologic material. While examples of traditional findings of infectious (bacterial, fungal, viral), inflammatory, and neoplastic disease involving the lower urinary tract are presented, special emphasis has been given to the criterion for recognition of entities associated with renal parenchymal disease. Much of the correlation has been achieved through evaluation of renal transplant biopsies and corresponding serial urine sediment examinations. Hopefully this atlas will provide a diagnostic approach for systematic evaluation of the urine sediment and its histologic bases.

G. Berry Schumann, M.D. and Mark A. Weiss, M.D.

ACKNOWLEDGMENTS

The authors are greatly appreciative of the supportive efforts of numerous individuals in preparing this atlas. Special thanks goes to David B. Jones, M.D., Margaret Palmieri, CT(ASCP) and other members of the Cytopathology Laboratory, Upstate Medical Center, State University of New York, Syracuse, New York, for their advice regarding numerous diagnostic cases. Special thanks goes to Janet L. Johnston, CT(ASCP) for her efforts in writing Chapter 1 on cytologic procedures and techniques, and for her superb work in photographing most of the cytologic illustrations. Many thanks goes to the Cytopathology Laboratory at the University of Cincinnati Medical Center for their support and patience while this atlas was being produced. We'd like to thank Sue Zaleski CT(ASCP) from the Carle Clinic Association, Urbana, Illinois, for donating her diagnostic case on transitional cell carcinoma. Also, our thanks goes to Linda Define, CT(ASCP), for donating her case of suspected urinary tract lymphoma. The authors are deeply grateful to Mrs. Barbara Boelscher, Ms. Georgia Zimmer, and Ms. Joyce Turner of the Document Processing Area, Department of Pathology, at the University of Cincinnati Medical Center, Cincinnati, Ohio, for their extreme patience and understanding in typing this manuscript. We wish to thank Consolidated Biomedical Laboratories for funding the technical cost of the photography in this atlas. Naturally, there are several other contributors whose names will not be mentioned in this limited space, but the authors wish to thank them for their sincere efforts in making this atlas possible.

ATLAS
OF
Renal and Urinary Tract Cytology and Its Histopathologic Bases

PART I

MATERIALS AND METHODS

1
CYTOLOGIC PROCEDURES AND TECHNIQUES

SPECIMEN COLLECTION

Urine sediment examinations are usually of spontaneously voided clean-catch urine or following instrumentation (catheterization). If catheterized urine is being submitted, it should be noted on a requisition form. Because excoriation of epithelial surfaces may occur during instrumentation, this notation is imperative. Detailed information regarding specimen collection is given in standard textbooks.

CELLULAR PRESERVATION

Morphological detail of cellular elements will be preserved if the specimen is submitted to the laboratory within a 2-hr period. Refrigeration (2–8°C) will minimize cellular degeneration up to 48-hr period. It should be noted that hyaline and granular casts will dissolve with prolonged unfixed storage. When it is anticipated that a period of longer than 48-hr will occur before the urine specimen is submitted to the laboratory, an appropriate fixative should be used. The use of equal volumes of urine and 50% alcohol, Saccomanno's or Mucolexx fixative is recommended. Alcohol-fixed urine sediments should be stored in the refrigerator until time for preparation. A major advantage of the Saccomanno and Mucolexx fixatives is that the specimen can be stored at room temperature.

PREPARATORY METHODS

The material and methods used for preparation of urine sediment are given in standard textbooks. Methods of cell recovery vary to improve cellular yield. The two techniques that have proved of most value are the cytocentrifugation and membrane filter techniques. Following notations of volume, color, appearance (turbidity or cloudiness), specific gravity, and reagent-strip testing, the urine is initially centrifuged 10 min at 1200 rpm. The visual examination of the urine specimen is useful in determining which technique should be utilized. If *no* sediment button is visible following centrifugation, the membrane filter technique is employed (Fig. 1-1). If a sediment button *is* observed, the supernatant should be carefully removed leaving a 1- to 3-ml total volume of supernatant and sediment. Cytocentrifugation is then performed (Fig. 1-2). Following

FIG. 1-1
MEMBRANE FILTER METHOD

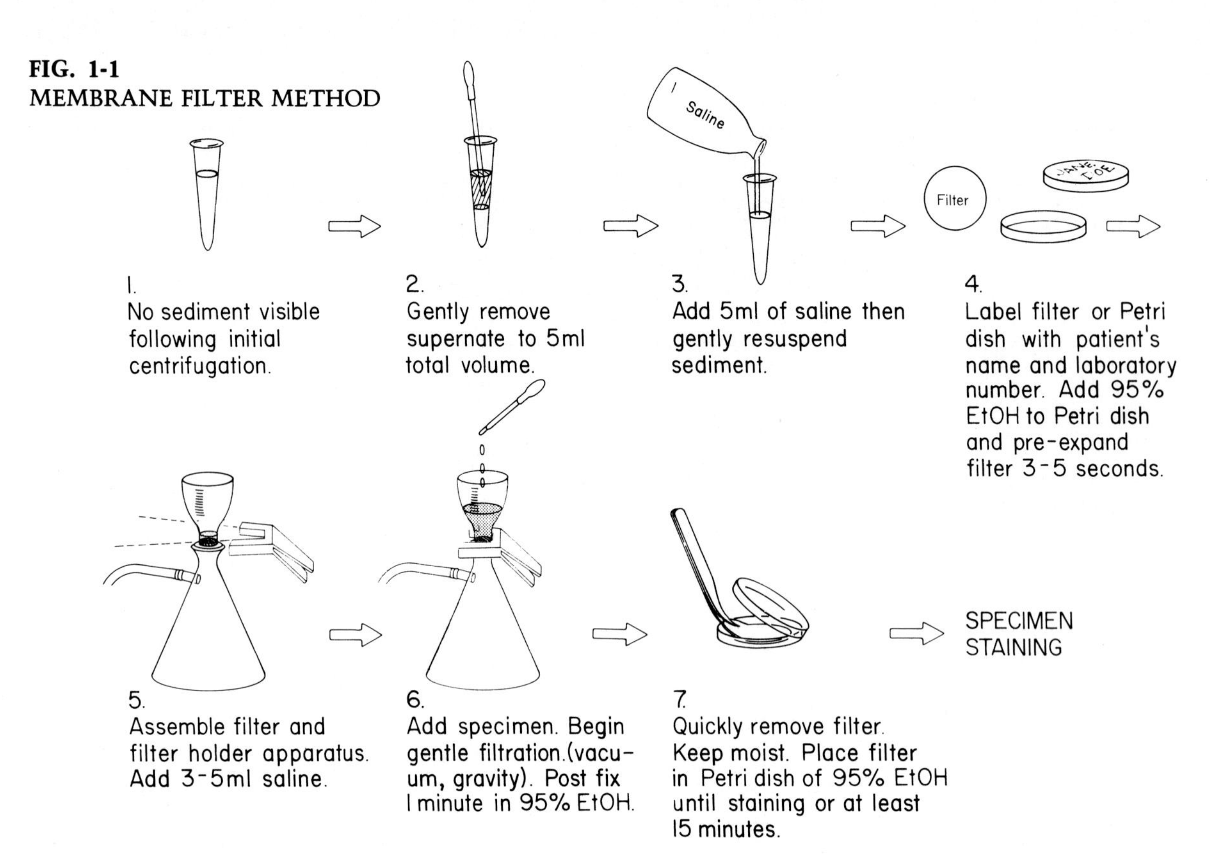

FIG. 1-2
CYTOCENTRIFUGATION METHOD

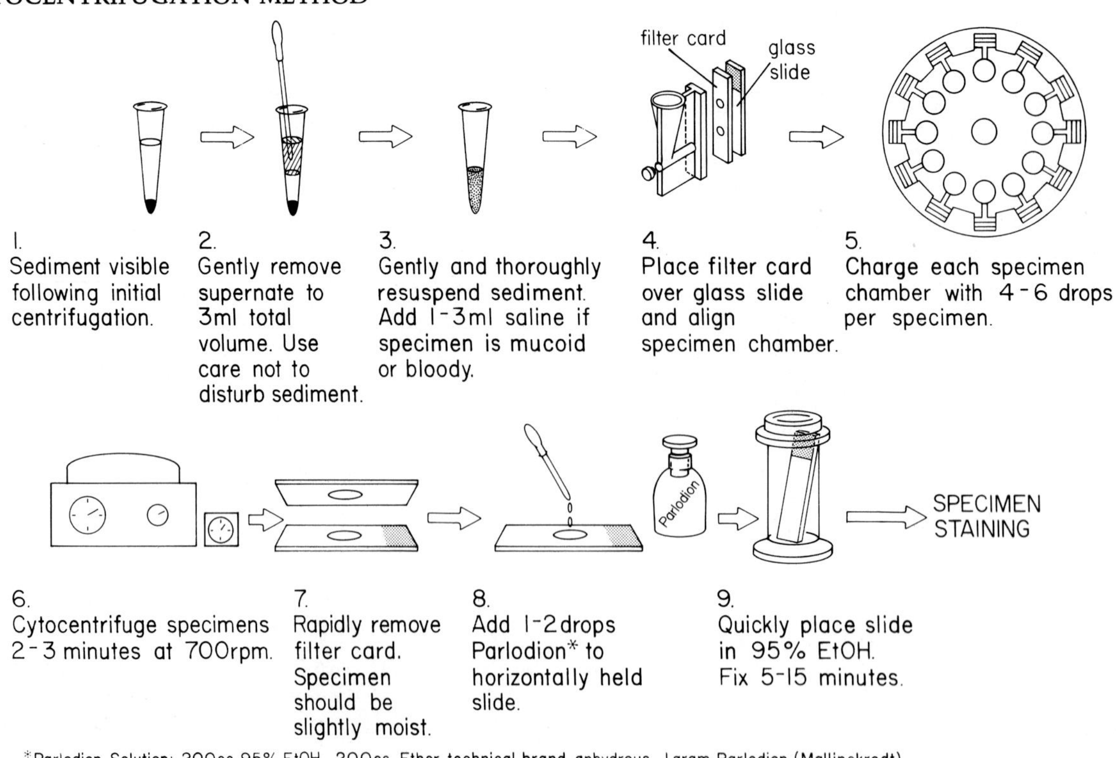

*Parlodion Solution: 200cc 95% EtOH, 200cc Ether technical brand anhydrous, 1 gram Parlodion (Mallinckrodt)

TABLE 1-1
STAINING METHOD OPTIONS FOR URINE SEDIMENT

STAIN	PURPOSE
PAPANICOLAOU STAIN (MANUAL, AUTOMATED, RAPID METHODS)	Routine diagnostic stain (multipurpose) – lends visibility to all of urine sediment content
TOLUIDINE BLUE	Supravital stain for nuclear: cytoplasmic ratio
METHYLINE BLUE	Supravital stain for nuclear: cytoplasmic ratio
PERIODIC ACID-SCHIFF (PAS)	Stain for fungi, glycogen, mucoproteins, and polysaccharides
GROCOTT'S SILVER METHENAMINE	Fungal organism stain
GRAM	Bacterial identification-classification
ACID-FAST	Identification of acid-fast bacilli
WRIGHT-GIEMSA	Blood cell identification Nuclear: cytoplasmic ratio evaluation
PEROXIDASE	Differentiation of leukocytes from epithelial cells
METHYL GREEN-PYRONINE	Cytoplasmic and nucleolar RNA evaluation
MUCICARMINE	Intracellular mucin production identification
OIL RED O (AIR DRY SLIDE)	Lipid identification
MELANIN (FONTANA-MASSON)	Intracytoplasmic melanin identification
PRUSSIAN BLUE	Iron and hemosiderin identification
GOMORI'S URATE	Identification of urate crystals

urine preparation, the sediment should be properly stained. Various staining methods are available (Table 1-1) and are extremely useful for the demonstration and differentiation of diagnostic entities.

STAINS USED FOR EVALUATING URINE SEDIMENT

Staining is used to further characterize morphological structures in urine sediment. The Papanicolaou staining method gives the best cellular detail. In general, staining of urine sediment allows one to (1) identify and classify organisms and crystals, (2) identify and type malignant cells, (3) differentiate hematopoietic cells from epithelial cells, (4) identify intra- or extracellular pigment (iron, melanin), and (5) differentiate cytoplasmic vacuoles (fat, mucin). Table 1-1 contains a summary of the staining reactions commonly used in evaluating urine sediments.

Although organisms can be identified in unstained urine sediment preparations, accurate identification is better achieved using staining procedures. With the Papanicolaou stain, bacterial organisms can be classified as coccal, bacillus, and pleomorphic forms, but the Gram stain is used to classify the organisms as gram-positive or gram-negative (Fig. 1-3). Although rarely occurring in urine sediments, tuberculous bacilli can be identified with the acid-fast stain (Fig. 1-4). Fungal organisms can be identified as spores (yeast) and hyphal forms with the Papanicolaou staining method. By using the periodic acid-Schiff (PAS) (Fig. 1.5) or Grocott's (Fig. 1-6) stains, fungi can be readily identified as obscuring inflammation.

Crystals are best appreciated using polarized microscopy. Cytochemical staining has limited value in the urine sediment examination, but special stains may be useful for the identification of crystalline material in histologic sections (Fig. 1-7).

Supravital (wet mount) stains, such as toluidine blue (Fig. 1-8) and methylene blue (Fig. 1-9), allow rapid identification of important cellular features of malignancy, particularly abnormal nuclear:cytoplasmic ratios. Nuclear enlargement, nuclear membrane indentations, hyperchromasia, chromatin clumping and clearing, and nucleoli can be appreciated with these supravital stains. Their use should be limited to the preliminary evaluation of urine specimens.

The Papanicolaou stain is the stain of choice in demonstrating pathologic casts, renal tubular epithelial cells, and malignant cells. This transparent stain is primarily used to better appreciate chromatin texture, nucleolar characteristics, and cytoplasmic differentiation (Fig. 1-10). The texture and cellular composition of casts are also readily identified.

Although the Wright and Wright-Giemsa stains can be used to identify malignant cells, they are most commonly used in the identification of hemopoietic cells such as lymphocytes, granulocytes, and lymphoreticular cells (Fig. 1-11). When a permanent stain, such as the Wright stain, cannot be employed, the peroxidase stain has been suggested as a method to distinguish nonsegmented neutrophils and monocytes from renal epithelial and urothelial cells (Fig. 1-12). Dark-brown peroxidase-positive granules help identify neutrophils and monocytic leukocytes. The methyl green-pyronine stain is used to identify lymphoreticular cells such as plasma cells and lymphocytes (Fig. 1-13).

Accurate identification of intra- and extracellular pigment may be of diagnostic importance. The prussian blue reaction (Fig. 1-14) is used to demonstrate cellular iron or hemosiderin. With the Fontana-Masson (melanin) stain, malignant melanoma cells can be identified and differentiated from undifferentiated carcinoma (Fig. 1-15).

Differentiation of cytoplasmic vacuoles containing fat and mucin from degenerative vacuoles is best appreciated using a special stain technique. The mucicarmine stain (Fig. 1-16) is valuable in distinguishing mucin containing vacuoles from those containing fat or resulting from hydropic degeneration. The oil red O fat stain is used to verify lipid-containing vacuoles (Fig. 1-17).

BIBLIOGRAPHY

Bradley M, Schumann GB, Ward PCJ: Examination of urine. Davidsohn I, Henry JB (eds): Todd-Sanford Clinical Diagnosis and Management by Laboratory Methods, 16th ed. Philadelphia, WB Saunders, 1979

Kunin CM: Detection, Prevention and Management of Urinary Tract Infections, 2nd ed. Philadelphia, Lea & Febiger, 1974

Schumann GB: The Urine Sediment Examination. Baltimore, Williams & Wilkins, 1980

Tweeddale DN: Urinary Cytology. Boston, Little, Brown, 1977

Voogt HJ, Rathert T, Beyer-Boon ME: Urinary Cytology. Phase Contrast Microscopy and Analysis of Stained Smears. New York, Springer-Verlag, 1977

FIG. 1-3
GRAM STAIN. Violet gram-positive cocci and red gram-negative rods. (× 1000).

FIG. 1-4
ACID-FAST STAIN. Red acid-fast positive tuberculous bacilli. (× 1000).

FIG. 1-5
PERIODIC ACID–SCHIFF (PAS) STAIN. Magenta-positive *Candida* species. Note neutrophilic background with PAS-positive cytoplasm. (× 1000).

FIG. 1-6
GROCOTT'S SILVER METHENAMINE STAIN. Dark brown to black organisms (*Candida* species). (× 1000).

FIG. 1-7
GOMORI'S URATE STAIN. Note black crystalline material. (× 400).

FIG. 1-8
TOLUIDINE BLUE SUPRAVITAL STAIN. Transitional cell carcinoma. (× 1000).

FIG. 1-9
METHYLENE BLUE SUPRAVITAL STAIN. Transitional cell carcinoma. (× 1000).

FIG. 1-10
PAPANICOLAOU STAIN. Transitional cell carcinoma. (× 1000).

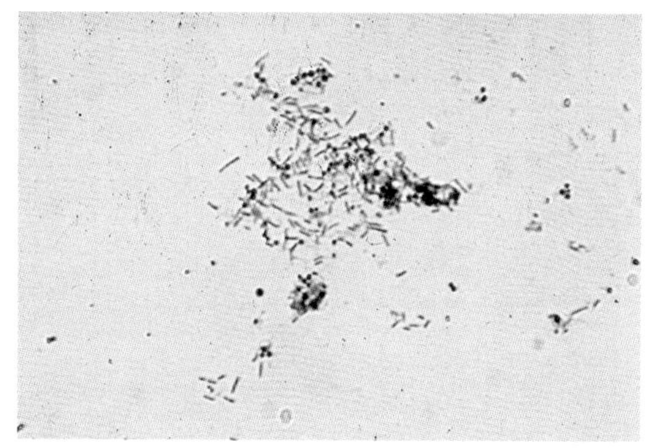

FIG. 1-3

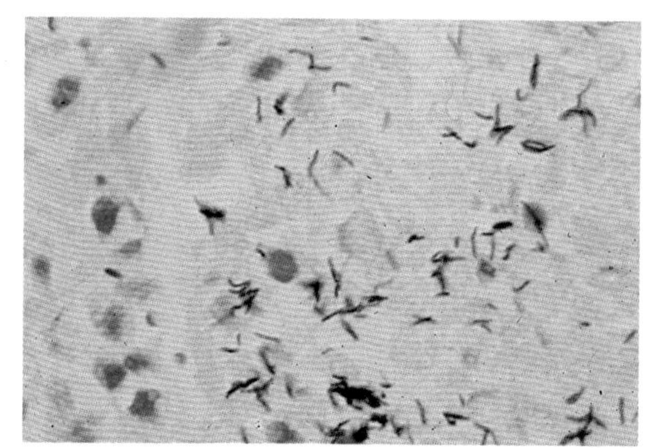

FIG. 1-4

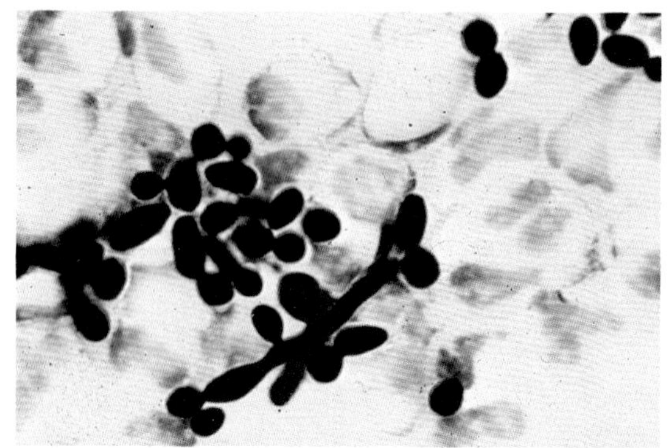

FIG. 1-5

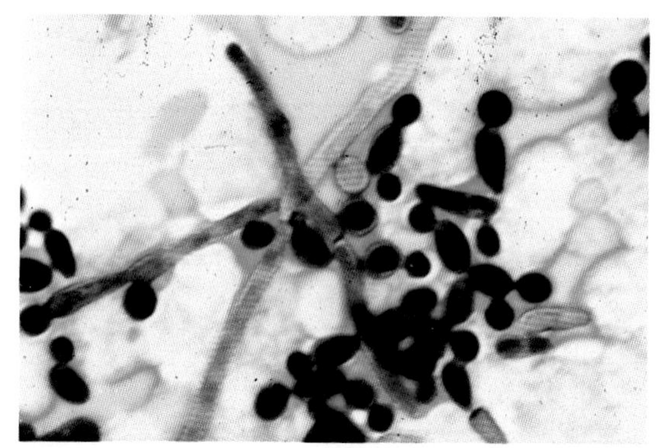

FIG. 1-6

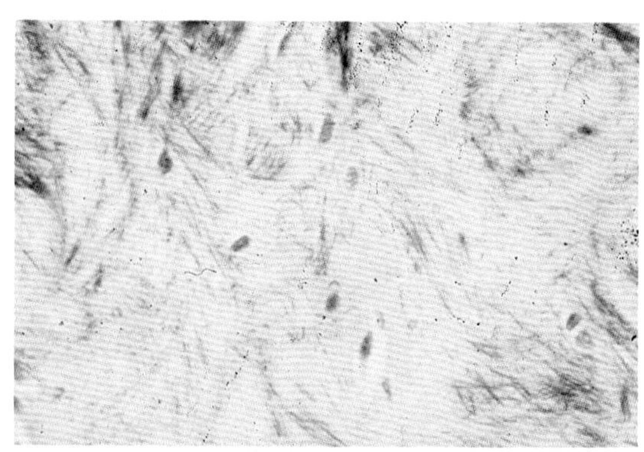

FIG. 1-7

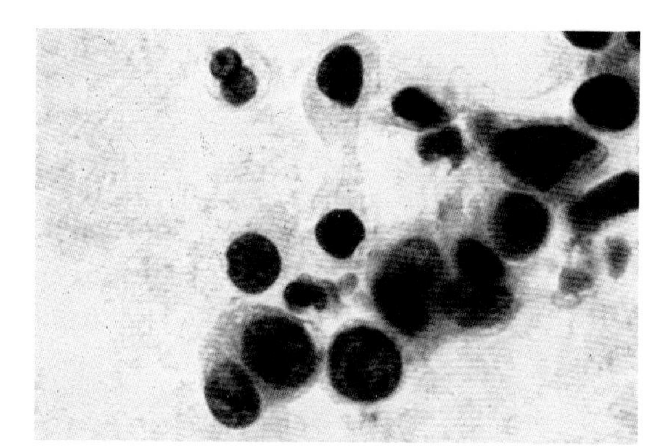

FIG. 1-8

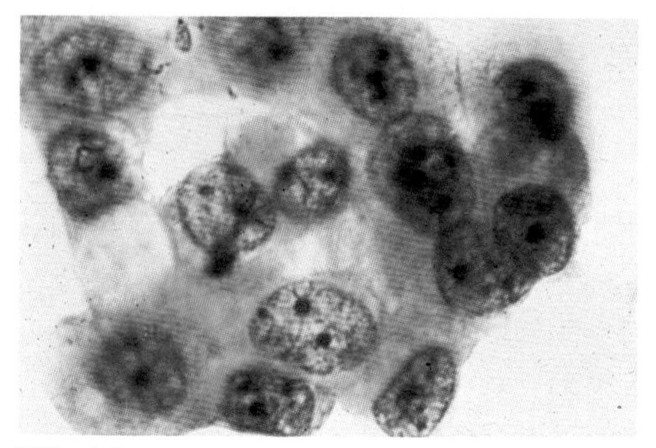

FIG. 1-9

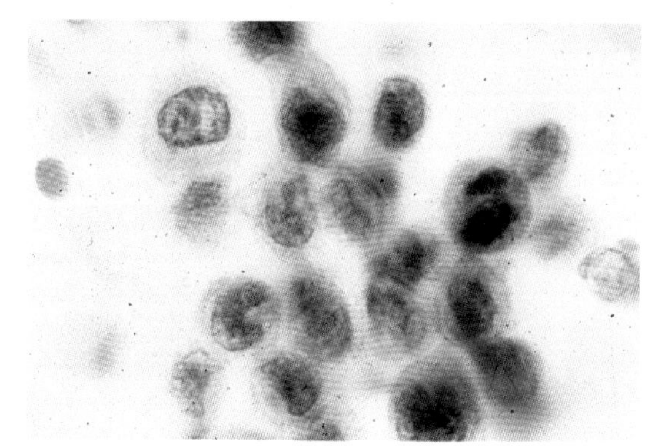

FIG. 1-10

FIG. 1-11
WRIGHT'S STAIN. Hematopoietic cells. (× 1000).

FIG. 1-12
PEROXIDASE STAIN. Peroxidase-positive cytoplasmic granules in neutrophils. (× 1000).

FIG. 1-13
METHYL GREEN-PYRONINE STAIN. Plasma cells and lymphocytes with magenta-positive cytoplasm. (× 1000).

FIG. 1-14
PRUSSIAN BLUE (FE) STAIN. Positive-blue iron granules in renal tubular epithelium. (× 100).

FIG. 1-15
FONTANA-MASSON (MELANIN) STAIN. Brown- to black-positive intracytoplasmic granules in malignant melanoma. (× 400).

FIG. 1-16
MUCICARMINE STAIN. Red-positive cytoplasmic mucin in epithelial cells. (× 400).

FIG. 1-17
OIL RED O (FAT) STAIN. Waxy casts containing renal cells with positive-red-orange cytoplasmic fat globules. (× 1000).

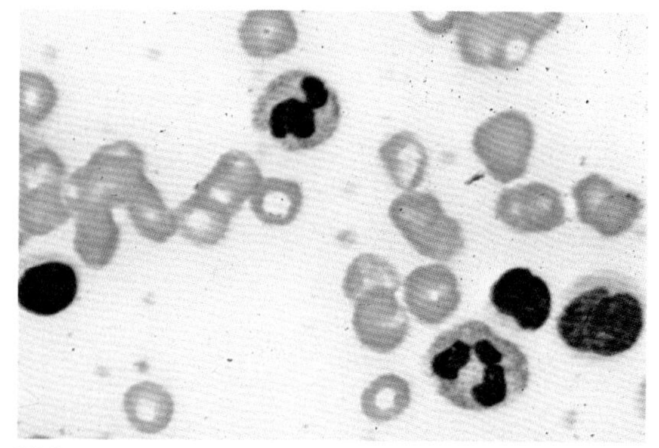

FIG. 1-11

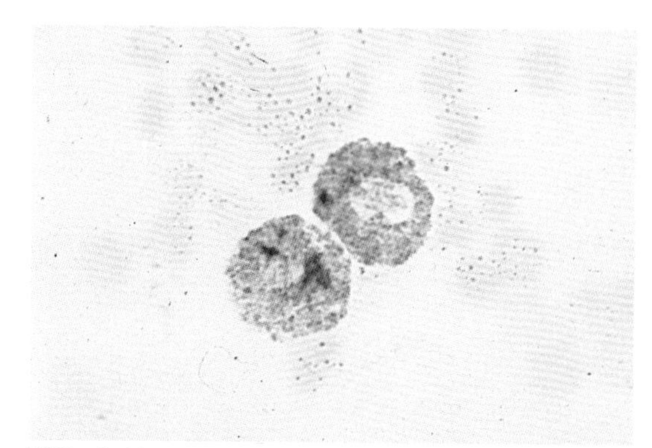

FIG. 1-12

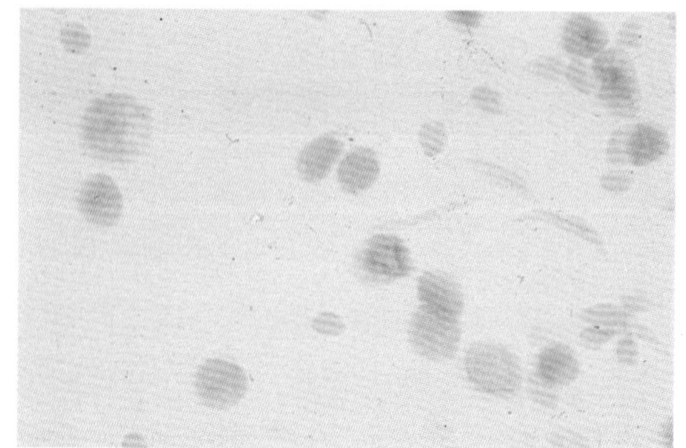

FIG. 1-13

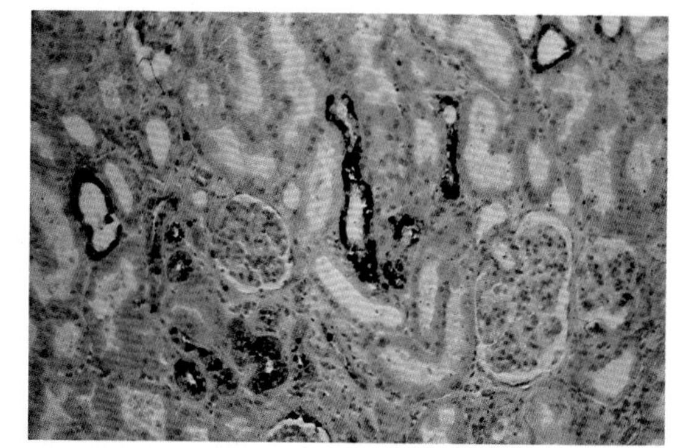

FIG. 1-14

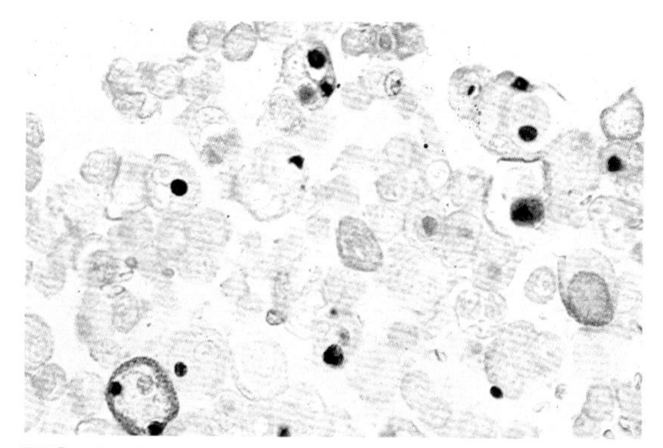

FIG. 1-15

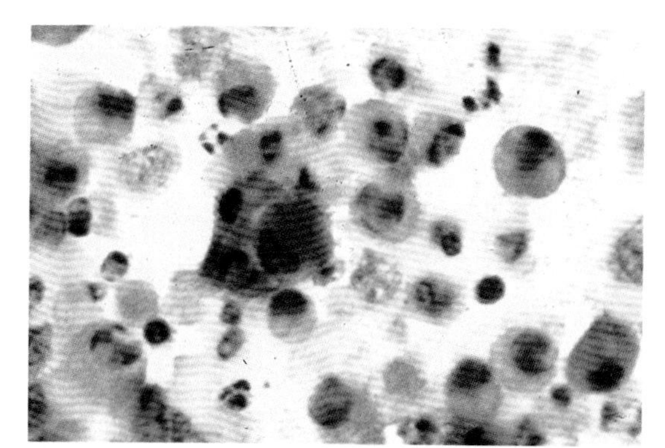

FIG. 1-16

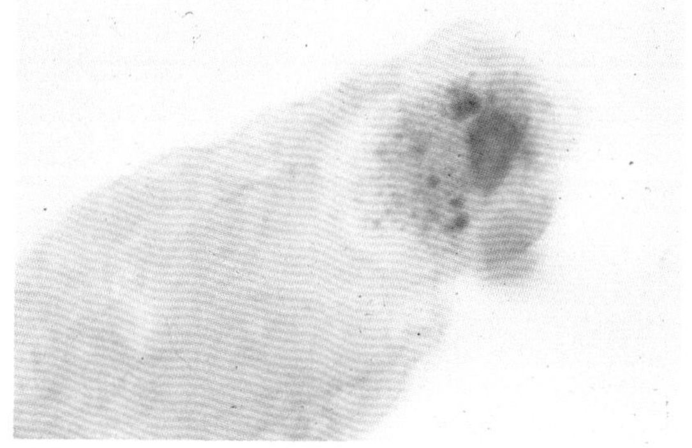

FIG. 1-17

2
HISTOPATHOLOGIC PROCEDURES AND TECHNIQUES

HANDLING AND PROCESSING FOR LIGHT MICROSCOPY
SPECIAL STAINS
IMMUNOFLUORESCENT TECHNIQUES
ELECTRON MICROSCOPIC TECHNIQUES

HANDLING AND PROCESSING FOR LIGHT MICROSCOPY

Histologic material presented in this atlas was obtained from specimens submitted to the surgical pathology laboratory. These specimens submitted for histological examination were routinely processed, cut, and stained in the general histopathology laboratory. Renal tissue was obtained by percutaneous needle biopsy in the majority of cases, although material from nephrectomy and autopsy specimens was also used, and provided a valuable source for tissue imprints. Needle biopsies were routinely divided for light microscopy, immunofluorescence, and electron microscopy. Renal tissue for light microscopy was fixed in Mossman's if obtained by needle biopsy. Nephrectomy and autopsy specimens, as well as urinary tract biopsies, were fixed in neutral buffered 10% formalin. Renal biopsies were sectioned at 2 to 3 μ.

SPECIAL STAINS

In addition to the standard hematoxylin and eosin stain, Masson's trichrome, Fraser-Lendrum, and Jones' methenamine silver stains were commonly used on renal tissue. The Masson's trichrome stain is useful for the evaluation of the interstitial compartment, particularly when assessing fibrosis and edema, the demonstration of deposits (immune complex and fibrin), and the accentuation and characterization of renal tubular casts, that is, hyaline, coarse versus fine granular, erythrocytic versus blood, and so forth. The staining reactions are as follows: collagen and basement membranes – blue; immune complexes and fibrin – red; cytoplasm and muscle fibers – red; and nuclei – black. With the Fraser-Lendrum stain, fibrin can be demonstrated within vascular and glomerular deposits as well as in casts. Fibrin, keratin, and some cytoplasmic granules stain red, and erythrocytes stain orange. Collagen and basement membranes stain blue. The Jones' methenamine silver stain delineates basement membrane material, including glomerular and tubular basement membranes and mesangial matrix. Basement membrane deformities, for example, thickening, splitting, and disruption, are more accurately defined with this technique. Basement membranes and reticulum fibers stain black, nuclei stain blue, and cytoplasm and collagen stain pink to orange. Additional information for light microscopic evaluation can be found in the bibliography list at the end of the chapter.

IMMUNOFLUORESCENT TECHNIQUES

Immunofluorescence was routinely performed on all renal biopsies, employing standard methodologies.[2] Fluorescein isothiocyanate (FITC) conjugated monospecific antisera to human immunoglobulin heavy chains (G, A, M); C3, and fibrin/fibrinogen were purchased commercially. Additional information detailing immunofluorescent techniques is given in standard textbooks.

ELECTRON MICROSCOPIC TECHNIQUES

Tissue for electron microscopy was fixed in 2% gluteraldehyde, rinsed in cacodylate buffer, postfixed in osmium tetroxide, dehydrated in graded alcohols, and embedded in plastic. Ultrathin sections were stained with uranyl acetate and lead citrate. Alternative methods for electron microscopy can be obtained in standard textbooks.

BIBLIOGRAPHY

Smith RD, Weiss M: Renal Biopsy: Technical Aspects of Light Microscopic Interpretation, Chicago, Educational Products Division, American Society of Clinical Pathologists, 1979

Smith RD, Weiss, M: Renal Biopsy: Technical Aspects of Immunofluorescence and Electron Microscopy. Chicago, Education Products Division, American Society of Clinical Pathologists, 1979

Williams G: Color Atlas of Renal Diseases, Chicago, Year Book Medical 1973

PART II

ENTITIES

3
BACKGROUND

As an initial step in the urine sediment evaluation, a low-power microscopic assessment of the background should be made, including cellular and noncellular entities. The terms ***cellularity*** *and* ***background*** *are not synonymous. Depending upon the amount of noncellular background, the urine is generally described as clear, slightly dirty, or dirty. The following list shows common noncellular entities:*

AMORPHOUS MATERIAL
- *Proteinaceous debris*
- *Pigmented debris*
- *Detritus*
- *Necrotic debris*

CRYSTALS
MUCUS
FIBRIN
CONCRETIONS
LIPID
ARTIFACTS

CELLULARITY

Cellularity includes hematopoietic, inflammatory, and epithelial cells and is graded mild, moderate, or marked. A clean background may be present with varying degrees of cellularity (Figs. 3-1 and 3-2), although with marked inflammation or epithelial exfoliation there is usually an associated dirty background (Fig. 3-3).

CELLULAR DEGENERATION, NECROSIS, AND DEBRIS

Cellular degeneration and necrosis contribute to the cellular and noncellular background. An amorphous, proteinaceous background should not be attributed to cellular degeneration or necrosis. This type of dirty background is frequently observed with proteinuria (Fig. 3-4).

As cells degenerate, the amount of cellular debris increases (Fig. 3-5). The presence of a clean or slightly dirty background with numerous degenerated or necrotic epithelial cells indicates rapid exfoliation following acute injury (Fig. 3-6). Dark, granular background material, representing stainable nuclear debris (most frequently disintegrating leukocytes), is termed **detritus** (Fig. 3-7). Detritus in the urine sediment is seen with marked acute inflammation or abscess formation. Necrotic debris is the result of lysis of epithelial and inflammatory cells and includes nuclear and cytoplasmic material. It is frequently present during marked inflammation, urothelial ulceration, and renal parenchymal ischemic necrosis (Fig. 3-8).

ORGANISMS

Organisms contribute to the characteristics or urinary sediment background. Bacterial organisms should be distinguished from amorphous debris or amorphous crystalline material (Fig. 3-9). Further discussion of urinary organisms is found in Chapter 5.

CRYSTALS

Amorphous urates and phosphates represent a common cause of particulate matter in sediment backgrounds (Figs. 3-10 and 3-11). The color of the amorphous material is usually the same color as the crystalline structures. Uric acid crystals are plate-like and golden-brown. Triple phosphate crystals have a coffin-lid appearance and are purple. Calcium oxalate crystals are characterized by their centrally located "X" formation (Fig. 3-12). The presence of calcium oxalate in the urine sediment is occasionally associated with severe renal parenchymal disease such as severe acute tubular necrosis, ethylene glycol poisoning, and methoxyfluorane exposure (Fig. 3-13).

FIBRIN

A poorly described entity in urine sediment are fibrils characteristic of fibrin. Fibrin is often associated with fresh erythrocytes and indicates a urinary system bleed. Aggregated fibrin will appear as a fibrillary network (Fig. 3-14). The identification of embedded epithelial cells allows localization of the site of bleeding (Fig. 3-15).

MUCUS

The significance of mucous threads in urine sediments is not well understood. They should be distinguished from fibrin (Fig. 3-16). Mucus commonly represents vaginal contamination but has also been associated with cystitis and degenerative or inflammatory conditions involving the renal parenchyma.

CONCRETIONS

Concretions are occasionally found in urine following prostatic massage. They are characterized by a laminar outer rim and a central core (Fig. 3-17). Concretions should not be confused with renal cast material.

BIBLIOGRAPHY

Bradley M, Schumann GB, Ward PCJ: Examination of urine. In Henry, JB (ed): Todd, Sanford, Davidsohn: Clinical Diagnosis and Management by Laboratory Methods, 16th ed. Philadelphia, WB Saunders, 1979

Monte-Verde D, Nosanchuk JS, Rudi MA, et al: Unknown crystals in urine. Lab Med 10:299–302, 1979

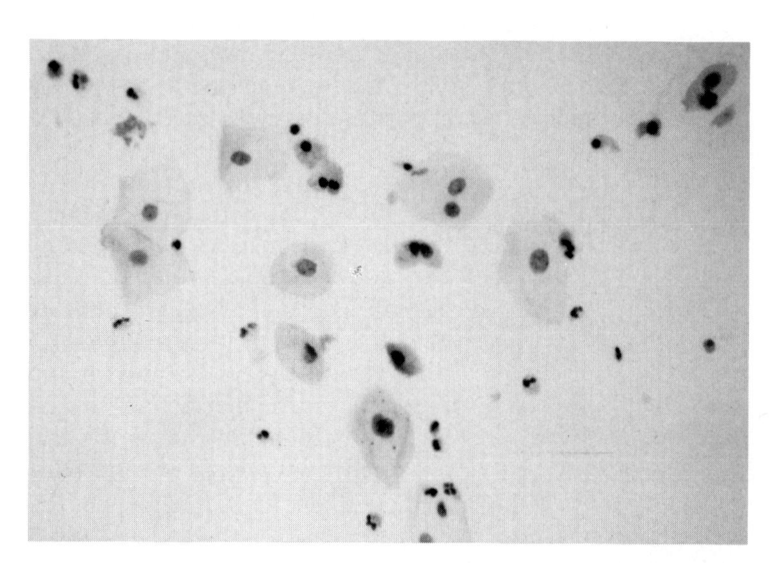

FIG. 3-1
URINE SEDIMENT WITH CLEAN BACKGROUND, MILD INFLAMMATION, AND UROTHELIAL CELLS. (× 400)

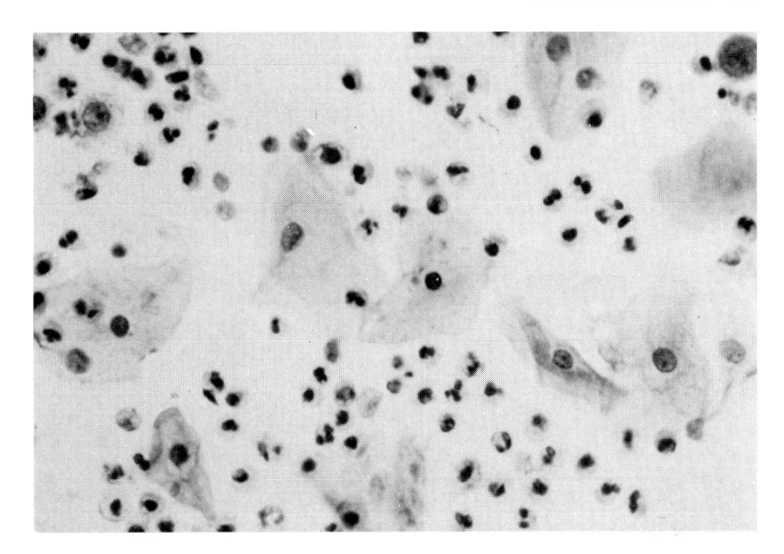

FIG. 3-2
URINE SEDIMENT WITH CLEAN BACKGROUND, MODERATE INFLAMMATION, AND SQUAMES. (× 400)

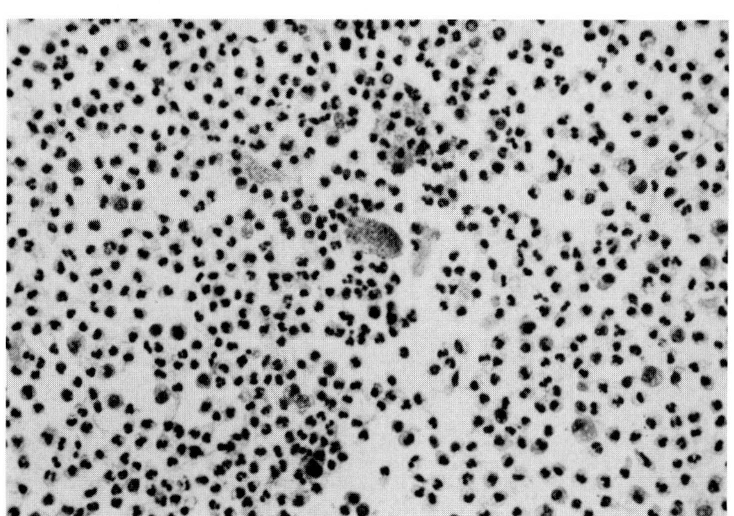

FIG. 3-3
URINE SEDIMENT WITH SLIGHTLY DIRTY BACKGROUND AND MARKED ACUTE INFLAMMATION. (× 250)

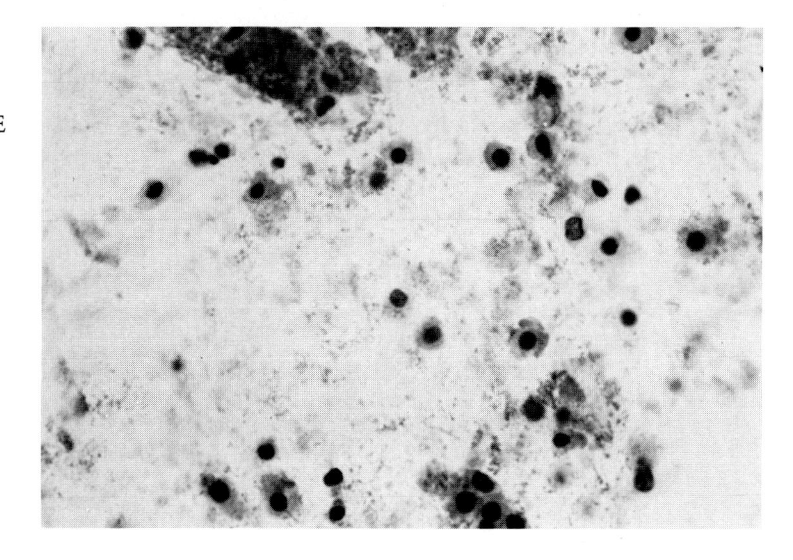

FIG. 3-4
URINE SEDIMENT WITH DIRTY PROTEINACEOUS BACKGROUND AND NUMEROUS SINGLE AND LOOSE CLUSTERS OF RENAL TUBULAR EPITHELIAL CELLS. (× 400)

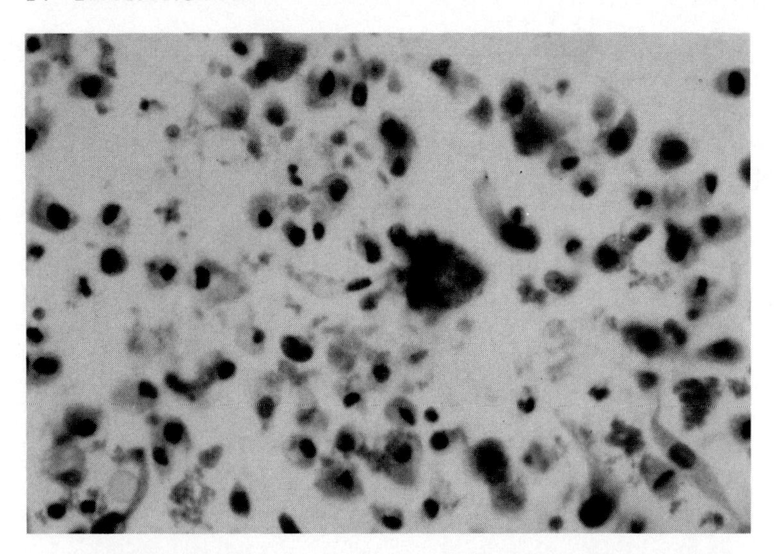

FIG. 3-5
URINE SEDIMENT WITH DIRTY BACKGROUND AND MARKED RENAL TUBULAR EPITHELIAL EXFOLIATION. Mild to moderate cell degeneration and cellular debris are present. (× 400)

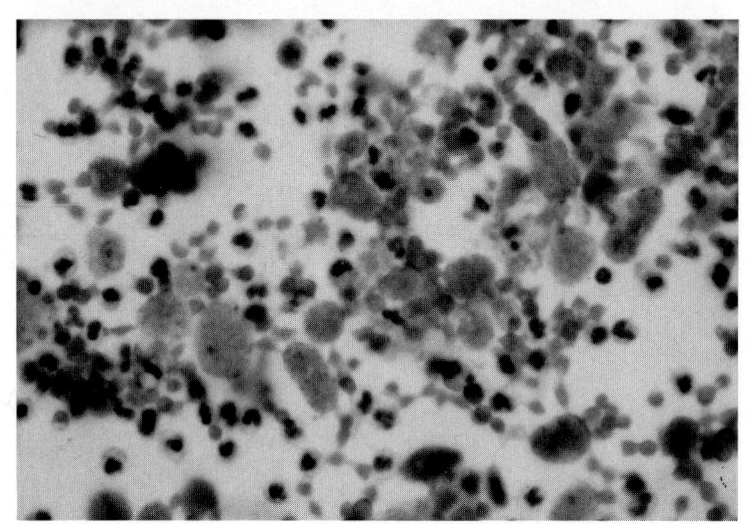

FIG. 3-6
URINE SEDIMENT WITH NUMEROUS NECROTIC ("GHOST") CELLS, INFLAMMATION, AND HEMATURIA. Ghost cells have unstainable nuclei but retain their size and shape. (× 400)

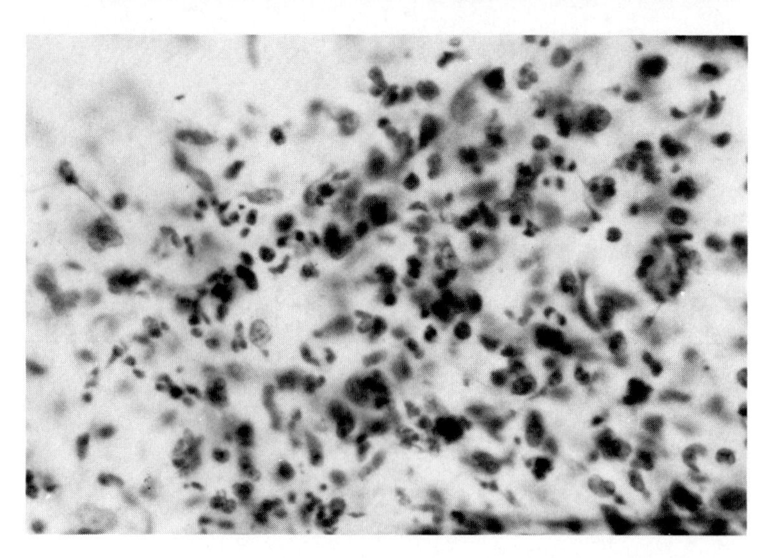

FIG. 3-7
URINE SEDIMENT WITH MARKED INFLAMMATION AND DETRITUS. Dark granular material in background represents stainable nuclear debris. (× 400)

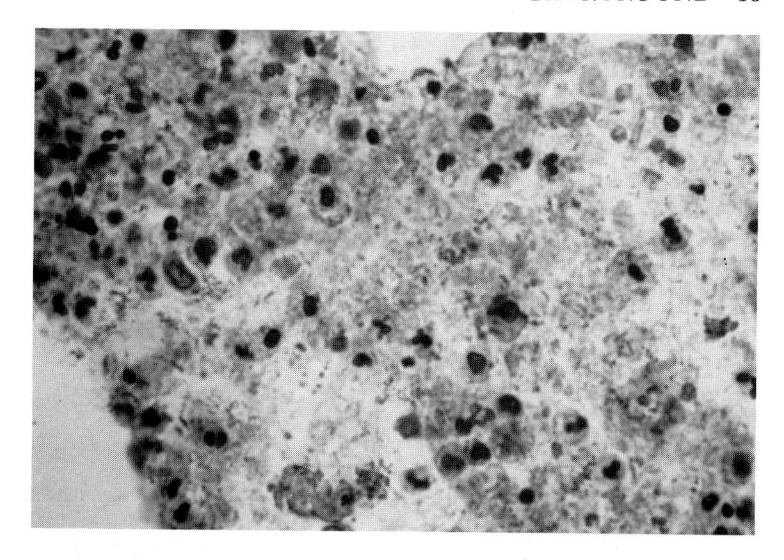

FIG. 3-8
URINE SEDIMENT WITH DIRTY BACKGROUND, ACUTE INFLAMMATION, AND RENAL TUBULAR EPITHELIAL EXFOLIATION. Note extensive necrotic debris in background. (× 400)

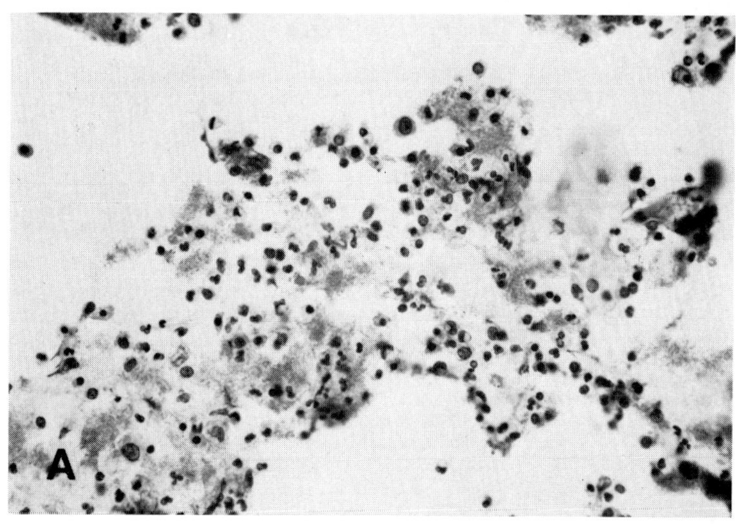

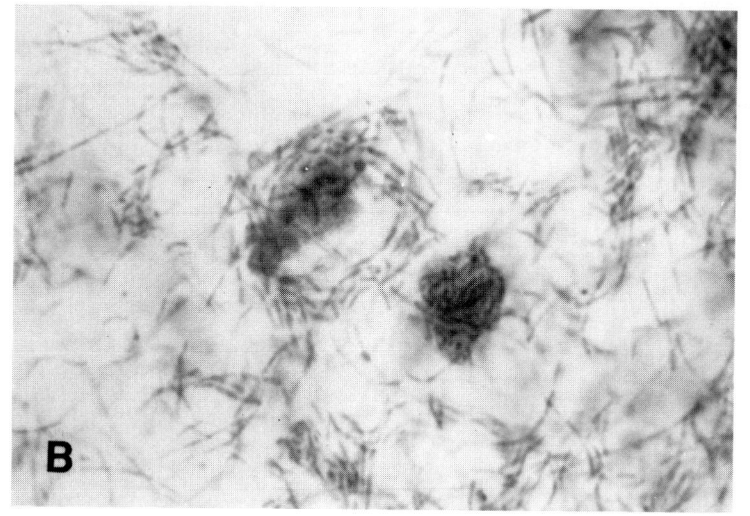

FIG. 3-9
BACTERIAL CYSTITIS. **(A)** Urine sediment with dirty background and marked acute inflammation. (× 250) **(B)** Amorphous dirty background is better characterized using higher magnification. Note numerous bacterial rods. (× 1000)

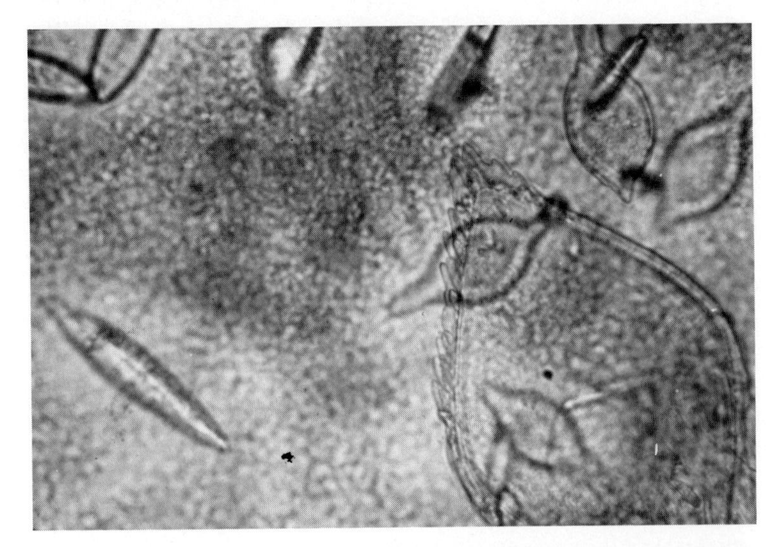

FIG. 3-10
URINE SEDIMENT WITH URIC ACID CRYSTALS AND GOLDEN BROWN AMORPHOUS URATES IN BACKGROUND. (× 400)

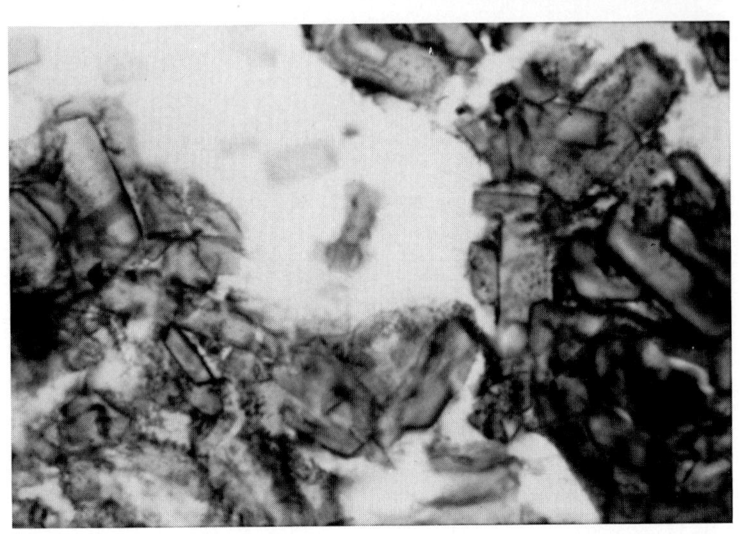

FIG. 3-11
URINE SEDIMENT WITH PURPLE AMORPHOUS PHOSPHATES SURROUNDING PACKED TRIPLE PHOSPHATE CRYSTALS. (× 400)

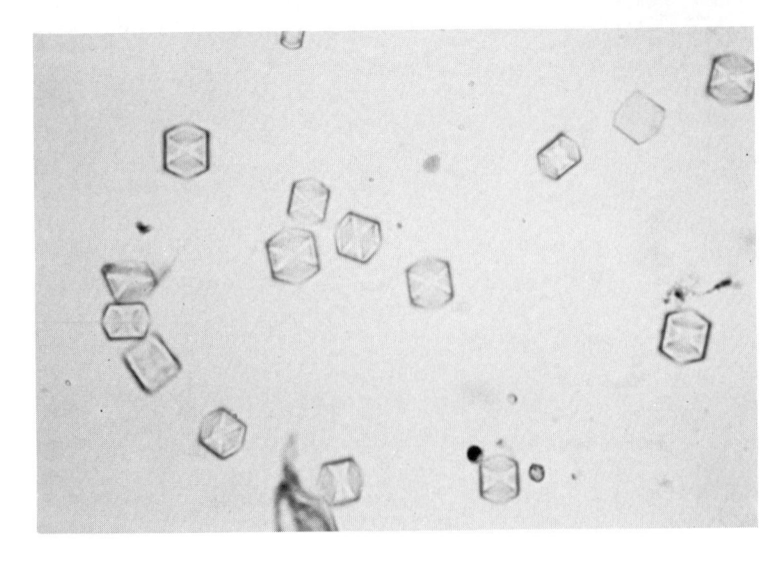

FIG. 3-12
URINE SEDIMENT WITH CALCIUM OXALATE CRYSTALS IN A CLEAN BACKGROUND. (× 1000)

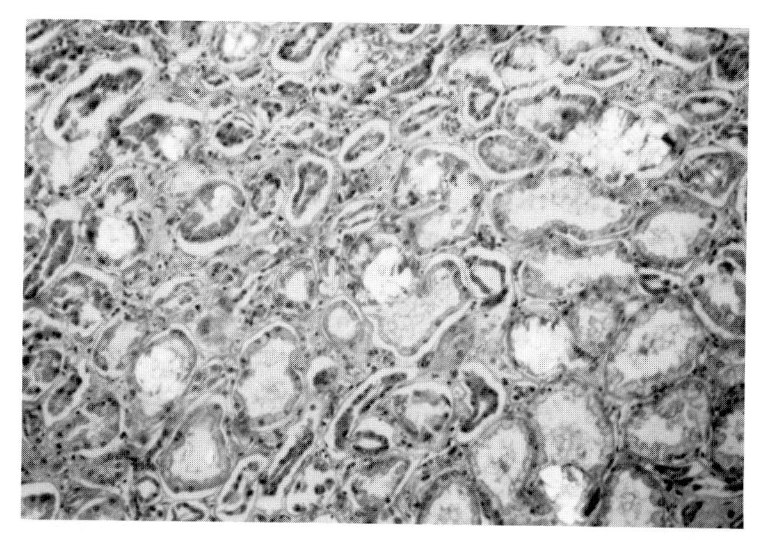

FIG. 3-13
HISTOLOGIC SECTION SHOWING ACUTE TUBULAR NECROSIS AND EXTENSIVE OXALATE DEPOSITION. (Polarized × 100)

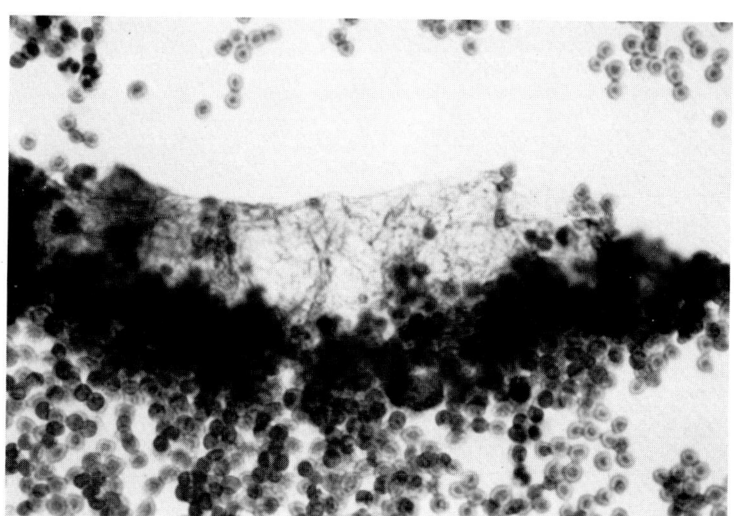

FIG. 3-14
MARKED HEMATURIA AND BACKGROUND CONTAINING FIBRILLAR NETWORK. (× 400)

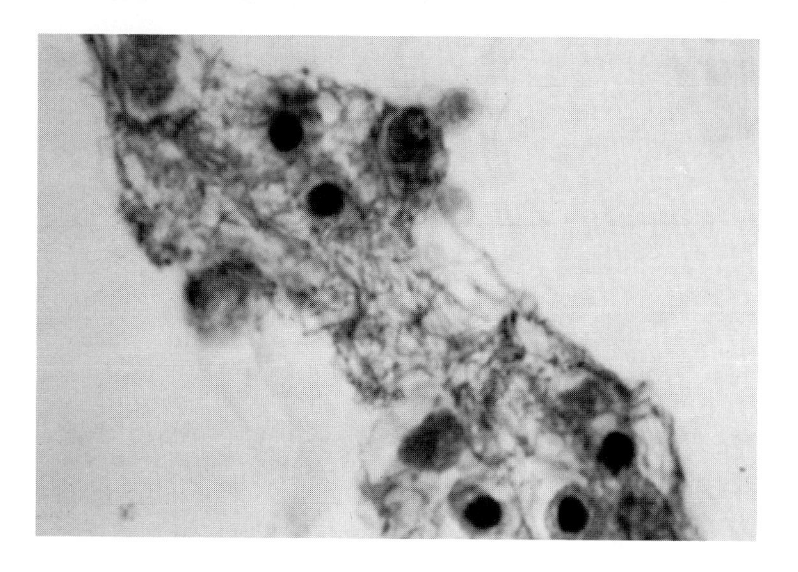

FIG. 3-15
FIBRILLAR THREADS CONTAINING RENAL TUBULAR EPITHELIAL CELLS. These strands, which are probably fibrin, are commonly seen when coagulation products leak into the urine. (× 1000)

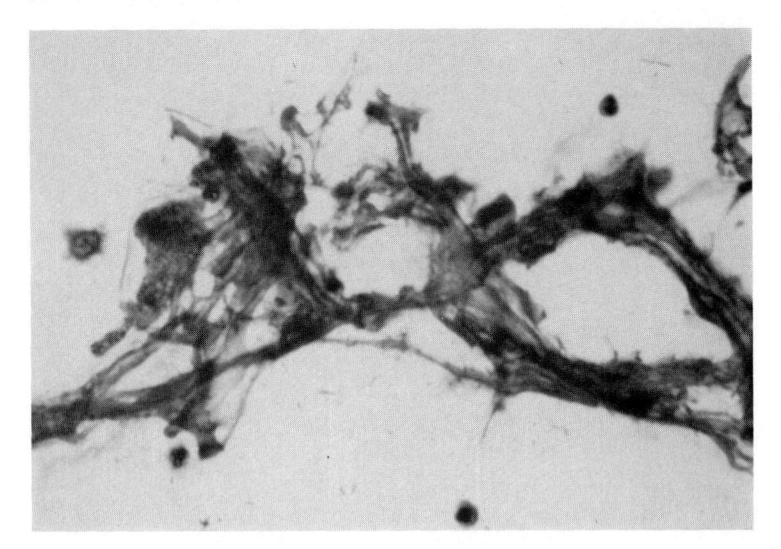

FIG. 3-16
URINE SEDIMENT WITH ABUNDANT THICK, TENACIOUS MUCOUS THREADS. Variability in length and width helps distinguish mucous threads from the delicate, thin threads of fibrin. (× 400)

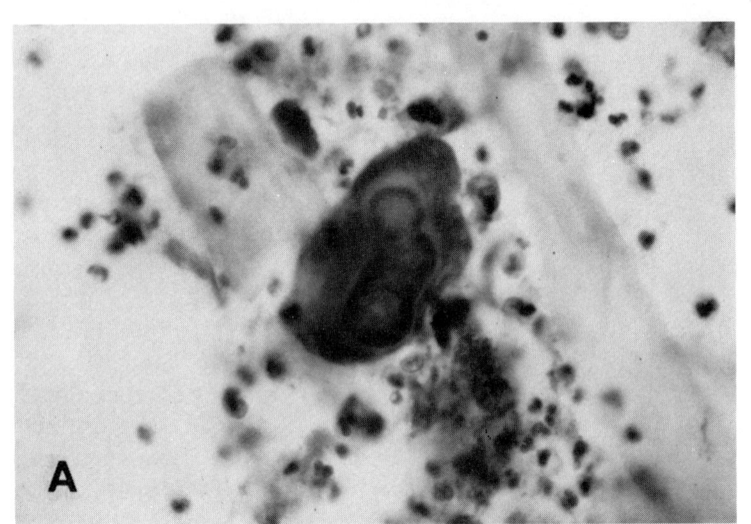

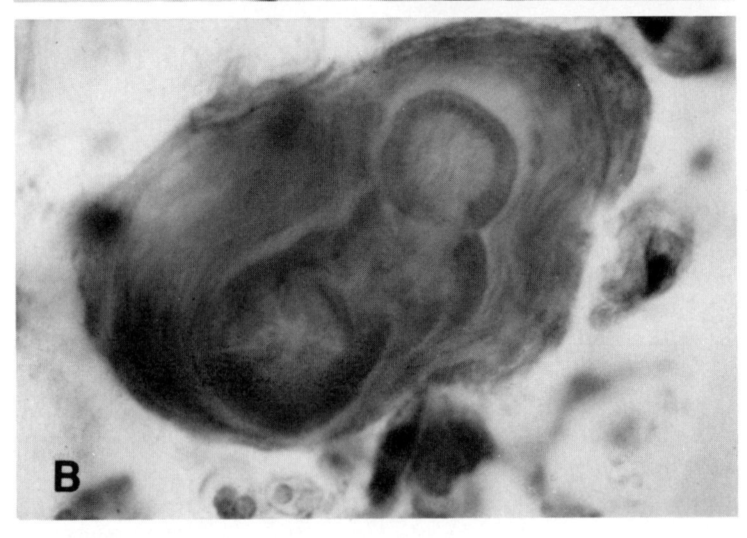

FIG. 3-17
(A) URINE SEDIMENT WITH CONCRETION. Note background containing inflammation and filter card artifact. (× 400) **(B)** The concretion has a laminated outer rim and a central crystalline core. Concretions are commonly seen following prostatic massage. (× 1000)

4
HEMATOPOIETIC-INFLAMMATORY CELLS

ERYTHROCYTES

Erythrocytes represent one of the most common types of hemopoietic cells found in the urinary sediment. The number of erythrocytes are usually recorded as none, trace, few, moderate, or marked. Fresh or hemoglobinized erythrocytes stain red and measure approximately 7 μ in greatest diameter (Fig. 4-1). Erythrocytes that have remained in urine for prolonged periods of time lose their hemoglobin content and are pale staining. Unstained erythrocytes may be mistaken for air bubbles and spore forms of fungi.

NEUTROPHILS

Neutrophils are commonly seen in urine following contamination as well as with hemorrhage or inflammatory responses. Their presence influences background cellularity, and their number is usually recorded as none, trace, few, moderate, or marked. With Papanicolaou-stained sediments, their appearance is similar to peripheral blood neutrophils (Fig. 4-2). In hypotonic urines, neutrophils may lose their nuclear segmentation and appear mononuclear. By unstained brightfield microscopy, these cells are termed **glitter cells** and are characterized by cytoplasmic granules showing brownian movement. In Papanicolaou-stained sediments, they need to be distinguished from lymphoid and epithelial cells.

The presence of neutrophils in clumps is evidence of urinary tract infection. Leukocytic or neutrophilic clumps accompanied by renal epithelial cell exfoliation and pathologic casts, particularly the leukocytic type, are necessary for a diagnosis of acute pyelonephritis (Fig. 4-3).

EOSINOPHILS

The presence of eosinophils in urine is termed **eosinophiluria.** Eosinophils are characterized by the orange granular cytoplasm and are mono- or binucleated (Fig. 4-4). Low numbers of eosinophils may be associated with a mixed inflammatory sediment and may be a component of a urinary inflammatory process. Eosinophiluria of greater than 5% of the inflammatory cells is significant and usually correlates with an allergic inflammatory response. The differential diagnosis includes such lower urinary tract diseases as eosinophilic cystitis (Fig. 4-5), eosinophilic granuloma, parasitic infection, and inflammation associated with bladder carcinoma. Eosinophiluria may also be present with allergic interstitial nephritis.

LYMPHOCYTES

The presence of increased numbers of lymphocytes in urine is termed **lymphocyturia** (> 10% of the inflammatory cells). These mononuclear cells have a dark-staining nucleus with a coarse chromatin pattern, a pale-staining thin rim of cytoplasm, and an average diameter similar to that of erythrocytes (Fig. 4-6). In comparison to mature lymphocytes, immature lymphocytes have nuclear enlargement, chromatin clearing, micronucleoli, and a more prominent blue-gray cytoplasm. Malignant lymphocytes, which rarely occur in urine sediments, are characterized by nuclear membrane irregularities and multiple irregular nucleoli. A mixed inflammatory background containing lymphocytes most frequently indicates chronic cystitis. The prominence of mature and immature lymphocytes suggests follicular cystitis (Fig. 4-7). Lymphocyturia may be seen during acute renal allograft rejection.

PLASMA CELLS

Stained permanent preparations are required for the accurate identification of plasma cells in urine. Plasma cells are characterized by their dense irregular chromatinic membrane (clock face), eccentric nucleus, and a cytoplasmic hof (pale zone) (Fig. 4-8). They are usually associated with other chronic inflammatory cells, such as lymphocytes and histiocytes. The methyl green-pyronine stain can be used to improve the identification of plasma cells. The appearance of plasma cells in urine is similar to that seen in peripheral blood smears and in histologic material (Fig. 4-9). The presence of plasma cells in the urinary stream usually represents chronic inflammation. Localization of the plama cell infiltrate is best accomplished by assessing other sediment parameters, such as renal casts, renal tubular epithelial cells, or urothelial cells. Myelomatous renal involvement has been associated with increased numbers of plasma cells in the urinary sediment.

HISTIOCYTES AND MACROPHAGES

Histiocytes are characteristically larger than neutrophils, have lobulated or kidney-bean-shaped nuclei, and abundant, finely vacuolated gray cytoplasm lacking phagocytized material (Figs. 4-10 and 4-11). If particulate matter is found in the cytoplasm of histiocytes, they are termed **macrophages.** Increased numbers of histiocytes in the urine sediment are usually present as a component of chronic inflammation. They are frequently found in large numbers in response to radiation injury.

MULTINUCLEATED GIANT CELLS

There are several sources of multinucleated cells in the urine sediment (Fig. 4-12). Multinucleated, superficial urothelial cells are frequently seen following bladder irrigation. Multinucleated histiocytic cells are characterized by a vacuolated or finely granular cytoplasm containing no particulate matter. They may be a prominent finding following radiation. An important cause of multinucleation and giant cell formation is viral infection. The diagnostic characteristics of virally induced multinucleated cells are given in Chapter 5. Another source of multinucleated epithelial cells is the renal tubule. Multinucleated syncytial epithelial cells can be seen with tubular regeneration, as a reactive response to crystalline material (Fig. 4-13**A**) and surrounding myeloma casts. Giant cells may be a component of a granulomatous infection, for example, renal tuberculosis (Fig. 4-13**B**).

BIBLIOGRAPHY

Koss LG: Diagnostic Cytology and Its Histopathologic Bases, 3rd ed. Philadelphia, JB Lippincott, 1979

Schumann GB: The Urine Examination. Baltimore, Williams & Wilkins, 1980

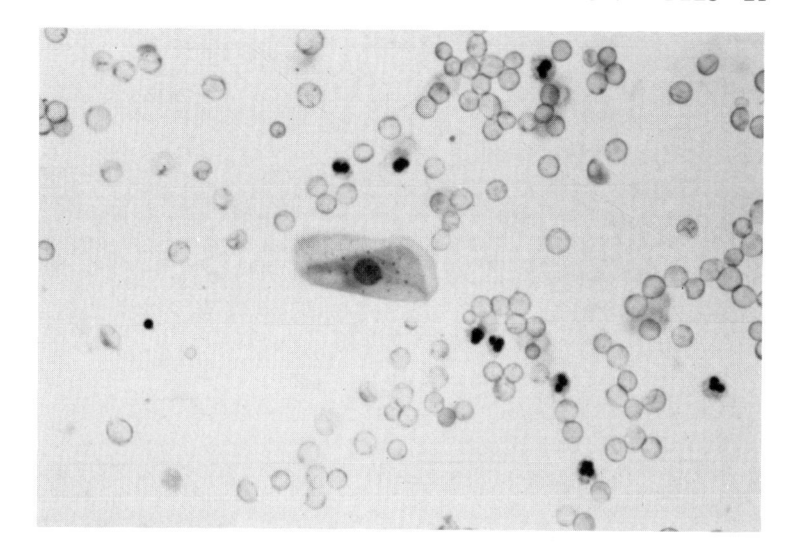

FIG. 4-1
URINE SEDIMENT WITH CLEAN BACKGROUND AND MODERATE HEMATURIA. The relatively few neutrophilic leukocytes present are consistent with fresh bleeding. Greater numbers of leukocytes are needed to indicate an inflammatory response. (× 400)

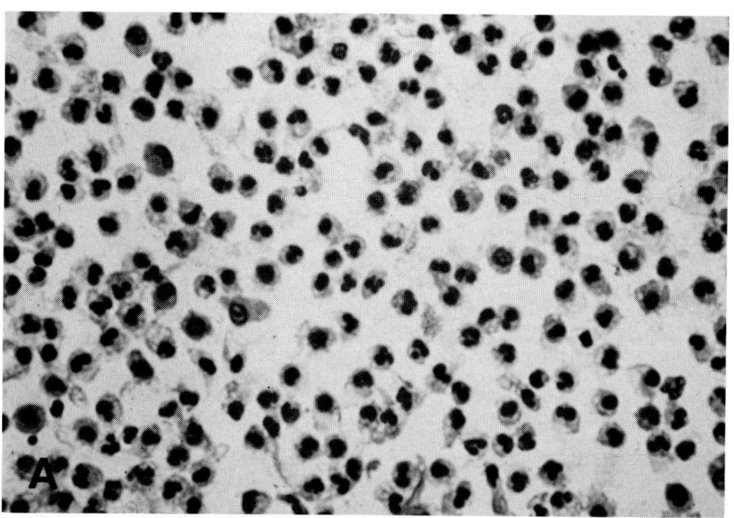

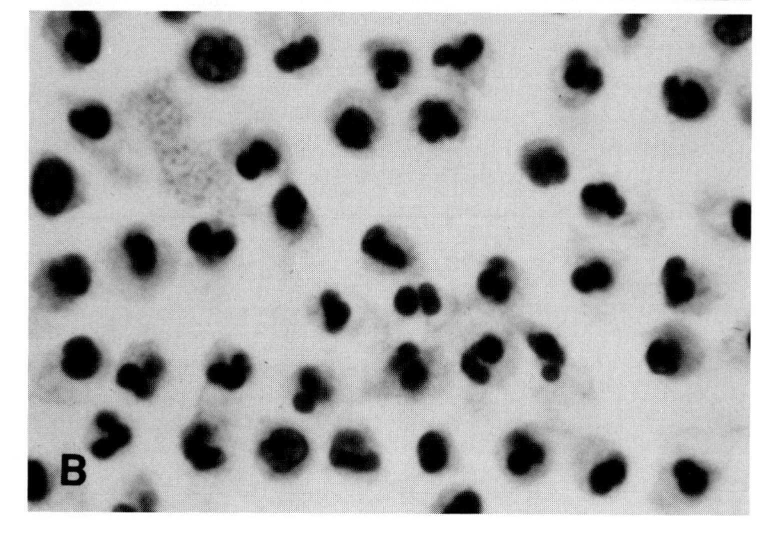

FIG. 4-2
(A) URINE SEDIMENT WITH MARKED NEUTROPHILIC EXUDATE AND SLIGHTLY DIRTY BACKGROUND INDICATING AN INFLAMMATORY RESPONSE. (× 400) **(B)** Characteristics of neutrophilic leukocytes are better appreciated using higher magnification. Note the nuclear segmentation similar to that seen in a peripheral blood smear. (× 1000)

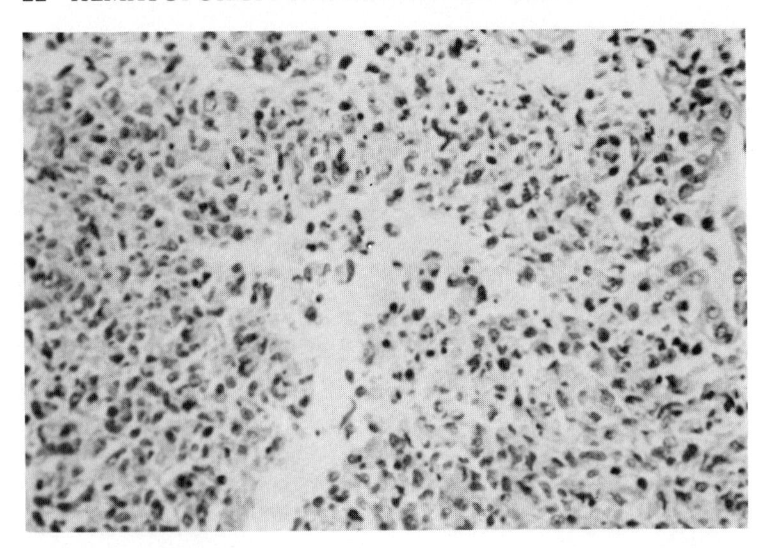

FIG. 4-3
NEUTROPHILIC (POLYMORPHONUCLEAR) LEUKOCYTES. Acute inflammatory destruction of renal medullary tubules with a neutrophil-rich abscess in a patient with acute pyelonephritis. (× 250)

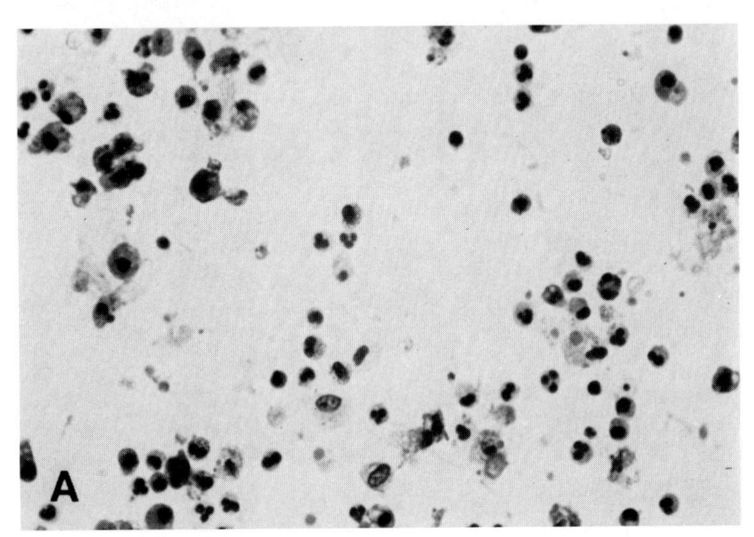

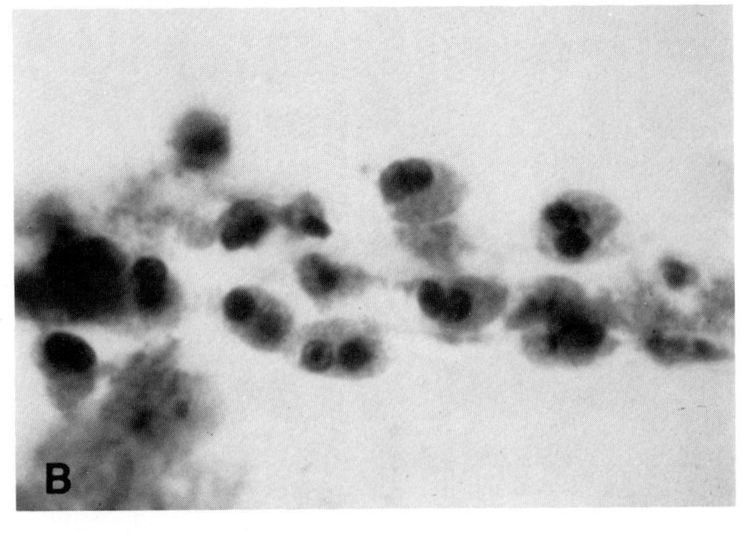

FIG. 4-4
(A) EOSINOPHILURIA. Slightly dirty background with numerous inflammatory cells present. (× 400) **(B)** Characteristics of the inflammatory cells are better appreciated using higher magnification. Note numerous bilobed eosinophilic leukocytes. Eosinophils are important in identifying allergic inflammatory responses. (× 1000)

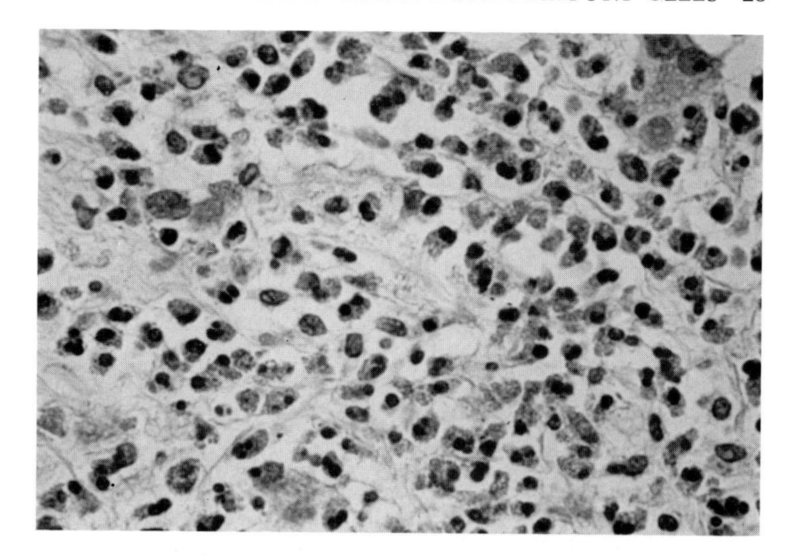

FIG. 4-5
EOSINOPHILS. The bladder lamina propria contains numerous mononucleated and binucleated eosinophils that can be easily recognized by their granular, brightly eosinophilic cytoplasm. Histiocytes are also present in this biopsy. The differential diagnosis would include eosinophilic cystitis, eosinophilic granuloma, parasitic infection, and associated bladder cancer. (× 400)

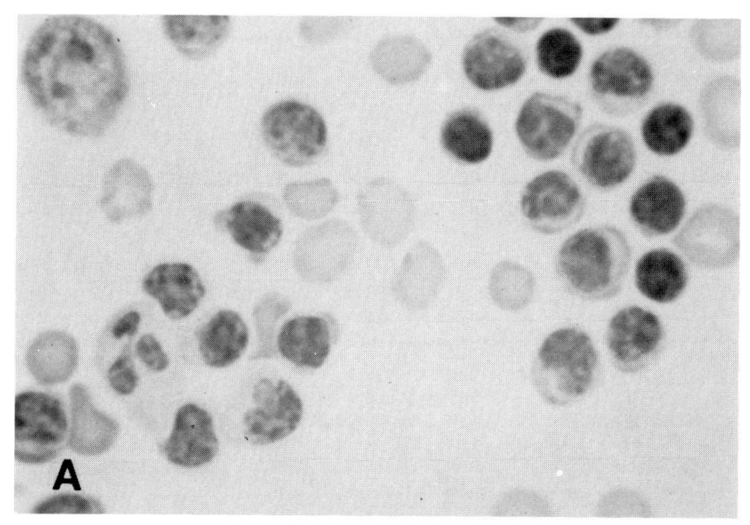

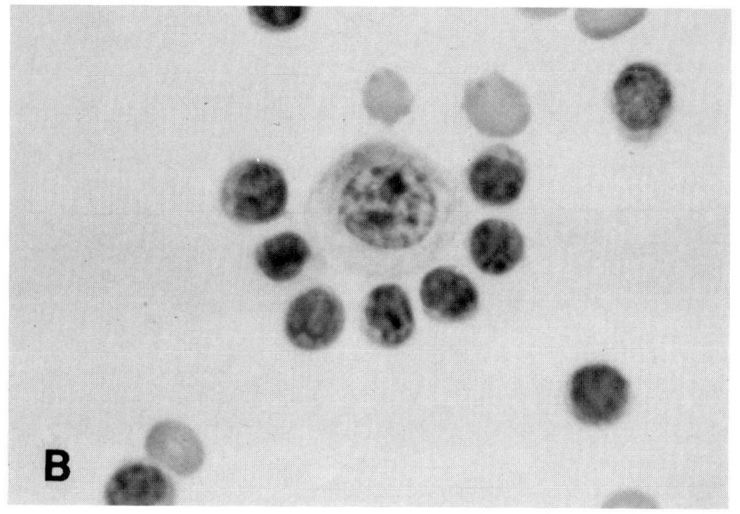

FIG. 4-6
(A) NUMEROUS LYMPHOCYTES, ERYTHROCYTES, AND A RARE NEUTROPHIL IN A CLEAN BACKGROUND. Note the similar size of lymphocytes and erythrocytes (approximately 7 μ in greatest diameter). This specimen was obtained from aspiration of the renal pelvic lymphocele. (× 1000) **(B)** Several mature lymphocytes surrounding an immature lymphoreticular cell, probably a lymphoblast. Note the dark nucleus and the thin rim of cytoplasm characteristic of small, mature lymphocytes. (× 1000)

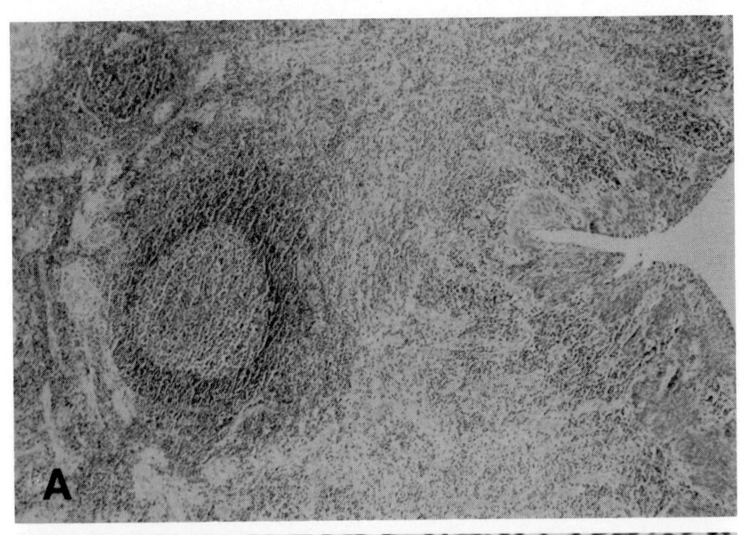

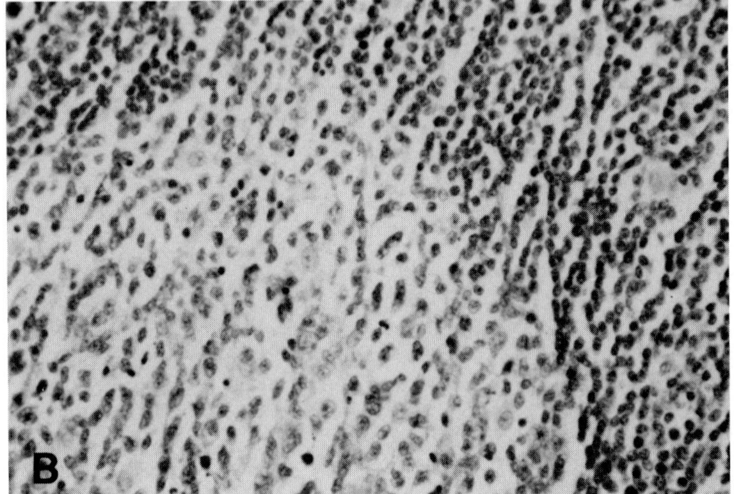

FIG. 4-7
LYMPHOCYTES AND GERMINAL CENTER CELLS. **(A)** A lymphoid follicle with a reactive germinal center is present in the lamina propria, which is edematous and expanded by a dense, chronic inflammatory infiltrate. (× 100) **(B)** The germinal center contains large and small lymphocytes and histiocytes with phagocytic activity and is rimmed by mature lymphocytes. (× 250)

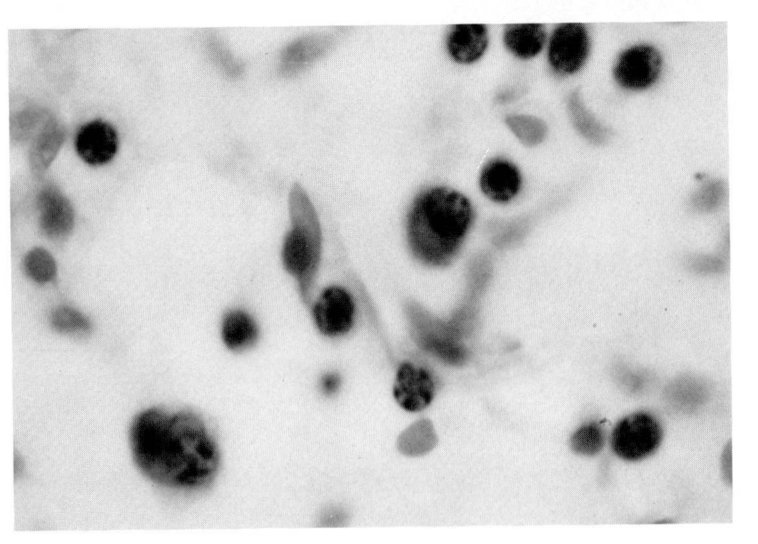

FIG. 4-8
EXUDATE CONTAINING SEVERAL PLASMA CELLS, ERYTHROCYTES, AND LYMPHOCYTES. Note a dense, irregular chromatinic membrane (clock face), an eccentric nucleus, and a cytoplasmic hof (pale) zone characteristic of plasma cells. (× 1000)

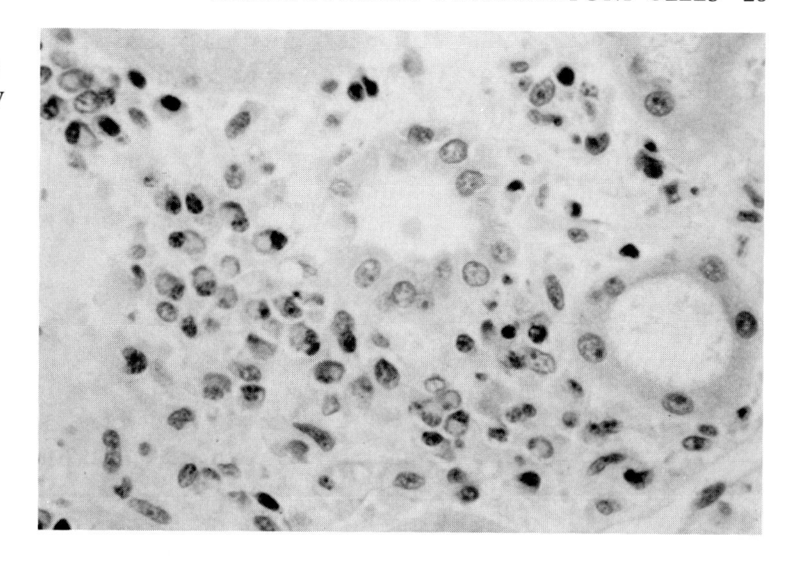

FIG. 4-9
PLASMA CELLS. Plasma cells within the renal interstitium are easily recognized by their eccentric nuclei, perinuclear halo, and cytoplasmic amphophilia. (× 400)

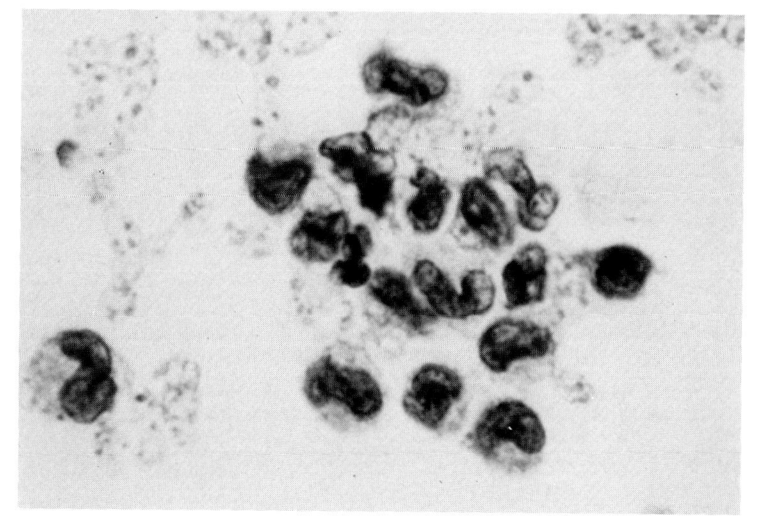

FIG. 4-10
URINE SEDIMENT WITH MARKED HISTIOCYTIC EXUDATE. Note kidney-bean shaped nuclei, and an indistinct cytoplasm lacking phagocytized particulate matter. (× 1000)

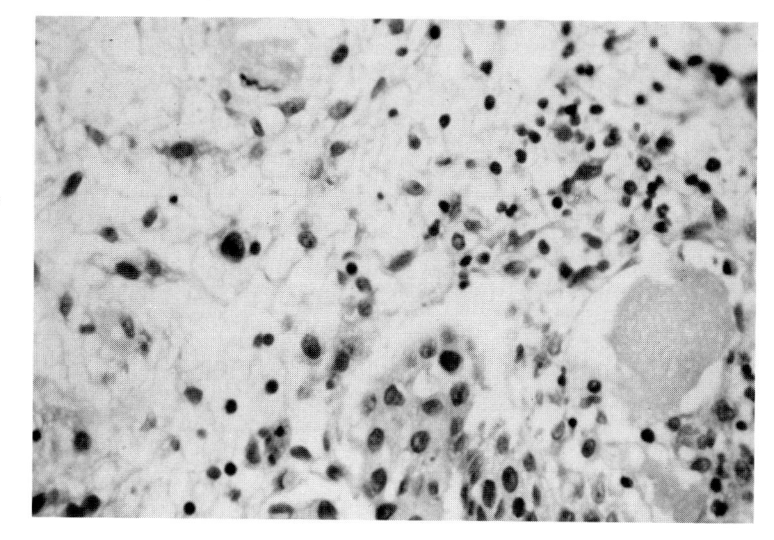

FIG. 4-11
HISTIOCYTES AND MACROPHAGES WITHIN BLADDER LAMINA PROPRIA. Histiocytes have indented or bean-shaped nuclei with a finely dispersed chromatin, a dot-like nucleolus, and indistinct cell borders. The nucleus is often eccentrically placed and cytoplasm is more prominent in macrophages, which contain phagocytized hemosiderin and cellular debris. (× 250)

FIG. 4-12
MULTINUCLEATED CELLS IN URINE SEDIMENT. **(A)** Multinucleated urothelial cell. These cells are often exfoliated during bladder irrigation. (× 400) **(B)** Multinucleated histiocytes with four to six eccentric nuclei. Note the lack of phagocytized material. (× 1000) **(C)** Multinucleated histiocyte with greater than ten nuclei. These cells are frequently seen in nonspecific chronic inflammation. (× 1000) **(D)** Multinucleated giant cell. Note numerous nuclei containing faint intranuclear inclusions characteristic of herpes simplex infection. (× 1000) **(E)** Multinucleated cell showing cytopathic effect suggestive of viral infection. Cytopathic effect is characterized by nuclear enlargement, variability in nuclear size, a chromatinic membrane, and a ground-glass nuclear appearance. Note the lack of an intranuclear inclusion (× 1000)

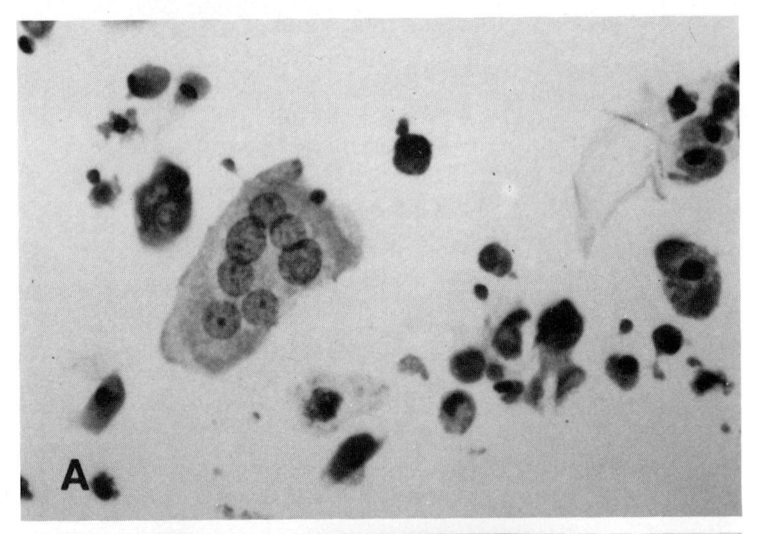

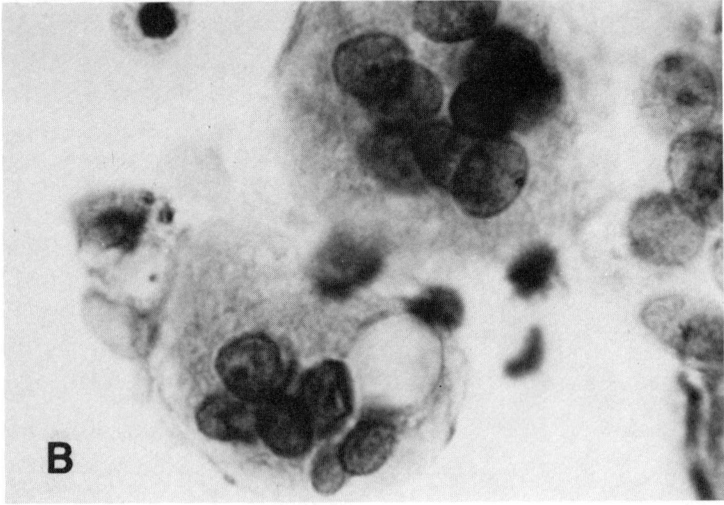

(continued)

FIG. 4-12 *(continued)*

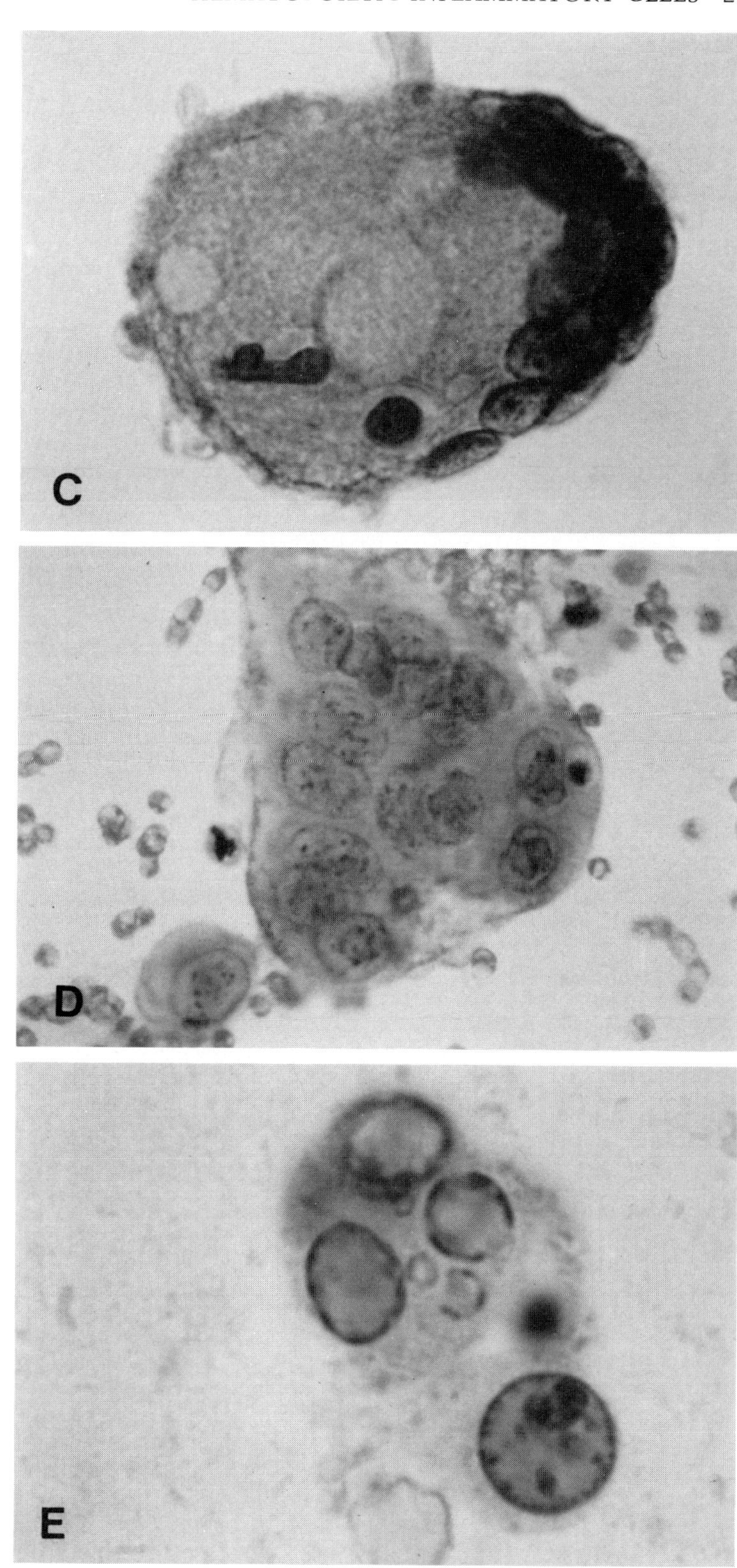

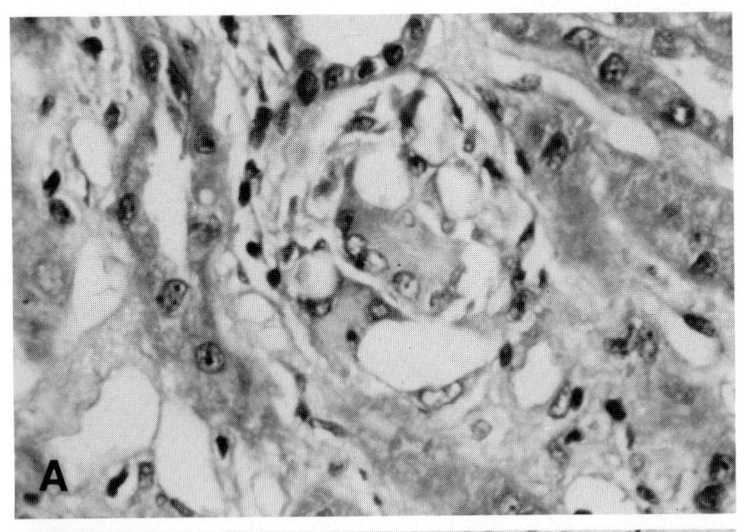

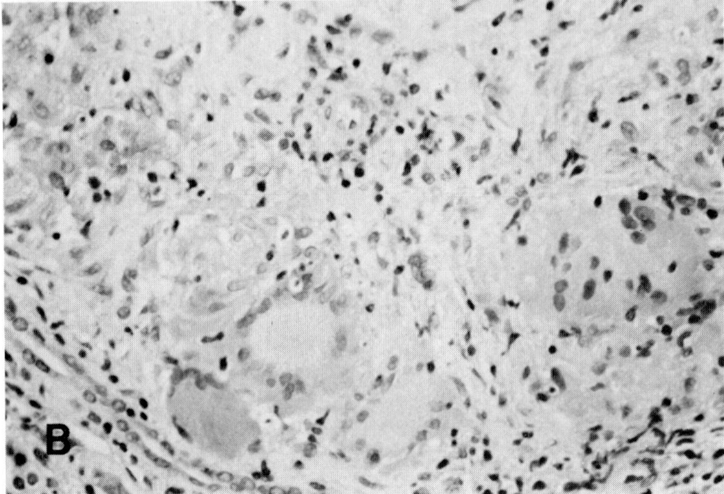

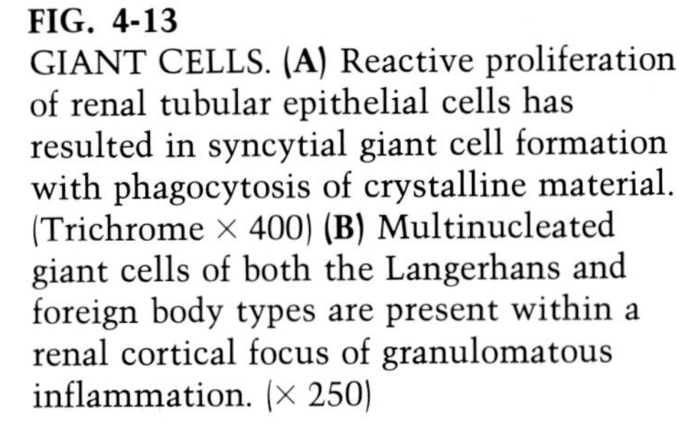

FIG. 4-13
GIANT CELLS. **(A)** Reactive proliferation of renal tubular epithelial cells has resulted in syncytial giant cell formation with phagocytosis of crystalline material. (Trichrome × 400) **(B)** Multinucleated giant cells of both the Langerhans and foreign body types are present within a renal cortical focus of granulomatous inflammation. (× 250)

5
ORGANISMS

EXTRACELLULAR ORGANISMS	INTRACELLULAR ORGANISMS
BACTERIA	BACTERIA
FUNGI	FUNGI
PROTOZOA	PROTOZOA
PARASITES	VIRUSES

Accurate identification of organisms aids in the differential diagnosis of urinary system infections. Table 5-1 lists the types of organisms seen in urine. In urine sediments, organisms are recognized as ***extracellular*** *or* ***intracellular*** *structures.*

EXTRACELLULAR ORGANISMS

BACTERIA

The presence of bacteria, usually extracellular, in urine is termed **bacteriuria.** Significant bacteriuria is present when greater than 10^5 organisms are found by urine culture count. In unspun urine, the presence of bacteria is clinically significant. In permanently stained preparations, the morphology of bacteria is readily appreciated; however, accurate classification requires microbiological techniques. The importance of stained preparations in the evaluation of bacteriuria is in the identification of associated inflammatory cells and the assessment of epithelial exfoliation and cast formation for purposes of localization. The lack of an inflammatory response may be misleading in immunosuppressed patients (Fig. 5-1).

FUNGI

Extracellular fungal organisms are frequently seen in urine. Yeast or spore forms must be distinguished from erythrocytes and cellular debris (Fig. 5-2). An important diagnostic feature is the presence of budding yeasts. Pseudohyphae are characteristic of invasive *Candida* species (Fig. 5-3). Urinary system candidiasis is commonly seen in immunosuppressed and diabetic patients.

Papanicolaou-stained sediments represent a rapid and accurate method of identifying fungal forms. Specific classification requires microbiological techniques. An additional advantage of the Papanicolaou technique is the identification of ascending infection with renal parenchymal involvement. The presence of certain fungi, such as *Cryptococcus,* usually represents hematogenous spread to the kidney (Fig. 5-4).

PROTOZOA

Trichomonas represents the most common type of protozoan lower urinary tract infection (Fig. 5-5). In males, *Trichomonas* may cause a bladder infection, and in females it usually represents vaginal contamination. If the infection has extended into the bladder, the exfoliation of urothelial cells with inflammation should be expected.

Amebae have been described in urine sediment. Their presence usually indicates fecal contamination.

PARASITES

Enterobius vermicularis (pinworm) and *Schistosoma* have been observed in urine sediment. Pinworm ova found in the urine is usually due to anal contamination (Fig. 5-6). In comparison to pinworm, schistosomes in the urine sediment may indicate bladder infection (*S. hematobium*). In countries with endemic schistosomiasis, there is an increased incidence of squamous cell carcinoma.

INTRACELLULAR ORGANISMS

BACTERIA

Phagocytized bacteria can be seen in neutrophils and macrophages (Fig. 5-7). Intracellular bacteria are often seen during an early acute bacterial urinary tract infection.

FUNGI

Both yeast and pseudohyphal forms can be ingested by neutrophilic leukocytes or macrophages (Fig. 5-8). *Candida* species represent the most common type of fungi found intracellularly. This finding usually indicates fungal urinary tract infection.

PROTOZOA

Toxoplasmosis may involve the urinary system. *Toxoplasma* are characterized by dark intracellular organisms found in macrophages. They have been recognized as the source of infection in immunosuppressed patients.

VIRUSES

Viral-induced cellular changes have been recognized with increased frequency in urine sediments from immunosuppressed patients. Cytomegalovirus (CMV), herpes simplex, and polyomavirus represent the most common types of urinary system viral infections (see Table 5-1).

CMV is frequently seen in the urine of infants with cytomegalic inclusion disease and in adults following immunosuppression (Fig. 5-9). CMV-induced viral cellular changes are characterized by nuclear enlargement, a chromatinic membrane, a basophilic intranuclear inclusion, and a peri-inclusion halo. Smaller basophilic intracytoplasmic inclusions may be identified and should be distinguished from eosinophilic, nonspecific degenerative droplets. These viral-induced changes may be present without cellular enlargement. The presence of CMV-infected cells in the urine sediment correlates with renal parenchymal localization. CMV-infected cells have been identified within glomeruli, parietal epithelium lining Bowman's capsule, and in convoluted tubular epithelium.

Herpes simplex infections usually occur in the lower urinary tract. The cytopathic effects are characterized by multinucleation, chromatinic membranes, ground-glass nuclear changes, and prominent eosinophilic intranuclear inclusions (Fig. 5-10).

Urinary system infection with polyoma virus typically manifests as marked exfoliation of degenerated epithelial cells (Fig. 5-11). These cells are characterized by nuclear enlargement, a homogeneous deep-staining chromatin, and a thin, almost indistinct peri-inclusion halo. These cells are similar to decoy cells and should not be misinterpreted as malignant urothelial cells (Fig. 5-12). Besides an abnormal nuclear:cytoplasmic ratio and hyperchromatism, the diagnosis of malignant cells requires identification and assessment of preserved chromatin detail.

TABLE 5-1
URINARY ORGANISMS

EXTRACELLULAR	INTRACELLULAR
Bacteria	Bacteria
Fungi: *Candida*	Fungi: *Candida*
Cryptococcus	
Protozoa: Trichomonads	Protozoa: *Toxoplasma*
Amoebae	
Parasites: Pinworm	Virus: Cytomegalovirus
Schistosoma	Herpes simplex
	Adenovirus
	Polyomavirus
	Measles

BIBLIOGRAPHY

Bradley M, Schumann, GB, Ward PCJ: Examination of urine. In Henry, JB (ed): Todd, Sanford, Davidsohn: Clinical Diagnosis and Management by Laboratory Methods, 16th ed. Philadelphia, WB Saunders, 1979

Koss LG: Diagnostic Cytology and Its Histopathologic Bases, 3rd ed. Philadelphia, JB Lippincott, 1979

Schumann GB: The Urine Sediment Examination. Baltimore, Williams & Wilkins, 1980

Tweeddale DN: Urinary Cytology. Boston, Little, Brown, 1977

Voogt HJ, Rathert P, Beyer-Boon ME: Urinary Cytology: Phase Contrast Microscopy and Analysis of Stained Smears. New York, Springer-Verlag, 1977

FIG. 5-1
BACTERIURIA. (**A**) A rare degenerated cell is surrounded by numerous bacterial rods. Note the lack of inflammation. These findings in a nonimmunosuppressed patient usually represent bacterial overgrowth. (× 1000) (**B**) Bacteriuria and acute inflammation in a patient with bacterial urinary tract infection. Note the relatively clean background. (× 1000)

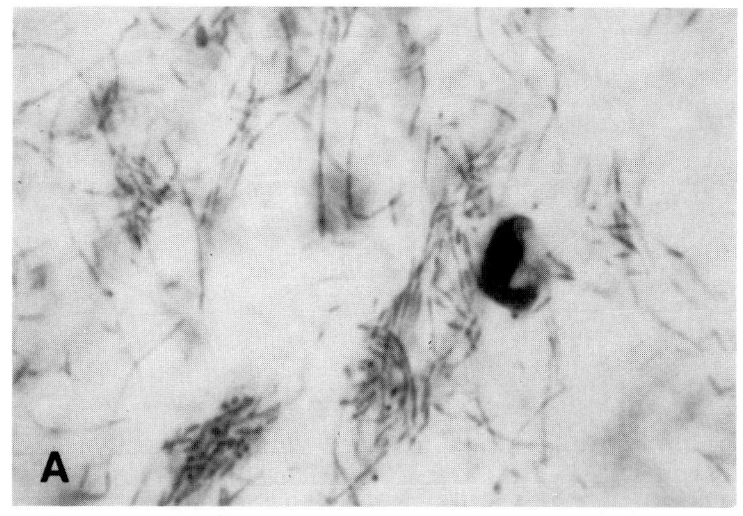

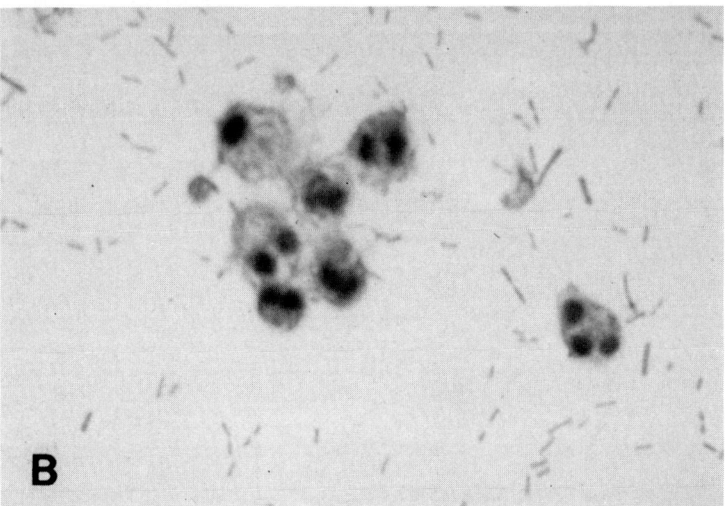

FIG. 5-2
EXTRACELLULAR FUNGAL ORGANISMS IN URINE SEDIMENT. Numerous yeast forms (spores) are present. Neutrophils and transitional cells are present in background. (× 400)

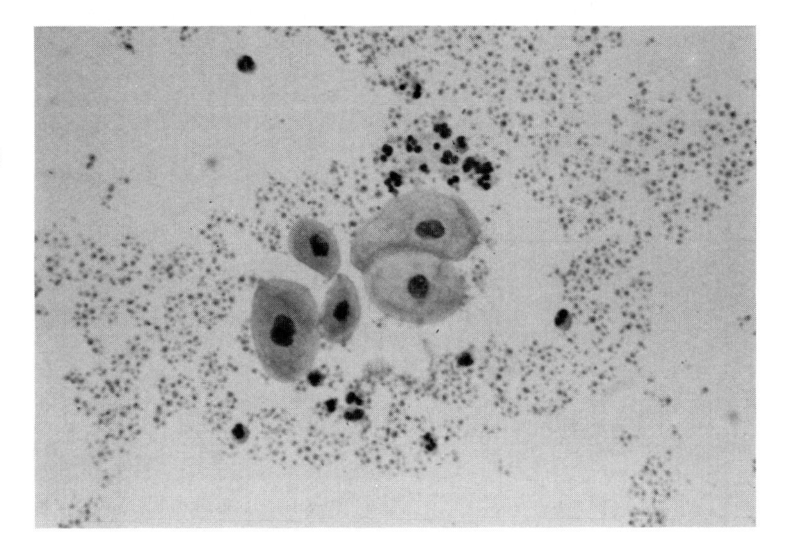

FIG. 5-3
CANDIDIASIS. **(A)** Glomerular abscess with yeast forms within Bowman's space, and the partially necrotic glomerulus is characteristic of septicemia with hematogenous source of renal infection. There is an associated neutrophilic and lymphocytic inflammatory infiltrate. (× 400) **(B)** Invasion of ureteral wall by yeast and pseudohyphal forms. There is a notable lack of inflammatory response in this immunosuppressed patient—(Grocott × 250) **(C)** Budding yeast is present in urine. The recognition of budding is important in distinguishing yeast from erythrocytes and other cellular debris. (× 1000) **(D)** Pseudohyphal form of *Candida* species. Note marked inflammation and epithelial cells. In unstained urine sediment preparations, intense inflammation may obscure identification of fungi (× 150) **(E)** Pseudohyphal form of *Candida* species. Note bacilli. Mixed organisms are frequently seen in immunosuppressed patients (× 1000)

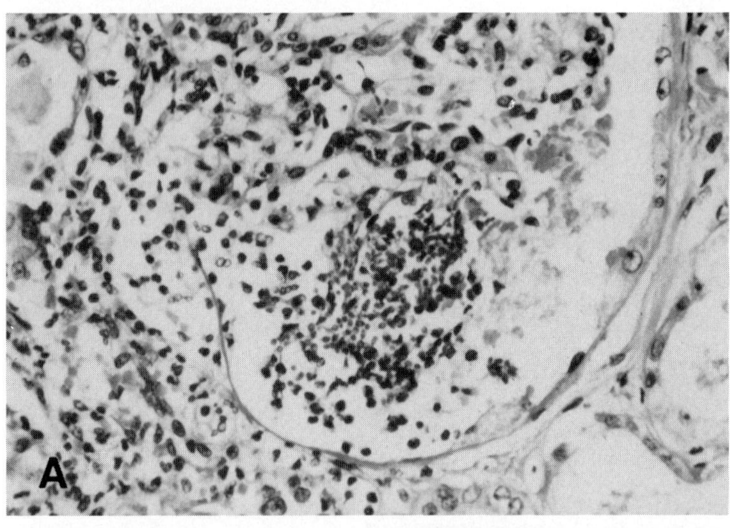

(continued)

FIG. 5-3 *(continued)*

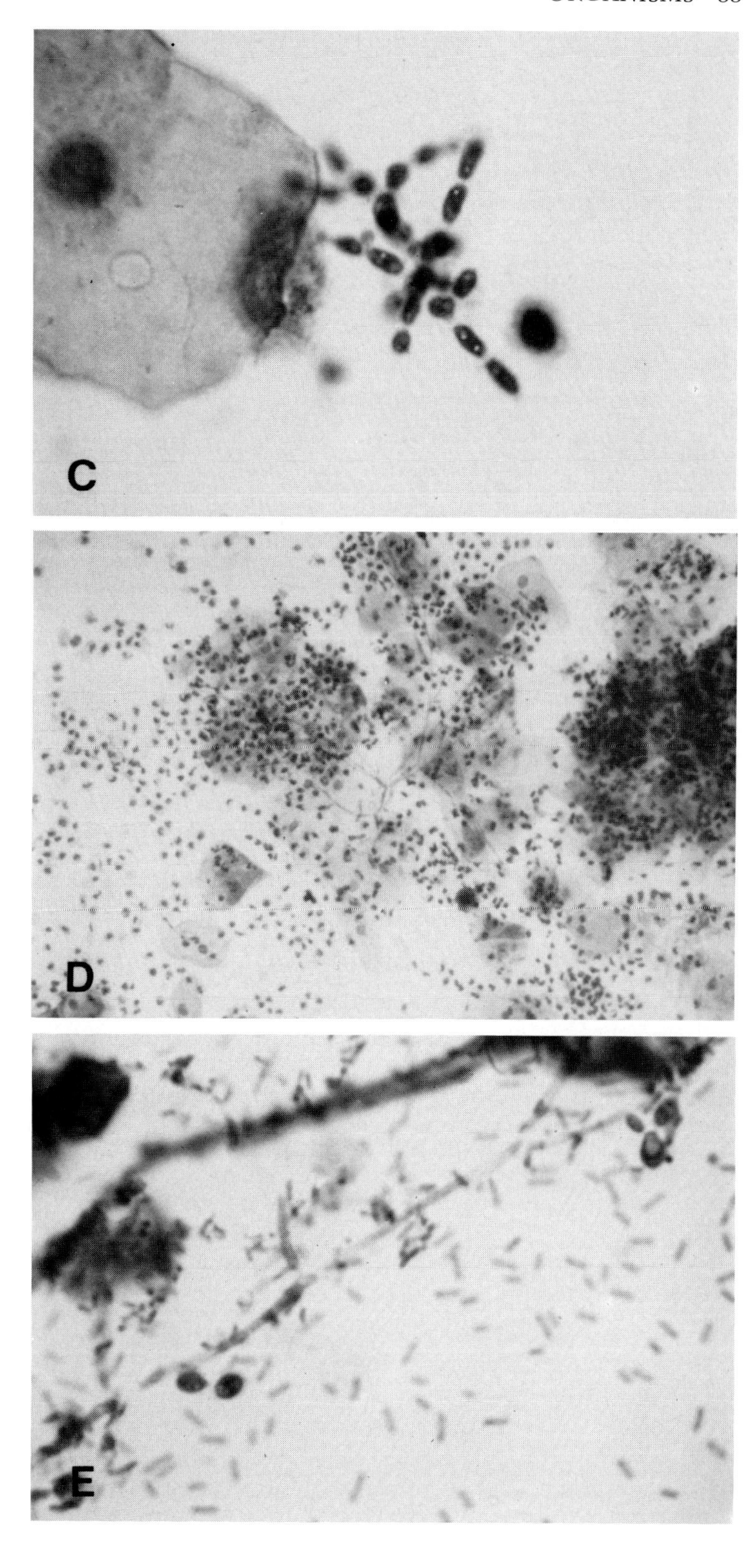

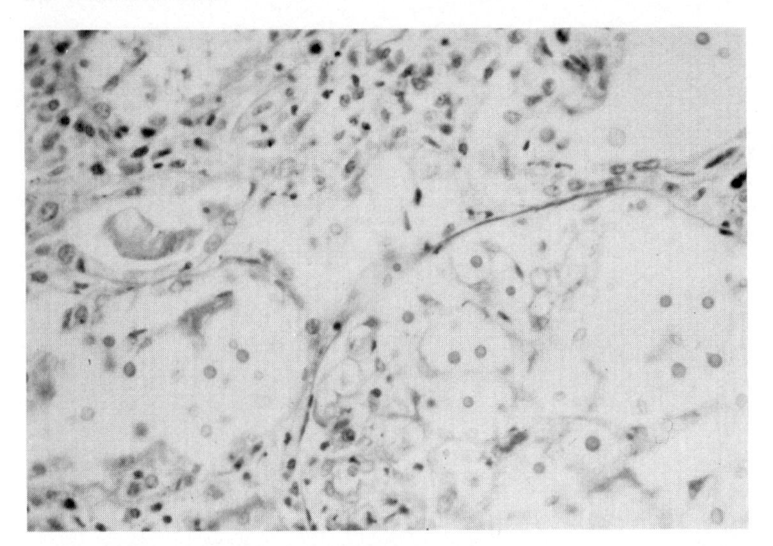

FIG. 5-4
CRYPTOCOCCOSIS. *Cryptococcus neoformans,* recognized as a yeast form with a surrounding halo representing the thick, unstained capsule, is present within the glomerulus and tubular lumens. (PAS × 250)

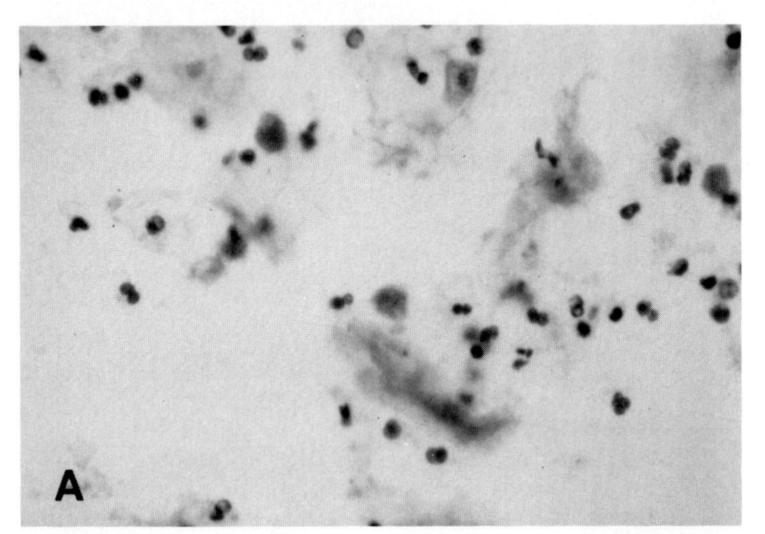

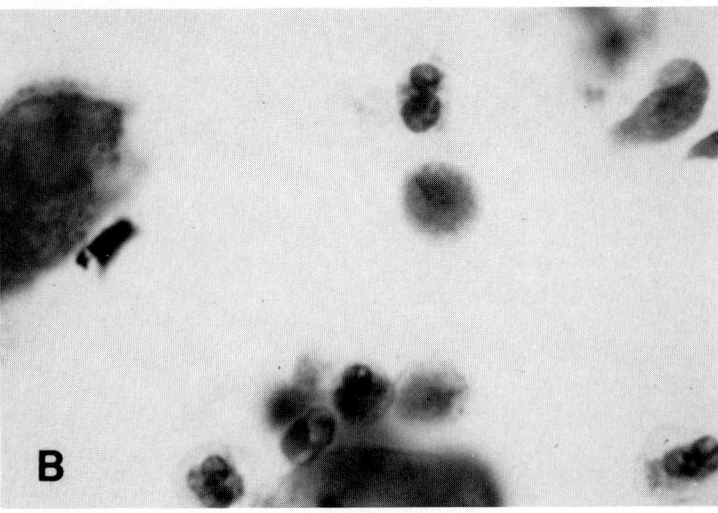

FIG. 5-5
TRICHOMONAS IN URINE SEDIMENT. **(A)** Organisms appear as nondescript oval structures. Note slightly dirty background with moderate acute inflammation. (× 400) **(B)** At higher magnification, trichomonads are recognized by their round to oval shape, pale nucleus, and vacuolated cytoplasm containing red granules. Flagella are not seen. Trichomonads must be distinguished from degenerating deep urothelial cells and renal tubular epithelial cells (× 1000)

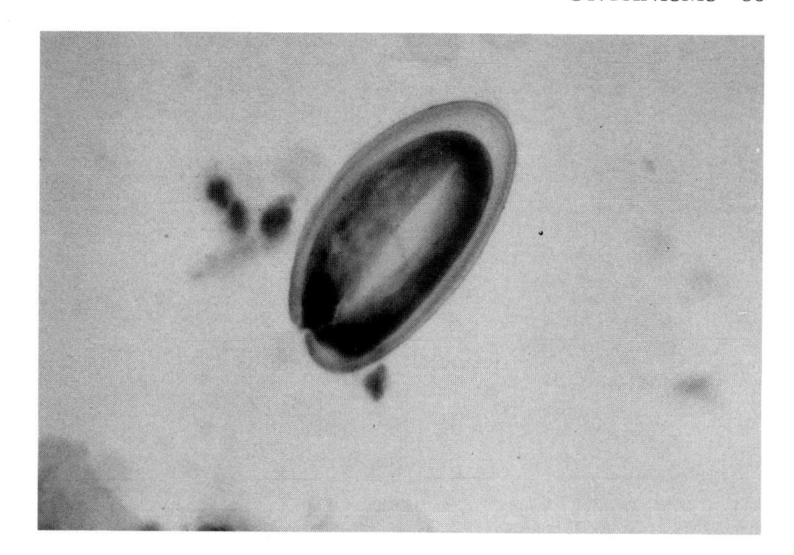

FIG. 5-6
PINWORM *(ENTEROBIUS VERMICULARIS)* OVUM IN URINE SEDIMENT RESULTING FROM ANAL CONTAMINATION. Note the thick continuous capsule. (× 1000)

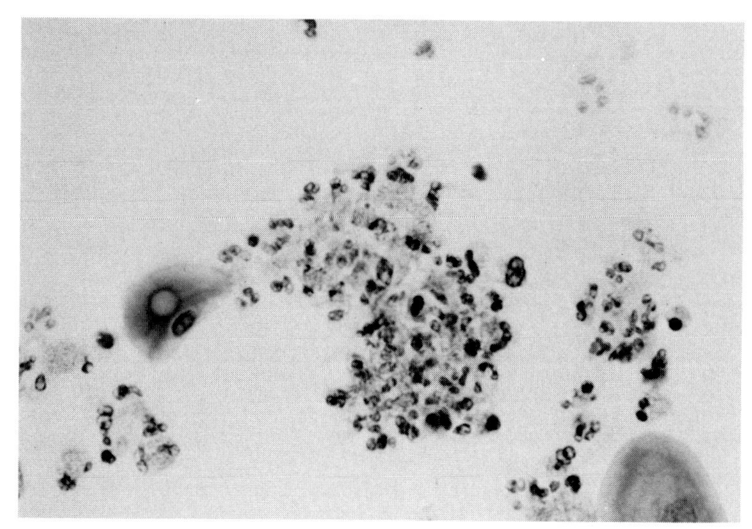

FIG. 5-7
BACTERIA PHAGOCYTIZED BY NEUTROPHILS. Note the loose neutrophilic clusters with ingested bacteria. This pattern is often seen during early acute bacterial urinary tract infections. (× 400)

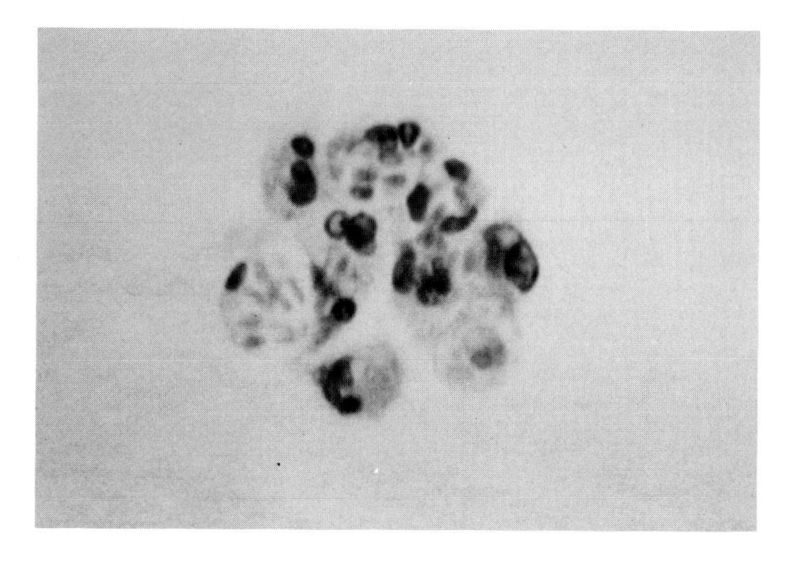

FIG. 5-8
FUNGI PHAGOCYTIZED BY NEUTROPHILS. Note discrete yeast forms. (× 1000)

FIG. 5-9
CYTOMEGALOVIRUS (CMV)-INDUCED INCLUSION-BEARING CELLS. **(A)** Nuclear enlargement, chromatinic membrane, basophilic intranuclear inclusions, and inclusion halo are characteristic features of a cytomegalovirus infection. (× 1000) **(B)** Both large intranuclear and a few smaller basophilic intracytoplasmic inclusions are present. Note large orange intracytoplasmic droplet representing a nonspecific degenerative change commonly seen in virally infected cells. (× 1000) **(C)** Giant cell with marked nuclear enlargement and a hyperchromatic nucleus found in the urine sediment of a transplant recipient with a CMV infection. Note the lack of intranuclear inclusions. A smaller cell with a viral inclusion (arrow) is present. (× 400) **(D)** CMV inclusions within glomerulus. Note cellular and nuclear enlargement with distinct basophilic intranuclear inclusions (arrows). Biopsy tissue from immunosuppressed renal allograft recipient. (× 250) **(E)** CMV inclusions within renal tubular epithelium. Classic CMV inclusions are present in attached and exfoliated cells. Note cellular and nuclear enlargement with basophilic intranuclear and faint cytoplasmic inclusions. (× 400).

FIG. 5-10
CYTOPATHIC EFFECTS OF A HERPES SIMPLEX INFECTION. **(A)** Multinucleation, chromatinic membranes and ground-glass nuclear changes are present in cells. Note the lack of distinct intranuclear inclusions. (× 1000) **(B)** Classic intranuclear inclusions are readily identified. (× 1000).

FIG. 5-11
(A) CYTOPATHIC EFFECT OF POLYOMAVIRUS INFECTION. Marked exfoliation of cells in the urine sediment during a polyomavirus infection. In comparison to CMV infections, urine sediments of patients infected with polyomavirus contain numerous degenerated epithelial cells. (× 400) **(B)** Note homogeneous dark nuclear change in polyomavirus-infected cell. (× 1000).

FIG. 5-12
DECOY CELLS IN URINE SEDIMENT. These cells have enlarged, homogeneous, hyperchromatic nuclei and a narrow rim of cytoplasm. Decoy cells should not be confused with malignant urothelial cells. Note morphological similarity to polyomavirus-infected cells. (× 1000).

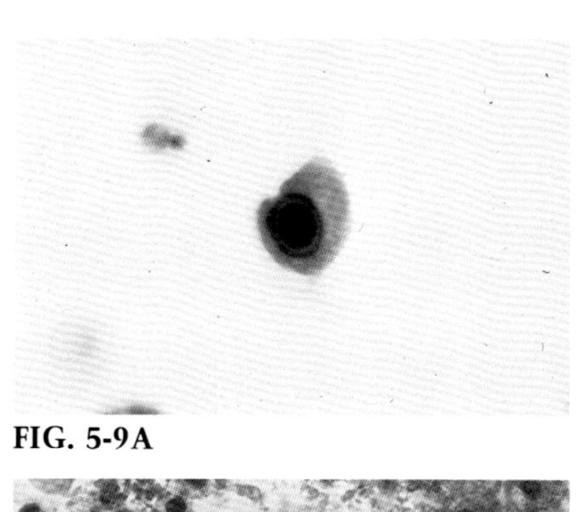
FIG. 5-9A

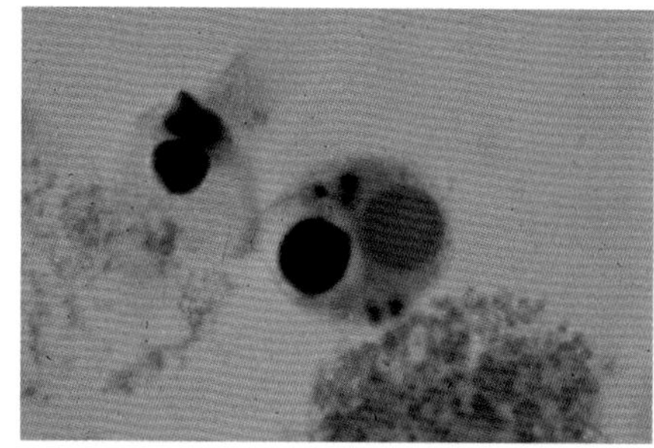
FIG. 5-9B

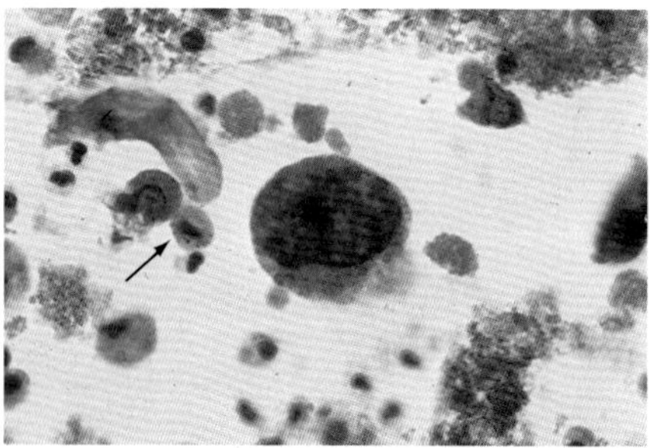
FIG. 5-9C

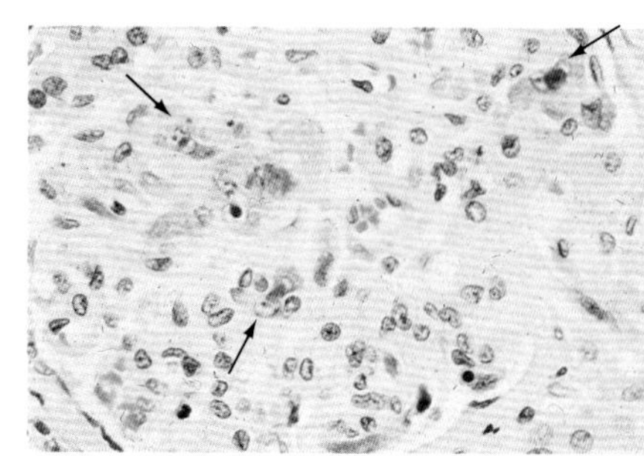
FIG. 5-9D

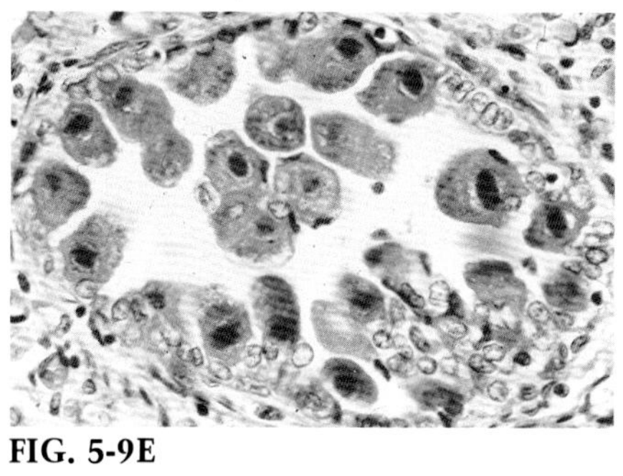
FIG. 5-9E

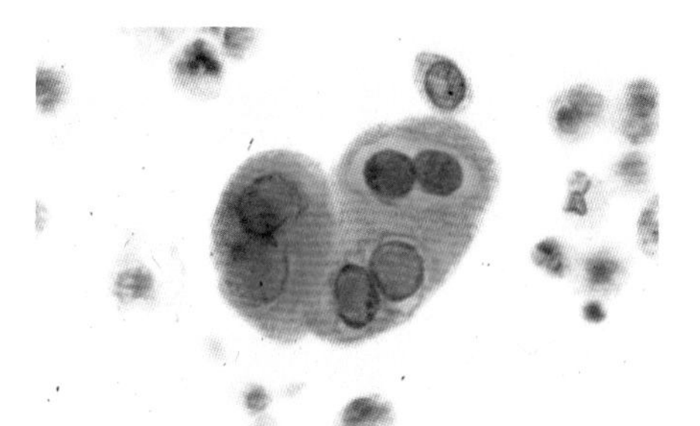
FIG. 5-10A

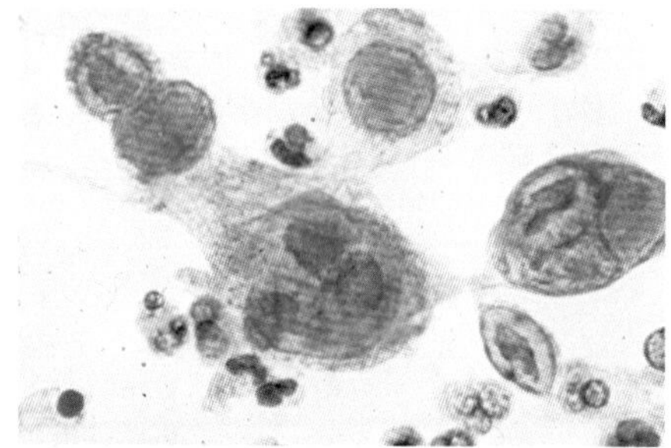
FIG. 5-10B

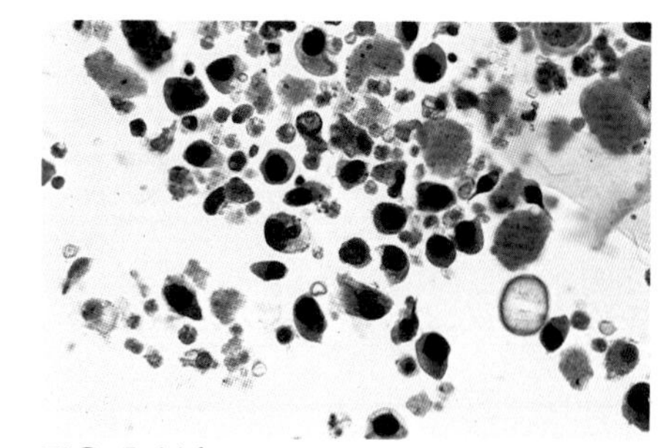
FIG. 5-11A

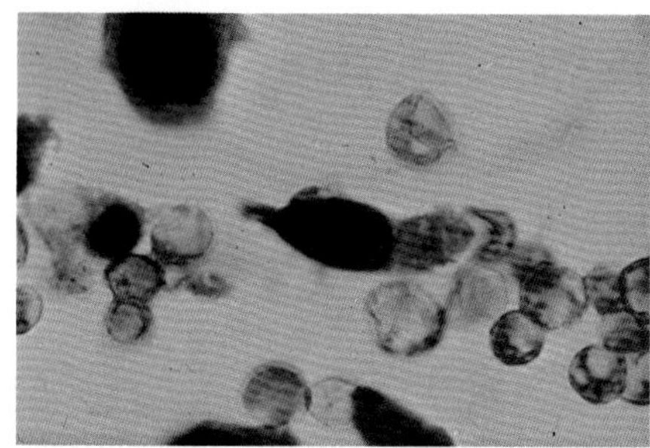
FIG. 5-11B

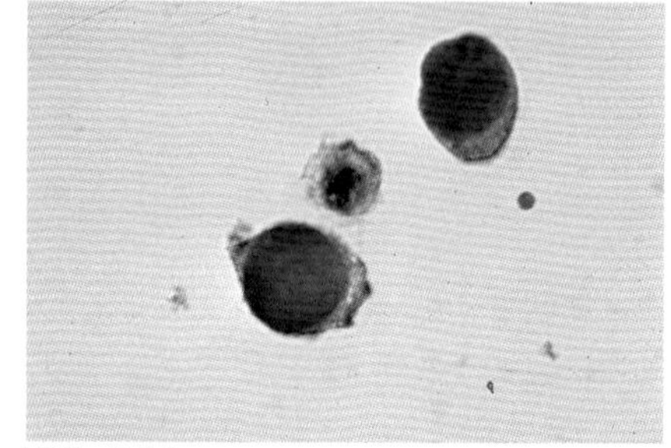
FIG. 5-12

6

EPITHELIUM

BENIGN LOWER URINARY TRACT EPITHELIAL CELLS

UROTHELIAL (TRANSITIONAL) CELLS

The surface of the lower urinary tract is lined by urothelial (transitional) cells. Few exfoliated urothelial cells are commonly seen in the urine sediment during the aging and maturation of the epithelium. Urinary tract diseases that alter urothelial surfaces will be reflected in the sediment by increased exfoliation of urothelial cells.

The urothelium is composed of superficial and deep cells (Fig. 6.1). Urothelial cells characteristically exhibit round to oval nuclei, dense cytoplasmic staining, and a pale outer ectoplasmic-endoplasmic rim (Fig. 6-2). In histologic sections, normal urothelium contains basal, intermediate, and superficial cells (Fig. 6-3). The average thickness of normal urothelium is five to seven cell layers. Brunn's nests occur most commonly in the trigone and may represent a variant of normal urothelium but are often regarded as the simplest form of proliferative cystitis (Fig. 6-4). Urine sediment changes would not be expected without an associated inflammatory lesion.

Urothelial cells are usually mononucleate, with the exception of superficial cells, which may have binucleated forms. Extreme examples of multinucleation, with superficial cells containing up to ten nuclei, are seen following stimulation such as bladder irrigation (Figs. 6-5 and 6-6).

SQUAMOUS CELLS

Squamous cells are frequently found in urine and can originate from the distal urethra in males and from the trigone area of the bladder in females. A properly obtained clean-catch specimen will minimize the amount of vaginal squamous epithelial contamination.

Squamous maturation represents an epithelial variant in the urinary bladder. Squamous cells have an appearance similar to those seen in gynecologic smears and are larger than urothelial cells (Fig. 6-7). Heavily glycogenated squames that are present in the urine of females may reflect cytohormonal stimulation (Fig. 6-8). The trigone in women of reproductive age frequently has variably glycogenated, nonkeratinizing, vaginal-type squamous epithelium (Fig. 6-9).

GLANDULAR CELLS

Glandular epithelial fragments are occasionally seen in urine sediment. (Figs. 6-10 and 6-11). These fragments are sometimes difficult to distinguish from exfoliated renal epithelial fragments. Exfoliated glandular epithelium may originate from aberrant surface epithelium, may represent contaminants from genital ducts, or may indicate proliferative cystitis. Occasionally, we have seen these exfoliated fragments in patients with cystitis cystica and glandularis. In proliferative cystitis, cystically dilated von Brunn's nests are present in the lamina propria (cystitis cystica). Glandular cells in the urine sediment are derived from columnar epithelium lining cysts (cystitis glandularis) (Fig. 6-12). There may be associated squamous metaplasia and variable inflammation (Fig. 6-13).

A less common glandular variant is a mucus-producing intestinal epithelium with goblet cells (Fig. 6-14). Many of the just-described conditions follow chronic irritation and inflammation. Colonic-type epithelium may be seen with proliferative cystitis, but is most often observed as aberrant epithelium in bladder exstrophy.

REGENERATIVE CELLS

Inflammatory conditions cause exfoliation of numerous reactive superficial and deep urothelial cells. Reactive or regenerative cellular changes are characterized by vesicular nuclei and prominent nucleoli (Figs. 6-15 and 6-16). Unlike dyskaryotic cells, the presence of nucleoli is not associated with hyperchromasia.

Sequential urine sediment examinations during chronic inflammatory conditions often show an increasing squamous epithelial component. These cells are characterized by small to pyknotic nuclei and abundant amphophilic or eosinophilic cytoplasm (Fig. 6-17). Large clusters of anucleated squames (Fig. 6-18) in urine sediment are consistent with hyperkeratosis and keratinizing squamous metaplasia. Contamination from external surfaces, for example, the vulva or vagina, must be excluded.

BENIGN RENAL EPITHELIAL CELLS

PROXIMAL AND DISTAL CONVOLUTED TUBULE CELLS

Besides exfoliation of epithelial cells from the lower urinary tract, an important potential source of epithelium in urine sediments is the renal parenchyma. Renal epithelial cells are rarely observed in urine sediment without renal parenchymal disease. Accurate identification usually requires a permanently stained preparation. Papanicolaou staining enhances cellular detail and provides the morphological information needed for identification and histologic correlation. The epithelium lining proximal convoluted tubules has abundant granular eosinophilic cytoplasm, indistinct cell borders, and uniform basally or centrally placed round nuclei (Fig. 6-19). Brush borders are occasionally identified as thin, dense lines along the luminal surfaces and can be delineated with the periodic acid-Schiff (PAS) stain.

The distal convoluted tubule has a wide lumen and is lined by cuboidal epithelium with pale, granular cytoplasm, prominent cell borders, and centrally placed nuclei, frequently with perinuclear cytoplasmic clearing (Fig. 6-20). One can appreciate more nuclei in cross sections of the distal convoluted tubules in comparison to proximal convoluted tubules.

Tissue imprints are useful for the cytologic characterization of renal tubular epithelium. Proximal convoluted tubular cells have round eccentric nuclei and eosinophilic granular cytoplasm with indistinct cell borders (Fig. 6-21).

Cells undergoing degeneration or necrosis lose their attachment to the underlying basement membrane and exfoliate into the urinary stream (Fig. 6-22). Convoluted tubular epithelial cells have only recently been recognized in urine sediment. Recognition of cells with prominent granular cytoplasm is important in accurate identification (Fig. 6-23). Cells with elongated shapes should not be misinterpreted as granular casts. Increased exfoliation of convoluted tubular cells indicates renal tubular injury.

COLLECTING DUCT CELLS

Early collecting duct epithelial cells are cuboidal, have homogeneous, finely granular cytoplasm, and distinct cell borders (Fig. 6-24). Similar features are present in tissue imprints (Fig. 6-25). In the urine sediment examination, these cells must be distinguished from other types of mononuclear cells.

Exfoliated cells of early collecting duct origin maintain a cuboidal shape (Fig. 6-26). With proper staining techniques one can distinguish the finely granular cytoplasm of collecting duct cells from the coarsely granular cytoplasm of convoluted tubular cells. Cells may appear in sediment singly or in loose clusters. Increased exfoliation of early collecting duct epithelium is frequently seen in cases of acute tubular necrosis and acute allograft rejection (Fig. 6-27).

With severe renal parenchyma ischemia, exfoliation of renal epithelium from convoluted tubules and early collecting ducts is increased and necrotic cells ("ghost cells") are prominent. In histologic sections and urine sediments, ghost cells have a distinctly granular cytoplasm without stainable nuclei or with nuclei undergoing karyolysis (Fig. 6-28). In urine sediments, renal epithelial ghost cells are typically seen with cellular debris, casts, and renal epithelial fragments (Fig. 6-29). Recognition of degrees of cellular degeneration and necrosis requires permanent stains.

The terminal collecting duct is lined by columnar epithelium with pale, finely granular cytoplasm and basally oriented nuclei (Fig. 6-30). Cell borders are distinct and give a typical "honeycomb" appearance to the epithelium, with tangential arrangements. The honeycomb arrangements of these columnar ductal cells can be recognized in tissue imprints of the lower renal medulla (Fig. 6-31).

EXFOLIATED BENIGN RENAL EPITHELIAL FRAGMENTS

Renal epithelial fragments in urine sediment have only recently been recognized and their origin delineated. Epithelial fragments are defined as structures containing at least three cells that maintain intercellular cohesiveness. In histologic sections, fragments have been observed to arise from collecting duct epithelium in association with basement membrane disruption (Fig. 6-32). They may appear as solid epithelial structures (Fig. 6-33), in sheets, or in a papillary configuration (Fig. 6-34). An additional type of renal epithelial fragment in histologic material is epithelial encasement of cast-like material (Fig. 6-35).

Fragments are usually composed of cuboidal to low columnar epithelial cells with finely granular cytoplasm and distinct cell borders. Cells may show regenerative changes, including cytoplasmic amphophilia and nucleolar prominence. These fragments have been identified with severe ischemic tubular necrosis. In this setting, disruption of convoluted tubules is commonly observed; however, the epithelium of convoluted tubules exfoliates as individual cells or loose clusters rather than as fragments.

Characteristics of renal epithelial fragments in urine sediments are summarized in Table 6-1.

Configurational characteristics of renal epithelial fragments in urine sediments are as follows: the renal epithelium is attached to or moulded by casts (Fig. 6-36**A**) and there is epithelial encasement of casts (Fig. 6-36**B**), cylindrical (sleeve-like) epithelium (Fig. 6-37), epithelial sheets with a "honeycomb" arrangement (Fig. 6-38), and epithelium containing spindle or elongated cells (Figs. 6-39, 6-40, and 6-41). Additional characteristics of renal epithelial fragments include intracytoplasmic pigment, for example, hemosiderin (Fig. 6-42), and lipochrome (Fig. 6-43) or crystals. These cytoplasmic inclusions are rarely, if ever, seen in urothelial cells.

The presence of renal fragments in urine sedi-

TABLE 6-1
CHARACTERISTICS OF RENAL EPITHELIAL FRAGMENTS

CONFIGURATIONAL CHARACTERISTICS
Epithelium attached to or moulded by casts
Epithelial encasement of casts
Cylindrical (sleeve-like) epithelium
Epithelial sheets with "honeycomb" arrangement
Epithelium containing spindle or elongate cells
ADDITIONAL CHARACTERISTICS
Epithelium with intracytoplasmic pigmentation, e.g. hemosiderin
Epithelium containing crystals

ment indicates ischemic necrosis and in the proper clinical setting one can suggest the possibility of acute tubular necrosis, renal infarction or cortical necrosis, or papillary necrosis. In urine sediment, renal fragments are usually accompanied by increased exfoliation of degenerated and necrotic renal tubular cells, pathologic casts, bleeding, and variable inflammation. With infarction of the papillary tip, exfoliation of large epithelial fragments occurs. Fragments may arise from collecting duct epithelium (Figs. 6-44 and 6-45) or urothelium lining the papillary tip or calyx (Figs. 6-46 and 6-47).

LOWER URINARY TRACT ATYPIAS

HISTOPATHOLOGY

Lower urinary tract atypias include degenerative cellular changes, such as those induced by chemotherapy and radiotherapy, and may be due to hyperplasia and dyskaryosis (vs. carcinoma *in situ* [CIS]). Accurate identification of urothelial atypias requires assessment of nuclear chromatin and cellular maturation.

Degeneration (Chemotherapy, Irradiation)

Chemotherapy and radiotherapy are common causes of urothelial atypia in urine sediment. Changes are characterized by varying degrees of nuclear enlargement, hyperchromasia, and altered cytoplasmic maturation. Further discussion and description of cellular changes induced by chemotherapy and radiotherapy as appears in Chapter 16.

Hyperplasia With or Without Atypia

Urothelial hyperplasia is defined as nonpapillary epithelium containing more than seven cell layers. The cells composing hyperplastic urothelium are indistinguishable from normal urothelium (Figs. 6-48 and 6-49). In atypical urothelial hyperplasia (Fig. 6-50), the thickened urothelium shows disturbed maturation and variable cytologic abnormalities.

Urothelial Atypia (Without Hyperplasia)

A similar spectrum of cytologic changes can occur in nonhyperplastic urothelium, and difficulty may arise in distinguishing severe atypia from CIS (Fig. 6-51). The diagnosis of CIS should be reserved for the finding of full-thickness changes characterized by complete lack of maturation, crowding of small cells with hyperchromatic nuclei, and minimal nuclear pleomorphism (Fig. 6-52). The histopathologic diagnosis is made independent of urothelial thickness.

This spectrum of urothelial hyperplasia and atypias, as well as CIS, is frequently found in patients with papillary transitional cell carcinoma. The incidence increases in the presence of high-grade papillary lesions. In addition, they may precede, coincide with, or follow the clinical presentation of a papillary transitional cell carcinoma. Therefore, in high-risk populations, serial urine cytology examinations and cystoscopy provide optimum followup and clinical management.

CYTOLOGIC DIFFERENTIAL DIAGNOSIS

Identification of urothelial atypias in urine sediment requires accurate assessment of nuclear chromatin content and texture and cytoplasmic maturation. Careful microscopic examination of exfoliated urothelial cells and fragments is essential.

Urothelial atypia may cause the exfoliation of increased numbers of cells and fragments. The persistence of urothelial fragments in a spontaneously voided urine is an important finding in establishing the diagnosis. Catheterized urine specimens often contain benign urothelial fragments (Fig. 6-53). Although there is cellular crowding, the cells have normochromatic nuclei. Careful microscopic examination may demonstrate cytoplasmic maturation at the periphery of the fragment. The lack of peripheral cytoplasmic maturation in a large urothelial fragment suggests exfoliation from a proliferative or hyperplastic lesion (see 16-1, Chap. 16).

Atypical urothelial cells are characterized by nuclear enlargement and irregularity, mild hyperchromasia with degenerative chromatin clumping, and an occasional micronucleolus (Fig. 6-54). These cells are seen in urine sediment in the following situations: during treatment with chemotherapy or radiation, following passage of stones, and with atypical hyperplasia or transitional cell carcinoma. Atypical urothelial cells with dyskaryotic features (Fig. 6-55) are characterized by nuclear enlargement and hyperchromasia with preserved chromatin texture and an increased nuclear-cytoplasmic ratio (Figs. 6-55 and 6-56). Attempts have been made to grade urothelial atypia as mild, moderate, or marked according to the degree of abnormality in the nuclear-cytoplasmic ratio. Markedly atypical urothelial cells cannot be distinguished from cells exfoliating from CIS. With CIS, however, one typically finds exfoliation of large numbers of markedly dyskaryotic cells often occurring in a syncytial arrangement (Fig. 6-57). The cellular features are similar to those of CIS cells from the cervix seen in gynecologic cytology smears. The recognition of markedly dyskaryotic cells, particularly in a clinical setting of negative results in a cystoscopic examination, is important. Because CIS may present as a multifocal or diffuse flat lesion, the cytologic diagnosis requires histologic confirmation.

LOWER URINARY TRACT NEOPLASIA

CARCINOMA *IN SITU*

CIS is a neoplastic lesion of nonpapillary urothelium with the biologic potential for invasion. For purposes of discussing differential characteristics, CIS is included in Table 16-1, Chapter 16.

PAPILLOMA

The transitional cell papilloma is not currently regarded as a malignant lesion. Because the urothelium is neither hyperplastic nor cytologically abnormal, a diagnosis from the urine sediment examination is not possible (Fig. 6-58).

PAPILLARY TRANSITIONAL CELL CARCINOMA, GRADES I TO III

The histologic grading of papillary transitional cell carcinomas, which was developed by the World Health Organization (WHO), is primarily based upon the degree of nuclear enlargement and hyperchromasia (Figs. 6-59, 6-60, and 6-61). In addition, high-grade lesions have pleomorphic nuclei and nucleoli with diminished intercellular cohesiveness, resulting in the loss of superficial cells and marked exfoliation of malignant cells (Fig. 6-62).

The cytologic diagnosis of malignant urothelial cells requires the identification of cells with pleomorphic, hyperchromatic nuclei and a high nuclear-cytoplasmic ratio. Exfoliated cells from a low-grade (grade I) papillary transitional cell carcinoma lack significant nuclear enlargement and hyperchromasia and, therefore, a diagnosis cannot be made. Exfoliated urothelial fragments from low-grade tumors are best categorized as atypical (Fig. 6-63). Differential diagnostic considerations should include chemotherapy, irradiation, and lithiasis, as well as low-grade transitional cell carcinoma.

Malignant nuclear features are more apparent in cells exfoliated from grade II papillary transitional cell carcinoma (Fig. 6-64). High-grade tumors (grades II–III) are characterized by increased epithelial exfoliation (Fig. 6-65). The presence of nucleoli in these malignant cells distinguishes high-grade from low-grade transitional cell carcinoma (Figs. 6-66 and 6-67). The presence of nucleoli in epithelial cells or fragments without hyperchromasia suggests a regenerative cell change rather than malignancy.

SQUAMOUS CELL CARCINOMA

Although high-grade transitional cell carcinomas may have malignant squamous component, lower urinary tract squamous cell carcinoma is composed almost entirely of malignant squamous cells (Fig. 6-68). Histologic grading of squamous cell carcinoma reflects the degree of keratinization. Exfoliated malignant cells derived from squamous cell carcinoma (Fig. 6-69) are characterized by pleomorphic pyknotic nuclei and abundant orangeophilic (keratinized) cytoplasm.

UNDIFFERENTIATED CARCINOMA

Undifferentiated carcinoma is composed of cells lacking glandular or squamous differentiation (Fig. 6-70). In addition, the tumor cytologically cannot be recognized as urothelial in origin.

ADENOCARCINOMA

Adenocarcinoma may arise from mesonephric remnants, aberrant colonic epithelium, and the urachus, or may be a component of high-grade papillary transitional cell carcinoma. Cystitis glandularis is not felt to be an origin of adenocarcinoma. The histologic appearance is variable depending upon the origin, although many resemble tumors of enteric or colonic origin (Fig. 6-71). Although the majority occur in the bladder trigone or dome, one rarely encounters a renal pelvic source (Fig. 6-72). Adenocarcinoma in the urine most often is metastatic in origin.

METASTATIC CARCINOMA

Secondary involvement of the urinary tract usually results from direct extension from a primary tumor or tumors in adjacent organs. Not uncommonly, prostatic carcinoma in males (Figs. 6-73 and 6-74) and endometrial and cervical carcinoma in females. (Fig. 6-75) can be diagnosed by urine cytology. Diffuse dissemination or widespread metastases from carcinoma of the breast (Figs. 6-76 and 6-77) or gastrointestinal tract, malignant melanoma, and lymphoma may also involve the urinary tract.

RENAL NEOPLASIA

RENAL CELL CARCINOMA

Renal cell carcinoma is a malignancy derived from the convoluted tubular epithelium (Fig. 6-78). The histologic appearance may be deceptively benign, especially in cystic, hemorrhagic lesions. A reproducible grading system has not been successfully developed. Urine cytology diagnosis usually requires the extension of tumor into the renal pelvis. With a clinically documented renal lesion and negative urine cytologic results, fine-needle aspiration is the recommended diagnostic procedure. Malignant cells derived from renal cell carcinoma (Fig. 6-79) have nuclear enlargement, slight hyperchromasia, a prominent, centrally located nucleolus, and abundant foamy cytoplasm.

WILMS' TUMOR

Wilms' tumor is a childhood malignancy derived from renal blastema and showing varying degrees of mesenchymal and abortive tubular differentiation. Fine-needle aspiration cytology may be helpful in establishing a diagnosis.

BIBLIOGRAPHY

Koss, LG: Diagnostic Cytology and Its Histologic Bases, 2nd ed. Philadelphia, JB Lippincott, 1979

Schumann, GB: The Urine Sediment Examination. Baltimore, Williams & Wilkins, 1980.

Tweeddale, DN: Urinary Cytology. Boston, Little, Brown, 1977

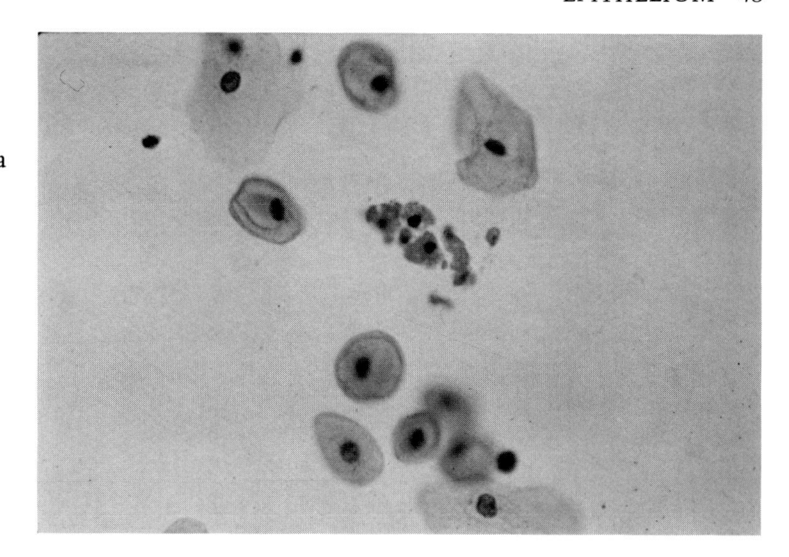

FIG. 6-1
NORMAL TRANSITIONAL (UROTHELIAL) CELLS IN URINE SEDIMENT. Superficial and deep urothelial cells are present in addition to a loose cluster of degenerated renal tubular epithelial cells. (× 400)

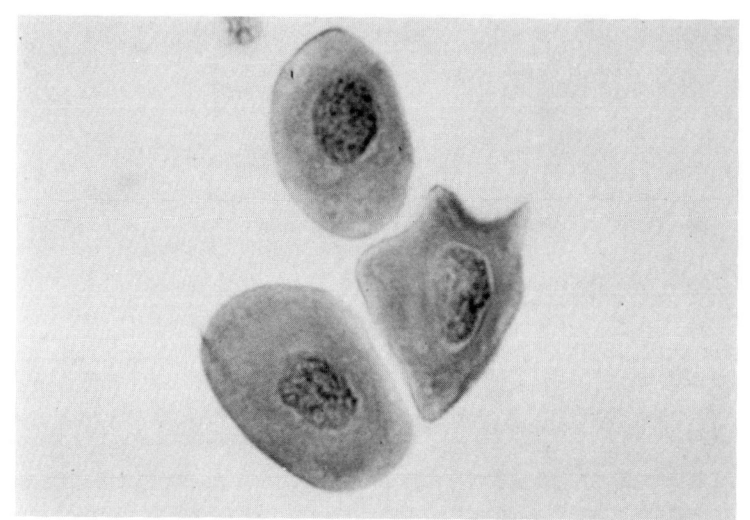

FIG. 6-2
TRANSITIONAL (UROTHELIAL) CELLS IN URINE SEDIMENT. Higher magnification of urothelial cells showing round to oval nuclei, dense cytoplasmic staining, and a pale-staining outer rim. (× 1000)

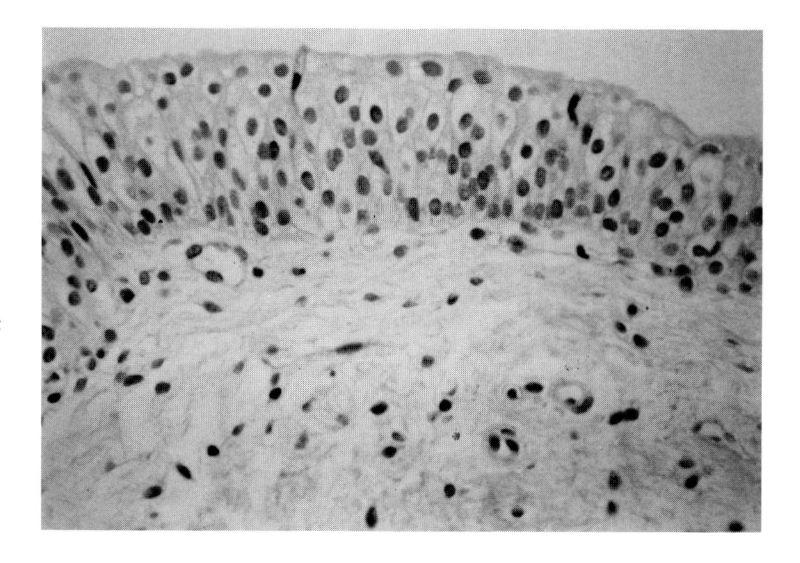

FIG. 6-3
NORMAL UROTHELIUM FROM AN UNDISTENDED BLADDER. Cuboidal superficial cells cover smaller polygonal intermediate cells in umbrella-like fashion. The average thickness of the urothelium, including basal cells, is five to seven cell layers; however, the thickness and prominence of the mononucleated or binucleated umbrella transitional cells will reflect the degree of bladder distension. (× 250)

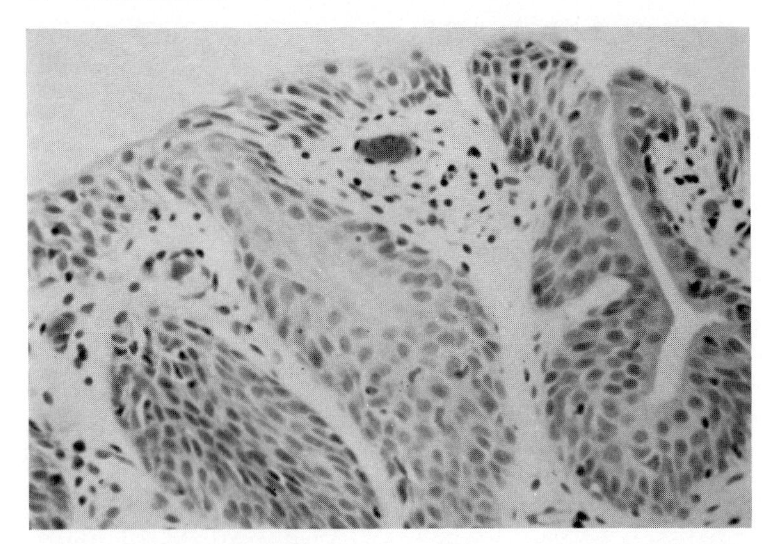

FIG. 6-4
BRUNN'S NEST. Normal surface urothelium with finger-like projections into lamina propria of small urothelial cells forming solid nests or having central lumens. (× 250)

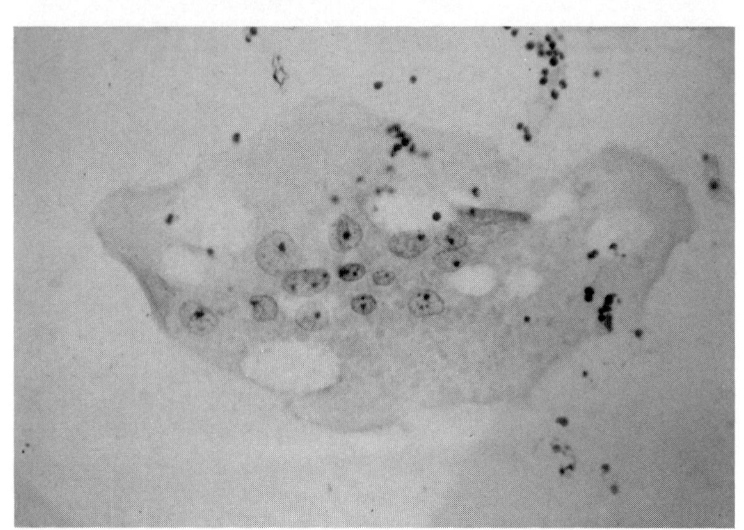

FIG. 6-5
MULTINUCLEATED UROTHELIAL CELL SEEN FOLLOWING BLADDER IRRIGATION. Note cellular enlargement and prominent cytoplasmic vacuoles. (× 400)

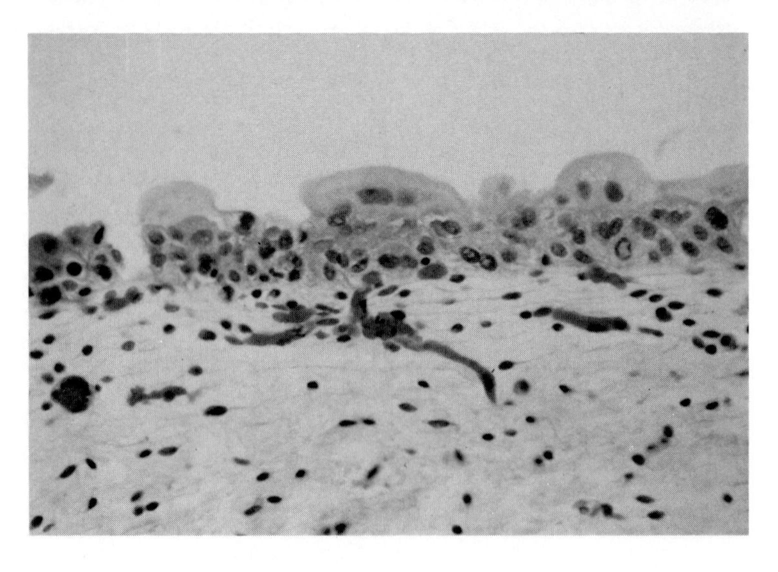

FIG. 6-6
NORMAL UROTHELIUM WITH PROMINENT UMBRELLA CELLS. In this bladder biopsy following distention and saline irrigation, umbrella cells are enlarged and multinucleated. Urothelium is three to four cell layers thick. (× 250)

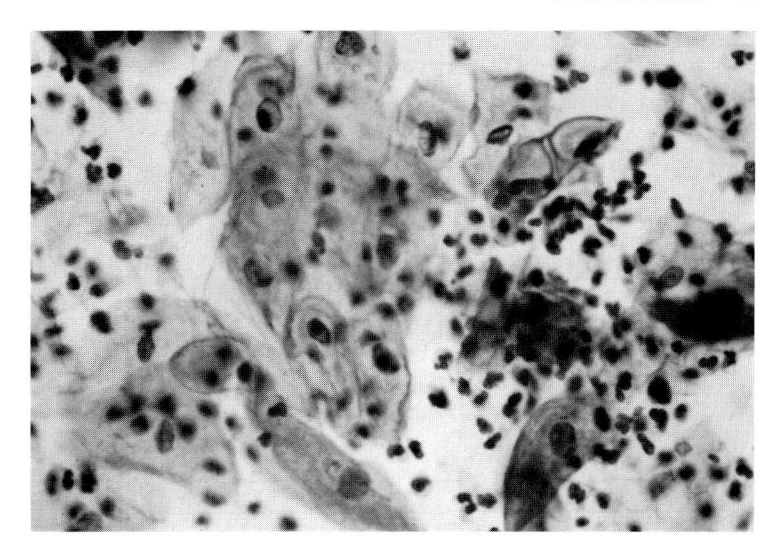

FIG. 6-7
EXTENSIVE ACUTE INFLAMMATION AND NUMEROUS SQUAMES IN URINE SEDIMENT. (× 400)

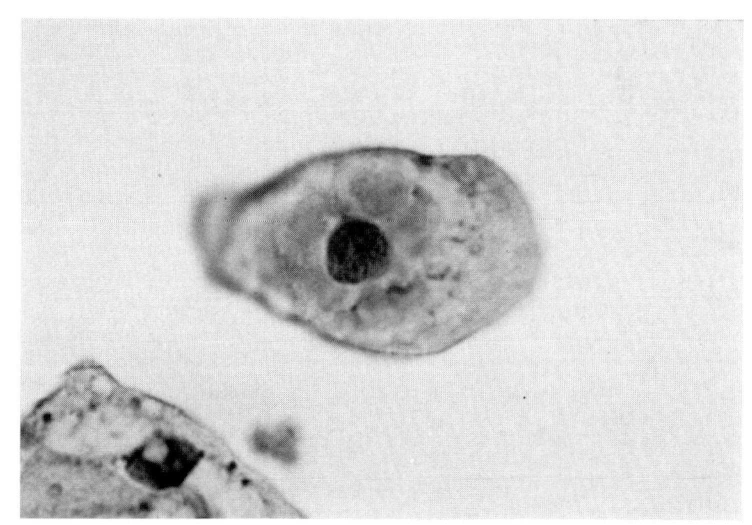

FIG. 6-8
HEAVILY GLYCOGENATED SQUAME IN URINE SEDIMENT. (× 1000)

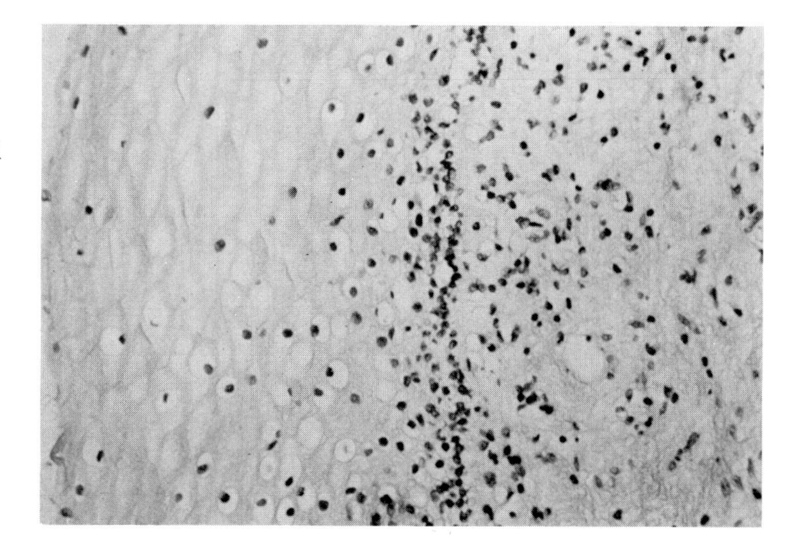

FIG. 6-9
EPITHELIAL VARIANT IN URINARY BLADDER. Nonkeratinizing vaginal-type squamous epithelium from trigone in woman of reproductive age. There is mild chronic inflammation in lamina propria. (× 160)

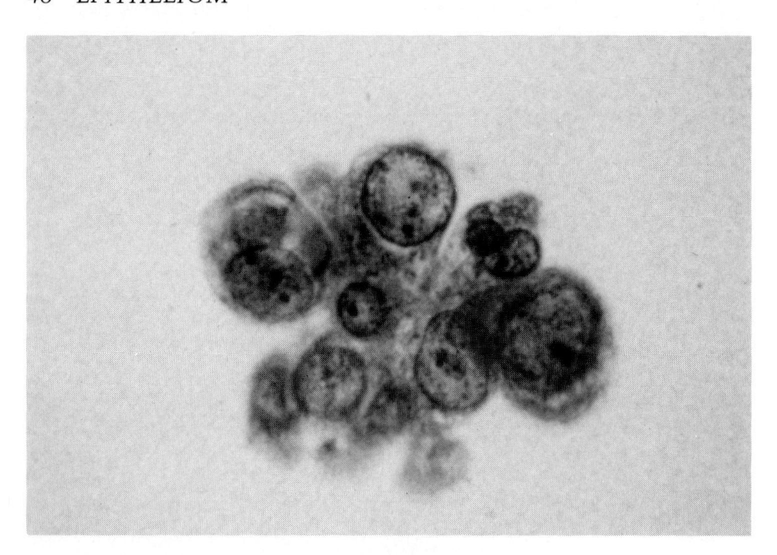

FIG. 6-10
GLANDULAR EPITHELIAL FRAGMENT IN URINE SEDIMENT. (× 1000)

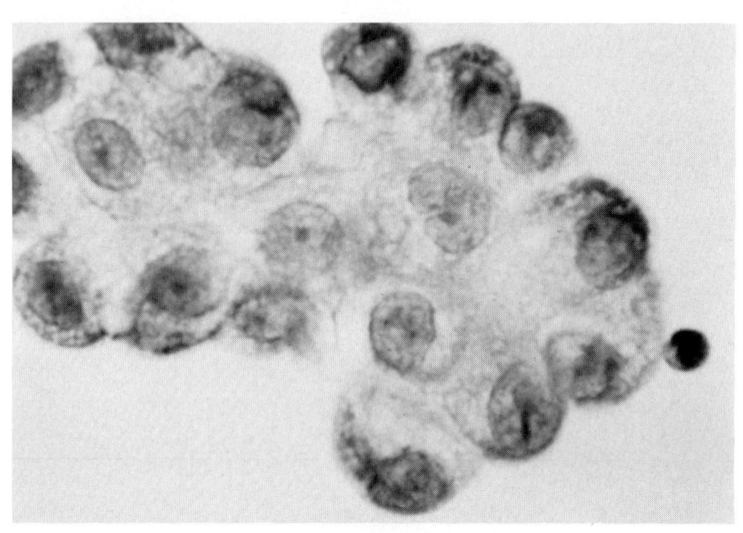

FIG. 6-11
GLANDULAR EPITHELIAL FRAGMENT IN URINE SEDIMENT. Cells have eccentric nuclei, prominent nucleoli, and abundant finely vacuolated cytoplasm. (× 1000)

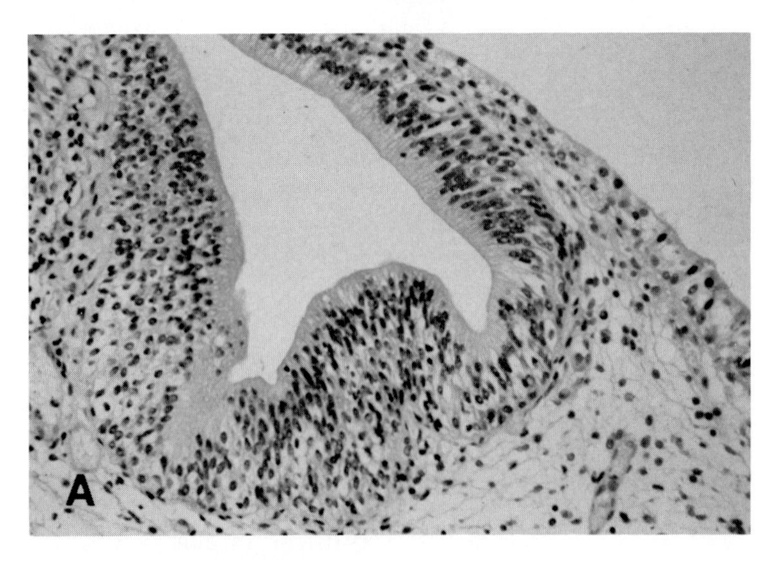

FIG. 6-12
CYSTITIS CYSTICA AND GLANDULARIS. **(A)** Cyst lies beneath normal urothelium within lamina propria. Although variably associated, there is edema and mild chronic inflammation in this biopsy. (× 160) **(B)** Note stratified columnar epithelial lining. (× 400)

(continued)

FIG. 6-12B (continued)

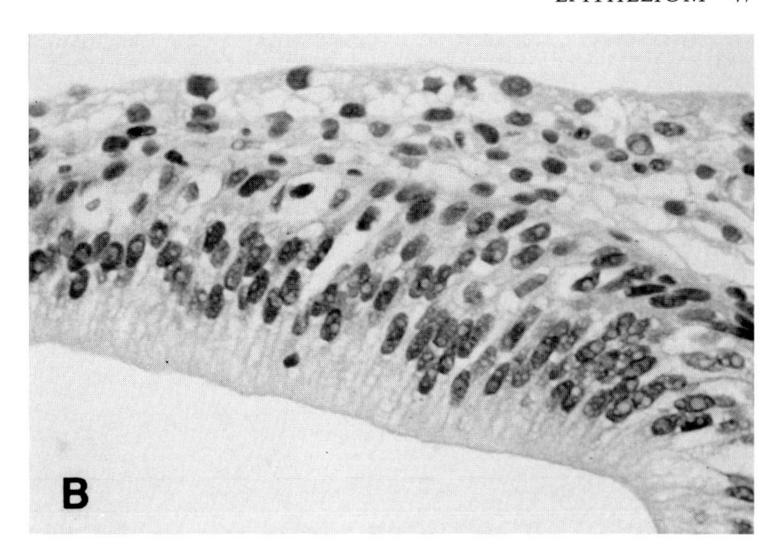

FIG. 6-13
CYSTITIS CYSTICA AND GLANDULARIS WITH SQUAMOUS METAPLASIA. **(A)** In addition to cystically dilated Brunn's nests, which contain mucus and focally have glandular metaplasia, one nest is filled with metaplastic squamous epithelium. There is an associated chronic inflammatory infiltrate in the lamina propria. (× 160) **(B)** The surface epithelium is also replaced by metaplastic nonkeratinizing squamous epithelium. (× 250)

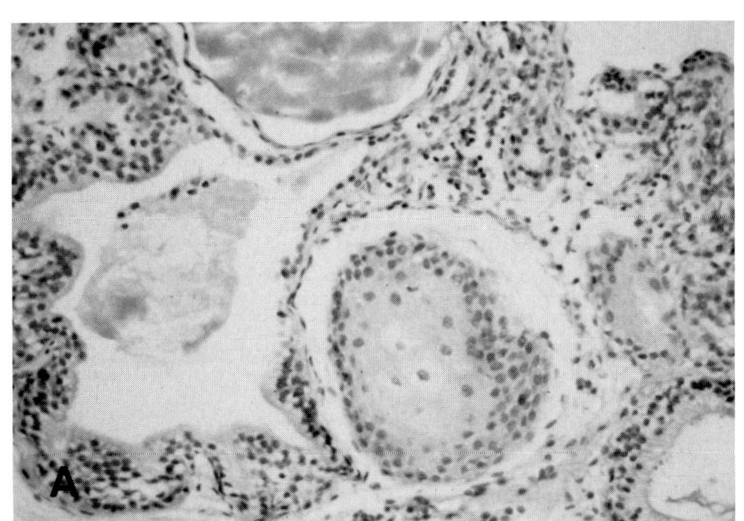

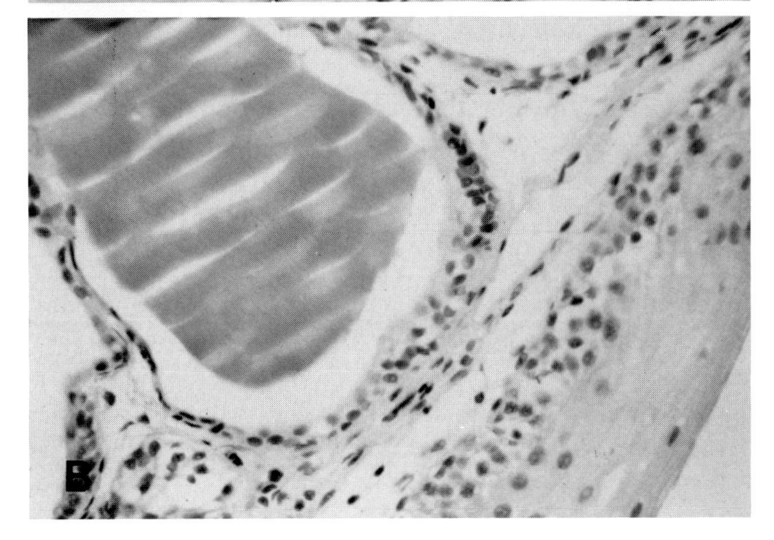

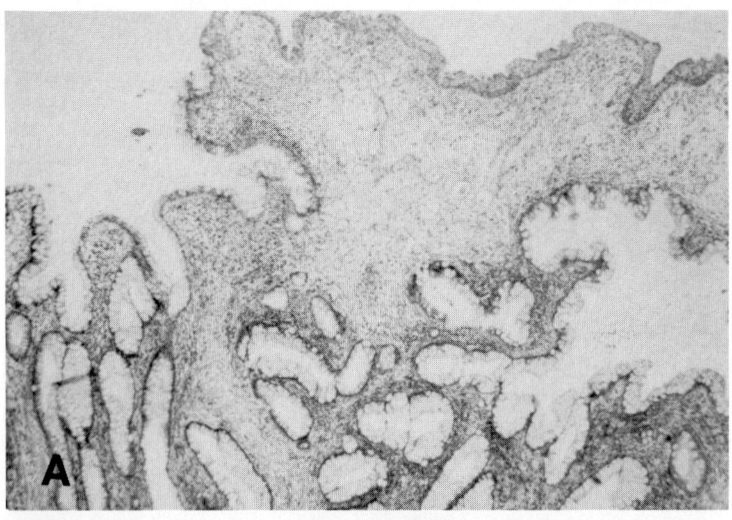

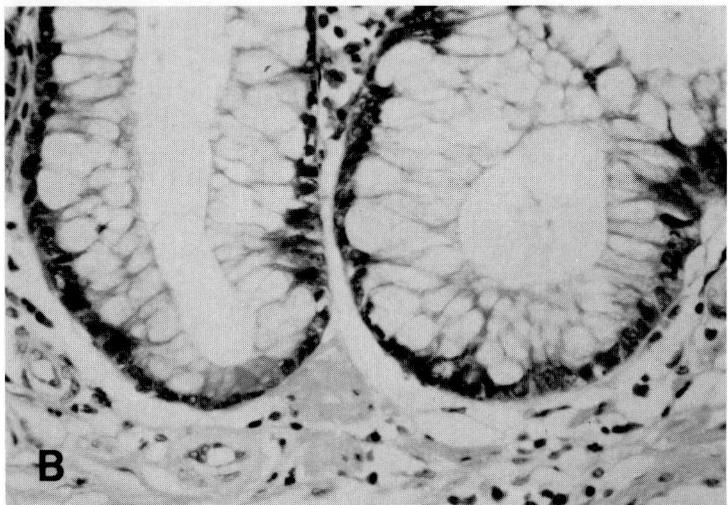

FIG. 6-14
EPITHELIAL VARIANT IN URINARY BLADDER. **(A)** The bladder surface epithelium and glands within the lamina propria are lined by mucus-producing intestinal epithelium with goblet cells. (× 25) **(B)** In addition to goblet cells, note the Paneth's cells at the base of the gland crypts. (× 250)

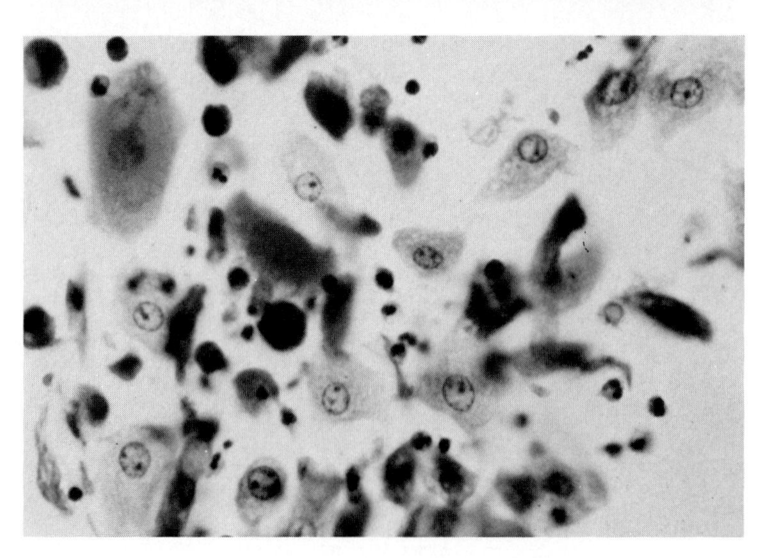

FIG. 6-15
NUMEROUS REACTIVE SUPERFICIAL AND DEEP UROTHELIAL CELLS IN URINE SEDIMENT WITH MODERATE INFLAMMATORY BACKGROUND. Reactive urothelial cells are characterized by vesicular nuclei containing prominent nucleoli. (× 400)

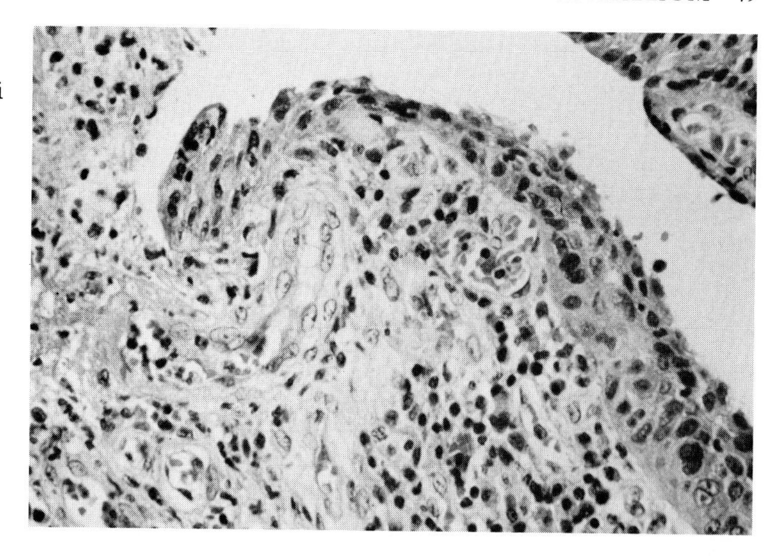

FIG. 6-16
BLADDER ULCER WITH REGENERATING UROTHELIUM. Nuclei vary in size, shape, and hyperchromatism. Nucleoli are frequently prominent. (× 250)

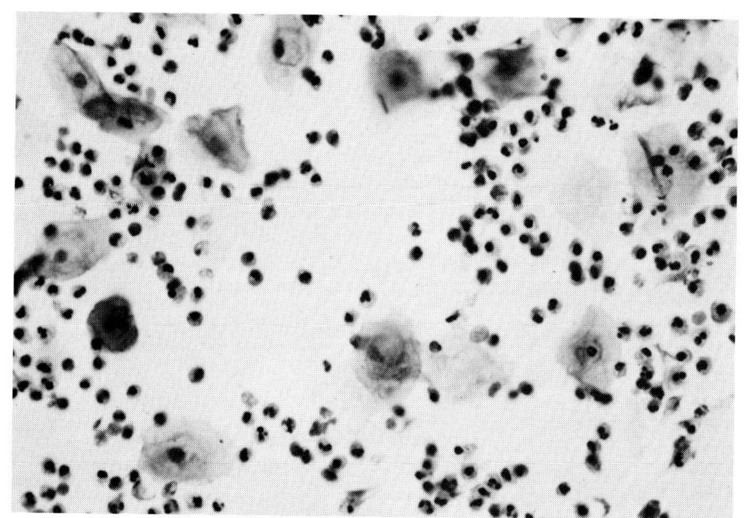

FIG. 6-17
EXTENSIVE INFLAMMATORY PATTERN WITH SEVERAL SQUAMES SEEN IN THE URINE SEDIMENT DURING CYSTITIS. Anucleated squames are consistent with hyperkeratosis or contamination. (× 400)

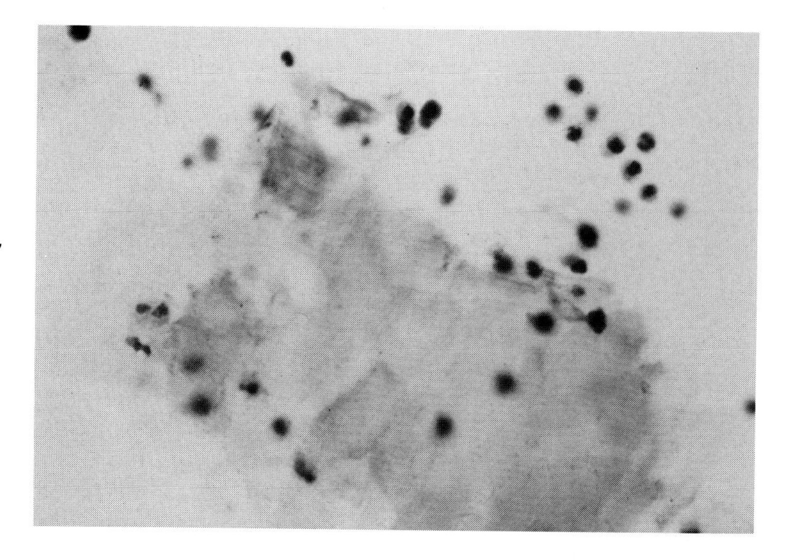

FIG. 6-18
LARGE CLUSTER OF ANUCLEATED SQUAMES IN URINE SEDIMENT CONSISTENT WITH HYPERKERATOSIS. Hyperkeratosis can be seen with keratinizing squamous metaplasia or can indicate urinary contamination from external surfaces (e.g., vulva or vagina). (× 400)

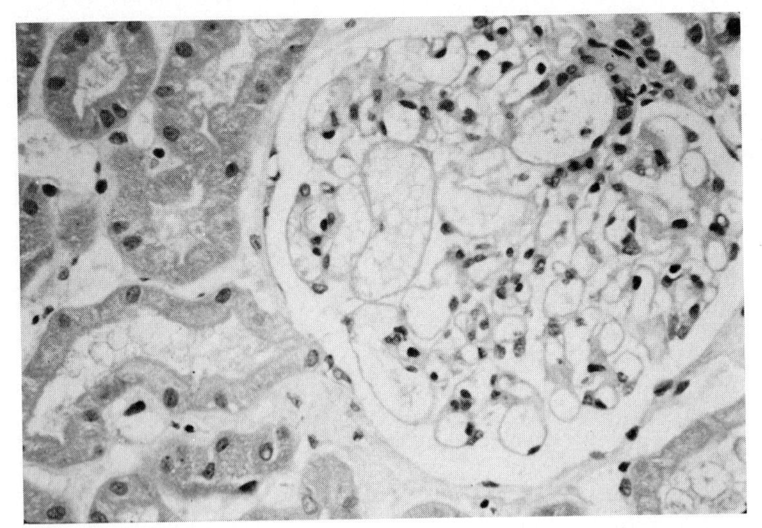

FIG. 6-19
NORMAL GLOMERULUS WITH SURROUNDING PROXIMAL CONVOLUTED TUBULES. Tubular epithelium has abundant granular eosinophilic cytoplasm, indistinct cell borders, and uniform basally or centrally placed round nuclei. Brush borders are occasionally identified as a thin, dense line along luminal surfaces. (× 250)

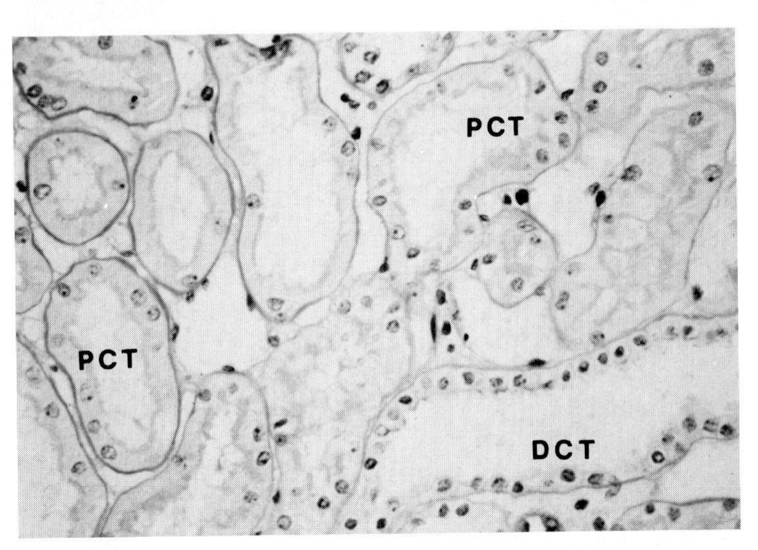

FIG. 6-20
PROXIMAL CONVOLUTED TUBULE (PCT) AND DISTAL CONVOLUTED TUBULE (DCT) EPITHELIUM. Note PCTs have PAS-positive brush borders. (PAS × 250)

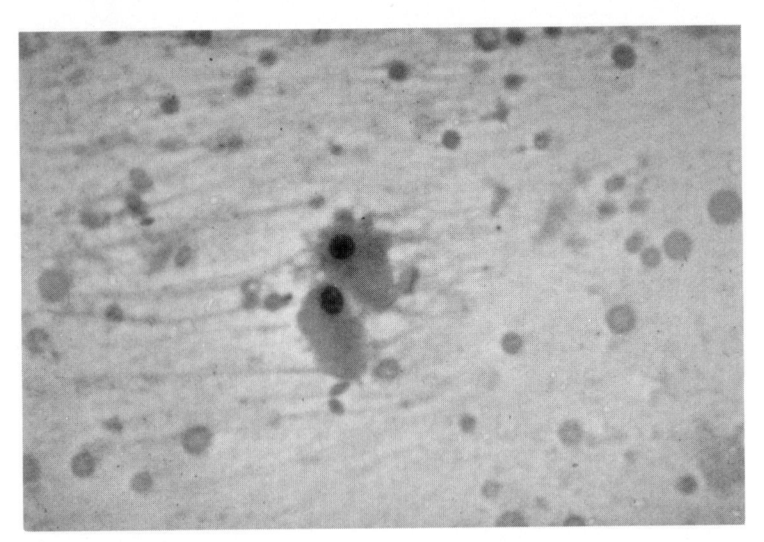

FIG. 6-21
IMPRINT OF RENAL CORTEX SHOWS TWO EPITHELIAL CELLS ORIGINATING FROM PROXIMAL CONVOLUTED TUBULES. Note the round eccentric nuclei and the eosinophilic granular cytoplasm with indistinct cell borders. (× 400)

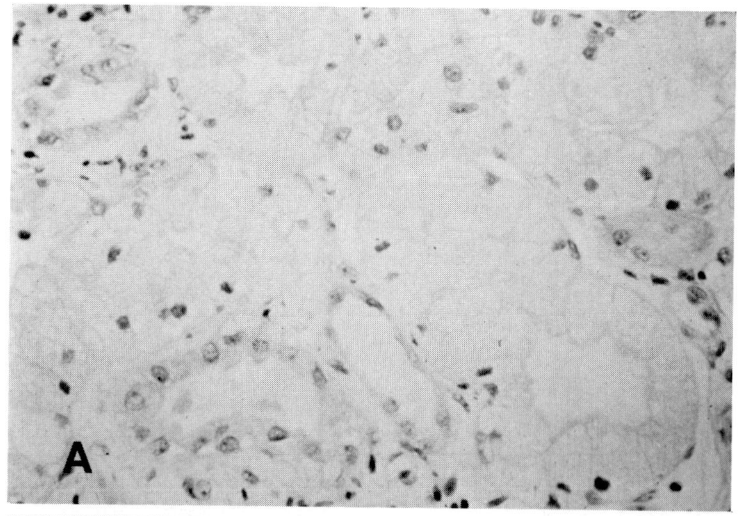

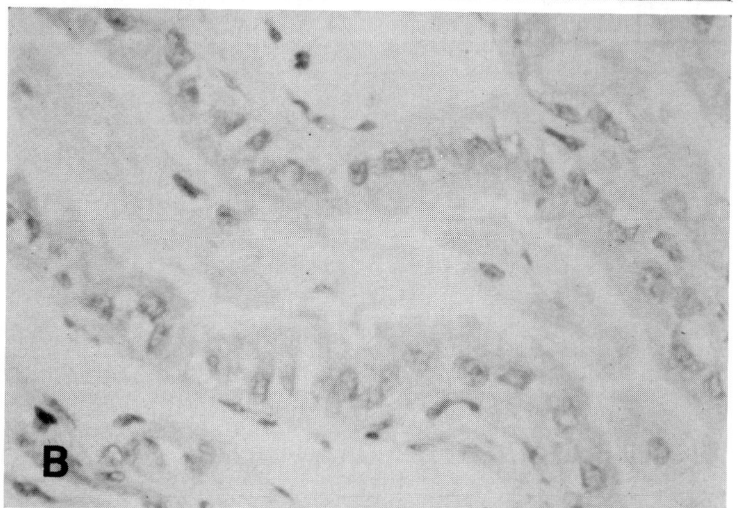

FIG. 6-22
EPITHELIAL DEGENERATIVE CHANGE WITH EXFOLIATION) **(A)** PCT epithelium has abundant foamy, slightly granular cytoplasm secondary to marked fatty change. (× 250) **(B)** Individual cells and cohesive sheets of PCT epithelium are present in the medullary portion of the biopsy within the lumen of a collecting duct. (× 400)

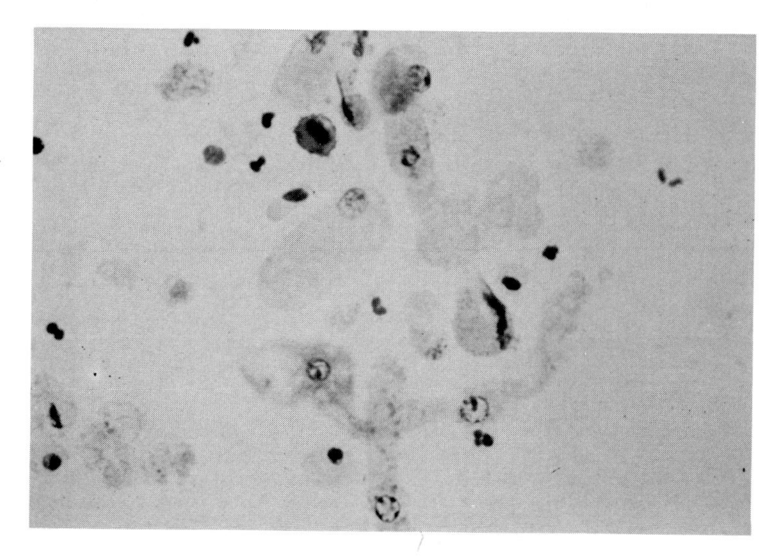

FIG. 6-23
CONVOLUTED TUBULAR CELLS IN URINE. Note numerous cells with reactive nuclei and abundant granular cytoplasm. Elongated granular cells should be distinguished from granular casts. (× 400)

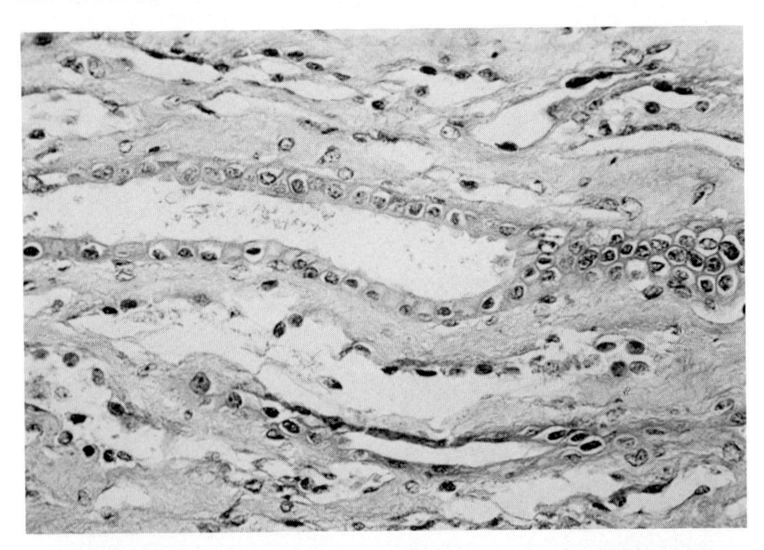

FIG. 6-24
EARLY COLLECTING DUCT WITH WIDE LUMEN AND MANY NUCLEI PRESENT IN CROSS SECTION. Note homogeneous finely granular cytoplasm and distinct cell borders. (× 250)

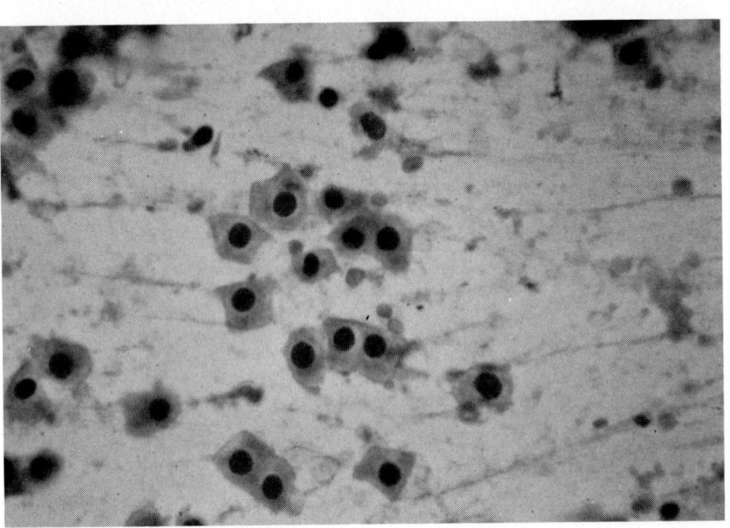

FIG. 6-25
IMPRINT OF RENAL MEDULLA CONTAINING NUMEROUS CUBOIDAL OR POLYGONAL EPITHELIAL CELLS OF EARLY COLLECTING DUCT ORIGIN. Note the round, centrally placed nucleus, perinuclear cytoplasmic clearing, and *finely* granular cytoplasm. (× 400)

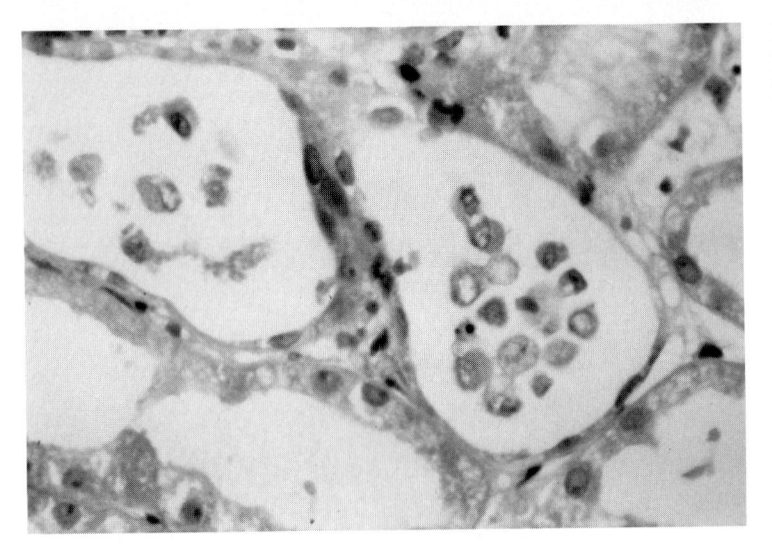

FIG. 6-26
EXFOLIATION OF RENAL EPITHELIUM IN CASE OF ACUTE TUBULAR NECROSIS. Cells maintain cuboidal appearance with variably dense, homogeneous, and finely granular cytoplasm and nuclei that are frequently eccentric. These features suggest exfoliation from the DCT or the early collecting duct epithelium. (× 400)

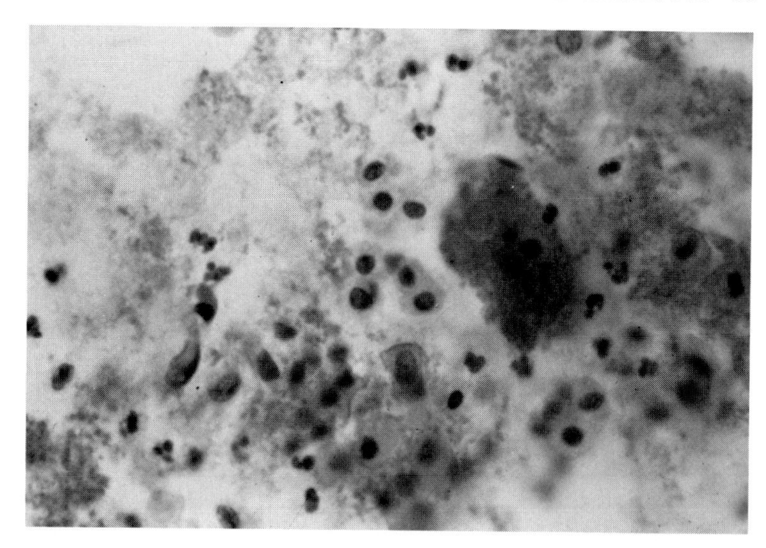

FIG. 6-27
MARKED RENAL EPITHELIAL CELL EXFOLIATION DURING ACUTE ALLOGRAFT REJECTION. Cells appear singly or in loose clusters. These intact exfoliated renal epithelial cells retain their cuboidal shape, have round to oval centrally placed nuclei, and have perinuclear clearing and finely granular cytoplasm with distinct cell borders. The majority of these renal epithelial cells originate from the early collecting duct. (× 400)

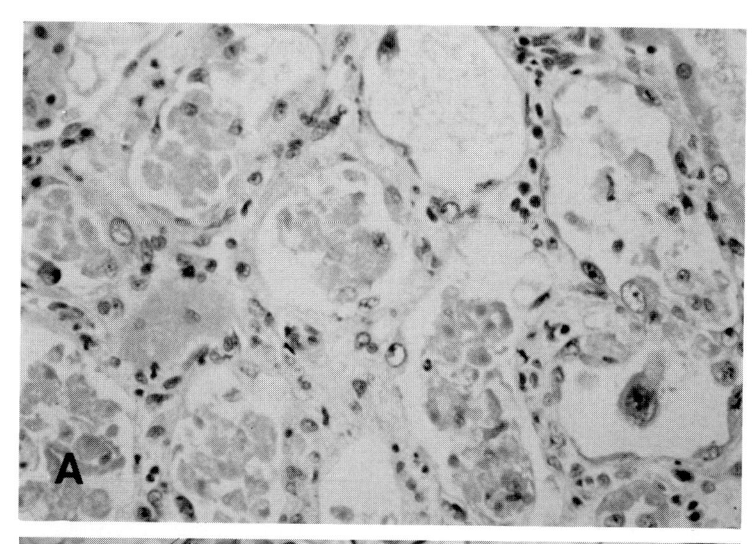

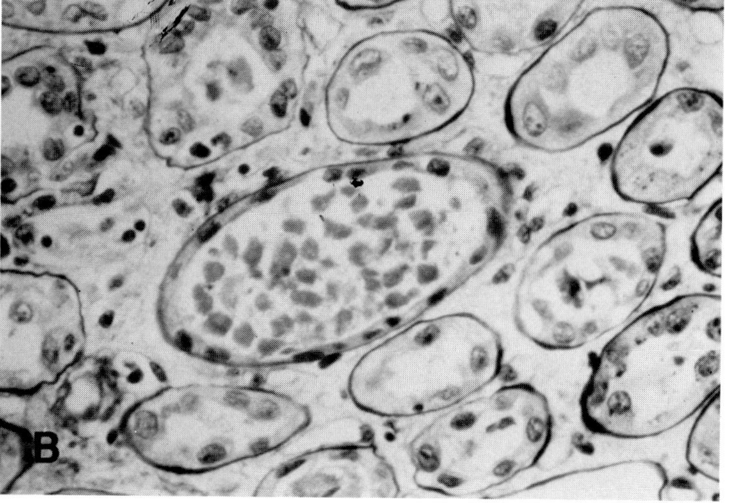

FIG. 6-28
EXFOLIATION OF NECROTIC RENAL EPITHELIUM DUE TO ISCHEMIC NECROSIS IN A RENAL TRANSPLANT. **(A)** Lumens contain aggregates of cells having eosinophilic granular cytoplasm without stainable nuclei ("ghost cells") or with nuclei undergoing karyolysis. Tubular basement membranes appear intact and are lined by an attenuated epithelium or more actively regenerating cells. (× 250) **(B)** Desquamated "ghost cells" are present distally in the lumen of a thin segment of convoluted tubule within the medullary portion of biopsy. (Jones × 250)

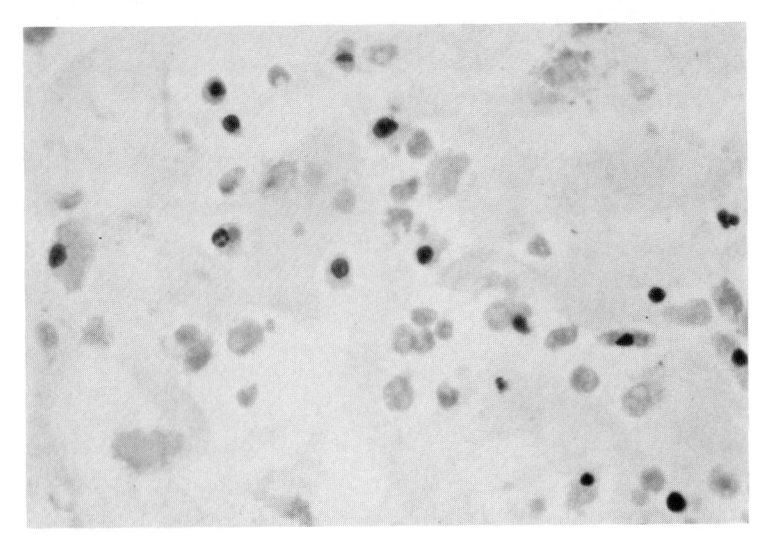

FIG. 6-29
URINE SEDIMENT FROM PATIENT WITH ACUTE TUBULAR NECROSIS CONTAINING NUMEROUS HYALINE CASTS AND DEGENERATED AND NECROTIC RENAL EPITHELIAL CELLS ("GHOST CELLS"). (× 400)

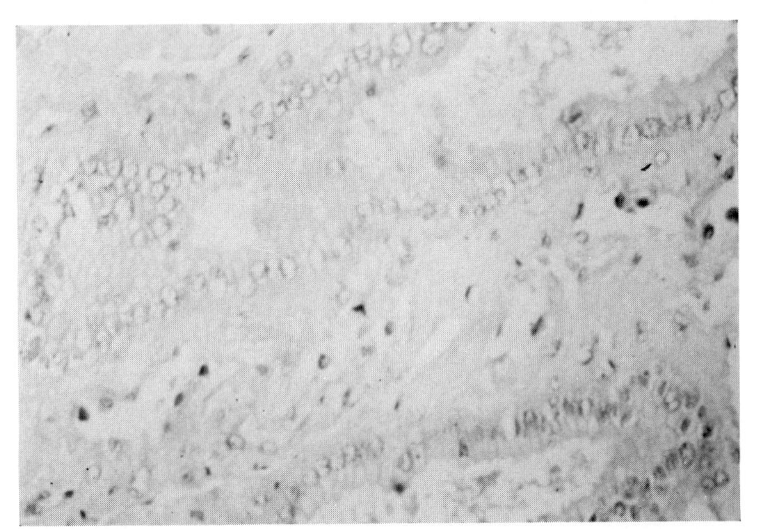

FIG. 6-30
TERMINAL COLLECTING DUCT. Note "honeycomb" appearance of epithelium when cut tangentially. (× 250)

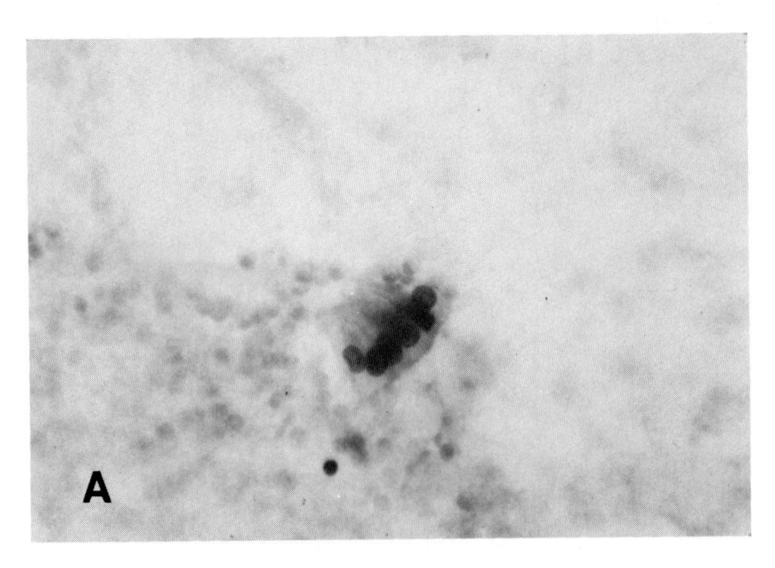

FIG. 6-31
TISSUE IMPRINT OF LOWER RENAL MEDULLA CONTAINING TERMINAL COLLECTING DUCT EPITHELIUM. **(A)** The cells have a round eccentric nucelus and tall, finely granular cytoplasm. (× 400) **(B)** This sheet of epithelial cells has the honeycomb arrangement characteristic of columnar ductal cells. (× 400)

(continued)

FIG. 6-31B *(continued)*

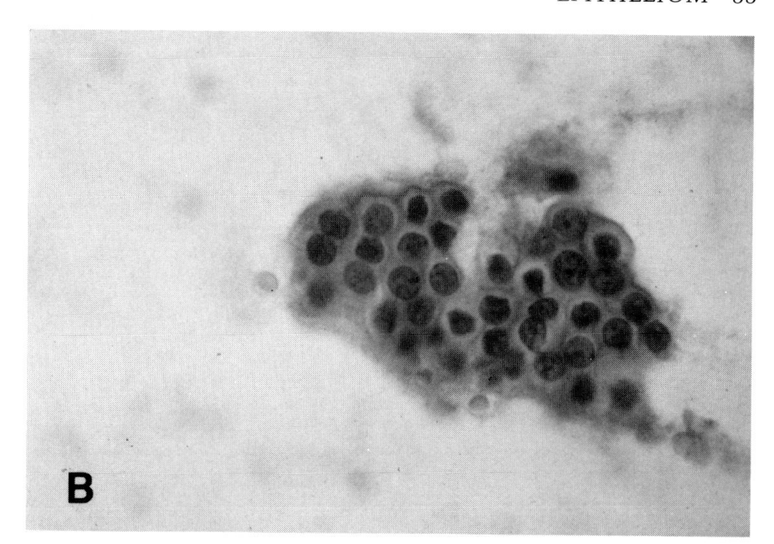

FIG. 6-32
ORIGIN OF RENAL EPITHELIAL FRAGMENTS. Inflamed interstitial connective tissue herniates through the ruptured basement membrane of the collecting duct and is covered by regenerating epithelium. Two small detached renal epithelial fragments (arrows) are present within the lumen. (× 250)

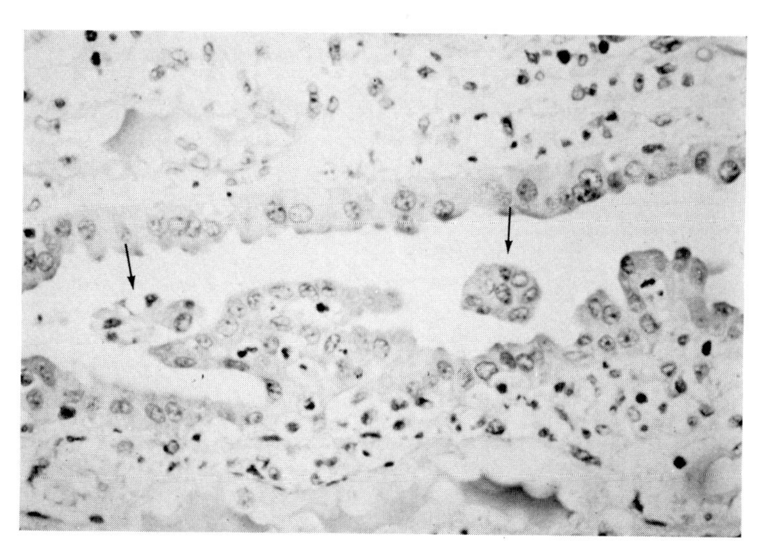

FIG. 6-33
RENAL EPITHELIAL FRAGMENTS SIMILAR TO THOSE IN FIGURE 6-32 ARE SEEN BETTER AT HIGHER MAGNIFICATION. Note solid epithelial fragment containing cells with enlarged nuclei and prominent nucleoli, characteristic of regeneration, and pale, slightly granular cytoplasm. This fragment, possibly of early collecting duct origin, is attached to fibrillar material similar to the fibrin cast in an adjacent tubule. (× 400)

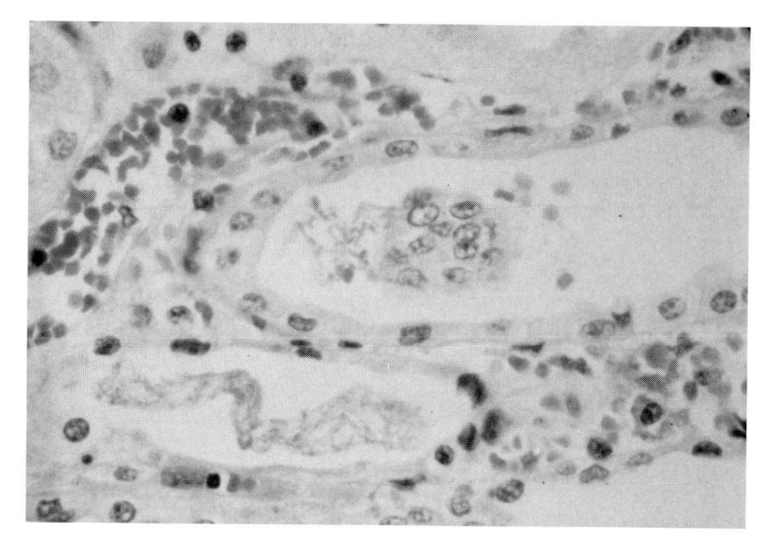

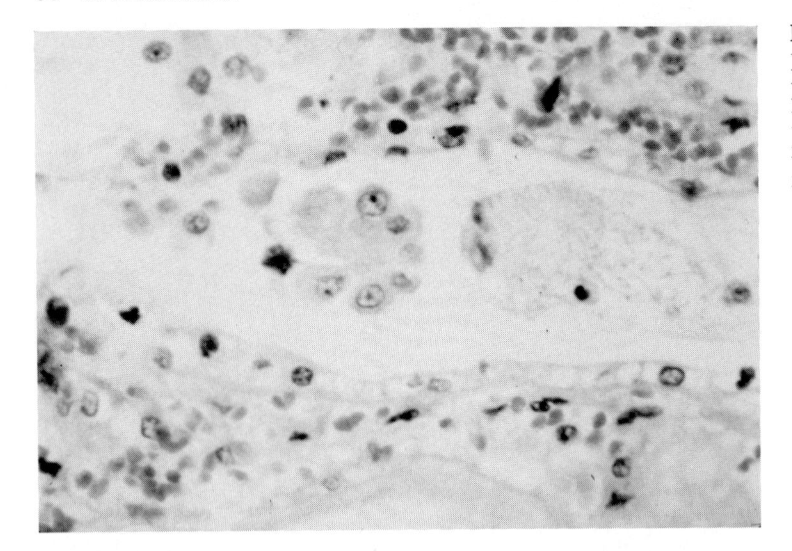

FIG. 6-34
FRAGMENT OF CUBOIDAL RENAL EPITHELIUM ARRANGED IN A PAPILLARY CONFIGURATION. Note finely granular eosinophilic cytoplasm and faint cell borders. (× 400)

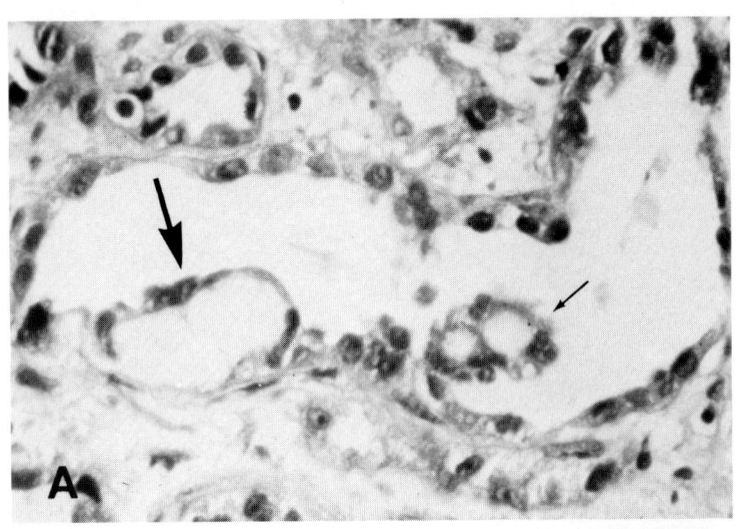

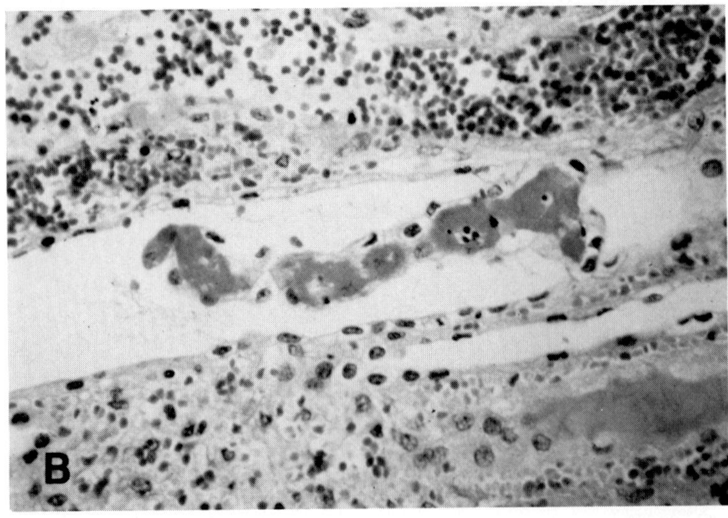

FIG. 6-35
RENAL FRAGMENT WITH EPITHELIUM SURROUNDING CAST-LIKE MATERIAL. **(A)** At site of healing tubular basement membrane breakage, dense hyaline material is covered by an attenuated, regenerating epithelium (large arrow). Similar material surrounded by regenerating epithelium protrudes into the tubular lumen but remains attached (small arrow). (Trichrome × 400) **(B)** Larger epithelial fragment with waxy, cast-like material encased by regenerating renal epithelium appears loosely attached. (× 250)

(continued)

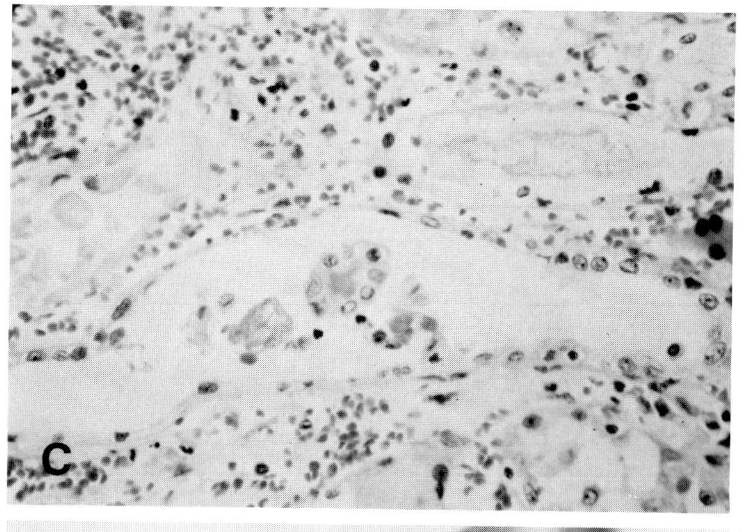

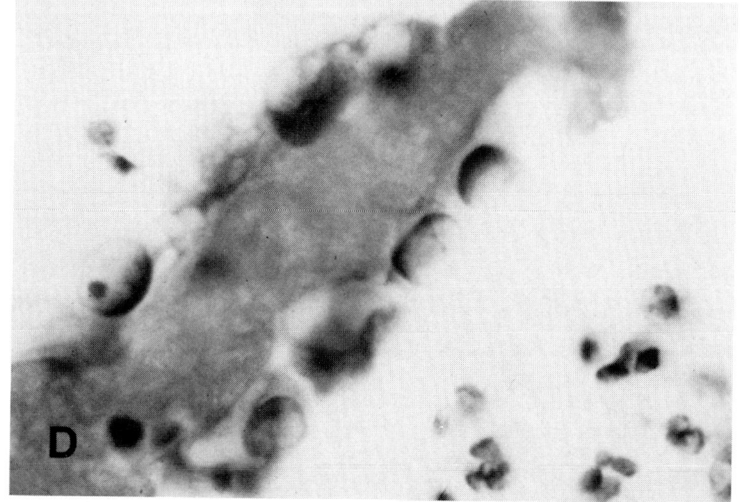

FIG. 6-35C, D *(continued)*
(C) Smaller detached renal epithelial fragment. (× 250) **(D)** Waxy, cast-like material encased by regenerating renal epithelium found in urine sediment during renal infarction. The epithelium has slightly enlarged, vesicular nuclei with prominent nucleoli and a thin rim of cytoplasm. (× 1000).

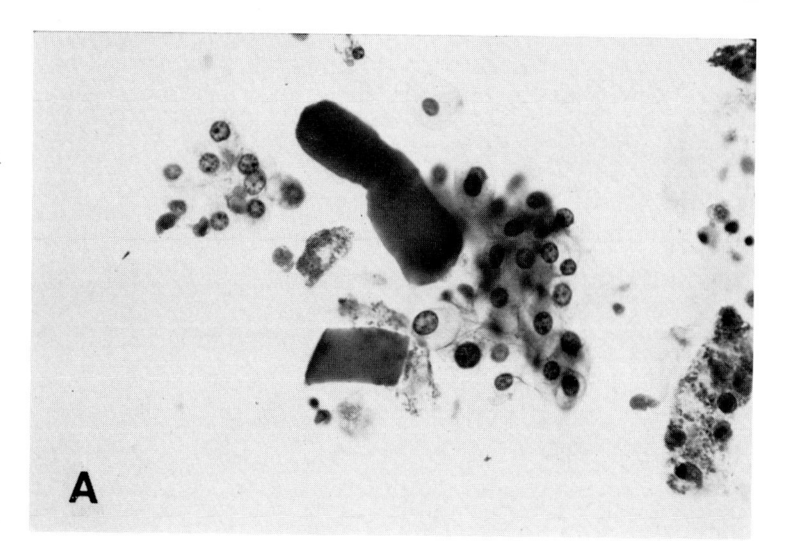

FIG. 6-36
RENAL EPITHELIAL FRAGMENTS FOUND IN URINE SEDIMENT. **(A)** Moulding by a renal cast represents an important configurational characteristic in identifying renal epithelium. (× 400)

(continued)

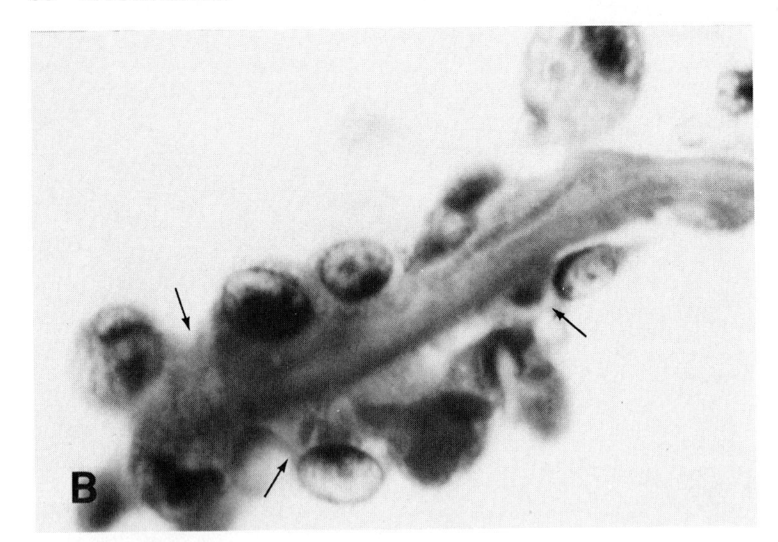

FIG. 6-36B (continued)
(B) Regenerative collecting duct epithelium encasing waxy, cast-like material. Waxy intracytoplasmic droplets are sometimes present in the cytoplasm of the lining epithelium (arrows). (× 1000)

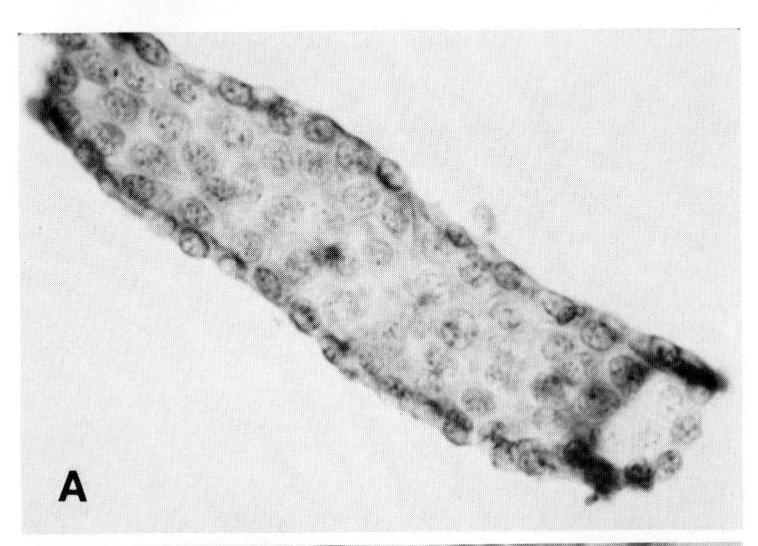

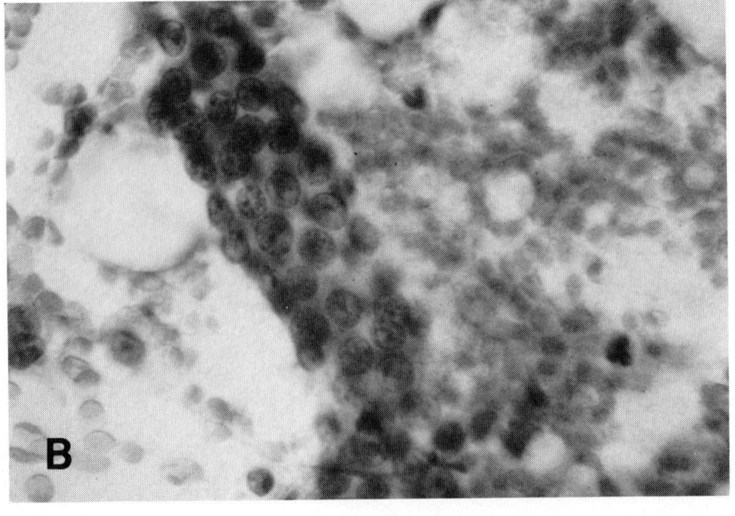

FIG. 6-37
CYLINDRICAL RENAL EPITHELIAL FRAGMENT. **(A)** A tissue imprint of renal medulla contains a sleeve-like sheet of epithelium. Note the luminal absence of cast material. (× 400) **(B)** Three-dimensional cylindrical renal epithelial fragment with hematuria. Microscopic focusing allows one to demonstrate the lack of luminal cast material. (× 400)

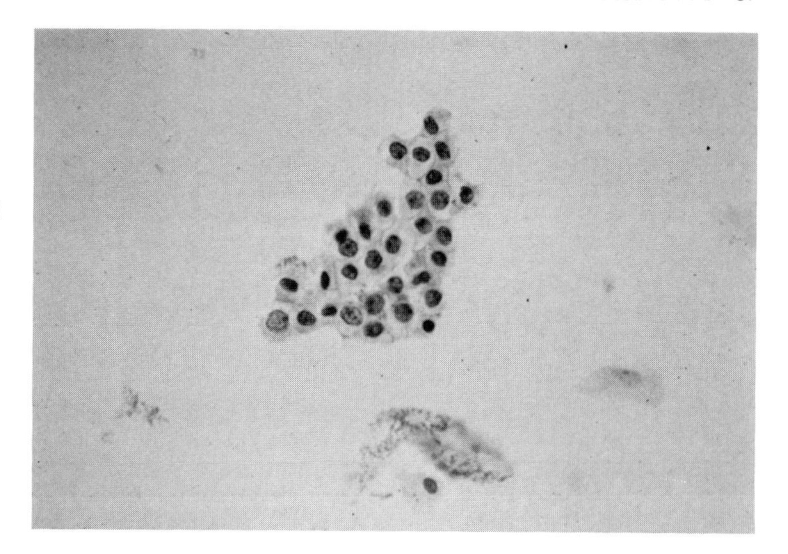

FIG. 6-38
RENAL EPITHELIAL FRAGMENT IN URINE SEDIMENT. The epithelial sheet contains numerous cells with round to oval nuclei and finely granular cytoplasm. The distinct cytoplasmic borders create a "honeycomb" appearance characteristic of collecting duct epithelium. (× 400)

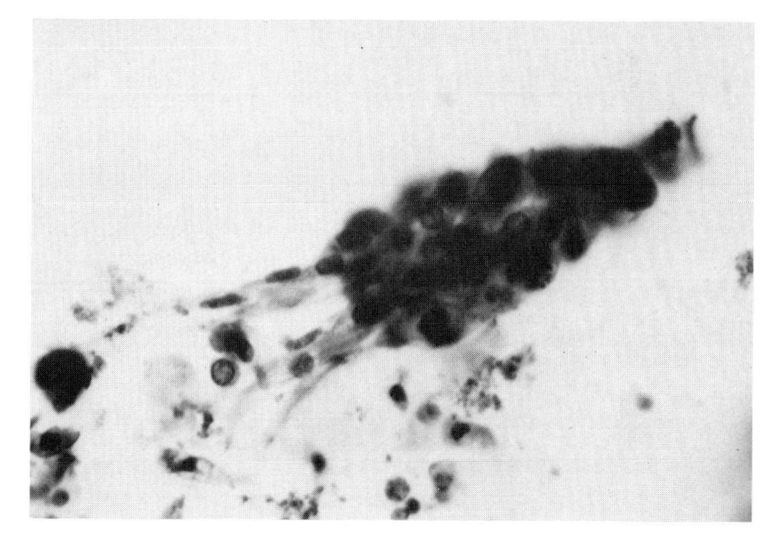

FIG. 6-39
RENAL EPITHELIAL FRAGMENT IN URINE SEDIMENT. The fragment has a three-dimensional cylindrical appearance and elongated or spindle-like epithelial cells at one end. (× 400)

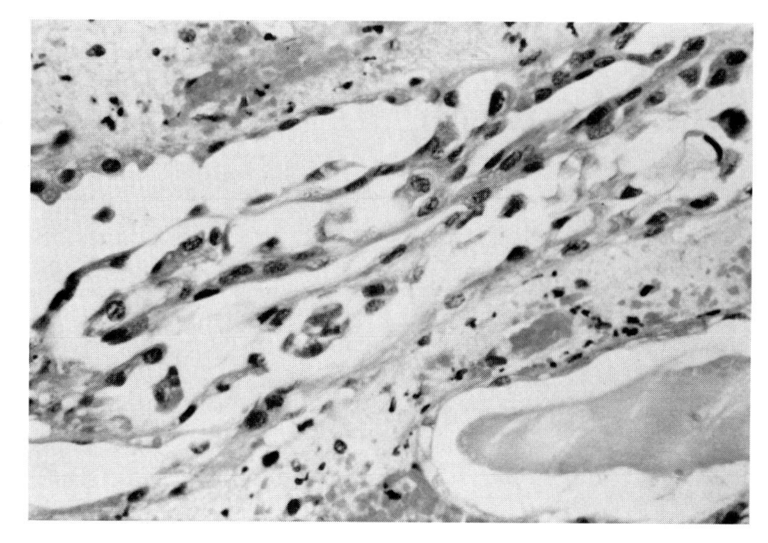

FIG. 6-40
RENAL EPITHELIAL FRAGMENT CONTAINING SPINDLE CELLS. A collecting duct contains a loosely cohesive sheet of elongated epithelial cells, which remain focally attached to the basement membrane. The spindle appearance of the cells probably reflects exfoliation of attenuated epithelium in an early phase of regeneration. (× 250)

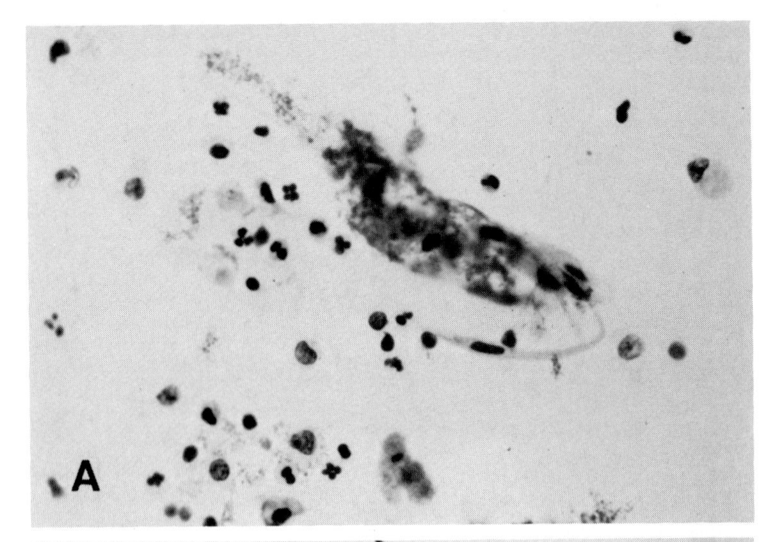

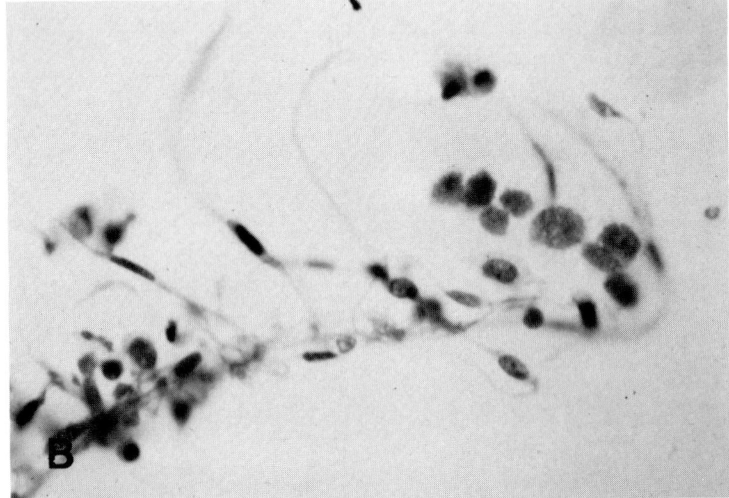

FIG. 6-41
SPINDLE RENAL EPITHELIAL CELLS IN URINE SEDIMENT. **(A)** Cells with elongated nuclei and tapered cytoplasm surround a granular cast. (× 400) **(B)** A loose cluster of spindle renal epithelial cells have barely visible stringy cytoplasm. Note numerous ghost cells in the background. (× 400)

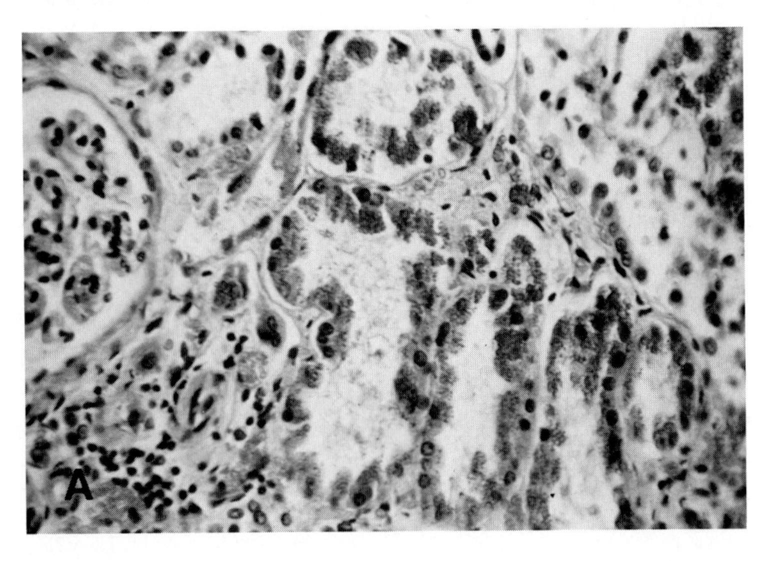

FIG. 6-42
PIGMENTED RENAL EPITHELIUM. Heavy deposits of hemosiderin in the cytoplasm of proximal convoluted tubular cells. **(A)** (× 150) **(B)** (Prussian blue × 150)

(continued)

FIG. 6-42B *(continued)*

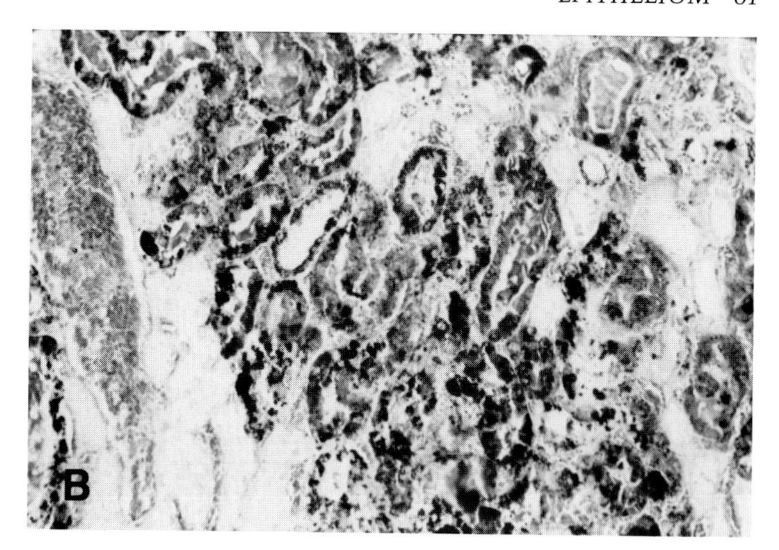

FIG. 6-43
PIGMENTED EPITHELIAL FRAGMENT IN URINE SEDIMENT. The epithelium has reactive nuclei containing prominent nucleoli and brown pigment granules in the cytoplasm. (× 400)

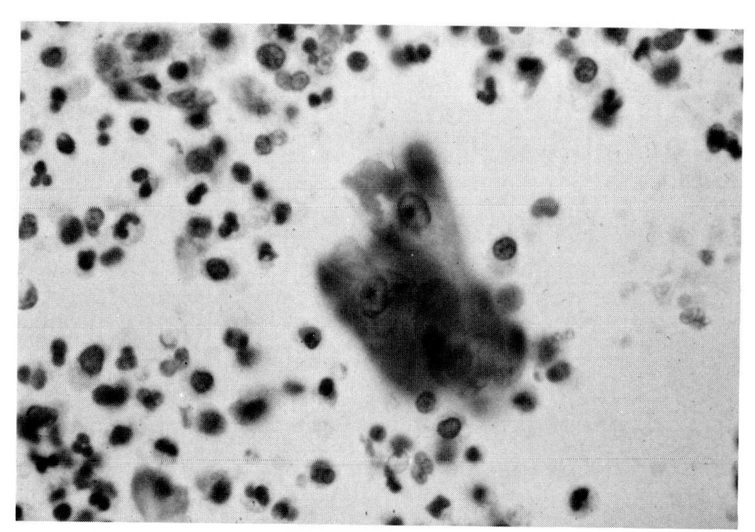

FIG. 6-44
INFARCTION OF PAPILLARY TIP WITH EXFOLIATION OF LARGE EPITHELIAL FRAGMENTS. Papillary configuration with columnar epithelium suggests origin from the collecting duct. (PAS × 250)

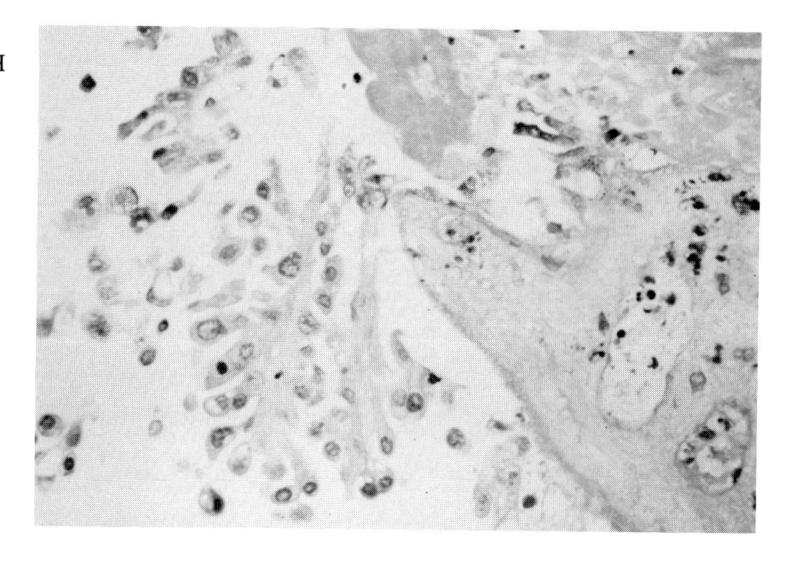

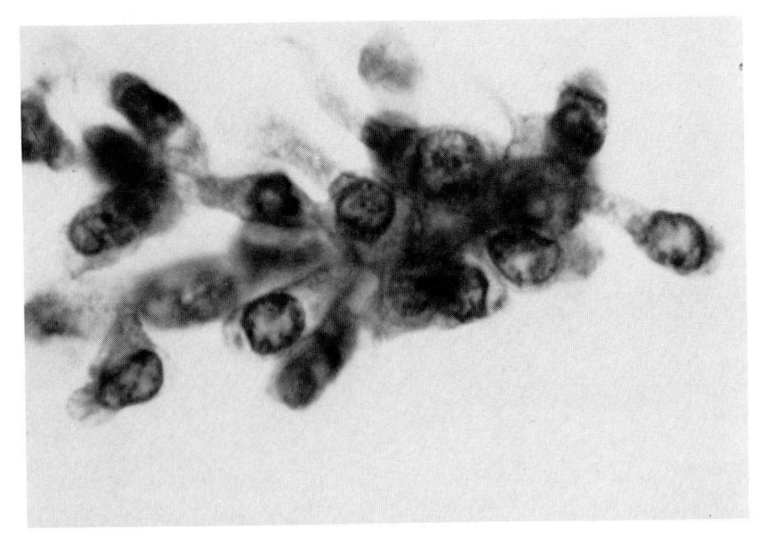

FIG. 6-45
RENAL EPITHELIAL FRAGMENT IN URINE SEDIMENT. Note eccentric nuclei and abundant, finely granular cytoplasm in these columnar-like cells, indicating terminal collecting duct origin. (× 1000)

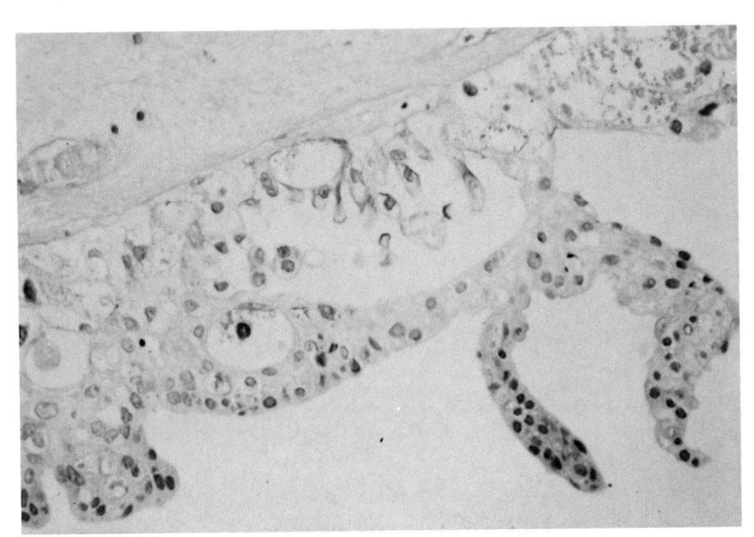

FIG. 6-46
INFARCTION OF PAPILLARY TIP. Superficial layers of urothelium are exfoliating as cohesive sheets of cells. Deeper layers are degenerating and are focally necrotic. (PAS × 250)

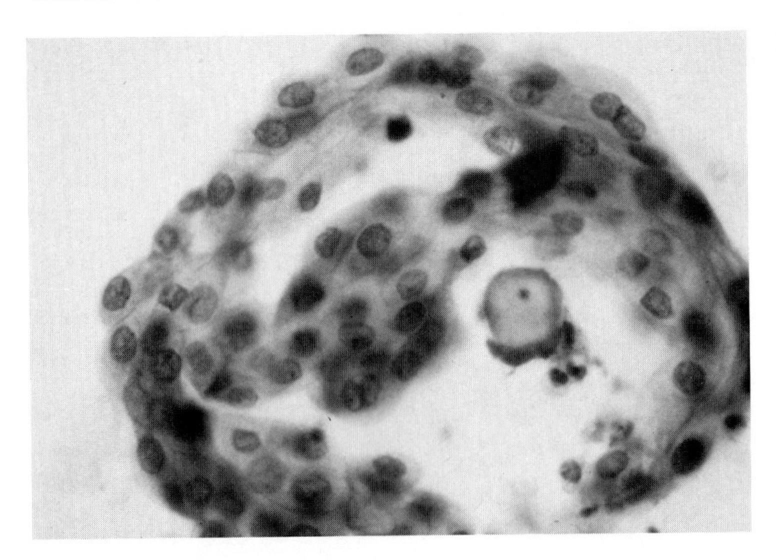

FIG. 6-47
UROTHELIAL FRAGMENT IN URINE SEDIMENT. The round to irregular nuclei and abundant dense cytoplasm are characteristic of urothelial cells. (× 400)

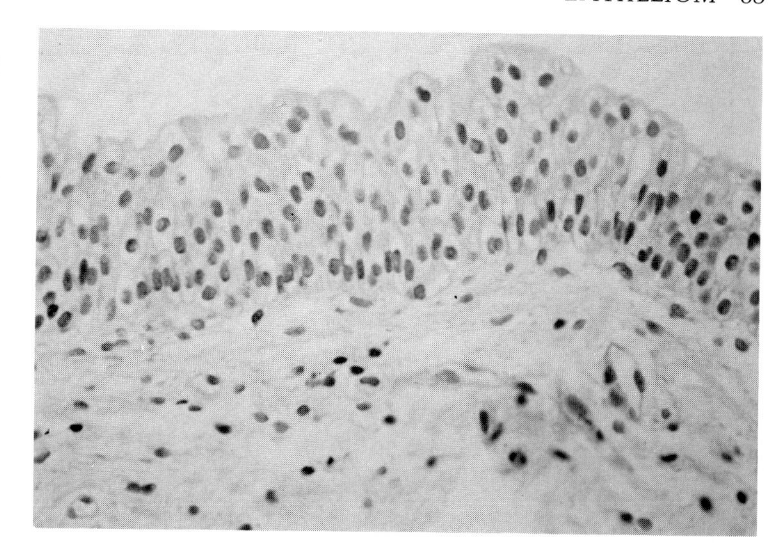

FIG. 6-48
NORMAL UROTHELIUM IN SLIGHTLY CONTRACTED BLADDER. (× 250)

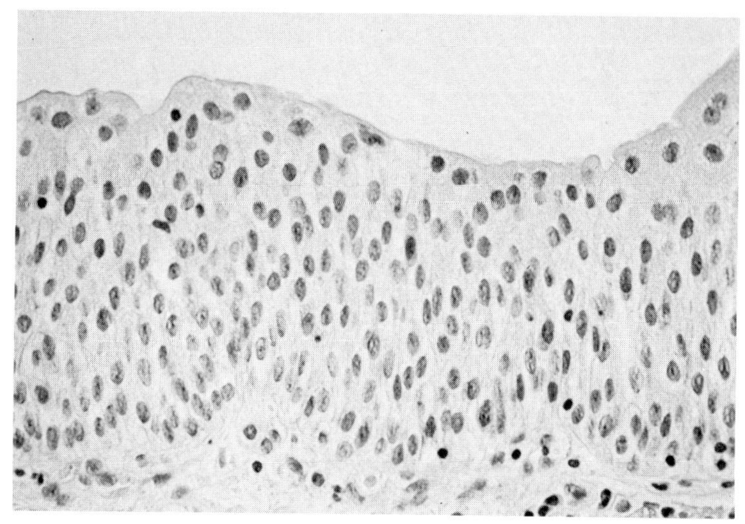

FIG. 6-49
UROTHELIAL HYPERPLASIA. The bladder epithelium is thickened, with more than seven cell layers present. Note there is no nuclear atypia, polarity is maintained, and superficial cells are preserved. (× 250)

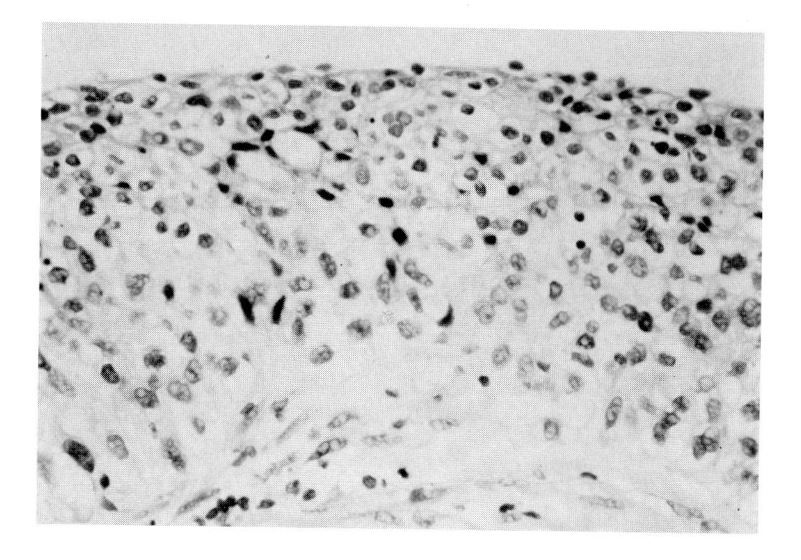

FIG. 6-50
ATYPICAL UROTHELIAL HYPERPLASIA. In addition to the more than seven cell layers in this thickened urothelium, there is loss of polarity and variability in nuclear size, shape, and hyperchromatism. Occasionally, flattened umbrella cells are identified with difficulty. These abnormalities were present adjacent to an invasive papillary transitional cell carcinoma, grade II. (× 250)

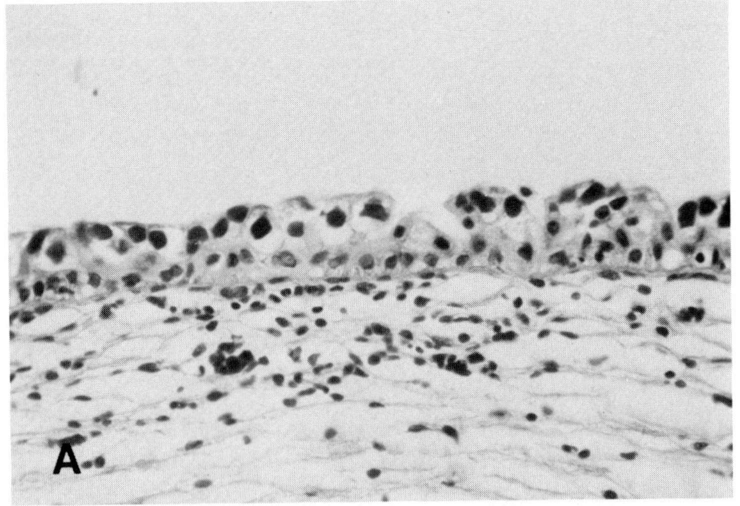

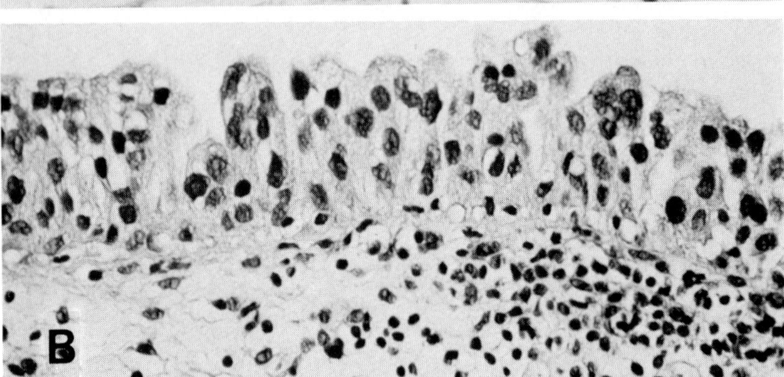

FIG. 6-51
UROTHELIAL ATYPIA WITHOUT HYPERPLASIA. **(A)** Superficial cells are enlarged and markedly hyperchromatic in comparison to normal-appearing basal cells. (× 250) **(B)** There is marked atypia bordering on carcinoma *in situ*. Cells with nuclear enlargement and hyperchromasia have abundant cytoplasm and are randomly dispersed throughout the urothelium with occasionally cytologically benign-appearing transitional cells and preserved superficial cells. (× 250) Bladder biopsies taken from a patient with invasive papillary transitional cell carcinoma, grade III.

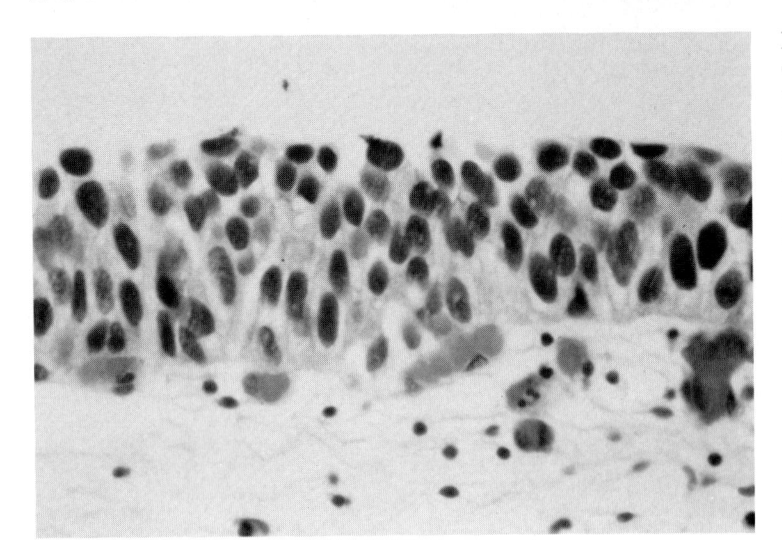

FIG. 6-52
CARCINOMA *IN SITU* OF THE BLADDER. Full-thickness changes are characterized by crowding of relatively small cells having enlarged, hyperchromatic nuclei. An adjacent focus of invasive nonpapillary transitional cell carcinoma was present. (× 400)

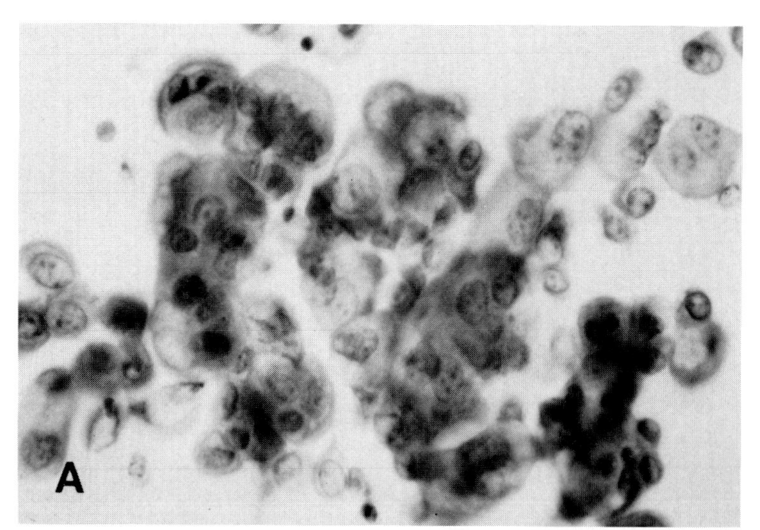

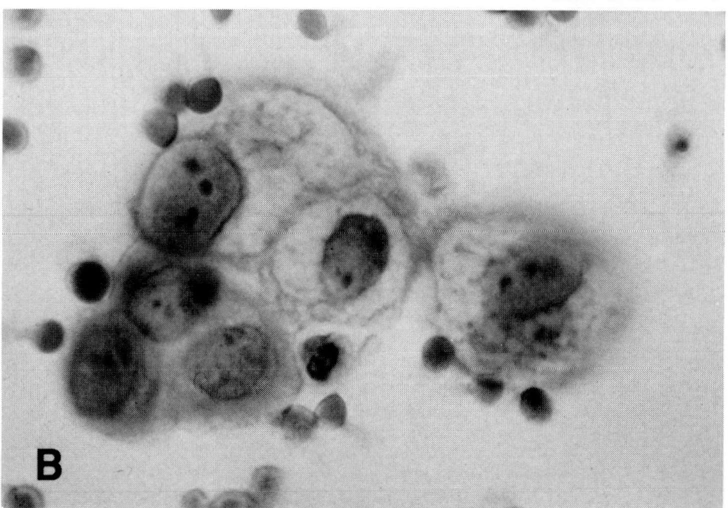

FIG. 6-53
CATHETERIZED URINE SPECIMEN WITH UROTHELIAL FRAGMENTS. **(A)** Although there is nuclear crowding, the nuclei are normochromatic. (× 400) **(B)** The normochromatic nuclei are round to oval, contain prominent chromocenters or micronucleoli, and have slightly irregular nuclear membranes. The cytoplasm is abundant and finely vacuolated. Note outer cytoplasmic rim characteristic of urothelial cells. (× 1000)

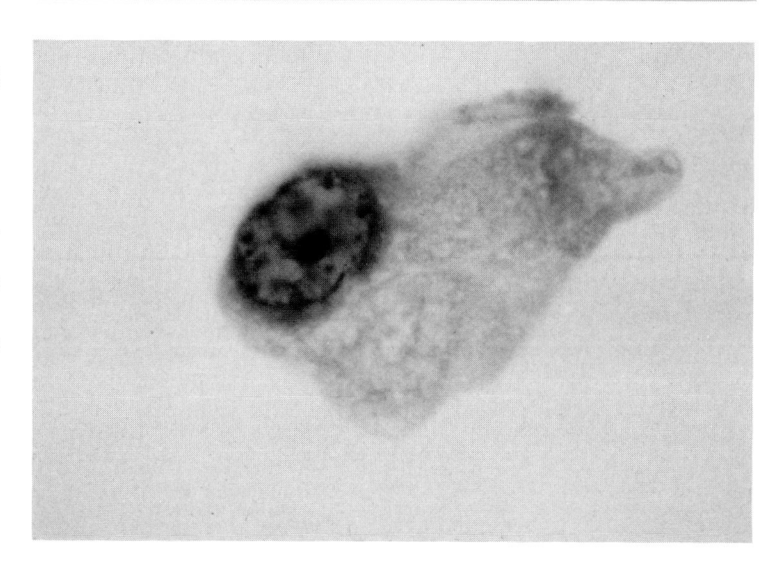

FIG. 6-54
ATYPICAL EPITHELIAL CELL IN URINE SEDIMENT. The nucleus is slightly enlarged, has an irregular membrane, is mildly hyperchromatic with chromatin clumping, and contains a prominent nucleolus. The cytoplasm is abundant and finely vacuolated. This type of atypical epithelial cell can be seen in the sediment in the following situations: after chemotherapy or irradiation, following the passage of a renal stone, or with atypical hyperplasia or transitional cell carcinoma. (× 1000)

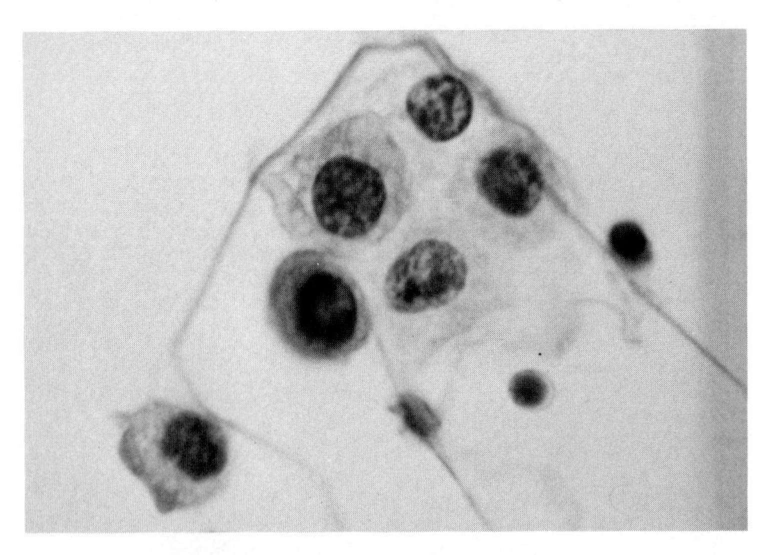

FIG. 6-55
ATYPICAL (DYSKARYOTIC) UROTHELIUM IN URINE SEDIMENT. Note enlarged nuclei with increased nuclear:cytoplasmic ratio and true hyperchromasia. (× 1000)

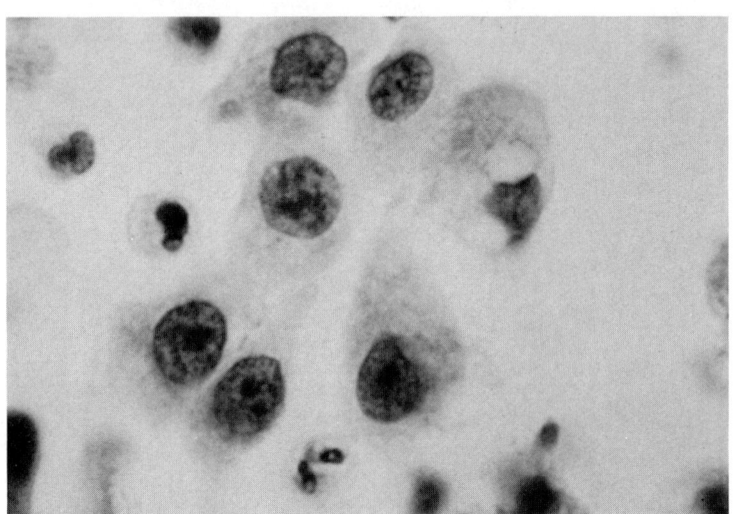

FIG. 6-56
MARKEDLY ATYPICAL (DYSKARYOTIC) UROTHELIUM IN URINE SEDIMENT. Nuclei are slightly enlarged and hyperchromatic with irregular chromatin clearing and irregular nuclear membranes and nucleoli. (× 1000)

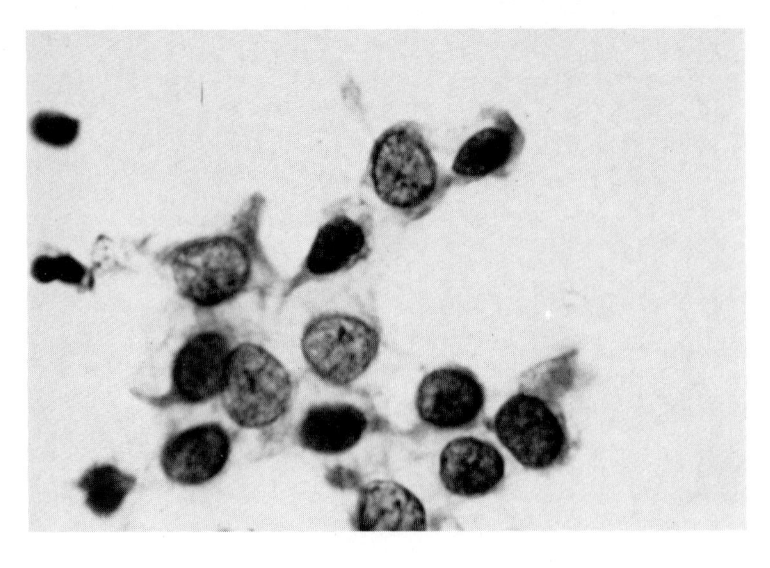

FIG. 6-57
MARKEDLY DYSKARYOTIC UROTHELIUM DERIVED FROM TRANSITIONAL CARCINOMA *IN SITU.* The syncytial arrangement of cells with round hyperchromatic nuclei and a high nuclear:cytoplasmic ratio is characteristic of carcinoma *in situ* (CIS). The cellular features are similar to CIS cells of the cervix seen on gynecologic cytology smears. (× 1000)

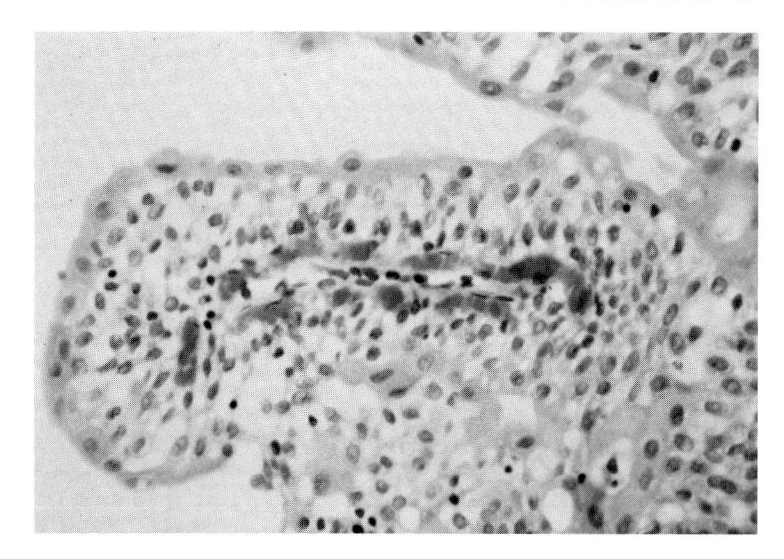

FIG. 6-58
TRANSITIONAL CELL PAPILLOMA. A delicate papillary frond is covered by urothelium of normal thickness and without nuclear abnormalities. (× 250)

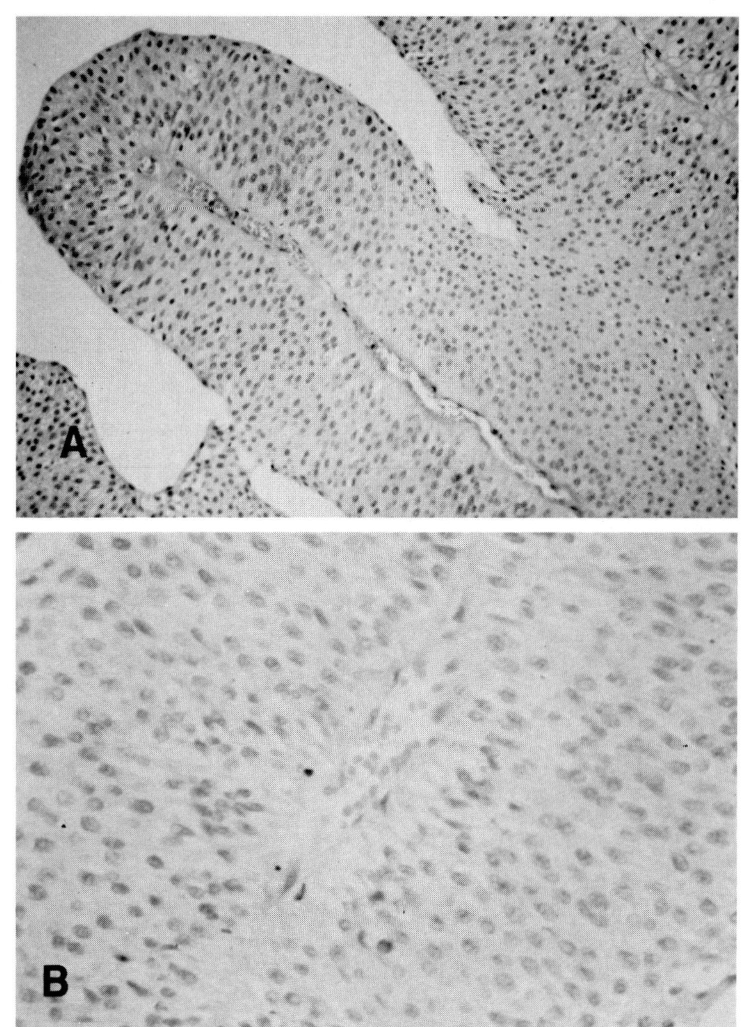

FIG. 6-59
PAPILLARY TRANSITIONAL CELL CARCINOMA, GRADE I. **(A)** Urothelium covering the papillary frond is thickened. Superficial cells are uniformly present. (× 100) **(B)** Nuclear enlargement is slight, and there is no significant hyperchromasia. (× 250)

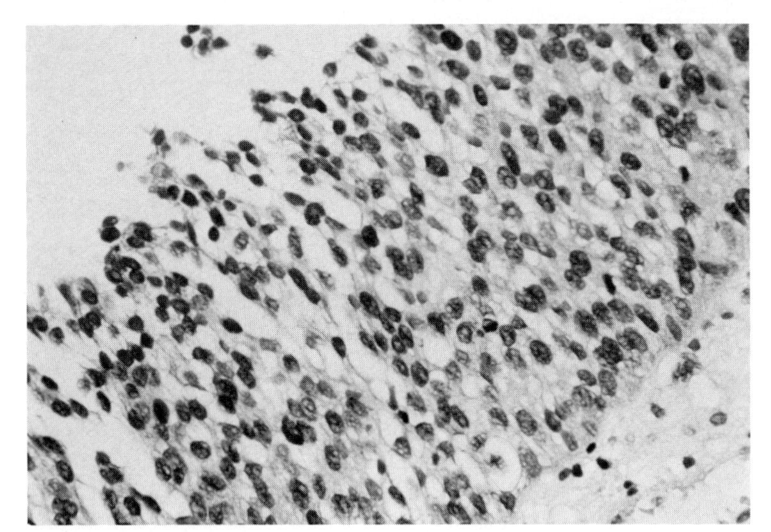

FIG. 6-60
PAPILLARY TRANSITIONAL CELL CARCINOMA, GRADE II. Cellular crowding and diminished intercellular cohesiveness with loss of superficial cells are present in this hyperplastic urothelium covering a papillary frond. Note moderate nuclear enlargement and hyperchromasia. (× 250)

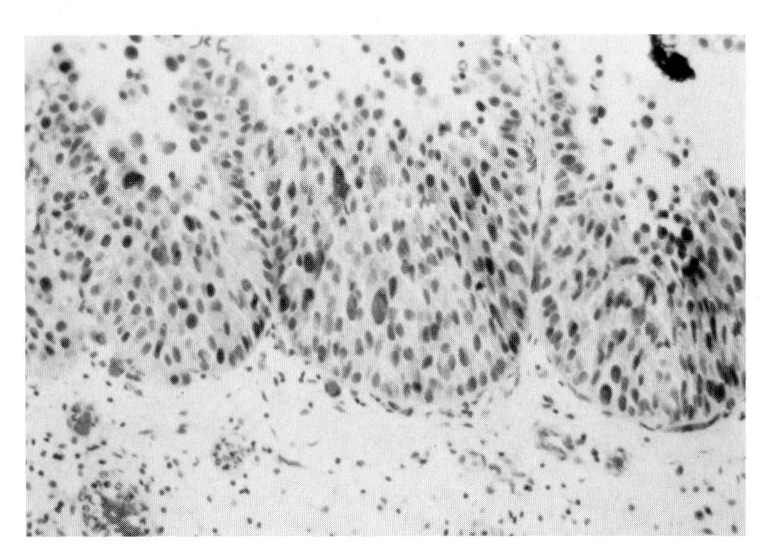

FIG. 6-61
PAPILLARY TRANSITIONAL CELL CARCINOMA, GRADE III. The poorly formed papillae appear broader and shorter due to markedly diminished intercellular cohesiveness with exfoliation of numerous malignant cells. (× 160)

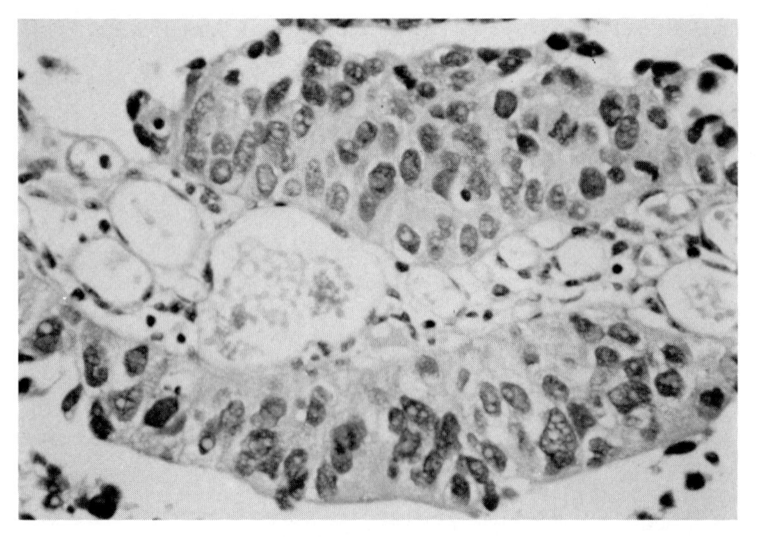

FIG. 6-62
PAPILLARY TRANSITIONAL CELL CARCINOMA, GRADE III. Disorderly epithelial growth and nuclear abnormalities are striking, with marked nuclear enlargement, pleomorphism, and hyperchromasia. Single and multiple nucleoli are frequent, and there are occasional mitoses. (× 250)

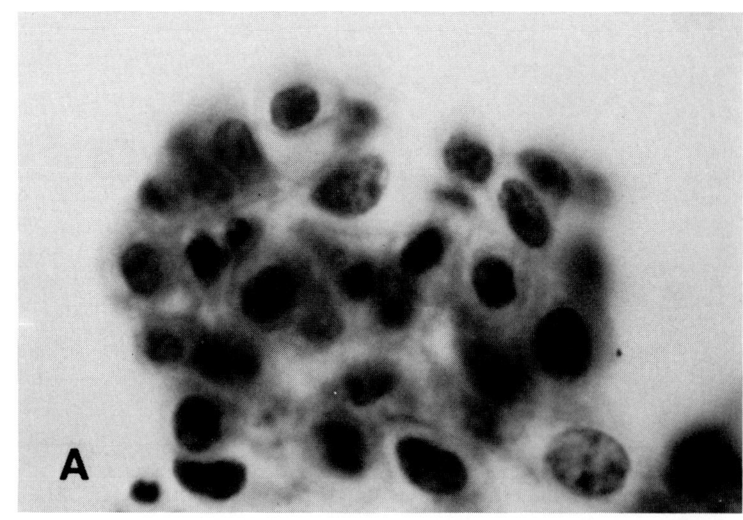

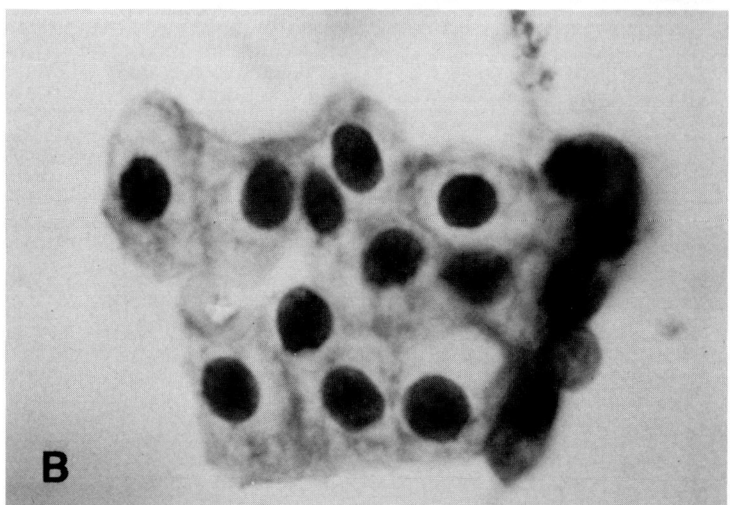

FIG. 6-63
ATYPICAL UROTHELIAL FRAGMENT IN URINE SEDIMENT. **(A)** Apparent nuclear hyperchromasia due to degeneration. Note variable nuclear size, wrinkled or shrunken nuclear membranes, and indistinct chromatin pattern. Cells such as these are frequently seen following chemotherapy, irradiation, lithiasis, and in patients having low-grade transitional cell carcinoma. (× 1000) **(B)** Atypical urothelial fragment from a patient with low-grade transitional cell carcinoma (grade I). The uniformity of nuclear size and hyperchromasia with lack of cytoplasmic maturation are more suggestive of a low-grade transitional cell carcinoma. (× 1000)

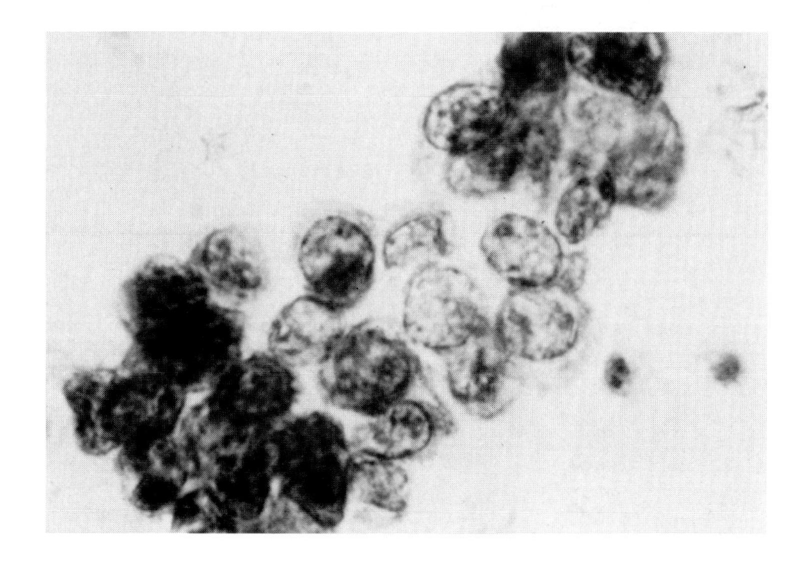

FIG. 6-64
MALIGNANT CELLS DERIVED FROM TRANSITIONAL CELL CARCINOMA (GRADE II). Note nuclear enlargement, increased nuclear:cytoplasmic ratio, and "true" hyperchromasia. (× 1000)

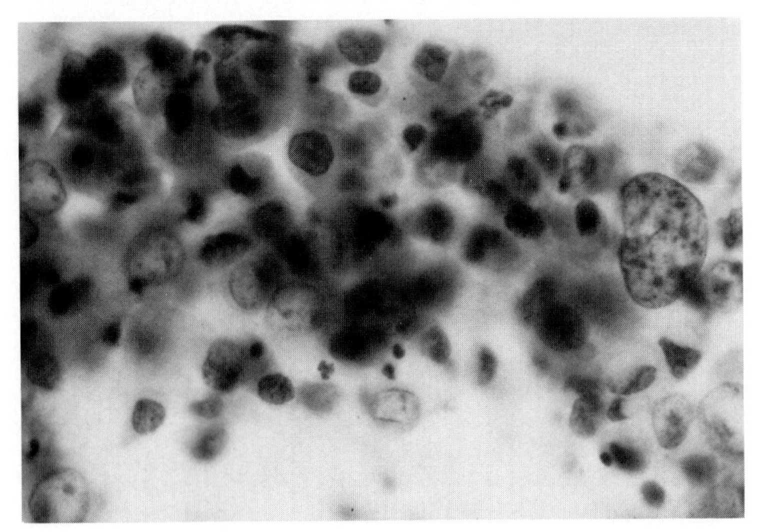

FIG. 6-65
MARKED EXFOLIATION OF MALIGNANT TRANSITIONAL CELLS (GRADE II–III). High-grade transitional cell carcinoma and transitional cell carcinoma *in situ* are characterized by increased epithelial exfoliation. (× 400)

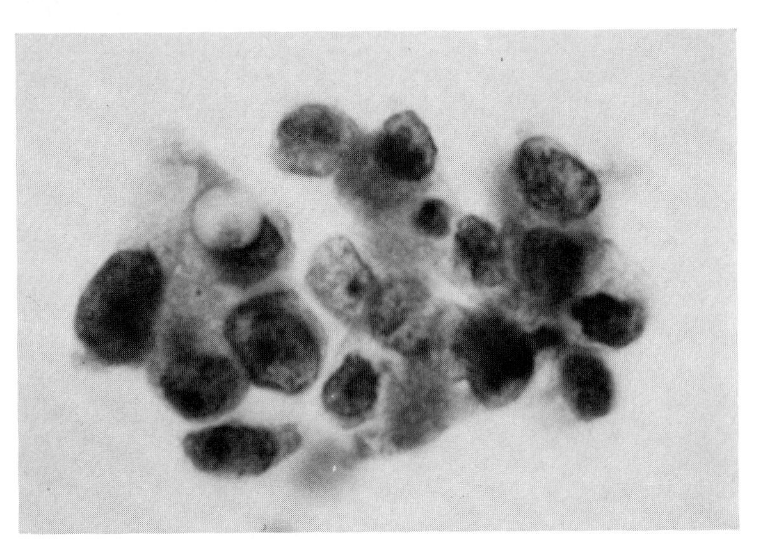

FIG. 6-66
MALIGNANT CELLS DERIVED FROM TRANSITIONAL CELL CARCINOMA (GRADE III). Note increased nuclear:cytoplasmic ratio and hyperchromasia. The presence of nucleoli in malignant cells distinguishes high-grade from low-grade transitional cell carcinoma. (× 1000)

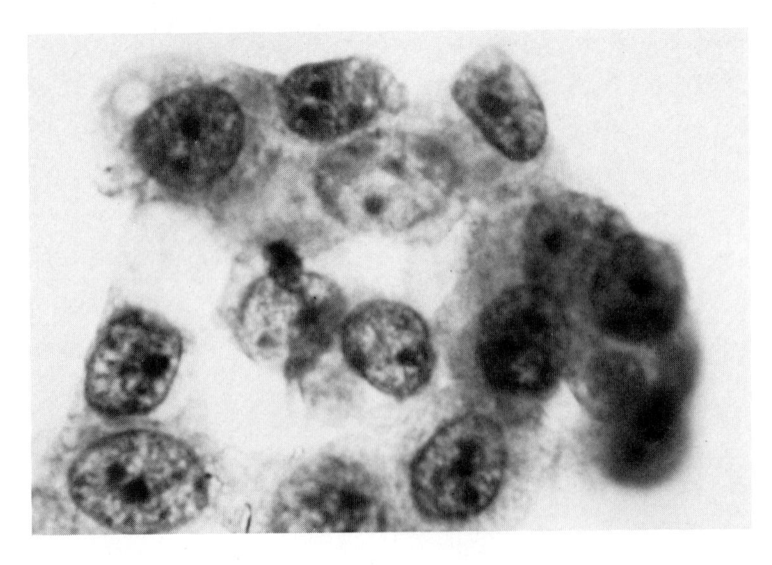

FIG. 6-67
MALIGNANT CELLS DERIVED FROM TRANSITIONAL CELL CARCINOMA (GRADE III). Note hyperchromasia, anisonucleosis, and prominent nucleoli. (× 1000)

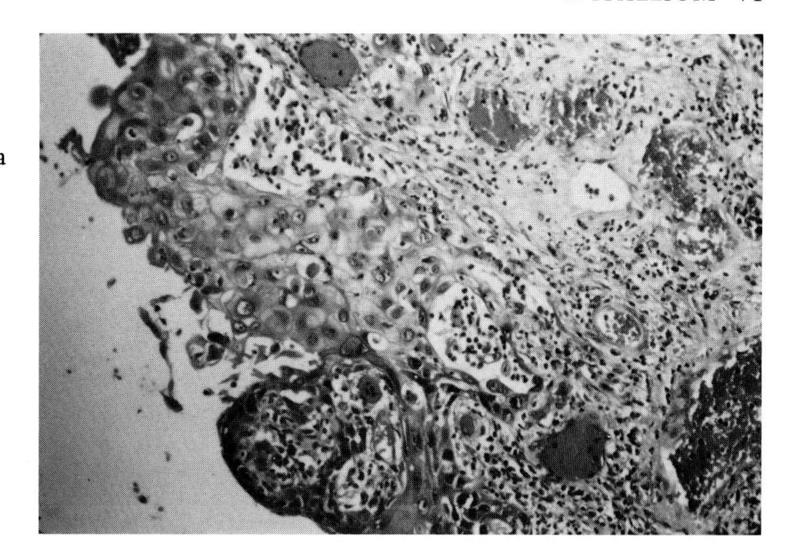

FIG. 6-68
SQUAMOUS CELL CARCINOMA OF THE BLADDER. Individual cell keratinization with rare pearl formation characterize this squamous cell carcinoma that invades the lamina propria. (× 150)

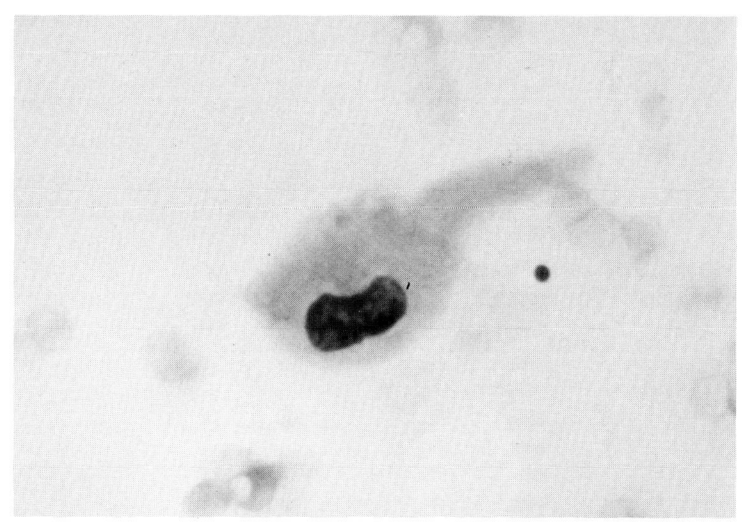

FIG. 6-69
MALIGNANT CELL DERIVED FROM SQUAMOUS CELL CARCINOMA. Note the enlarged pyknotic nucleus and abundant orangeophilic (keratinized) cytoplasm. (× 1000)

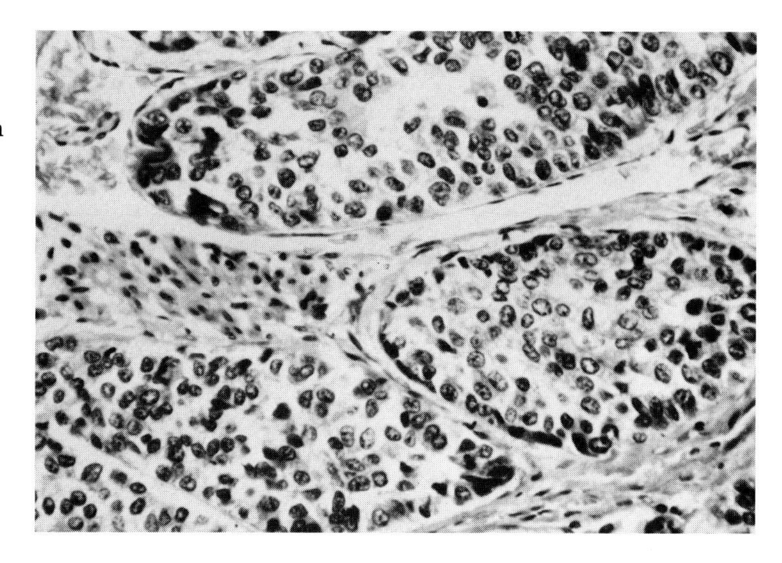

FIG. 6-70
FOCUS OF UNDIFFERENTIATED CARCINOMA IN BLADDER. Poorly cohesive nests of infiltrating tumor within the lamina propria contain small cells with scant cytoplasm. Hyperchromasia is marked and nucleoli are present. This primary bladder tumor was composed predominantly of keratinizing squamous cell carcinoma. (× 250)

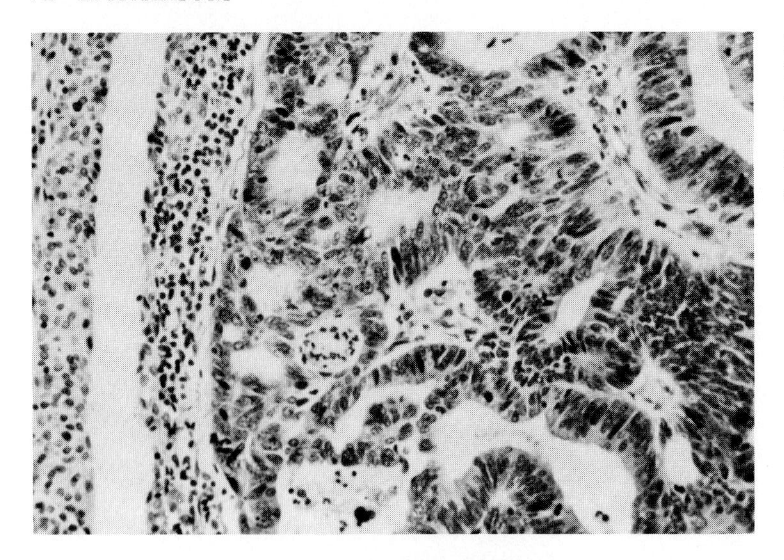

FIG. 6-71
URACHAL ADENOCARCINOMA. This tumor, which originally arose along the urachus, is shown within the wall of the bladder dome and resembles an adenocarcinoma of enteric origin. (× 160)

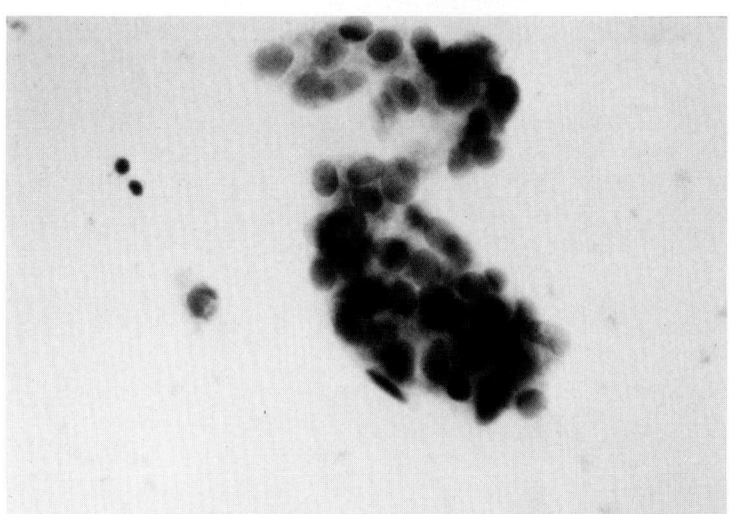

FIG. 6-72
URINE SEDIMENT WITH MALIGNANT CELLS DERIVED FROM ADENOCARCINOMA OF THE RENAL PELVIS. Note the loose glandular arrangement of these cells having hyperchromatic nuclei and high nuclear: cytoplasmic ratios. (× 400)

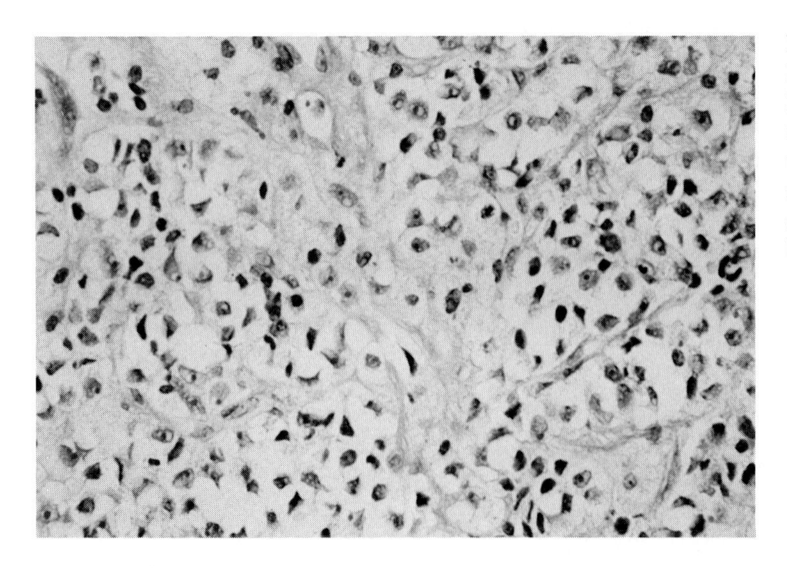

FIG. 6-73
METASTATIC CARCINOMA FROM PROSTATE. Small nests of prostatic carcinoma infiltrate the lamina propria of the bladder. Cells have irregularly shaped nuclei and conspicuous nucleoli. There is abundant vacuolated cytoplasm, but cell borders are indistinct. (PAS × 250)

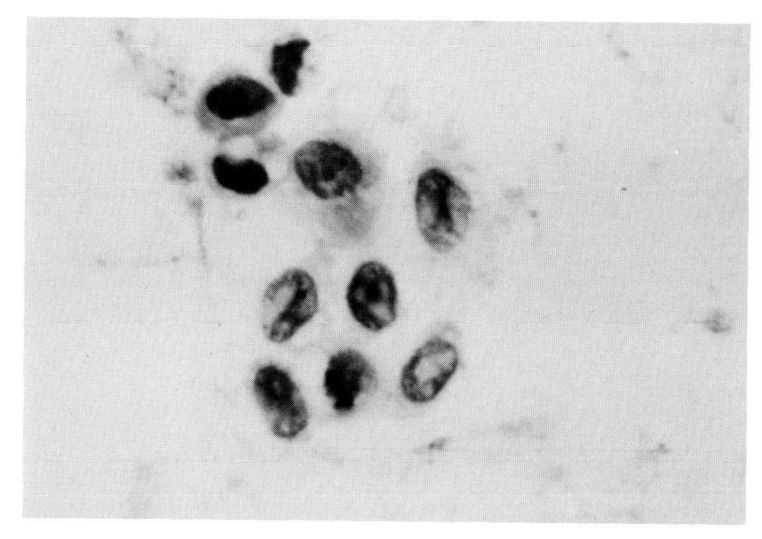

FIG. 6-74
URINE SEDIMENT WITH MALIGNANT EPITHELIAL CELLS DERIVED FROM PROSTATIC ADENOCARCINOMA. The glandular-like fragment contains cells with hyperchromatic nuclei, prominent nucleoli, and an increased nuclear: cytoplasmic ratio. (× 1000)

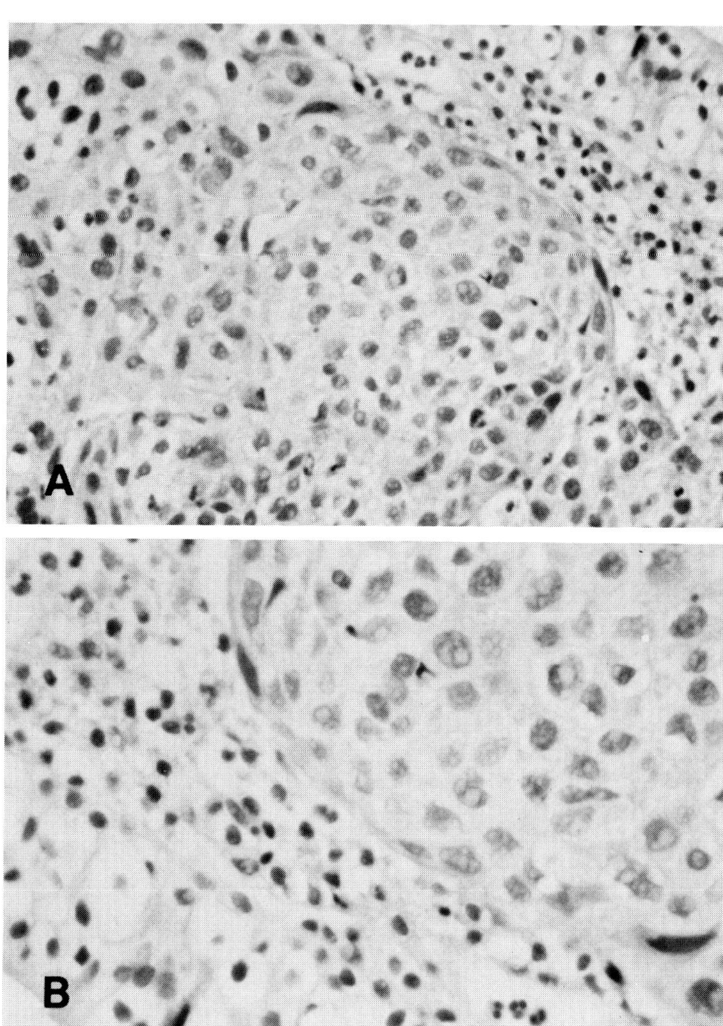

FIG. 6-75
METASTATIC CARCINOMA FROM UTERINE CERVIX. **(A)** Beneath the urothelium is a nest of large cell nonkeratinizing squamous cell carcinoma. (× 160) **(B)** Intercellular bridges are present (× 250)

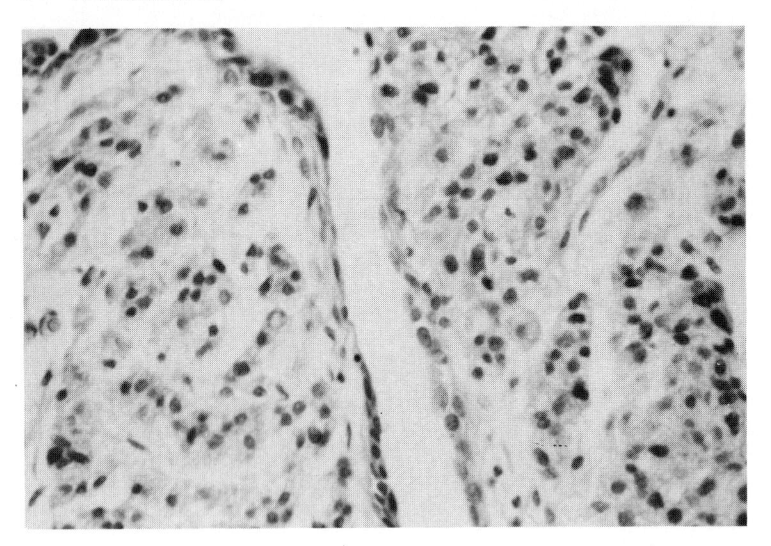

FIG. 6-76
METASTATIC CARCINOMA FROM THE BREAST. Infiltrating nests and cords of small, uniform malignant cells are present beneath the urothelium of the ureter. (× 250)

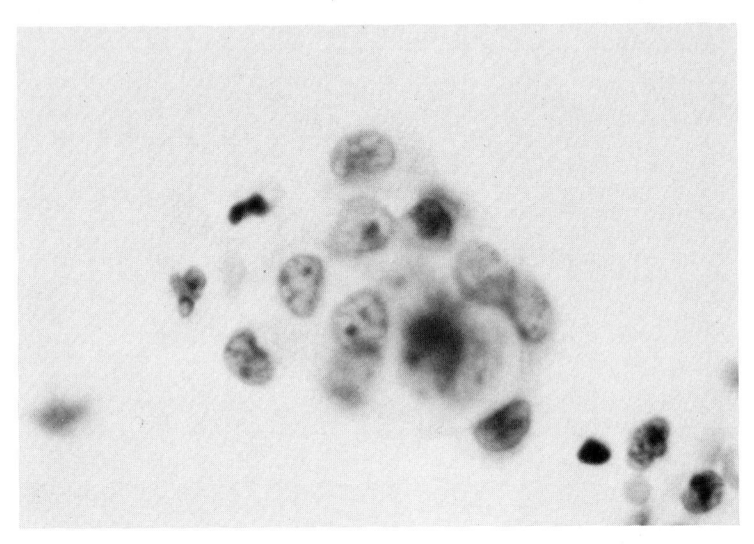

FIG. 6-77
URINE SEDIMENT WITH MALIGNANT EPITHELIAL FRAGMENT DERIVED FROM METASTATIC BREAST ADENOCARCINOMA. The acinar fragment contains cells with enlarged, hyperchromatic nuclei and prominent nucleoli. (× 1000)

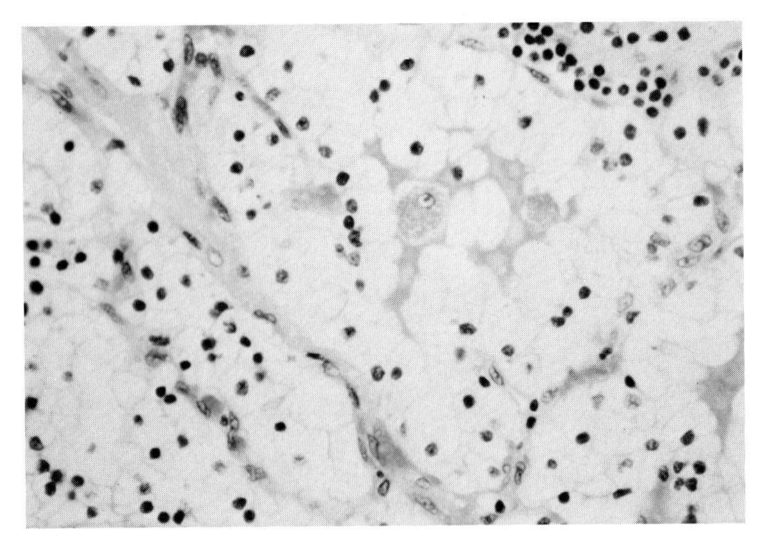

FIG. 6-78
RENAL CELL CARCINOMA. Cells have centrally placed hyperchromatic nuclei and abundant clear cytoplasm. (× 250)

FIG. 6-79
MALIGNANT CELL DERIVED FROM RENAL CELL CARCINOMA. Note slight nuclear enlargement, slight hyperchromasia, prominent, centrally located nucleolus, and abundant foamy or finely vacuolated cytoplasm. (× 1000)

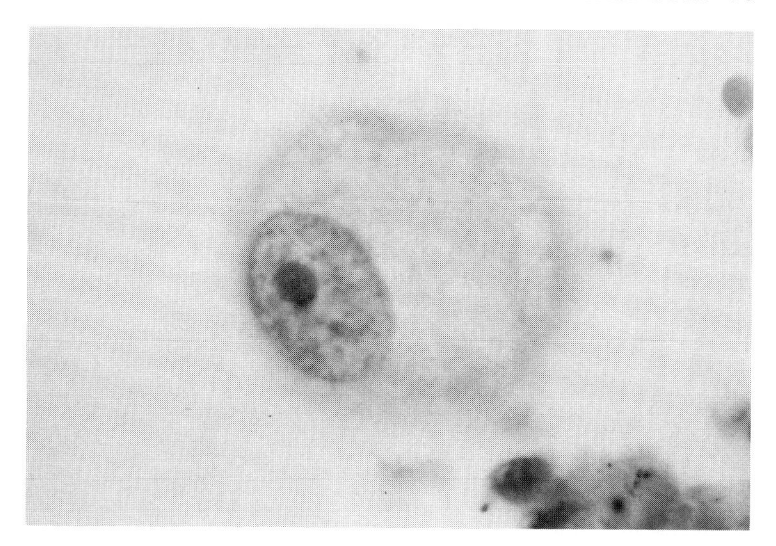

7
RENAL CASTS

Renal casts are composed of mucoprotein (Tamm-Horsfall) and are cylindrical. Casts are formed during stasis and are influenced by various salt concentrations and the pH of the environment. They reflect the configuration of tubular lumens and represent a significant localizing entity. Casts are further characterized by the appearance of the matrix and by the cellular or particulate matter embedded in the matrix.

Renal casts are classified as ***physiologic*** *or* ***pathologic*** *for diagnostic purposes.*

Physiologic Renal Casts

Hyaline
Finely granular

Pathologic Renal Casts

Coarsely granular
Finely granular
Waxy
Erythrocytic
Blood
Bile
Leukocytic
Renal epithelial
Fatty
Fibrin
Broad (any of the above types)
Crystal Inclusion
Mixed

Phsyiologic casts (hyaline and finely granular types) may be seen with other types of casts with renal parenchymal disease, but as isolated entities they are insufficient evidence of disease. The presence of both hyaline and finely granular casts increases following exercise and during dehydration.

The term ***pathologic casts*** *is used because these casts reflect renal parenchymal damage. Structural changes in glomeruli or renal tubules produce these morphological entities. The specific type or types of pathologic casts is dependent upon the severity and chronicity of the renal lesions and may reflect a specific disease process.*

HYALINE CASTS

Hyaline casts are transparent cylinders that can be easily overlooked in tissue sections and urine sediment (Figs. 7-1 and 7-2). They lack embedded particulate matter and cells. These casts may dissolve in hypotonic urine or in urine with a high pH. The potential loss of hyaline casts using the Papanicolaou staining method will not affect the diagnostic accuracy of the urine sediment examination.

GRANULAR CASTS

Granular casts contain discrete granules embedded in a hyaline matrix. The presence of finely particulate material characterizes finely granular casts (Fig. 7-3). In comparison, coarse granular casts (Figs. 7-4 and 7-5) contain granular material with increased particulate size and amount. Finely granular casts are produced by the accumulation of serum proteins embedded in a cast matrix or from degradation of coarsely granular casts. Coarsely granular casts are produced by increased accumulation of serum proteins and other material with urinary stasis or by transformation from a cellular cast to a coarsely granular cast through cell degeneration and lysis.

ERYTHROCYTIC CASTS

Erythrocytic casts are defined as red blood cells with intact cell membranes present within a hyaline cast matrix (Fig. 7-6). Erythrocytic casts in urine sediment (Fig. 7-7) indicate glomerular leakage (e.g., glomerulonephritis) or intrarenal parenchymal hemorrhage.

BLOOD CASTS

Blood casts are disrupted erythrocytes and heme pigment embedded in a hyaline matrix. In comparison to the erythrocytic casts, red blood cell membranes cannot be appreciated. In histologic sections, transition from erythrocytic casts to blood casts is frequently observed (Fig. 7-8). In Papanicolaou-stained sediment, blood casts are typically bright red (Fig. 7-9). In comparison, granular casts are usually pale red to light red-orange, resulting from the eosin in the Papanicolaou staining method. Because of the degeneration of erythrocytes in urine, blood casts are more commonly seen than erythrocytic casts in sediment.

BILE CASTS

Bile casts represent bile pigment embedded in a hyaline matrix. In histologic sections, the coarsely granular bile pigment may have a variable color (Fig. 7-10). In urine sediment (Fig. 7-11), bile casts are usually characterized by coarse greenish-yellow granules.

LEUKOCYTIC CASTS

The presence of intact neutrophils embedded in a hyaline matrix is characteristic of leukocytic casts. In histologic sections (Fig. 7-12) and in urine sediments (Fig. 7-13) accurate identification requires the demonstration of segmented neutrophils. The presence of leukocytic casts indicates inflammation within the collecting duct system. Urine sediments that contain leukocytic casts are frequently seen with tubulointerstitial disease (e.g., pyelonephritis) and are often seen with severe forms of glomerulonephritis associated with glomerular inflammation and basement membrane disruption.

RENAL EPITHELIAL CASTS

Renal epithelial casts usually contain more than three renal tubular epithelial cells embedded within a cast (Fig. 7-14). In urine sediment, the renal epithelium may have degenerative changes, but the polygonal or cuboidal cells retain their finely granular cytoplasm (Fig. 7-15). Renal epithelial casts are commonly seen in tubulointerstitial diseases, acute allograft rejection, and tubular necrosis. Renal epithelium may also be necrotic (Fig. 7-16). This type of cast should not be confused with a coarse granular cast. The presence of necrotic renal epithelial casts commonly indicates acute tubular necrosis.

WAXY CASTS

Waxy casts are formed by the degeneration of cellular material or particulate matter (Fig. 7-17). The disintegration of coarse granular casts to waxy casts (Fig. 7-18) is a reflection of chronicity. The presence of numerous waxy casts in tubules indicates chronic renal disease (Fig. 7-19). Waxy casts in histologic sections and urine sediments (Fig. 7-20**A**) are characterized by their dense homogeneous texture. The cracks present in waxy casts reflect their brittle nature (Fig. 7-20**B**). Unlike hyaline casts, waxy casts do not dissolve with a delay in processing urine specimens.

FATTY CASTS

The presence of fat globules or lipid-laden cells embedded in a cast matrix is termed **fatty cast.** Fat stains can be used to confirm the presence of lipid within casts. Lipid is dissolved by alcohol solutions in the Papanicolaou staining method, but distinct vacuoles are retained (Fig. 7-21). These casts are commonly seen with lipiduria associated with the nephrotic syndrome.

FIBRILLAR (FIBRIN) CASTS

The presence of delicate threads or a meshwork incorporated into a hyaline matrix is termed a **fibrillar cast.** These fibrils frequently have the morphological appearance and staining characteristics of fibrin (Fig. 7-22). The delicate appearance of these fibrils is maintained in urine sediments (Fig. 7-23). The cytologic examination of fibrinous exudates demonstrates similar fibrils. Fibrillar (fibrin) casts are found following leakage of coagulation proteins into the urine. Occasionally, "thready" material (Fig. 7-24) is found in histologic sections and in urine sediment and may represent alteration or aggregation of fibrin.

MIXED CASTS

The term **mixed cast** is used when a dominant cellular component is lacking, for example, when there are equal mixtures of leukocytes and erythrocytes or leukocytes and renal tubular epithelial cells. Mixed casts may also refer to a unipolar or bipolar distribution of cells (Figs. 7-25 and 7-26). The term should not be used in describing casts undergoing nonuniform transformation, for example, cellular to granular casts or granular to waxy casts. The diagnostic usefulness often depends on the cellular elements.

BROAD CASTS

A cast at least twice the usual width is termed a **broad cast.** All the aforementioned types of casts can appear in sediment in broad forms (Figs. 7-27, 7-28, and 7-29). In urine sediment, broad waxy casts are frequently referred to as Addis renal failure casts. In histologic sections, broad casts occur in zones containing dilated, frequently atrophic, tubules. Their presence indicates prolonged parenchymal urinary stasis or chronic renal disease with loss of functioning nephrons, or both.

CRYSTAL CASTS

Crystal casts represent a type of inclusion cast and are characterized by the presence of crystals in a cast matrix (Fig. 7-30). These casts indicate intratubular formation of crystals with associated urinary stasis and suggest tubular dysfunction with or without parenchymal damage.

PSEUDOCASTS

Pseudocasts are aggregations or accumulations of cells, fibrillar material, and particulate matter that lack a cylindrical configuration or mucoprotein matrix. The following may comprise pseudocasts: bacterial colonization, fungal aggregation, crystal aggregation, mucus strands, filter paper fibers, radiographic material, hair, and other artifacts. They may appear globular or rectangular. Pseudocasts may represent artifacts and urinary contaminants, form secondary to bladder disease, or may reflect renal parenchymal disease.

CELLULAR PSEUDOCASTS

Bacterial colonization is a common pseudocast and may represent a contaminant, overgrowth in storage, or severe infection (Fig. 7-31). Fungal pseudocasts in urinary sediments (Fig. 7-32) are frequently seen with lower urinary tract *Candida* infections. Leukocytic pseudocasts (Fig. 7-33) represent aggregated neutrophilic leukocytes and amorphous debris and are commonly seen with lower urinary tract infections. They must be distinguished from a true leukocytic cast, which indicates acute tubulointerstitial disease. Aggregation of erythrocytes is another common pseudocast and is associated with urinary system bleeding.

FIBRILLAR (MUCOUS) PSEUDOCASTS

Mucous material or threads are commonly seen in urine sediment (Fig. 7-34). Mucous threads are irregular and appear thicker than the delicate fibrils characteristic of fibrin. The presence of mucus in sediment usually indicates vaginal contamination or an inflammatory process involving the urinary system.

MUCOPROTEIN MATERIAL

Mucoprotein material in histologic sections (Fig. 7-35) and urine sediments (Fig. 7-36) often appears vacuolated and has scalloped edges with variable clefts. Renal tubules containing this material are frequently dilated, indicating urinary stasis. In urine sediments a cylindrical configuration typical of casts is lacking.

Renal interstitial pseudocasts are a source of globular mucoprotein material in urine sediment. They occur when there has been a rupture of tubular basement membranes (Fig. 7-37). In urine sediment, the globular material has a dense hyaline matrix but does not have the cylindrical shape of a hyaline cast (Fig. 7-38).

ARTIFACTS AND CONTAMINANTS

Concretions (Fig. 7-39), lubricants (Fig. 7-40), hair, and filter paper threads (Fig. 7-41) are commonly observed in urine sediment. These artifacts and contaminants can be reduced by careful urine specimen collection and processing.

BIBLIOGRAPHY

Schumann, GB: The Urine Sediment Examination. Baltimore, Williams & Wilkins, 1980.

FIG. 7-1
HYALINE CASTS. In tissue sections, these delicate-appearing transparent casts can be easily overlooked. Tubules are focally lined by an attenuated, flattened epithelium. They also have active regeneration with cells possessing an amphophilic cytoplasm and enlarged nuclei with prominent eosinophilic nucleoli. (× 250)

FIG. 7-2
HYALINE CAST IN URINE SEDIMENT. Note pale-blue, transparent matrix. (× 400)

FIG. 7-3
HYALINE AND FINELY GRANULAR CASTS IN URINE SEDIMENT. Note finely particulate material embedded in a hyaline matrix. (× 400)

FIG. 7-4
GRANULAR CAST IN COLLECTING DUCT. Granularity is accentuated in this trichrome-stained section. (× 250)

FIG. 7-5
COARSE GRANULAR CAST IN URINE SEDIMENT. Note increase in particulate size and amount in comparison to finely granular cast. (× 400)

FIG. 7-6
ERYTHROCYTIC (RED BLOOD CELL) CAST IN TUBULE. **(A)** Erythrocytes with intact cell membranes are present within hyaline cast matrix. (× 400) **(B)** The hyaline matrix and intact erythrocytes are accentuated with the trichrome stain. (× 250)

FIG. 7-7
ERYTHROCYTIC CASTS IN URINE SEDIMENT. Note intact erythrocytic cell membranes. (× 400)

FIG. 7-8
BLOOD (PIGMENTED) CAST IN CONVOLUTED TUBULES. In comparison to the erythrocytic cast, red blood cell membranes are disrupted with release of heme pigment, forming a granular heme-pigmented cast. (× 150)

FIG. 7-9
BLOOD (PIGMENTED) CAST IN URINE SEDIMENT. Note bright red color. Granular casts are usually pale red to light red-orange or blue with Papanicolaou staining. An attenuated epithelial layer is present along the top of the cast structure. (× 400)

FIG. 7-10
BILE CAST IN TUBULES. **(A)** Note intraluminal greenish, coarsely granular pigmented material in this section of renal medulla. (× 250) **(B)** A distal convoluted tubule contains an intraluminal aggregate of coarsely granular pigmented (orange-brown) material typical of a bile cast. (× 250)

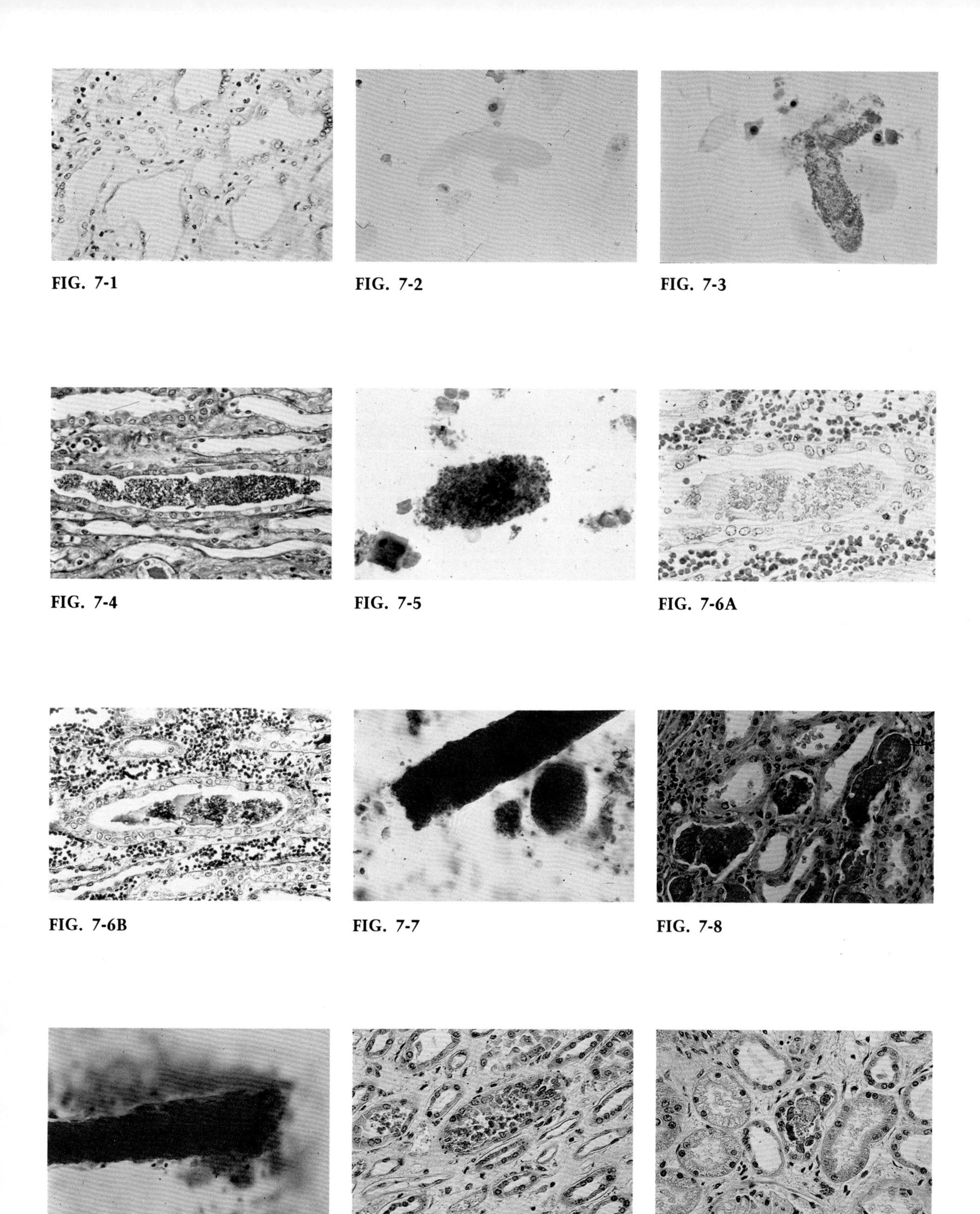

FIG. 7-1

FIG. 7-2

FIG. 7-3

FIG. 7-4

FIG. 7-5

FIG. 7-6A

FIG. 7-6B

FIG. 7-7

FIG. 7-8

FIG. 7-9

FIG. 7-10A

FIG. 7-10B

FIG. 7-11
BILE CAST IN URINE SEDIMENT. Note greenish-yellow granules. (× 400)

FIG. 7-12
LEUKOCYTIC CAST IN TUBULE. Note neutrophilic leukocytes in a hyaline matrix. Tubular epithelium is flattened and partially desquamated. The edematous interstitium contains an inflammatory infiltrate. (× 250)

FIG. 7-13
LEUKOCYTIC CAST IN URINE SEDIMENT. Segmented neutrophils are readily identified. (× 1000)

FIG. 7-14
RENAL EPITHELIAL CAST IN TUBULE. Exfoliated renal epithelial cells are partially incorporated into hyaline cast matrix. (× 250)

FIG. 7-15
RENAL EPITHELIAL CAST IN URINE SEDIMENT. Note mildly degenerated polygonal renal epithelial cells in a hyaline matrix. (× 400)

FIG. 7-16
NECROTIC RENAL EPITHELIAL CAST IN URINE SEDIMENT. Necrotic renal (ghost) cells in a hyaline matrix can be confused with a coarse granular cast. (× 400)

FIG. 7-17
DISINTEGRATING COARSE GRANULAR CAST IN URINE SEDIMENT. With *in vivo* intraparenchymal degeneration, cellular casts (renal epithelial, leukocytic, and erythrocytic) change to coarsely granular to finely granular and then finally to homogeneous waxy casts. (× 400)

FIG. 7-18
DISINTEGRATING COARSE GRANULAR CAST IN TUBULE. As a reflection of chronicity, a granular cast is undergoing transition to a waxy form. (Trichrome × 250)

FIG. 7-19
WAXY CAST IN TUBULES. Numerous shrunken renal tubules contain dense, amphophilic, hard-appearing casts, with cracks resulting from their brittle nature. These casts have been called "colloid-like" and the tubular alteration has been designated "thyroidization." (× 100)

FIG. 7-20
WAXY CASTS IN URINE SEDIMENT. **(A)** Note dense homogeneous texture. The convolutions found in this structure have no diagnostic significance. (× 1000) **(B)** Crack reveals the brittle nature of waxy cast. (× 400)

FIG. 7-21
FATTY CAST IN URINE SEDIMENT. Note distinct vacuoles. (× 400)

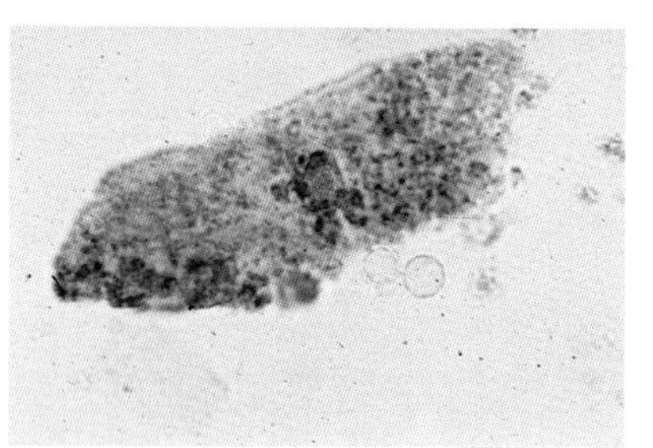

FIG. 7-11

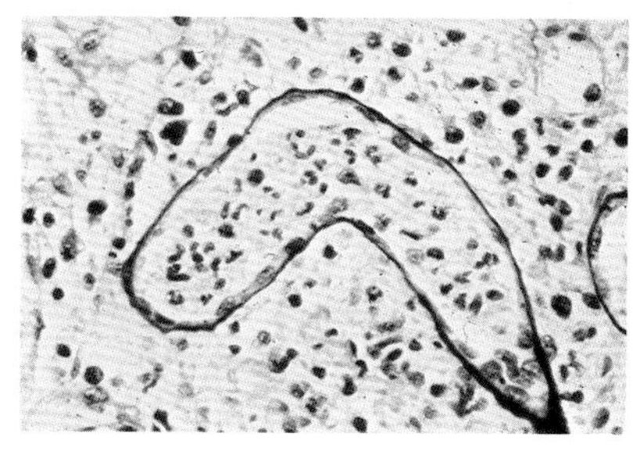

FIG. 7-12

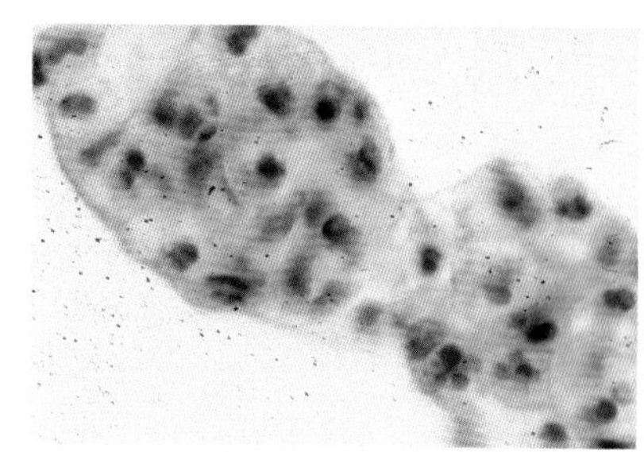

FIG. 7-13

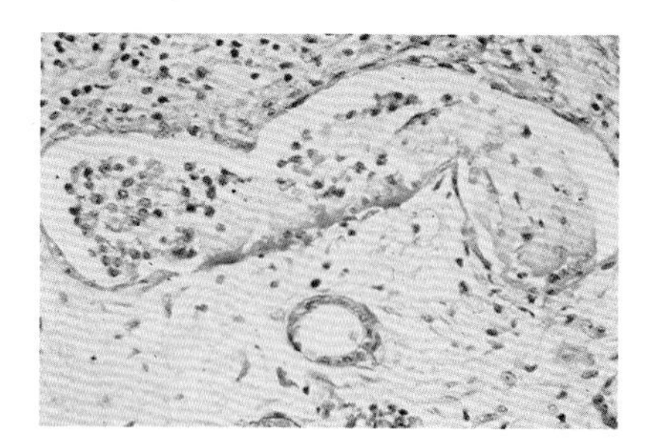

FIG. 7-14

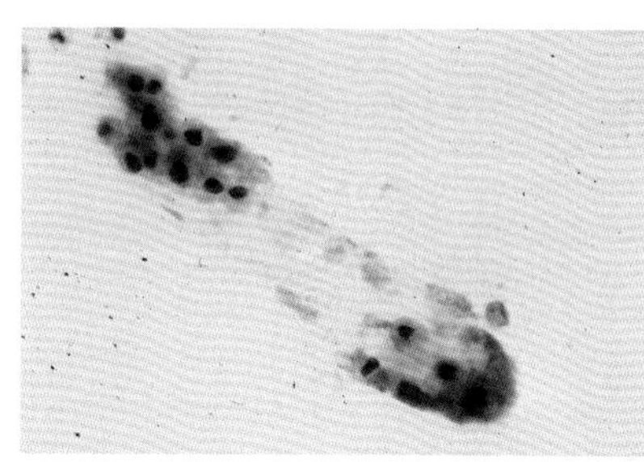

FIG. 7-15

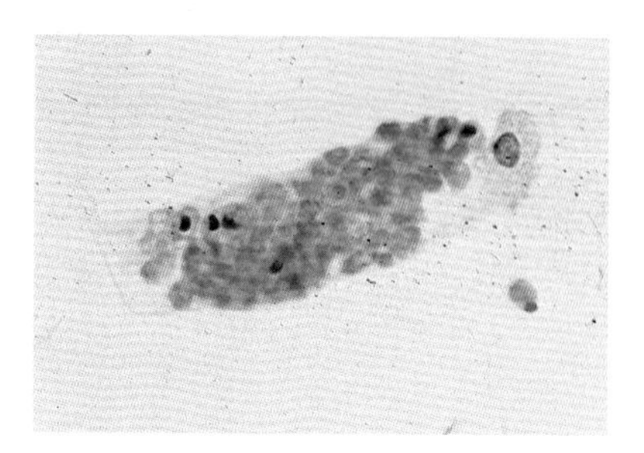

FIG. 7-16

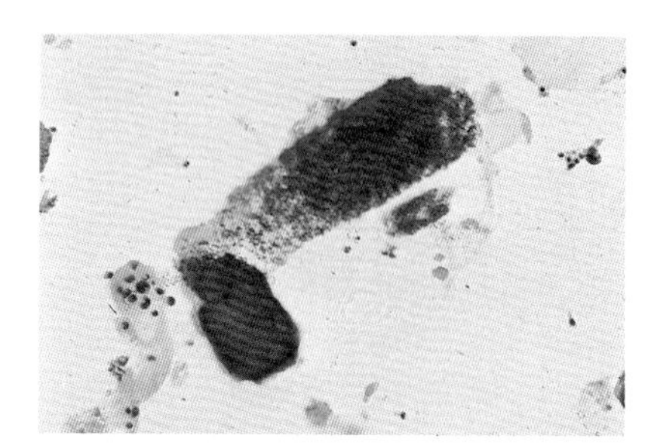

FIG. 7-17

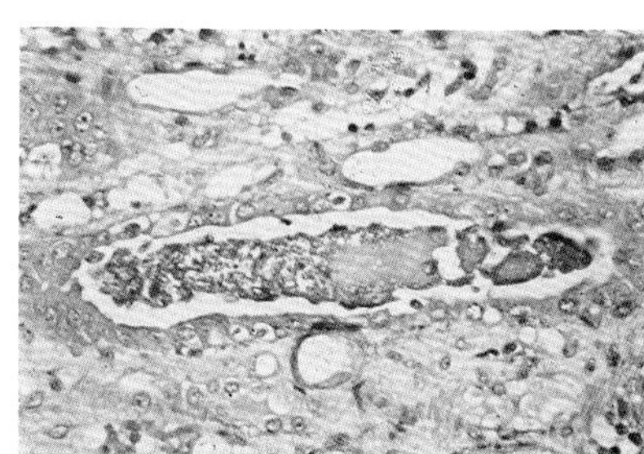

FIG. 7-18

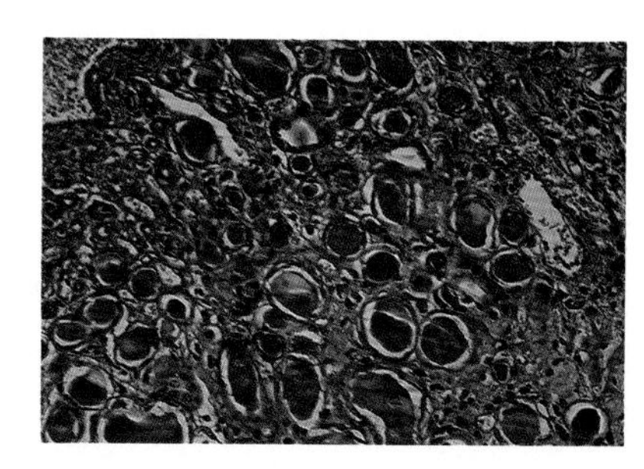

FIG. 7-19

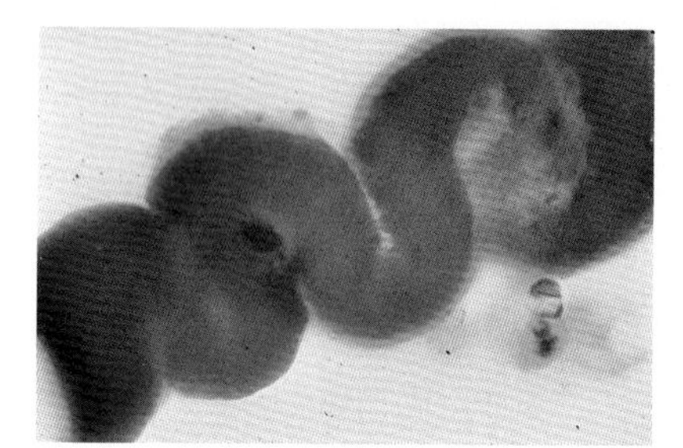

FIG. 7-20A

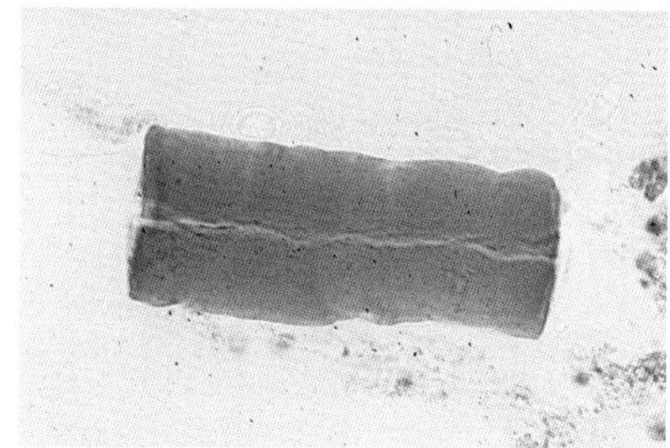

FIG. 7-20B

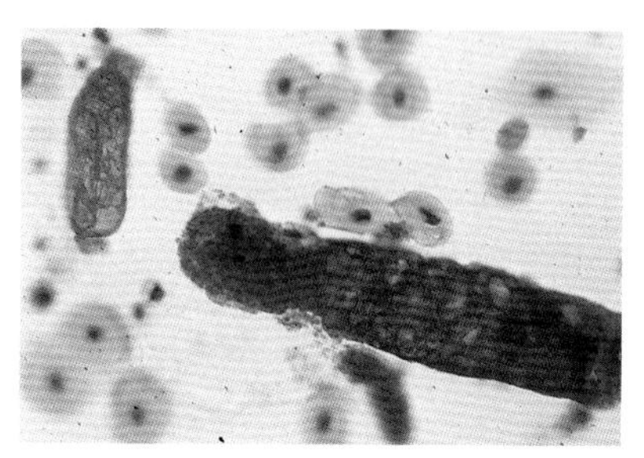

FIG. 7-21

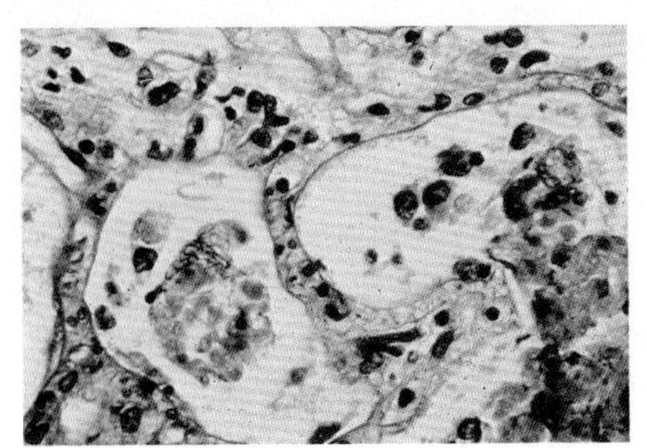

FIG. 7-22A

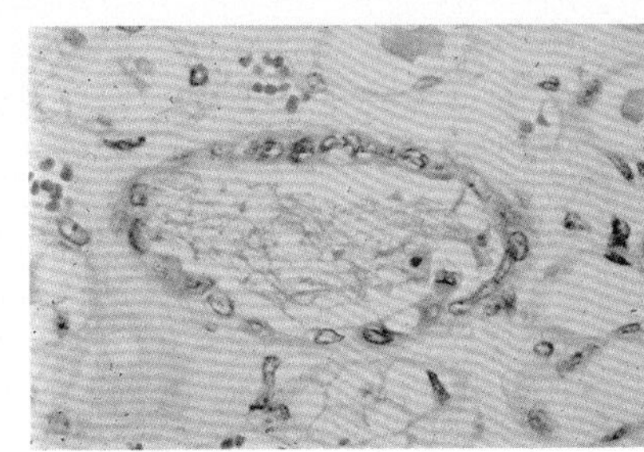

FIG. 7-22B

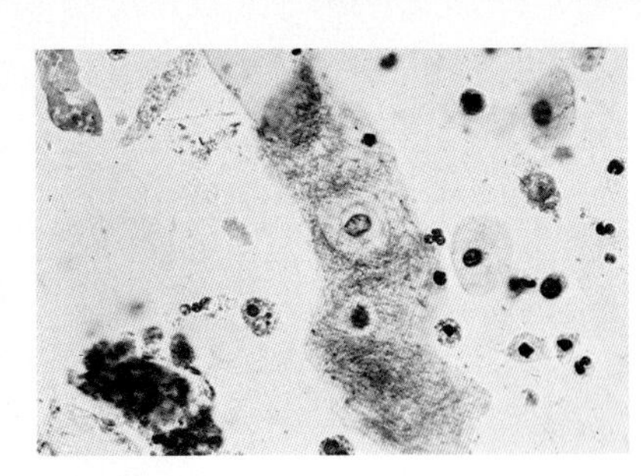

FIG. 7-23

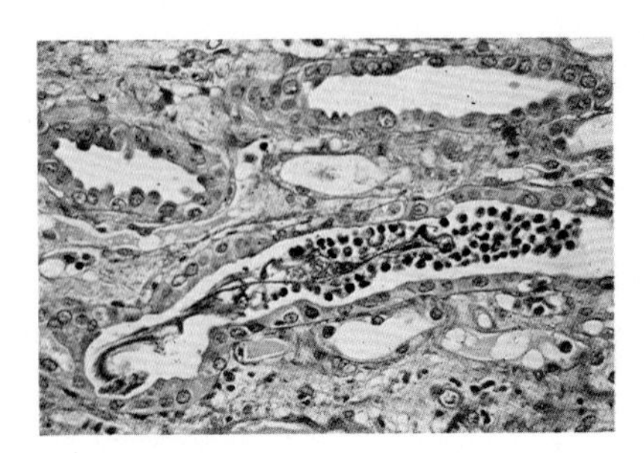

FIG. 7-24A

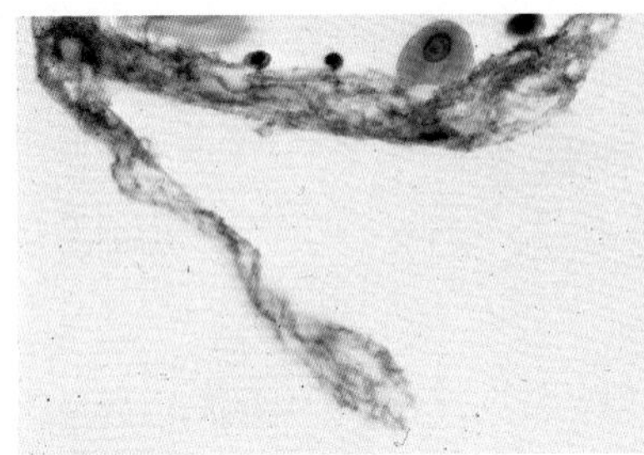

FIG. 7-24B

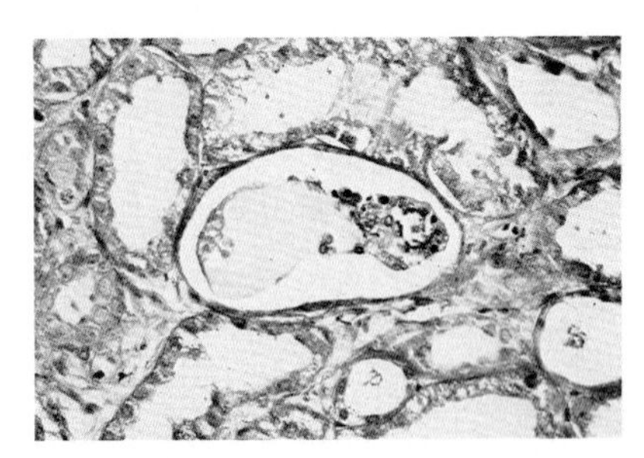

FIG. 7-25

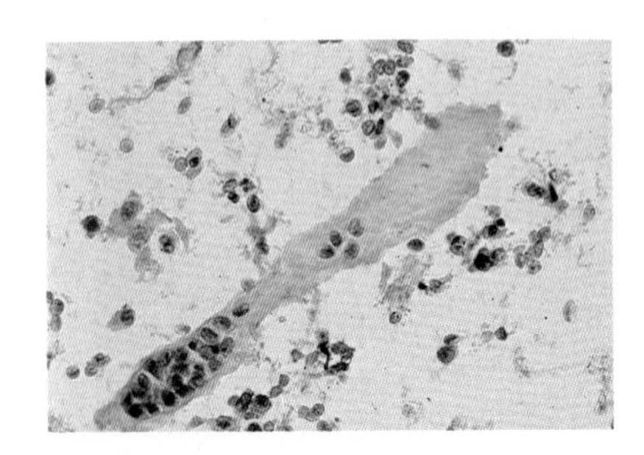

FIG. 7-26

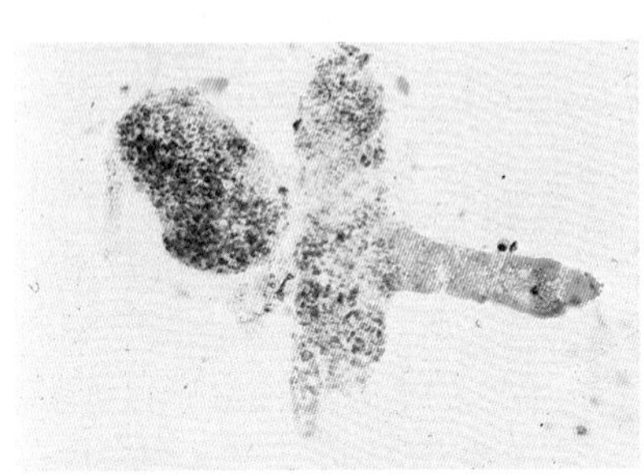

FIG. 7-27

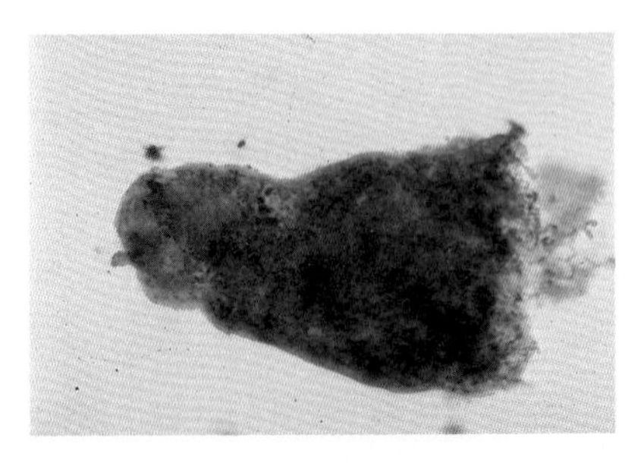

FIG. 7-28

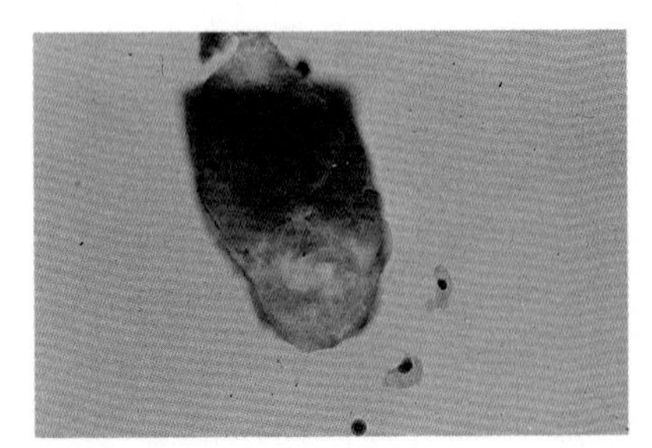

FIG. 7-29A

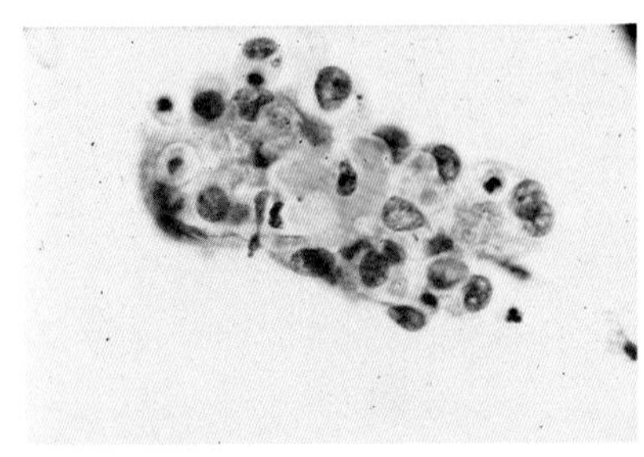

FIG. 7-29B

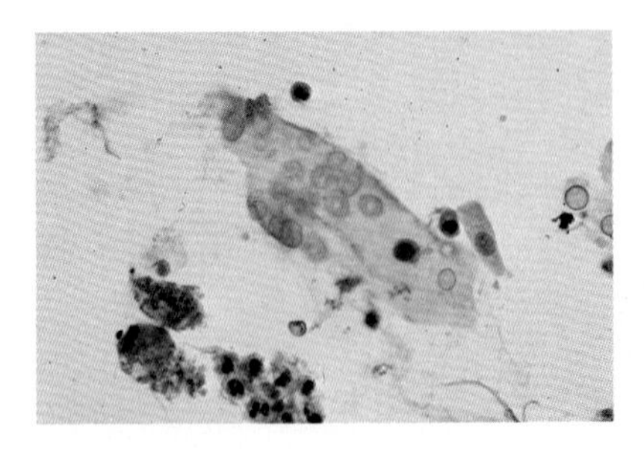

FIG. 7-30

FIG. 7-22
FIBRILLAR (FIBRIN) CASTS. **(A)** Two proximal convoluted tubules contain desquamated, necrotic renal epithelium mixed with delicate fibrils of fibrin. (Lendrum × 400) **(B)** Similar fibrils form a meshwork within hyaline cast matrix in the thin limb of a convoluted tubule. (× 400)

FIG. 7-23
FIBRIN CAST IN URINE SEDIMENT. Note delicate fibrils and rare epithelial cells in hyaline matrix. Fibrin casts are found following leakage of coagulation products into the urinary stream. (× 400)

FIG. 7-24
"THREADY" MATERIAL. **(A)** Nonfuchsinophilic, thick-appearing "thready" material is surrounded by numerous exfoliated renal epithelial cells, and may represent altered fibrin or mucoprotein. (Trichrome × 250) **(B)** Similar material in urine sediment. (× 400)

FIG. 7-25
MIXED CAST IN TUBULE. A dilated tubule contains a broad hyaline cast in continuity with a degenerating renal tubular epithelial cast. (Trichrome × 250)

FIG. 7-26
MIXED CAST IN URINE SEDIMENT. Note dense hyaline matrix containing renal tubular epithelial cells at one end. (× 400)

FIG. 7-27
GRANULAR AND BROAD GRANULAR CASTS IN URINE SEDIMENT. (× 400)

FIG. 7-28
BROAD GRANULAR CAST IN URINE SEDIMENT. (× 400)

FIG. 7-29
BROAD CASTS IN URINE SEDIMENT. (A) Broad waxy cast. (× 400) **(B)** Broad mixed cast. (× 400)

FIG. 7-30
CRYSTAL CAST IN URINE SEDIMENT. Note crystals embedded in a hyaline matrix. (× 400)

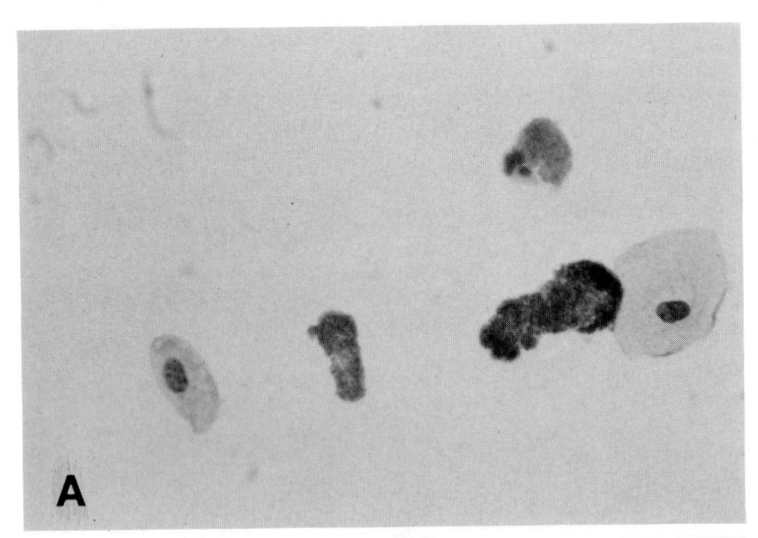

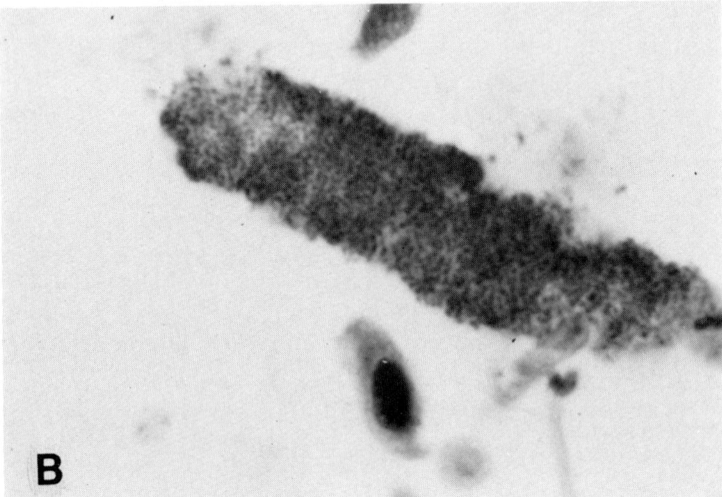

FIG. 7-31
BACTERIAL PSEUDOCASTS IN URINE SEDIMENT. **(A)** At low magnification bacterial pseudocasts can resemble granular casts. (× 250) **(B)** At higher magnification, discrete organisms can be recognized. (× 1000)

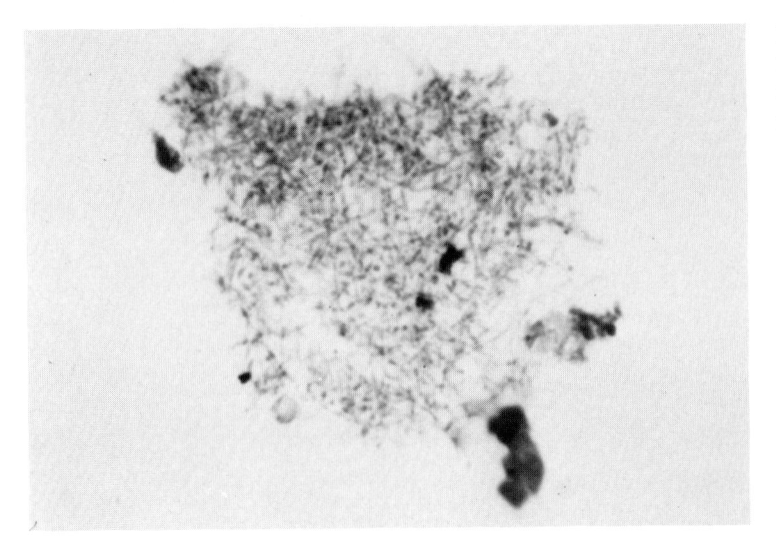

FIG. 7-32
FUNGAL PSEUDOCASTS IN URINE SEDIMENT. Note yeast and pseudohyphal forms of *Candida* species. (× 400)

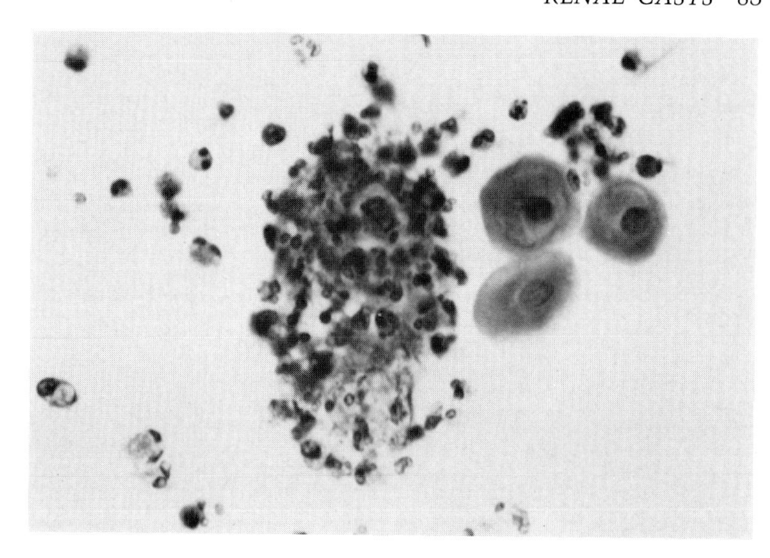

FIG. 7-33
LEUKOCYTIC PSEUDOCAST IN URINE SEDIMENT. Aggregated neutrophilic leukocytes and amorphous debris. (× 400)

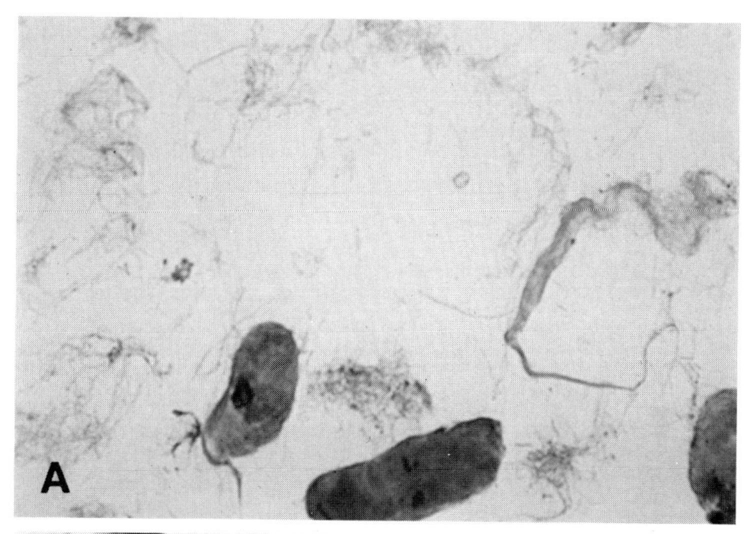

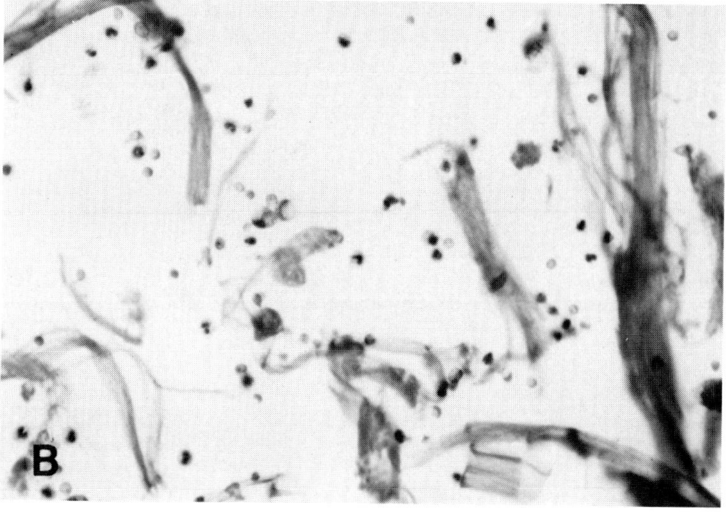

FIG. 7-34
MUCOUS THREADS IN URINE SEDIMENT. **(A)** Dense hyaline-like cast material and numerous mucous threads are present. (× 400) **(B)** Mucous threads are irregular and appear thicker than the delicate fibrils characteristic of fibrin. (× 400)

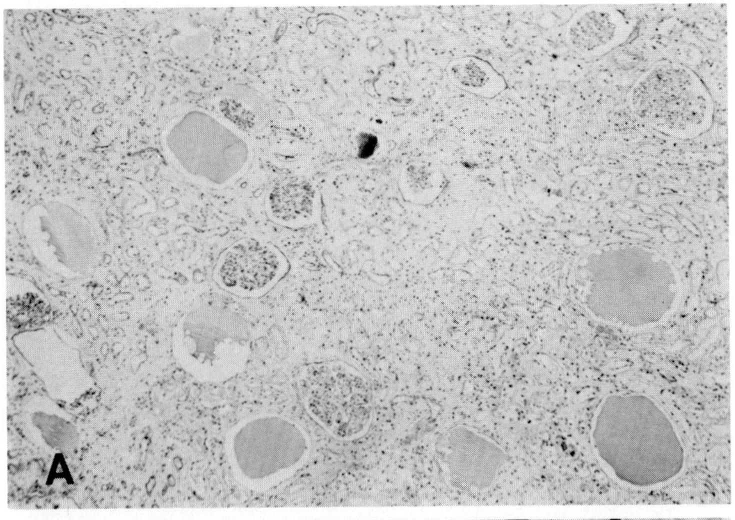

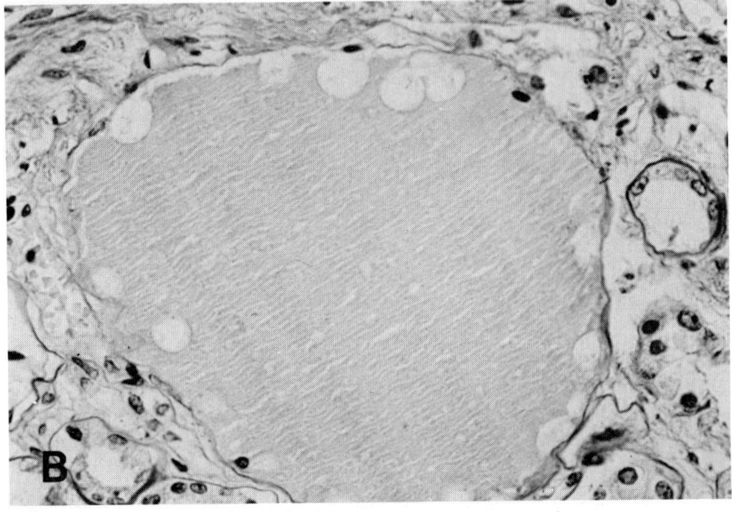

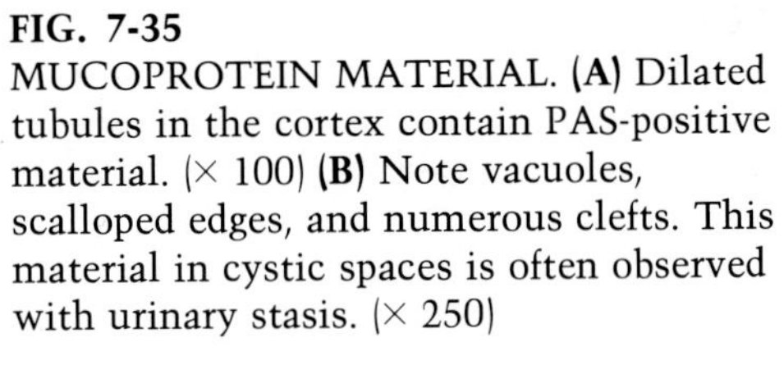

FIG. 7-35
MUCOPROTEIN MATERIAL. **(A)** Dilated tubules in the cortex contain PAS-positive material. (× 100) **(B)** Note vacuoles, scalloped edges, and numerous clefts. This material in cystic spaces is often observed with urinary stasis. (× 250)

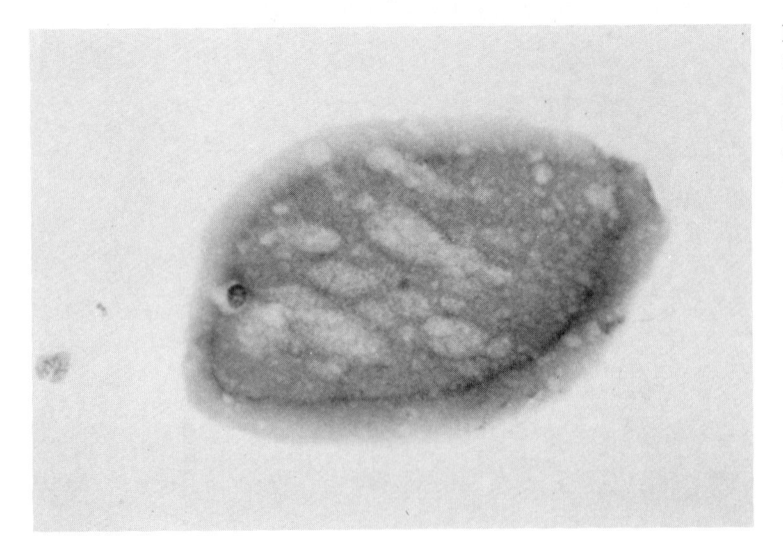

FIG. 7-36
GLOBULAR MATERIAL IN URINE SEDIMENT. Note absence of cylindrical configuration and dense hyaline appearance with vacuoles and clefts. (× 1000)

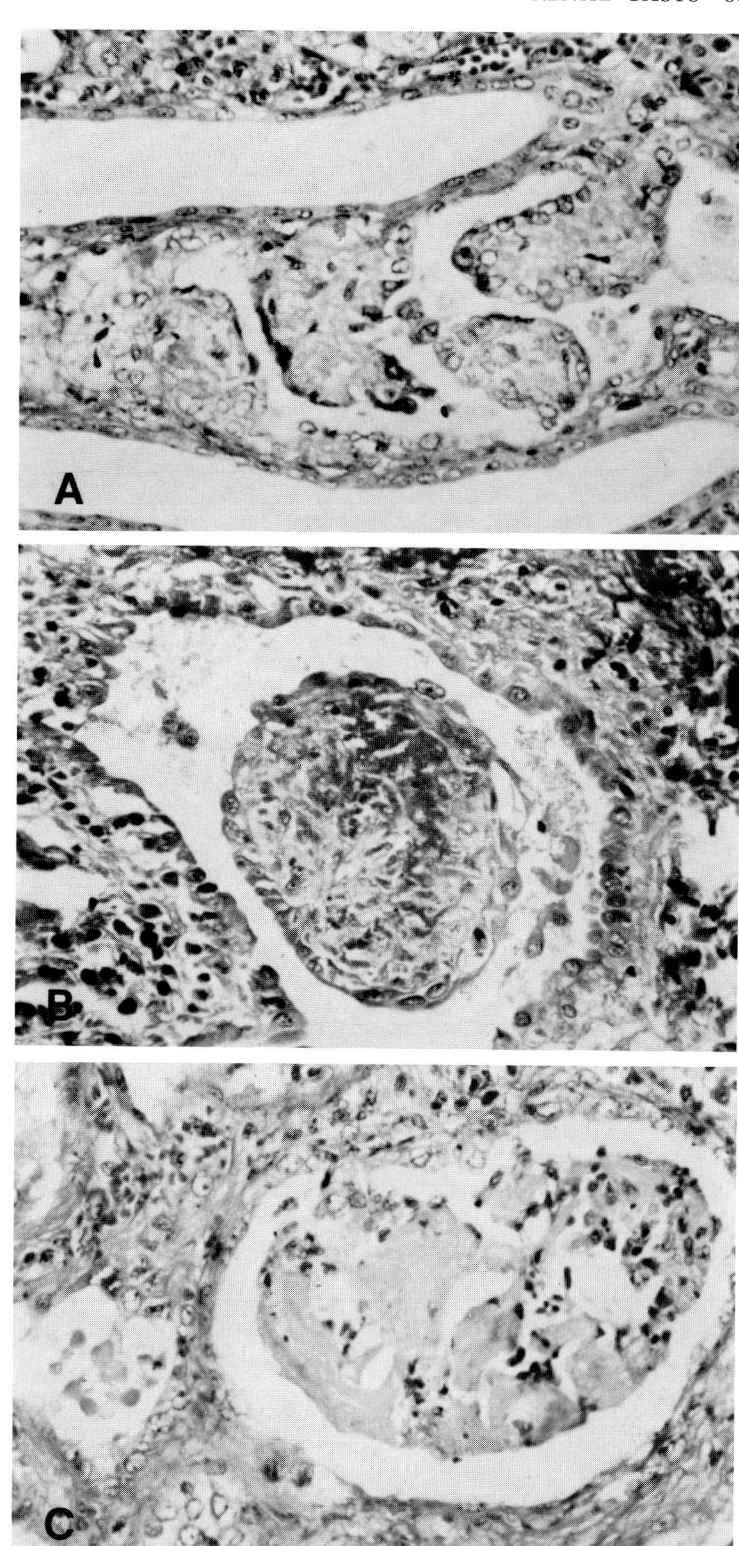

FIG. 7-37
RENAL INTERSTITIAL PSEUDOCAST. **(A)** Interstitial tissue has herniated through multiple ruptures in the tubular basement membrane. Regenerating epithelium covers these polypoid masses. (Lendrum × 400) **(B)** When the plane of section is parallel to the site of rupture, an apparently detached intraluminal mass is observed. This pseudocast may contain fibrin, fibroblasts, and a mixture of inflammatory cells, or may appear as a hypocellular hyaline mass. (Trichrome × 250) **(C)** With severe or repeated ischemic parenchymal damage, this mass may detach and exfoliate into the urine as a fragment, or may resemble a dense hyaline or waxy cast. (Lendrum × 250)

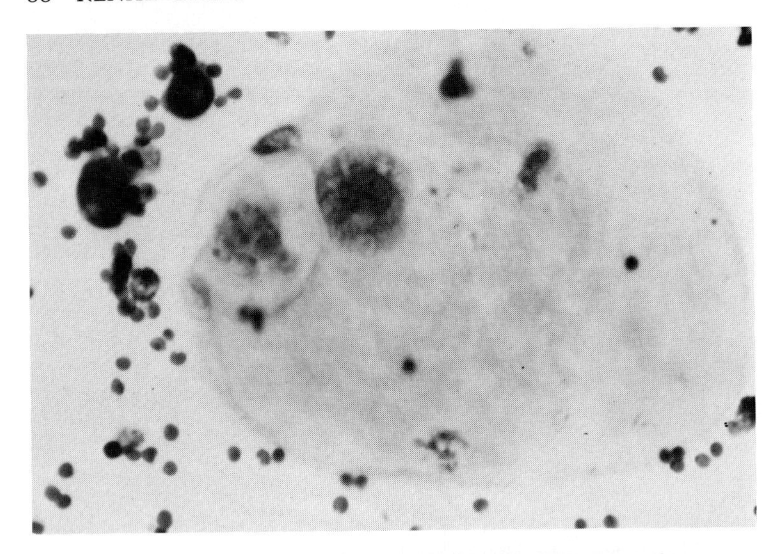

FIG. 7-38
GLOBULAR MATERIAL IN URINE SEDIMENT. Note dense hyaline appearance and incorporated cells. (× 1000)

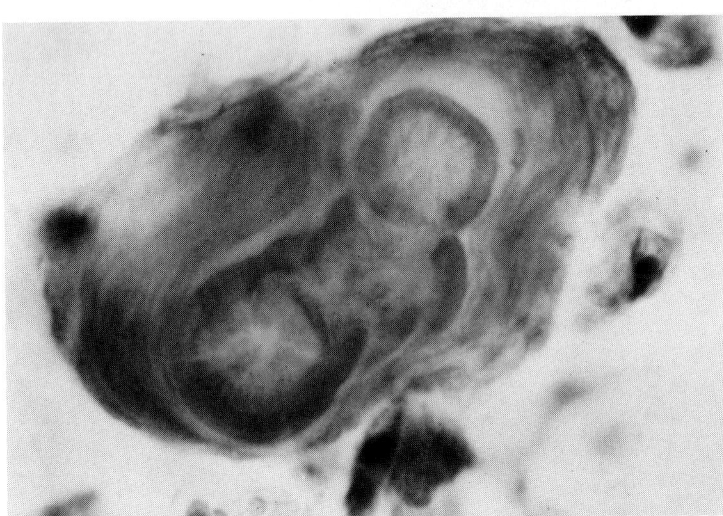

FIG. 7-39
LAMINATED CONCRETION IN URINE SEDIMENT. (× 1000)

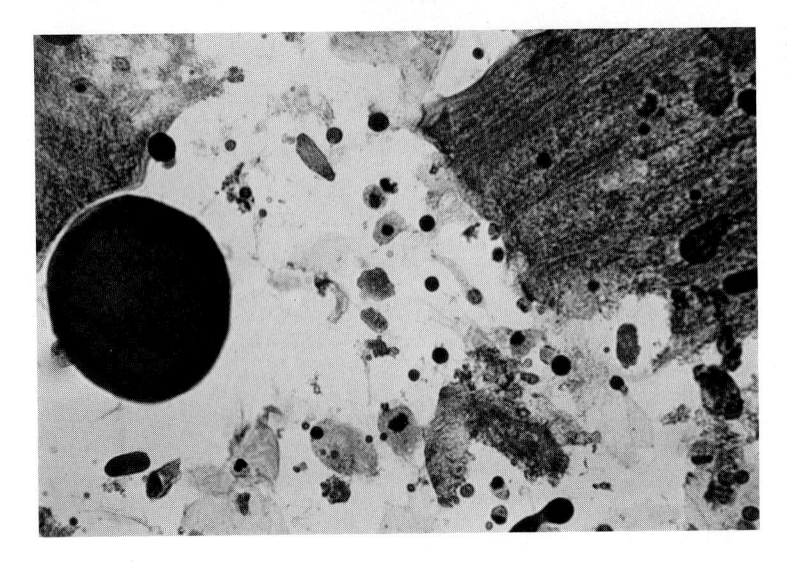

FIG. 7-40
PROBABLE LUBRICANT ARTIFACT (ARROW) IN URINE SEDIMENT. (× 400)

FIG. 7-41
FILTER PAPER ARTIFACT. Note irregular refractile appearance. (× 400)

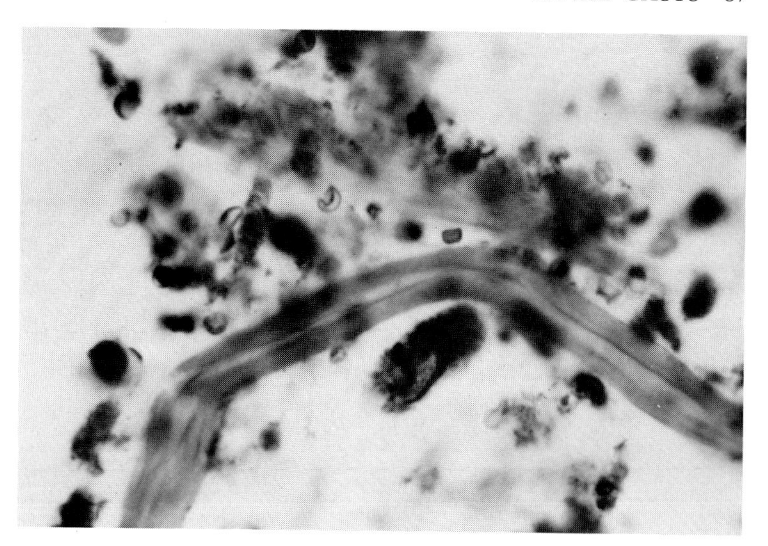

8

INCLUSION-BEARING CELLS

Inclusion-bearing cells in urine sediment should be categorized as ***intranuclear*** *or* ***intracytoplasmic*** *(Table 8-1). Accurate identification of inclusion-bearing cells requires a permanently stained preparation.*

INTRANUCLEAR INCLUSION-BEARING CELLS

NONSPECIFIC DEGENERATIVE-REGENERATIVE TYPE

Nonspecific degenerative nuclear changes in regenerating cells are a common source of intranuclear inclusion-bearing cells (Fig. 8-1). Chromatin crowding with deposition along nuclear membranes (chromatinic membrane) and nuclear pyknosis give the appearance of viral intranuclear inclusions. Regenerating renal tubular epithelium may have "pseudoviral" intranuclear inclusions (Fig. 8-2). These regenerative changes are characterized by nuclear enlargement, chromatin clearing, prominent irregular eosinophilic macronucleoli, and abundant amphophilic cytoplasms. These intranuclear inclusion-bearing cells are typically seen in urine sediment following renal ischemic necrosis and radiation-induced cystitis.

HEAVY METAL-INDUCED TYPE

Heavy metal exposure is a well-recognized cause of renal epithelial inclusion-bearing cells in urine sediment. Although studies suggest that the inclusions contain the heavy metal, this point is controversial. Regardless of whether these inclusions contain the suspected metal, the presence of increased numbers of exfoliated renal epithelial inclusion-bearing cells following acute or chronic heavy metal exposure indicates a related tubular injury.

Lead-induced inclusion-bearing cells in urine sediment have a dark, discrete intranuclear inclusion and ballooning, granular cytoplasm containing nonspecific eosinophilic droplets (Fig. 8-3). Lead-induced intranuclear inclusions in histologic sections occur primarily in the epithelium of proximal convoluted tubules and loops of Henle (Fig. 8-4). These eosinophilic intranuclear inclusions have a characteristic acid-fast staining property.

Recent studies on patients exposed to cadmium have shown exfoliation of numerous intranuclear inclusion-bearing cells in urine. The highly granular cytoplasm characterizes these cells as convoluted tubular in origin (Fig. 8-5). The lack of renal biopsy material from patients with heavy

TABLE 8-1
INCLUSION-BEARING CELLS

INTRANUCLEAR	INTRACYTOPLASMIC
Nonspecific degenerative-regenerative	Endogenous storage products:
Heavy metal–induced:	Glycogen
Lead	Altered organelles:
Cadmium	Eosinophilic degenerative droplets
Drug-induced:	Neutrophilic toxic granulation
Gentamicin	Michaelis-Gutmann bodies
Viral-induced:	Increased fusion of cytoplasmic granules (Chediak-Higashi Syndrome)
Cytomegalovirus	Phagocytized organisms:
Herpes simplex	Bacteria
Adenovirus	Bacteria-like: chlamydial Agent
Polyomavirus	Fungi: *Histoplasma, Candida*
	Protozoa: *Toxoplasma*
	Phagocytized cells:
	Erythrophagocytosis
	Phagocytized Material:
	Cast material
	Crystal
	Pigmentation:
	Lipochrome
	Hemosiderin
	Melanin
	Viral-Induced
	Cytomegalovirus
	Measles

metal exposure, such as lead and cadmium, has prevented cytohistologic correlations. Intranuclear inclusion-bearing renal cells often appear in urine before detectable renal function changes.

DRUG-INDUCED TYPE

Intranuclear inclusion-bearing renal cells may appear in urine sediment following exposure to nephrotoxic drugs, such as gentamicin (Fig. 8-6). Discrete basophilic intranuclear inclusions are found in granular cells, indicating a convoluted tubular origin. These patients frequently have temporally related renal dysfunction.

VIRAL-INDUCED TYPE

Viral-induced intranuclear inclusion-bearing cells in the urine sediment are frequently reported. Herpes simplex (Fig. 8-7), cytomegalovirus (CMV) (Figs. 8-8 and 8-9), polyomavirus and adenovirus may produce exfoliation of intranuclear inclusion-bearing cells and are important to exclude in immunosuppressed patients. Further discussion of viral-induced cellular changes is given in Chapter 5.

INTRACYTOPLASMIC INCLUSION-BEARING CELLS

A wide variety of cellular changes can be associated with intracytoplasmic inclusions (see Table 8-1). These inclusions represent accumulated storage products, altered organelles, and phagocytized foreign material, cells, and organisms, or may be virally induced. The inclusions may occur in epithelial and inflammatory cells. Morphologically, they may be granular, globular, or droplet-like and discrete bodies.

ENDOGENOUS STORAGE PRODUCTS

Material normally produced by urinary system lining cells may accumulate in the cytoplasm (Fig. 8-10). This may reflect overproduction or the inability to break down or release the material.

ALTERED ORGANELLES

A common cytoplasmic alteration associated with cellular degeneration is the formation of droplet-like inclusions. These eosinophilic bodies are frequently observed in histologic material (Fig. 8-11) and in urine sediments (Fig. 8-12). They probably represent enlarged lysosomes or phagosomes. They are commonly seen in the urine sediment in increased frequency in virally infected cells and following heavy metal exposure and renal parenchymal ischemia.

Granular cytoplasmic inclusions in neutrophils found in urine sediment correspond to toxic granulation observed in peripheral blood smears (Fig. 8-13). They should not be misinterpreted as intracellular bacteria.

Michaelis-Gutmann bodies, diagnostic of malakoplakia, are dark, discrete intracytoplasmic inclusions within macrophages (Fig. 8-14). In histologic sections, the inclusions appears eosinophilic and are variably calcified (Fig. 8-15). These inclusions represent fused phagolysomes that ultrastructurally contain fragmented bacterial walls.

PHAGOCYTIZED ORGANISMS, CELLS, AND MATERIAL

Common phagocytized organisms include bacteria and fungi (Fig. 8-16). Accurate interpretation requires morphological identification, that is, cocci, bacilli, and spores.

Erythrophagocytosis has been observed in urine sediment (Fig. 8-17). Intact erythrocytes may be ingested by neutrophils or macrophages. The significance of this finding is not known.

Phagocytized material may occur in inflammatory cells or epithelial cells. Exfoliated renal epithelial cells with ingested crystals or waxy cast-like material have been observed (Fig. 8-18).

PIGMENTATION

Intracytoplasmic pigments that can be identified in urine sediment include lipochrome (Figs. 8-19 and 8-20), hemosiderin (Figs. 8-21 and 8-22), and melanin. Special stains are required for accurate identification. Common sources of pigmented cells are macrophages, renal tubular epithelial cells, and seminal vesicle cells.

VIRAL-INDUCED TYPE

Cells infected with CMV frequently contain basophilic intracytoplasmic inclusions along with diagnostic intranuclear inclusions. Although rarely seen in urine sediments, the numerous intracytoplasmic inclusions present in measles-infected cells is a diagnostic feature.

BIBLIOGRAPHY

Koss, LG: Diagnostic Cytology and Its Histologic Bases, 3rd ed. Philadelphia, JB Lippincott, 1979

Schumann, GB: The Urine Sediment Examination. Baltimore, Williams & Wilkins, 1980

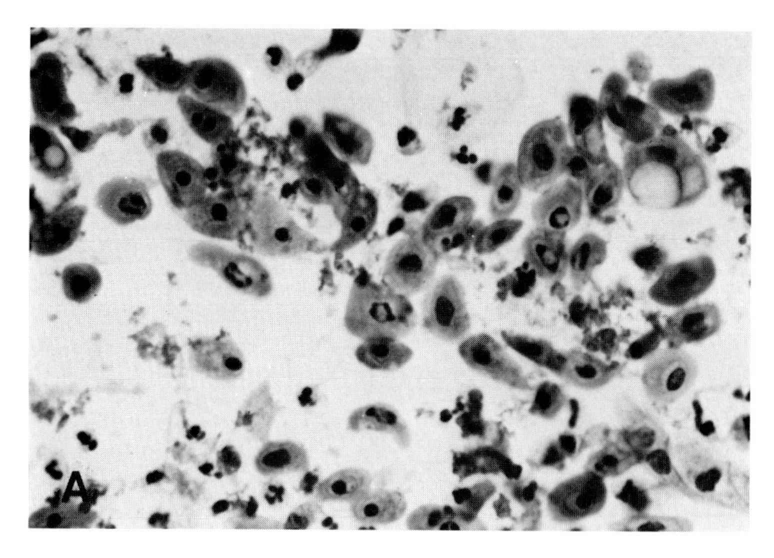

FIG. 8-1
DEGENERATIVE NUCLEAR CHANGES IN EXFOLIATED UROTHELIUM. **(A)** Note numerous degenerating urothelial cells with inflammation and cellular debris. (× 400)

(continued)

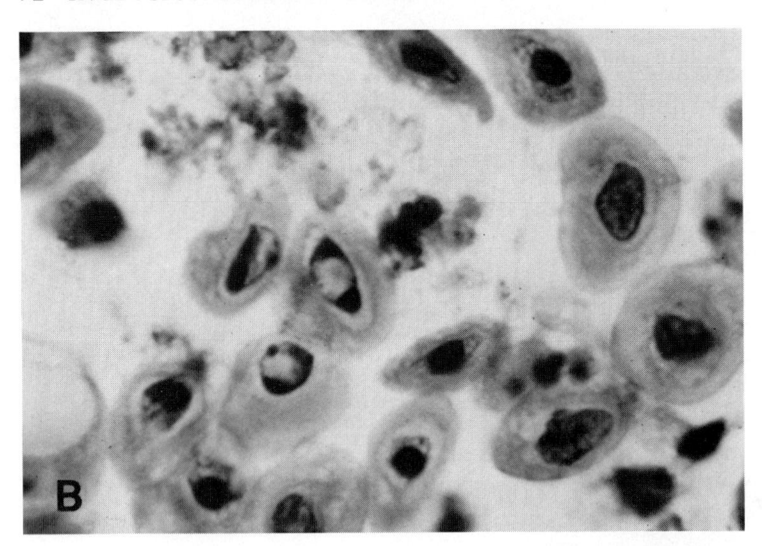

FIG. 8-1B *(continued)*
(B) Chromatin clearing and deposition along nuclear membrane (chromatinic membrane) and nuclear pyknosis give appearance of intranuclear inclusions. (× 1000)

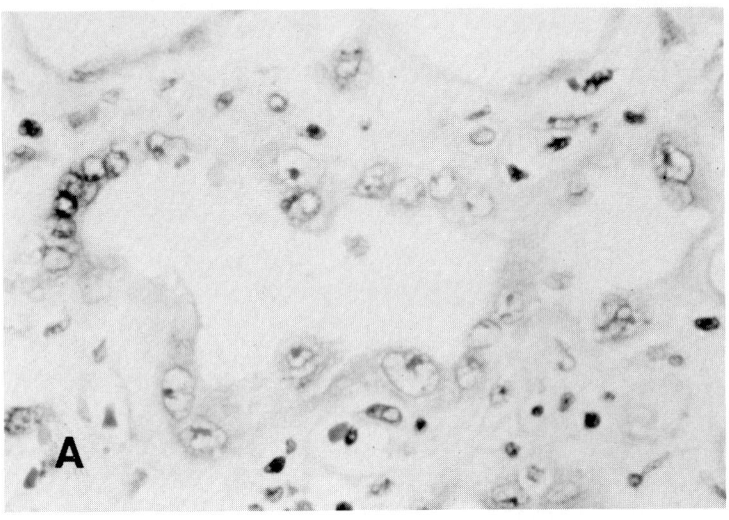

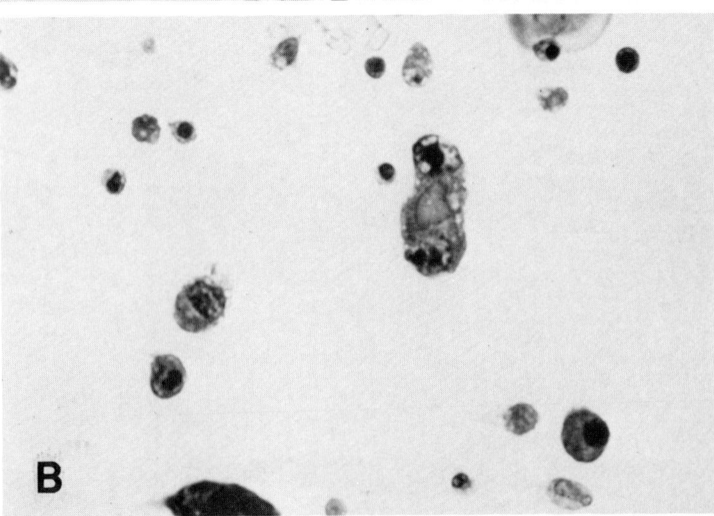

FIG. 8-2
PSEUDOVIRAL INCLUSIONS IN REGENERATING TUBULAR EPITHELIUM. **(A)** Nuclear enlargement and pleomorphism, chromatin clearing, prominent eosinophilic nucleoli, and abundant amphophilic cytoplasm characterize regenerating epithelium. Irregularity of nucleoli distinguishes them from viral inclusions. Piling up and crowding of regenerating cells give syncytial appearance. (× 400) **(B)** Exfoliated degenerative-regenerative cell in urine sediment. The nucleus is decreased in size, is wrinkled, and contains a dark, irregular intranuclear inclusion surrounded by a clear halo. These inclusions are commonly seen with ischemic necrosis. Note nonspecific eosinophilic intracytoplasmic droplet. (× 400)

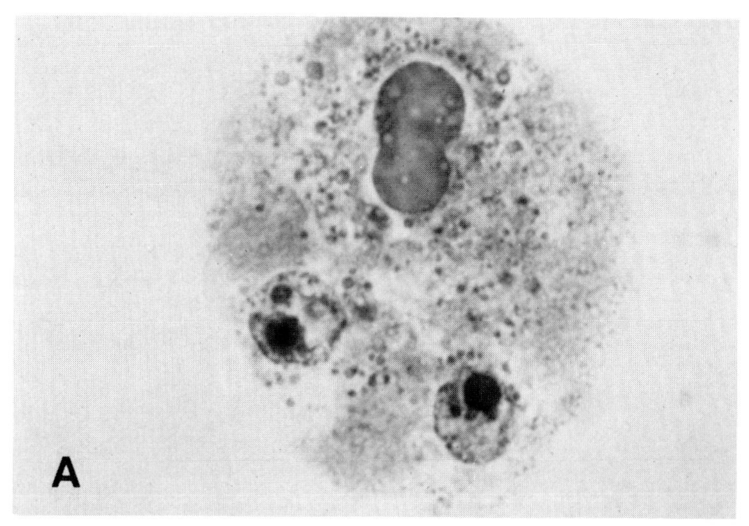

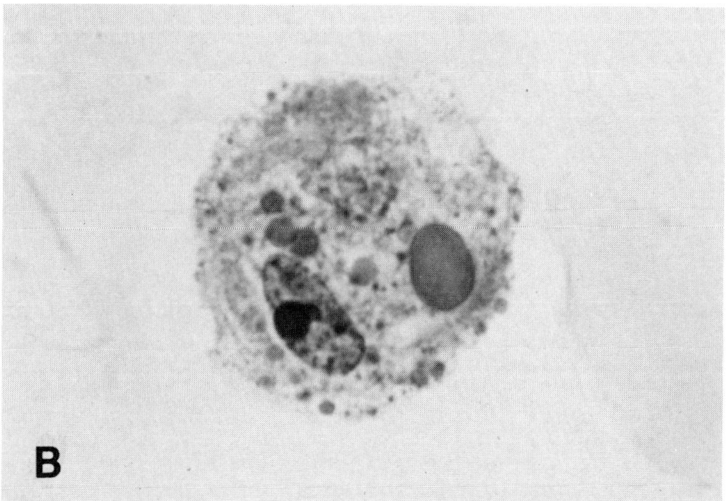

FIG. 8-3
LEAD-INDUCED INCLUSION-BEARING CELLS IN URINE SEDIMENT. **(A)** Large binucleated proximal convoluted tubular epithelial cell with intranuclear inclusions and ballooning granular cytoplasm containing nonspecific eosinophilic droplets. (× 1000) **(B)** Note cell with a prominent dark intranuclear inclusion and ballooned cytoplasm containing numerous nonspecific eosinophilic droplets. (× 1000)

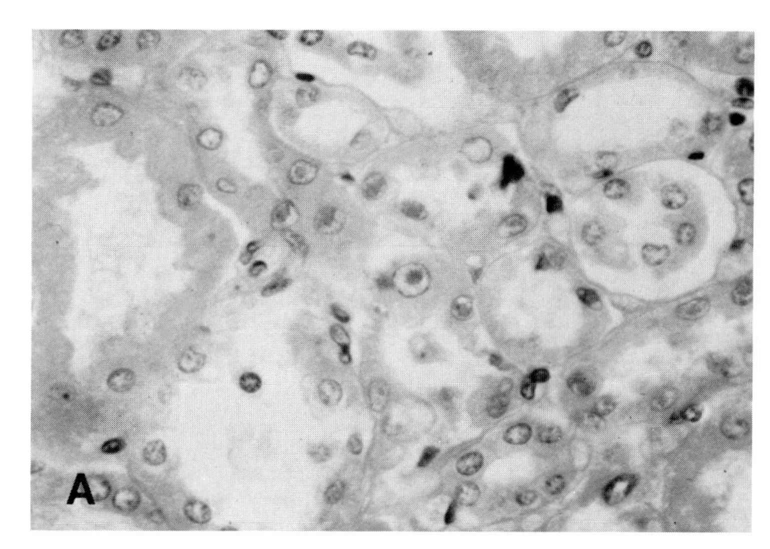

FIG. 8-4
LEAD-INDUCED INTRANUCLEAR INCLUSIONS. **(A)** Epithelium of proximal convoluted tubules contain numerous eosinophilic intranuclear inclusions. Occasional nuclei have chromatin clearing, but there is no chromatinic membrane and no nuclear or cellular enlargement. (× 400)

(continued)

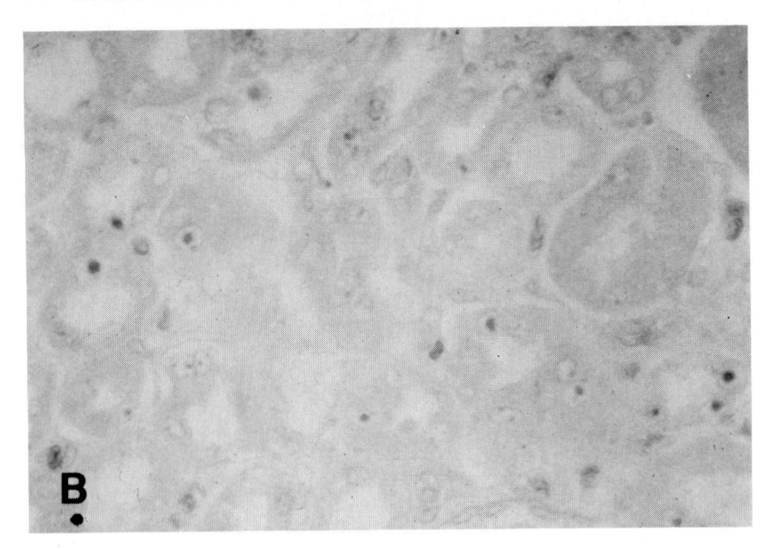

FIG. 8-4B *(continued)*
(B) Inclusions, also present in cells lining thin segments, have characteristic acid-fast-positive staining. (× 250)

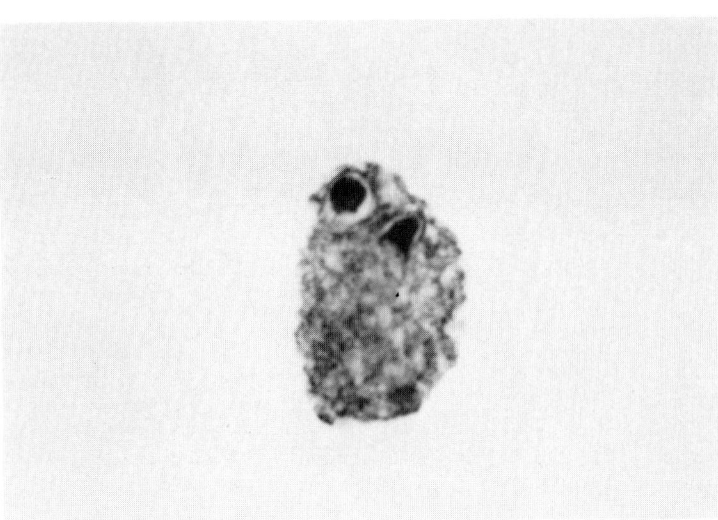

FIG. 8-5
INTRANUCLEAR INCLUSION-BEARING CELL IN URINE SEDIMENT. Note binucleated cell with dark intranuclear inclusions and abundant granular cytoplasm. In contrast to viral-induced inclusion-bearing cells, there is no nuclear enlargement. Urine specimen from patient with cadmium exposure. (× 1000)

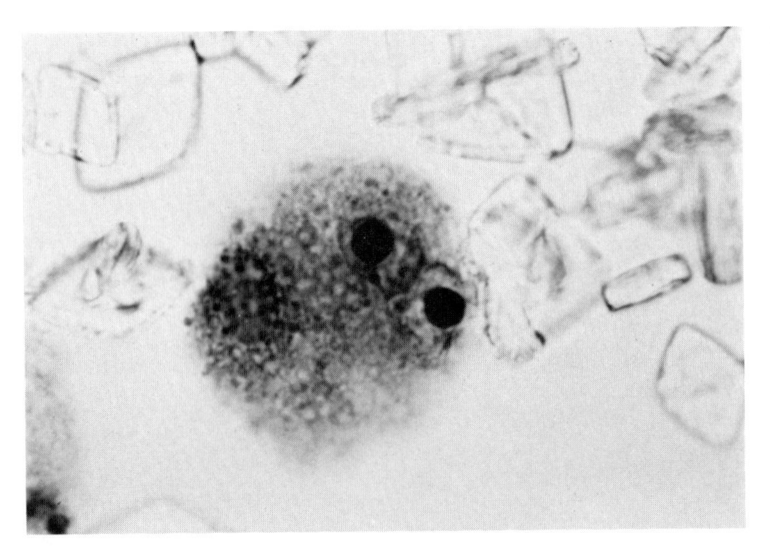

FIG. 8-6
INCLUSION-BEARING CELL IN URINE SEDIMENT. Binucleated renal convoluted tubular epithelial cell with intranuclear inclusions. Note similarity to the lead-induced inclusion-bearing cells. Numerous uric acid crystals are in background. Urine specimen obtained after gentamicin therapy. (× 1000)

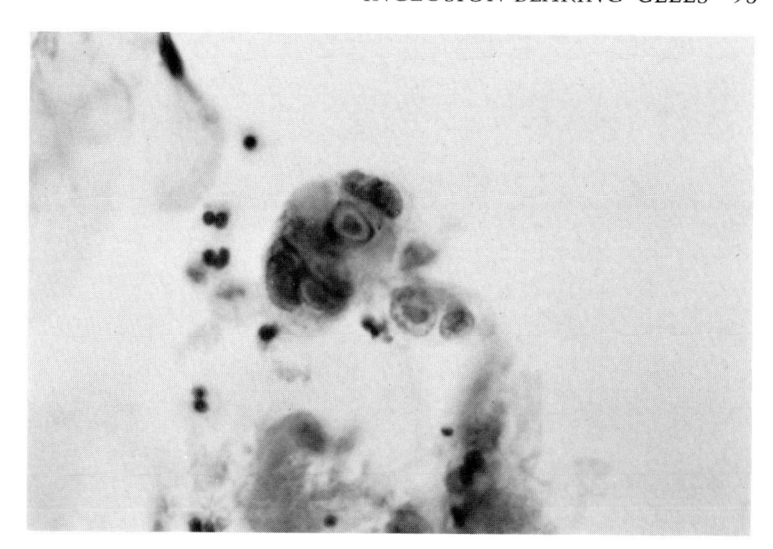

FIG. 8-7
EXFOLIATED HERPETIC INCLUSION-BEARING CELLS. Note nuclear enlargement and faint eosinophilic intranuclear inclusions. Multinucleated forms are characteristic for herpes simplex infection. (× 1000)

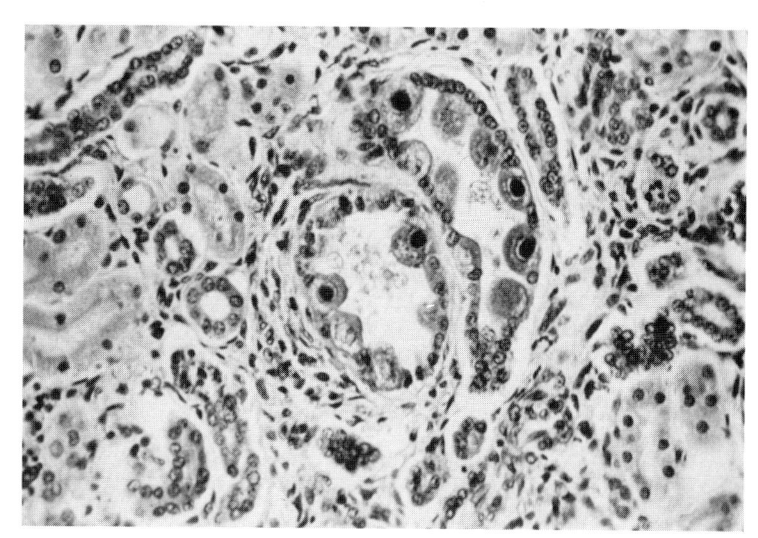

FIG. 8-8
CYTOMEGALIC INCLUSION DISEASE. Renal tubular epithelial cells infected with CMV are enlarged and contain dense basophilic intranuclear inclusions. (× 400)

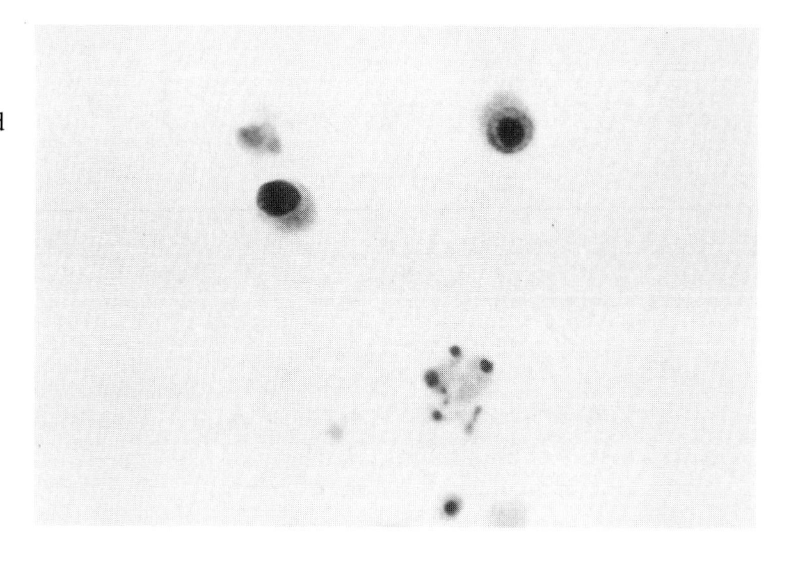

FIG. 8-9
EXFOLIATED CMV-INDUCED INTRANUCLEAR INCLUSION-BEARING CELLS. One of two degenerated epithelial cells contains a distinct intranuclear inclusion. (× 400)

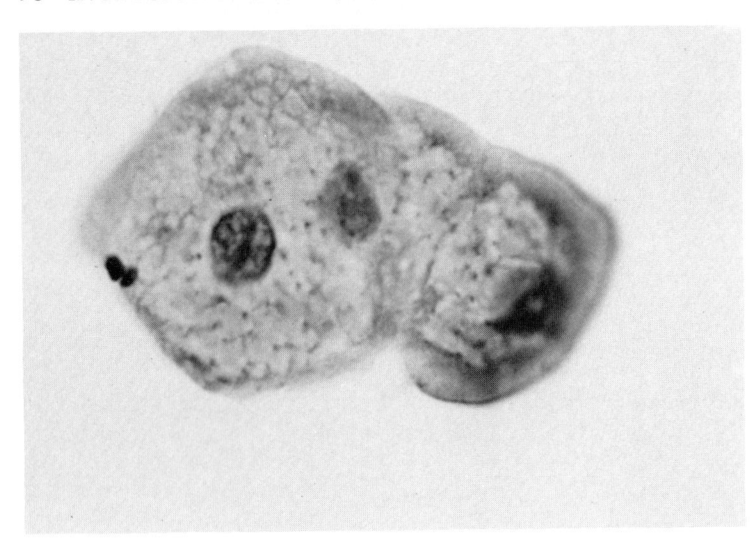

FIG. 8-10
GLYCOGENATED UROTHELIAL CELL IN URINE SEDIMENT. Note golden cytoplasmic granules. (× 1000)

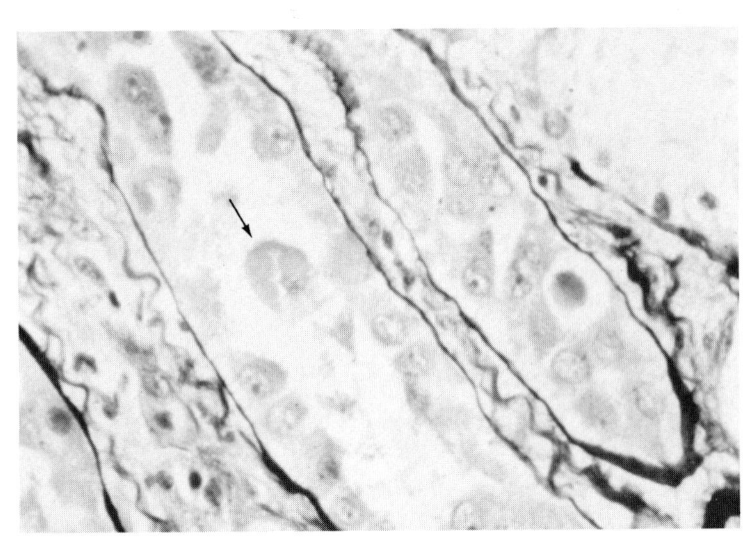

FIG. 8-11
DEGENERATIVE INTRACYTOPLASMIC INCLUSIONS. A tubule lined by regenerating epithelium contains an exfoliated epithelial cell with multiple, discrete, round eosinophilic (hyaline) bodies, probably representing enlarged lysosomes (arrow). In the adjacent tubule, a necrotic cell lacking a nucleus and having dense, contracted eosinophilic cytoplasm is surrounded by a halo and must be distinguished from an intracytoplasmic viral inclusion. (× 400)

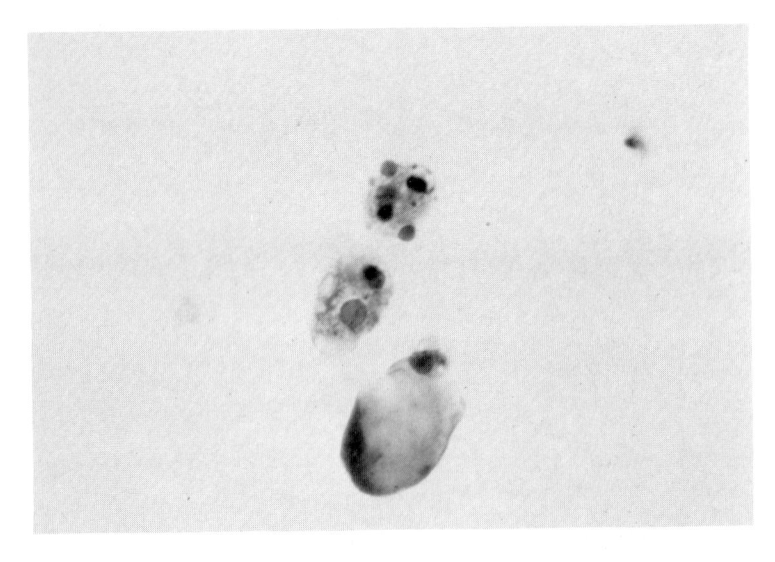

FIG. 8-12
EXFOLIATED CELL CONTAINING NONSPECIFIC INTRACYTOPLASMIC EOSINOPHILIC BODY. The inclusion body should not be confused with erythrophagocytosis. (× 400)

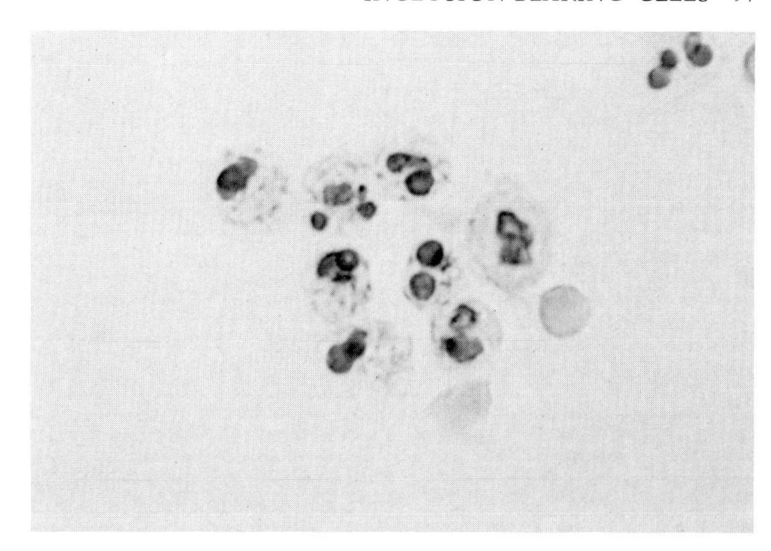

FIG. 8-13
NEUTROPHILS WITH TOXIC GRANULATION IN URINE SEDIMENT.
Distinct dark cytoplasmic granules are similar to toxic granulation seen in neutrophils on peripheral blood smears. Toxic granulation should not be confused with phagocytized bacteria. (× 1000)

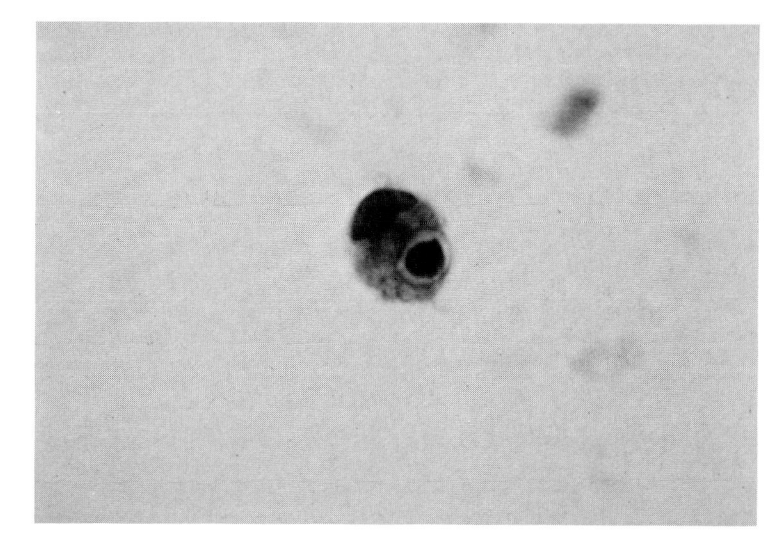

FIG. 8-14
EXFOLIATED CELL CONTAINING MICHAELIS-GUTMANN BODY.
Macrophage with distinct dark intracytoplasmic inclusion. (× 1000)

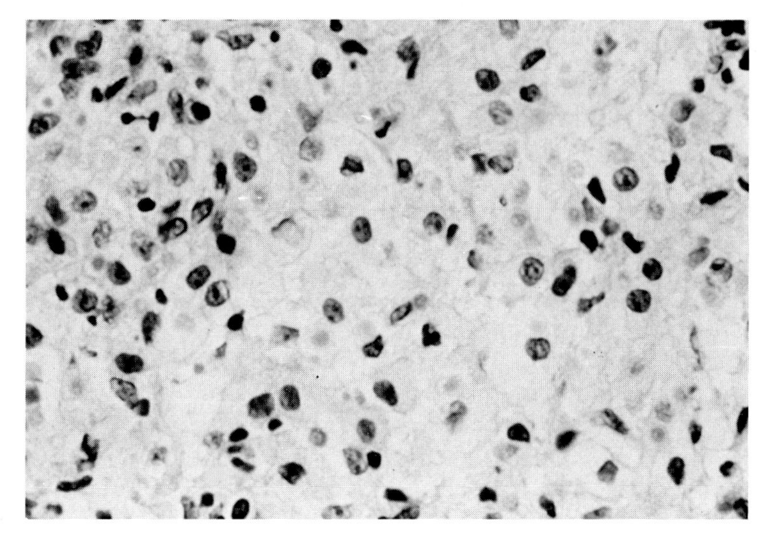

FIG. 8-15
MALAKOPLAKIA OF BLADDER.
Numerous eosinophilic inclusions (Michaelis-Gutmann bodies) within foamy, slightly granular cytoplasm of macrophages is diagnostic of malakoplakia. Slight basophilia noted in occasional inclusions is due to the presence of calcium. (× 400)

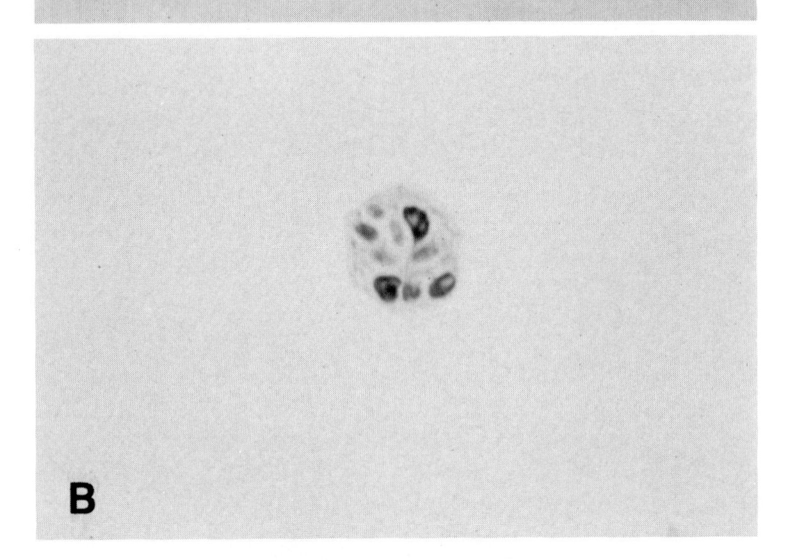

FIG. 8-16
INTRACELLULAR FUNGI IN URINE SEDIMENT. **(A)** Neutrophils with phagocytized yeast and pseudohyphal forms of *Candida* species. (× 400) **(B)** Permanently-stained urine sediment. preparations allow for accurate identification of intracytoplasmic yeast. (× 1000)

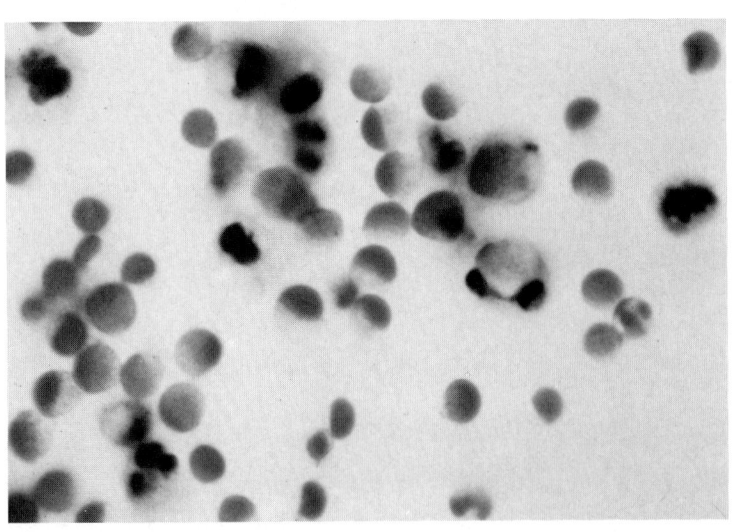

FIG. 8-17
URINARY ERYTHROPHAGOCYTOSIS. Note engulfment of erythrocytes by neutrophils. (× 1000)

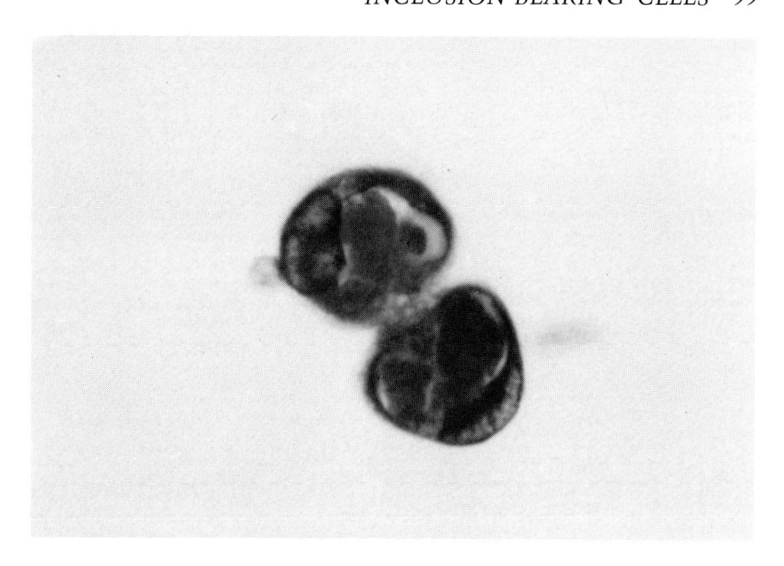

FIG. 8-18
EXFOLIATED EPITHELIAL CELLS WITH HOMOGENEOUS, WAXY INTRACYTOPLASMIC MATERIAL. These cells often accompany waxy cast. (× 1000)

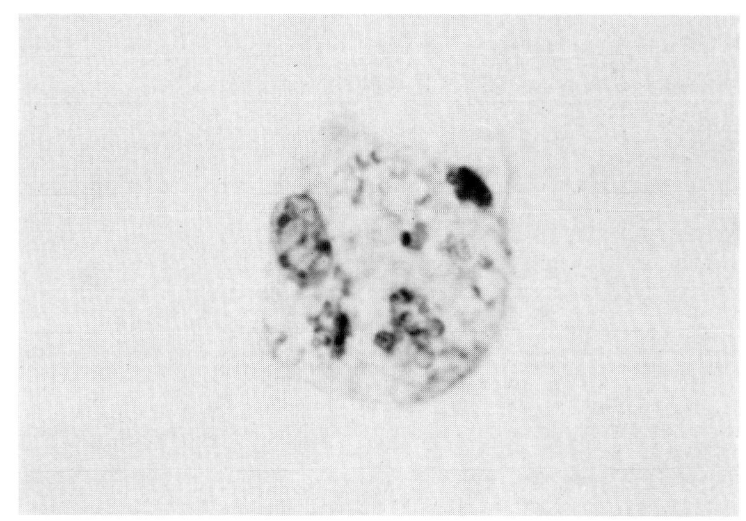

FIG. 8-19
EXFOLIATED CELL WITH CYTOPLASMIC PIGMENT. Yellow-brown appearance suggests lipochrome. (× 1000)

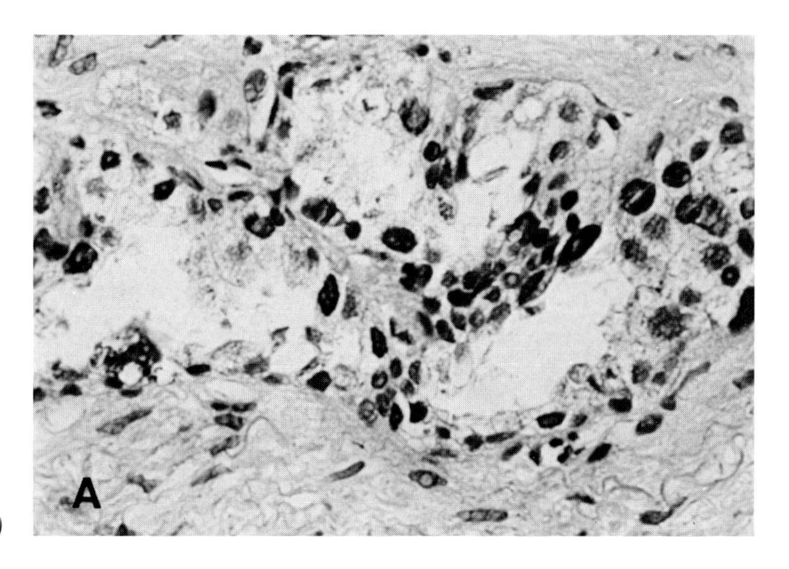

FIG. 8-20
INTRACYTOPLASMIC LIPOCHROME PIGMENT. (**A**) Seminal vesicle. A potential source of lipochrome-containing epithelium, particularly following prostatic massage, is the seminal vesicle. Nuclear hyperchromatism and pleomorphism as well as pigment characteristically increase with age. (× 400)

(continued)

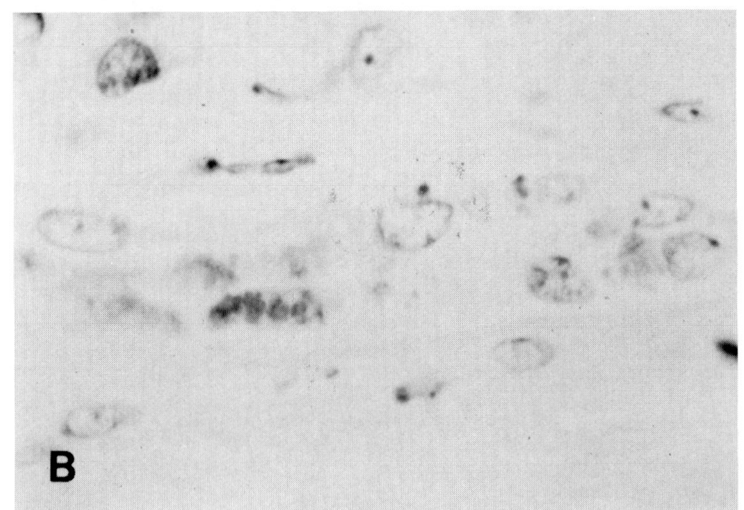

FIG. 8-20B *(continued)*
(B) Renal epithelium. Yellow-brown lipochrome pigment within the cytoplasm of epithelium lining the thin segment of convoluted tubules is commonly seen following anoxic or toxic tubular damage. (× 400)

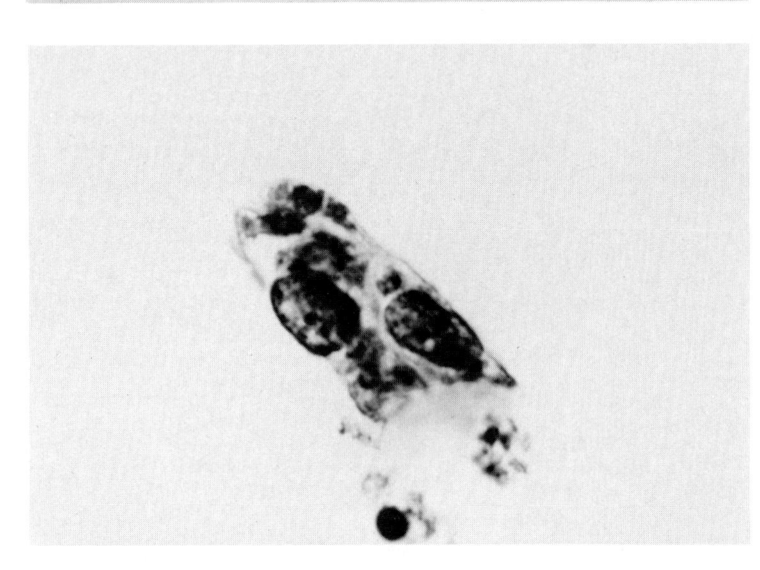

FIG. 8-21
EXFOLIATION OF PIGMENTED REACTIVE EPITHELIUM. (× 1000)

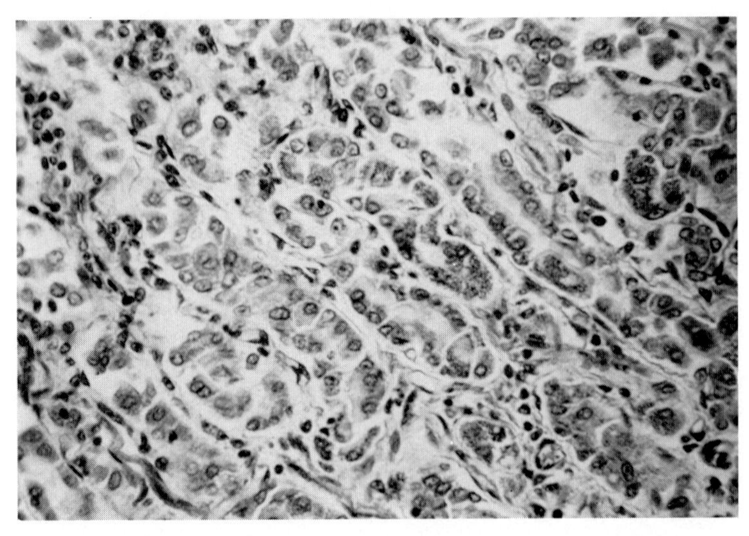

FIG. 8-22
RENAL HEMOSIDEROSIS. Note renal tubular epithelium with large amount of brown cytoplasmic pigment. (× 160)

PART III

RENAL PARENCHYMAL DISEASES

9
TUBULAR INJURY

PATHOGENETIC MECHANISMS

Tubular injury may be the result of a disturbance in metabolic function or direct toxicity induced by fever, anoxia, infection, drugs, and heavy metals. This disturbance leads to irreversible cellular damage with resulting exfoliation of epithelial lining cells. Renal epithelium lining proximal convoluted tubules is the most metabolically active part of the nephron and is, therefore, most susceptible to injury.

Structural integrity of the nephron is not necessarily altered during tubular injury. As long as the capability of cellular regeneration is present or the lesion or insult is not severe, renal dysfunction may not be clinically apparent. Severe lesions characterized by extensive tubular epithelial necrosis, often with disruption of tubular basement membranes, will be clinically overt and will give urine sediment findings of acute tubular necrosis.

CYTOHISTOLOGIC CRITERIA FOR DIAGNOSIS

The urinary cytodiagnosis of tubular injury requires the accurate identification of various types of renal epithelial cells. Exfoliated renal epithelial cells are persistently elevated in semiquantitative counts. A spectrum of cellular degenerative changes can be seen, including intranuclear inclusion-bearing cells. Although pathologic casts and necrotic renal epithelial cells may be observed with severe forms of tubular injury, their presence should not be relied upon for diagnosis of milder forms of tubular injury.

The urine sediment examination in patients with heavy metal exposure, typically without associated renal dysfunction, shows exfoliated convoluted epithelial cells. These cells have eccentric and prominent granular cytoplasm characteristic of convoluted tubular epithelium. Cellular changes induced by both lead and cadmium exposure include darkly staining discrete intranuclear inclusions and degenerative intracytoplasmic droplets. Large granular cells, exfoliated following lead exposure (Fig. 9-1), have the cytologic appearance of proximal convoluted epithelium. Cytoplasmic ballooning is a characteristic feature. Although the acid-fast stain is useful in histologic sections for the identification of lead-induced intranuclear inclusions, it is an unrealiable method in urine sediment. More sophisticated techniques are required for detecting the presence and quantitating the amount of lead present in inclusion-bearing cells.

In contrast to lead exposure, exposure to cadmium results in a significantly higher semiquantitative count of exfoliated renal epithelial cells.

In addition, the presence of smaller intranuclear inclusion-bearing granular cells suggests exfoliation from distal convoluted tubules (Fig. 9-2). These exfoliated cells are commonly accompanied by a background containing mucus, crystals, and hyaline and granular casts. Occasionally, an inflammatory component with neutrophils and leukocytic casts is present and suggests a more severe lesion, that is, tubulointerstitial nephritis. For diagnosis and documentation of early, potentially reversible lesions, renal biopsy would be useful. Chronic cadmium exposure has been reported to lead to interstitial fibrosis and nephron atrophy.

The urine sediment from patients during gentamicin therapy often has increased renal epithelial cells, many of which contain intranuclear inclusions (Fig. 9-3). These inclusion-bearing cells are similar in appearance to those seen following lead exposure. Using semiquantitative methods, the number of these cells is usually higher, and there is often renal dysfunction. Although many drugs, including antibiotics, are thought to be toxic to tubular epithelium because of clinical correlations, a direct relationship is difficult to prove in patients who are often febrile, septic, or in shock.

Certain drugs may produce more severe forms of tubular injury (Fig. 9-4). The urine sediment findings include marked exfoliation of degenerated and necrotic renal epithelial cells of convoluted tubular and collecting duct origin as well as pathologic casts. This sediment pattern is more consistent with acute tubular necrosis. Renal biopsy is more often performed because of associated renal dysfunction and provides confirmatory evidence of severe tubular injury (Fig. 9-5).

BIBLIOGRAPHY

Dunhill MS: A review of the pathology and pathogenesis of acute tubular necrosis. J Clin Pathol 28:2–13, 1974

Goyer RA, Rhyne BC: The pathological effects of lead. Int Rev Exp Pathol 12:1–71, 1973

Landing BH, Nakai, H: Histochemical properties of renal lead inclusions and their demonstrations in urinary sediment. Am J Clin Pathol 31:499–503, 1959

Schumann GB: The Urine Sediment Examination. Baltimore, Williams & Wilkins, 1980.

FIG. 9-1
LEAD-INDUCED INCLUSION-BEARING CELLS IN URINE SEDIMENT. **(A)** Exfoliated cell with amphophilic intranuclear inclusion. Note small nuclear size and ballooned granular cytoplasm containing multiple degenerative droplets. (× 1000) **(B)** Binucleated inclusion cell. Note granular cytoplasm characteristic of convoluted tubular cells. (× 1000) **(C)** Inclusion-bearing cell with suspicious inclusion. (× 1000) **(D)** Disintegrating exfoliated inclusion-bearing cell with characteristic granular cytoplasm containing a prominent degenerative droplet. Frequently, *anucleated* granular cells are present in the urine of patients exposed to lead. (× 1000) **(E)** Lead nephropathy. Note prominent eosinophilic inclusions in proximal convoluted tubular epithelium. (× 250) **(F)** Note characteristic acid-fast positive inclusions. (Acid-fast × 250)

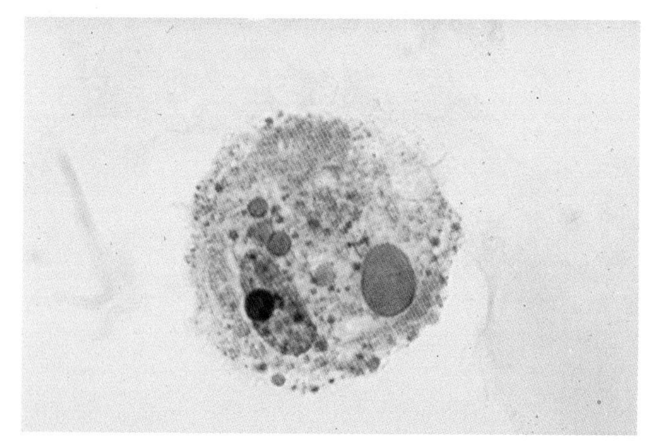

FIG. 9-1A

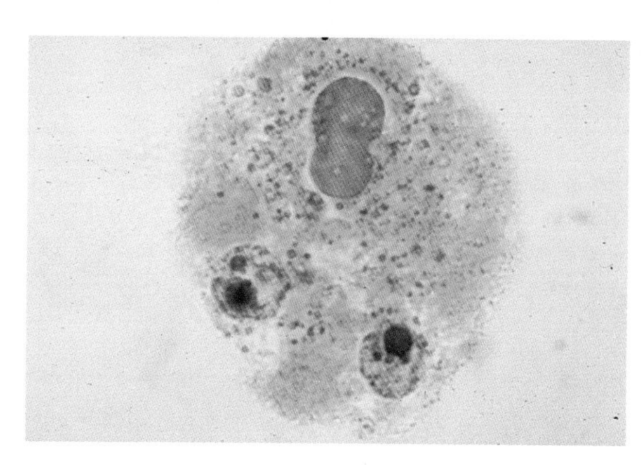

FIG. 9-1B

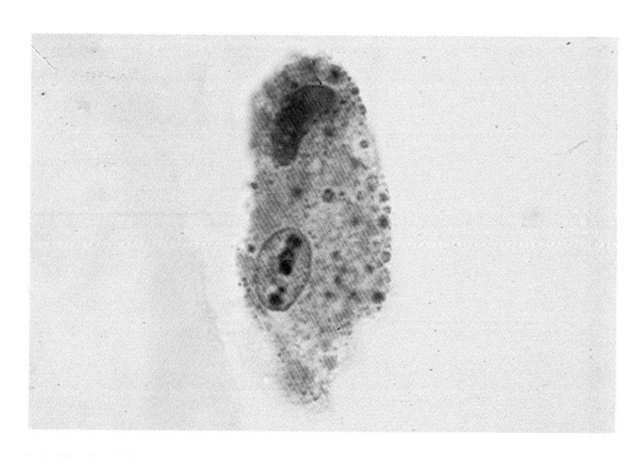

FIG. 9-1C

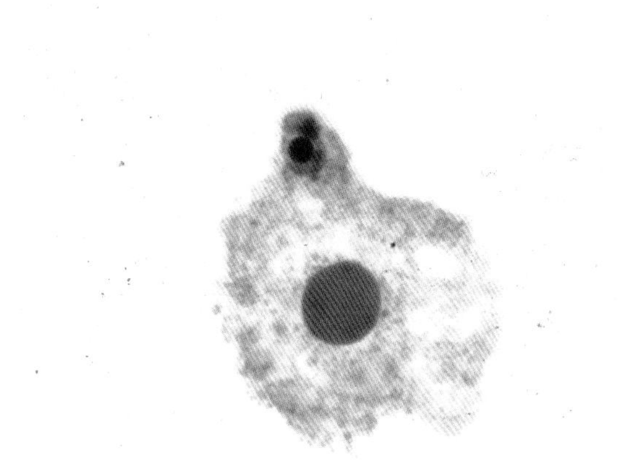

FIG. 9-1D

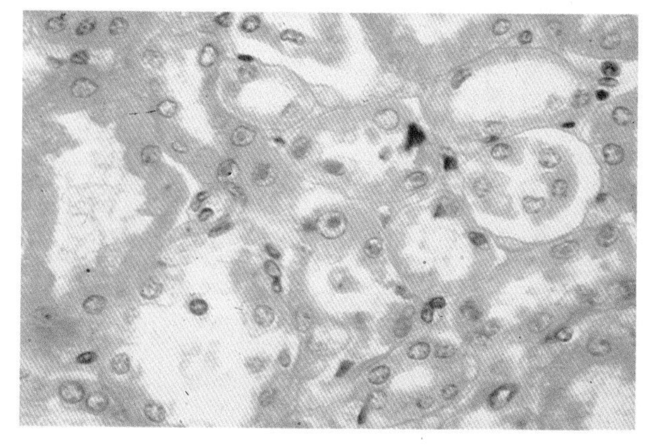

FIG. 9-1E

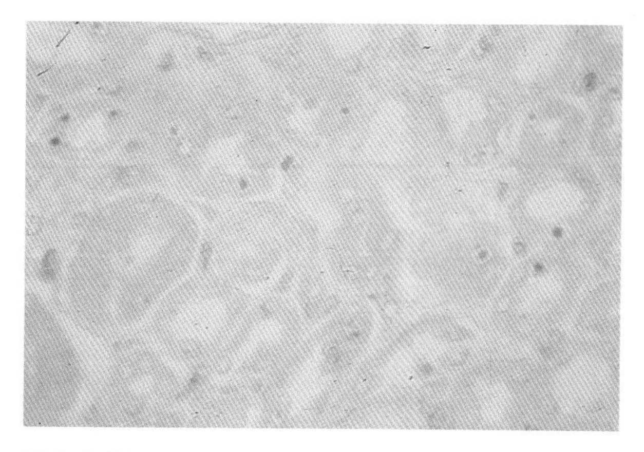

FIG. 9-1F

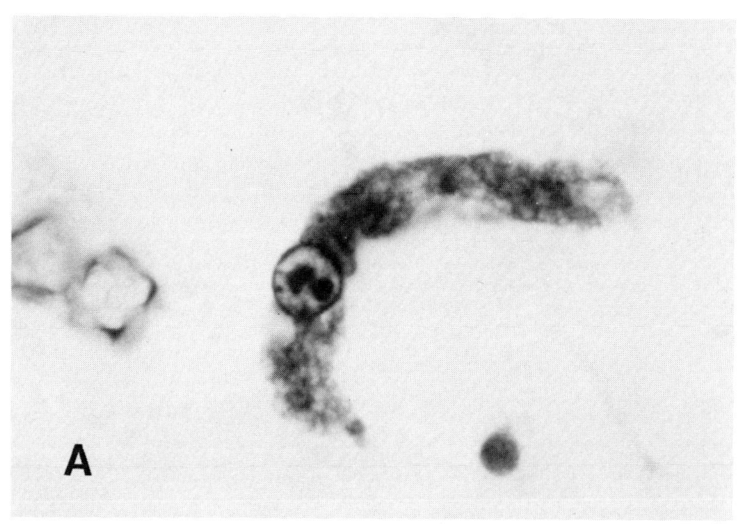

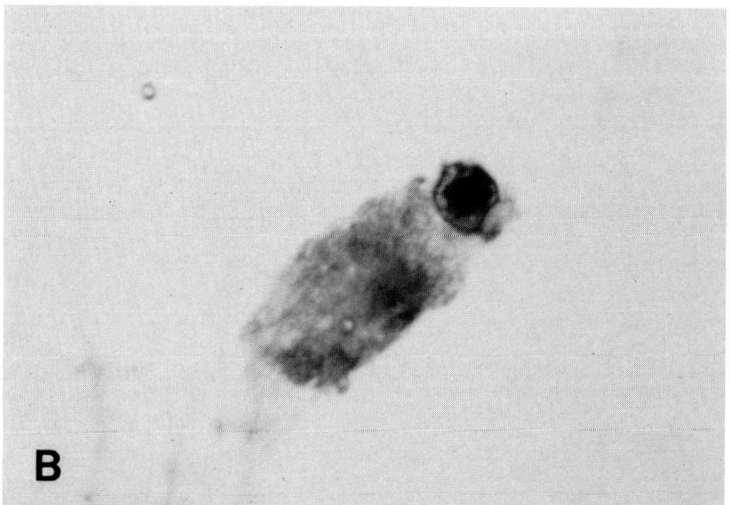

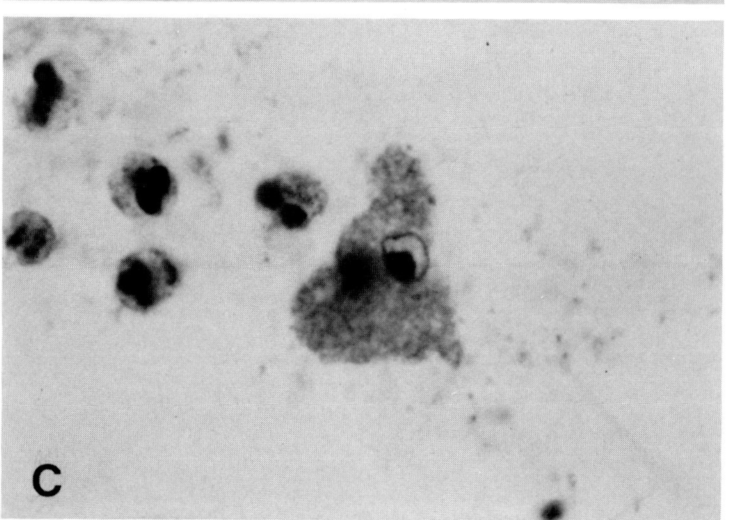

FIG. 9-2
CADMIUM-INDUCED INCLUSION-BEARING CELLS IN URINE SEDIMENT. **(A)** Elongated cell with basophilic intranuclear inclusion and prominent granular cytoplasm suggestive of convoluted epithelium. (× 1000) **(B)** Elongated cell with prominent inclusion. The granular cytoplasm should not be confused with a granular cast. (× 1000) **(C)** Triangular-shaped inclusion-bearing cell. Note neutrophils in background. Acute inflammation and leukocytic casts are occasionally found in cadmium-exposed individuals, indicating tubular-interstitial inflammation. (× 1000)

(continued)

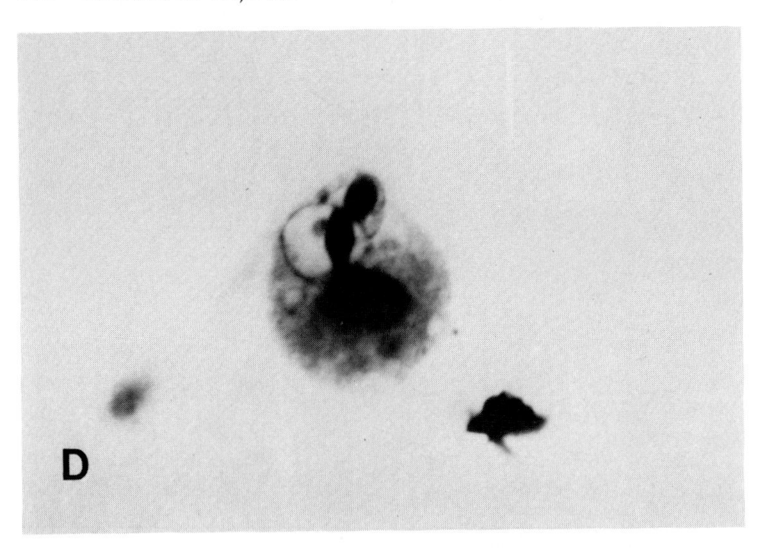

FIG. 9-2D *(continued)*
(D) Convoluted renal cell with prominent basophilic intranuclear inclusion. (× 1000)

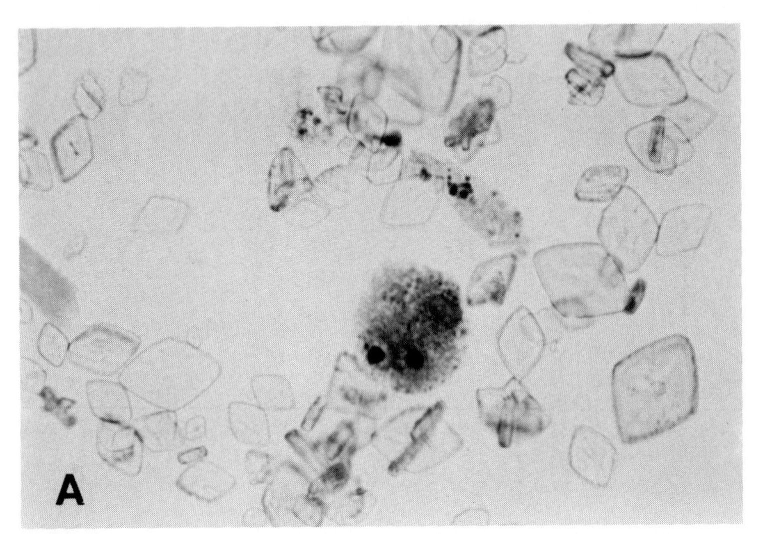

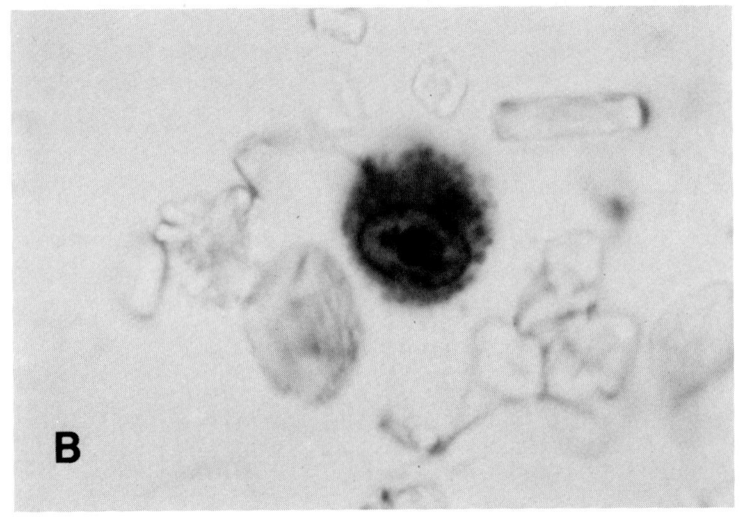

FIG. 9-3
DRUG-INDUCED INCLUSION-BEARING CELLS IN URINE SEDIMENT DURING GENTAMICIN THERAPY. **(A)** Binucleated intranuclear inclusion-bearing cell and background containing numerous uric acid crystals. Granular cytoplasm suggests convoluted tubular origin. (× 400) **(B)** Basophilic intranuclear inclusion-bearing cell. (× 1000)

(continued)

FIG. 9-3C, D *(continued)*
(C) Trinucleated inclusion-bearing cell with large intracytoplasmic degenerative eosinophilic droplet. (× 1000) **(D)** Binucleated inclusion-bearing cell with ballooned cytoplasm. Note numerous small eosinophilic droplets. (× 1000)

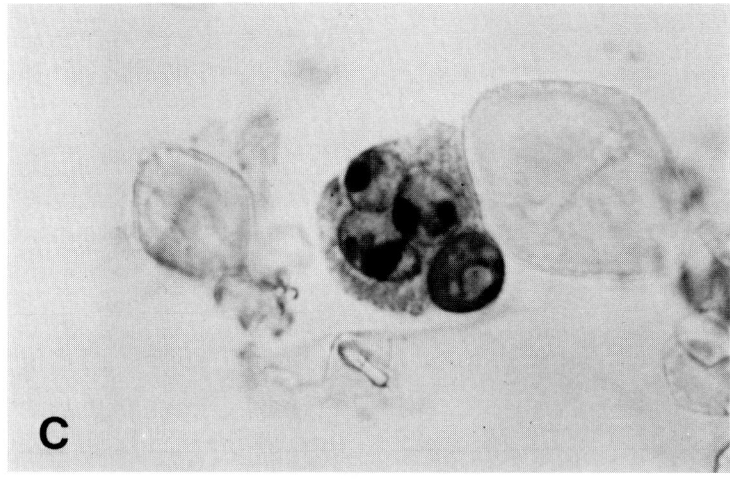

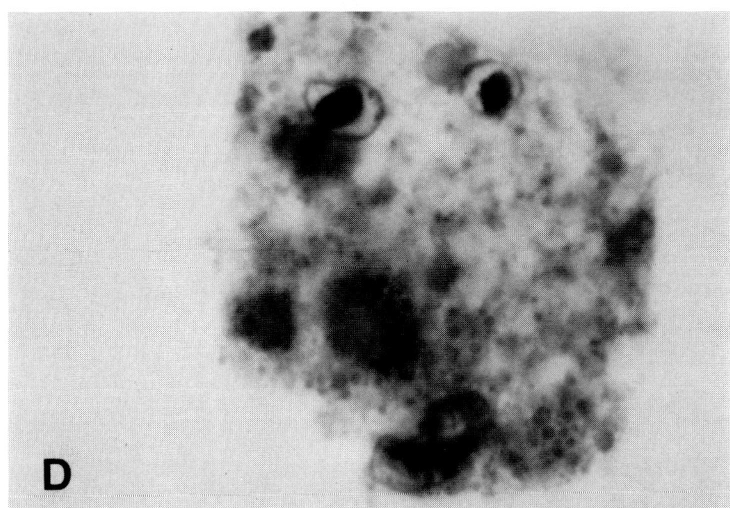

FIG. 9-4
TUBULAR INJURY IN LEUKEMIC PATIENT RECEIVING CHEMOTHERAPY. **(A)** Marked renal tubular epithelial exfoliation. Note necrotic renal epithelial cell cast (right). (× 400)

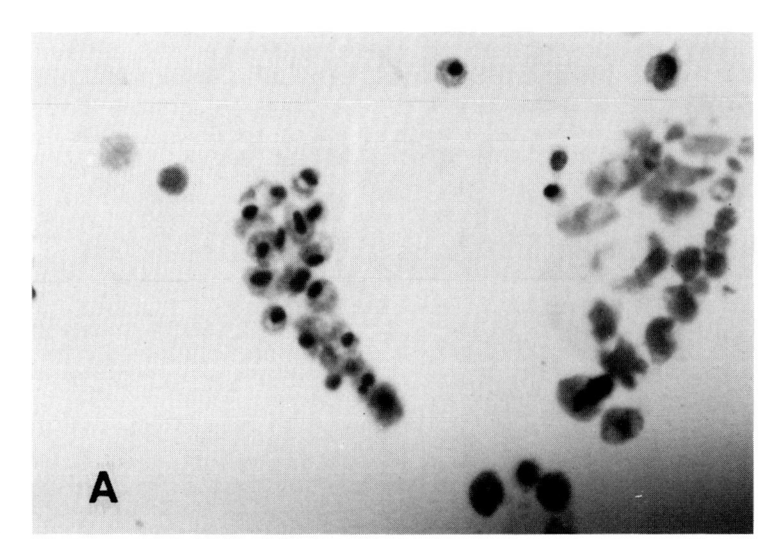

(continued)

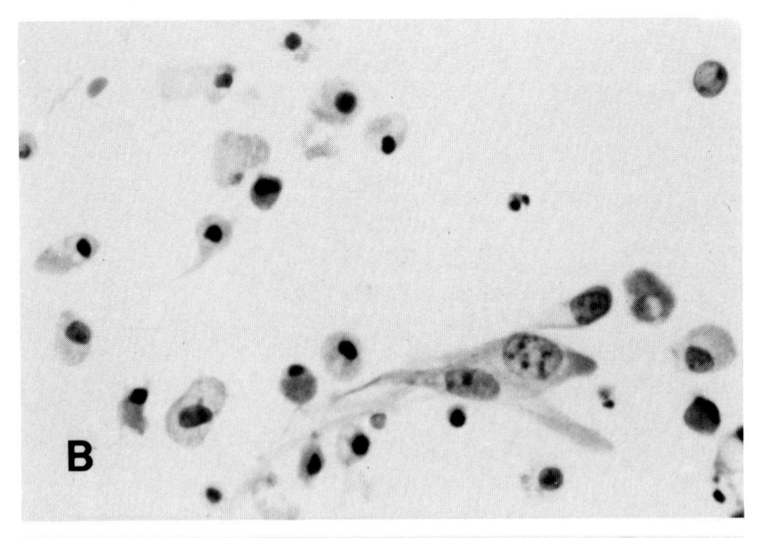

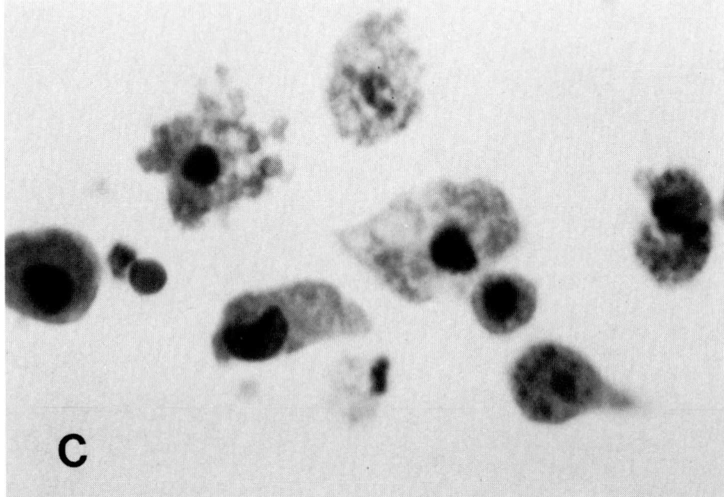

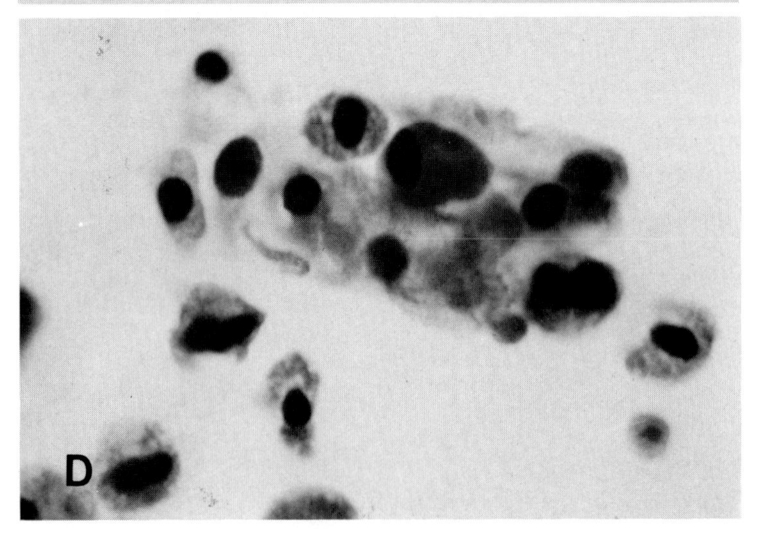

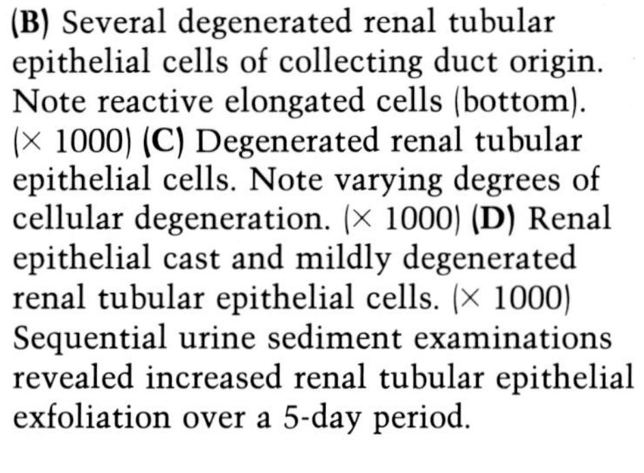

FIG. 9-4B-D *(continued)*
(B) Several degenerated renal tubular epithelial cells of collecting duct origin. Note reactive elongated cells (bottom). (× 1000) **(C)** Degenerated renal tubular epithelial cells. Note varying degrees of cellular degeneration. (× 1000) **(D)** Renal epithelial cast and mildly degenerated renal tubular epithelial cells. (× 1000) Sequential urine sediment examinations revealed increased renal tubular epithelial exfoliation over a 5-day period.

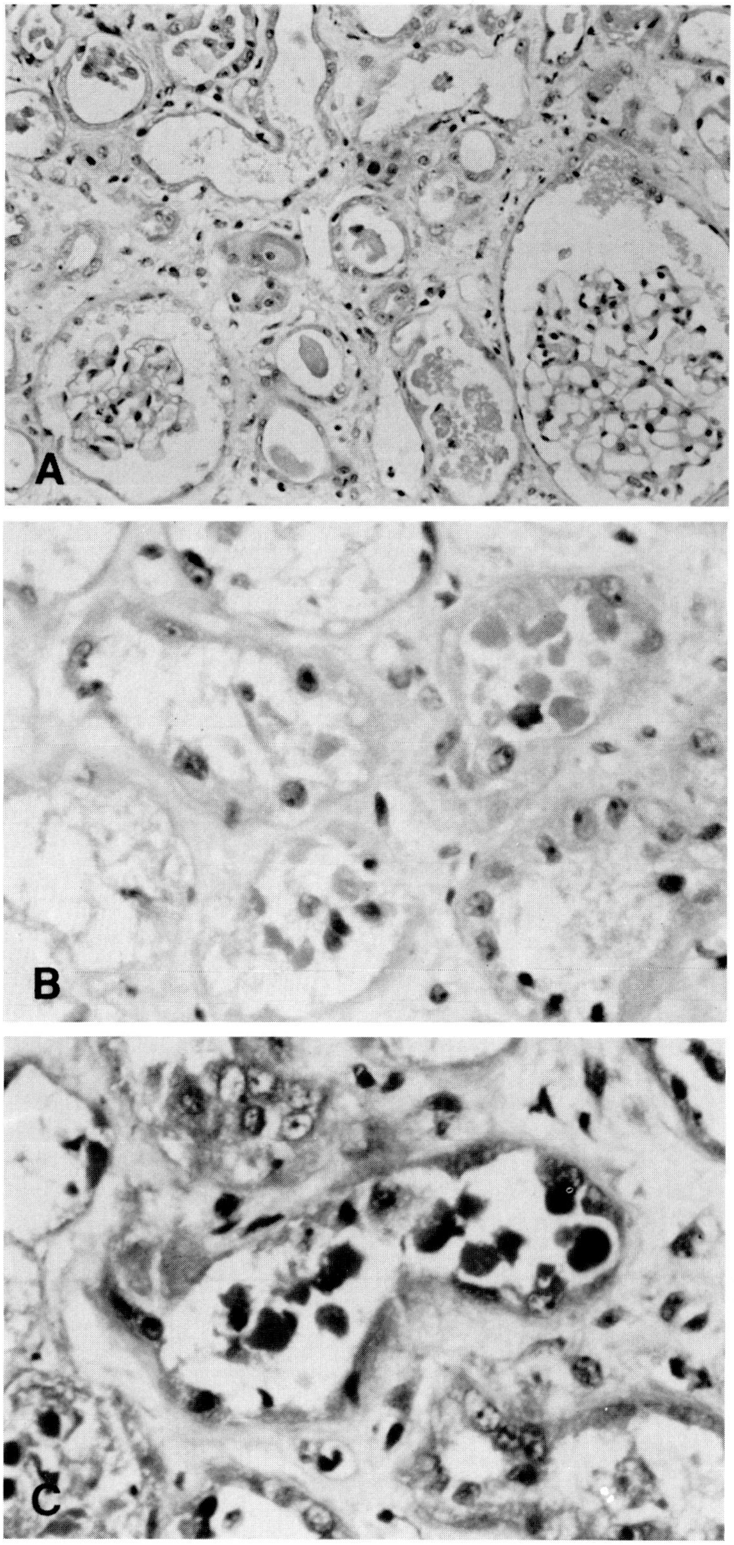

FIG. 9-5
TOXIC ACUTE TUBULAR NECROSIS. **(A)** Note tubular dilatation, epithelial flattening, and casts. (× 150) **(B)** Tubules contain necrotic renal epithelial cells. Note the lack of nuclear staining in the necrotic ("ghost") cells. (× 400) **(C)** Renal tubule contains numerous ghost cells. (Trichrome × 400)

(continued)

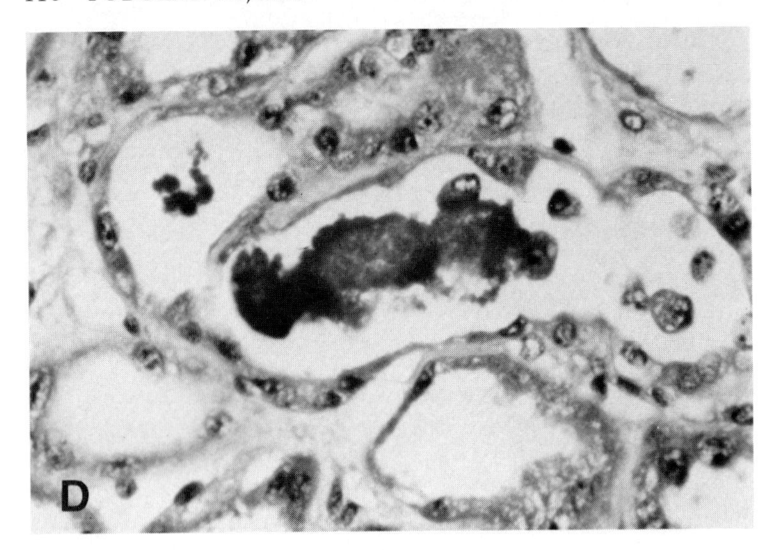

FIG. 9-5D *(continued)*
(D) Dilated tubule contains coarse granular cast and degenerated epithelial cells. (Trichrome × 400)

10

ISCHEMIC (TUBULAR) NECROSIS

In the preceding chapter, a spectrum of tubular injury was discussed. Emphasis was given to the semiquantitative assessment of renal epithelial exfoliation and to the recognition of cellular degenerative changes. With severe forms of tubular injury, recognized clinicopathologically as acute tubular necrosis (ATN), the urine sediment contains exfoliated necrotic renal tubular epithelial cells. The presence of necrotic renal cells in the sediment does not allow etiologic discrimination of toxic, that is, drug-related, tubular necrosis from an ischemic mechanism. Occasionally, however, ATN, secondary to severe renal parenchymal ischemia, may produce renal epithelial fragments in the urine sediment. Exfoliated renal epithelial fragments are a hallmark of ischemic necrosis.

CYTODIAGNOSTIC CRITERIA

Ischemic necrosis is a urinary cytodiagnostic term based upon the identification of exfoliated renal epithelial fragments. Morphological characteristics of renal epithelial fragments are given in Chapter 6. Other sediment evidence of ischemic necrosis includes a dirty background containing amorphous and cellular debris, exfoliation of necrotic ("ghost") renal epithelial cells, and pathologic cast formation.

CYTOHISTOLOGIC CORRELATIONS

There are a variety of renal pathologic conditions that may be reflected in the urine sediment as ischemic necrosis. These conditions include ATN, renal infarction, cortical necrosis, and papillary necrosis.

In ATN, the prominence of "ghost" cells in the urine sediment (Fig. 10-1) is produced by massive exfoliation of necrotic renal tubular epithelium (Fig. 10-2). Disruption of tubular basement membranes are commonly observed and represent the likely cause for exfoliation of epithelial fragments. Tubular epithelial regeneration, which is often striking, may be the only histologic evidence of previous epithelial exfoliation. Calcium oxalate crystals are frequently seen with ATN. The association of oxalate crystals with ATN is poorly understood.

With renal infarction and cortical necrosis, exfoliated renal epithelial fragments of collecting duct origin are numerous and persist in sequential urine sediment examinations. Case examples are given in Chapter 13.

Papillary necrosis may complicate urinary tract infections with obstruction, may occur in diabetic patients as a cause of acute renal failure, or may be related to analgesic abuse. The inflammatory component will vary depending upon the underlying etiology (Figs. 10-3 and 10-4). A case example of papillary infarction in a renal allograft recipient is given in Chapter 13.

BIBLIOGRAPHY

Heptinstall RN: Pathology of the Kidney, 2nd ed. Boston, Little, Brown, 1974

Schumann GB: The Urine Sediment Examination. Baltimore, Williams & Wilkins, 1980

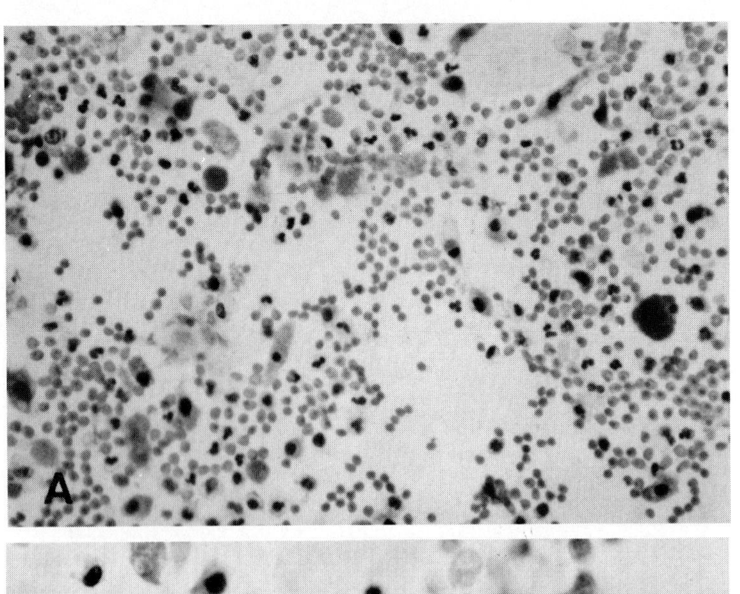

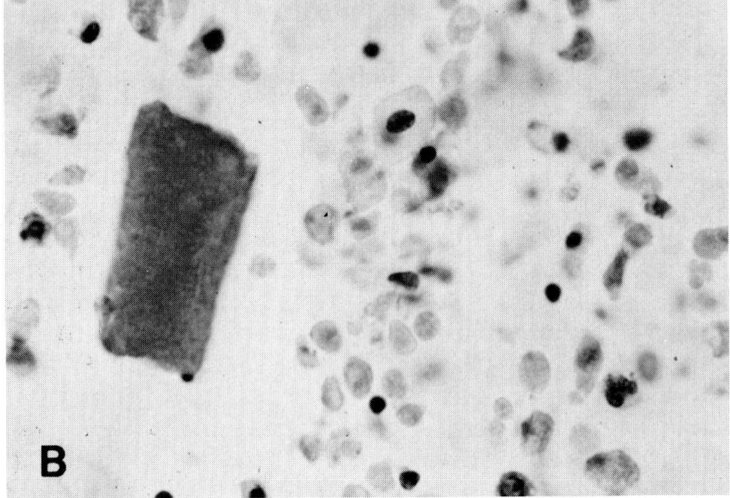

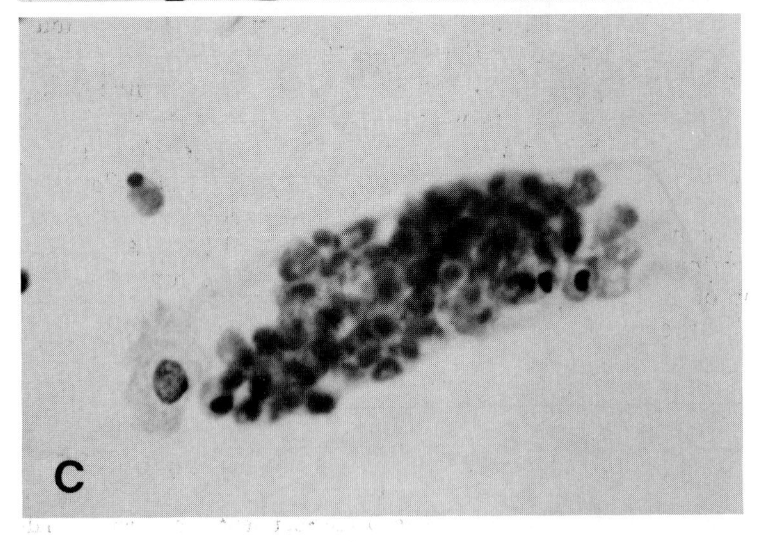

FIG. 10-1
"GHOST" CELLS IN URINE SEDIMENT. **(A)** Bloody background and numerous necrotic cells. (× 250) **(B)** Marked exfoliation of necrotic ("ghost") cells and a waxy cast. (× 400) **(C)** Necrotic renal epithelial cast (× 400)

(continued)

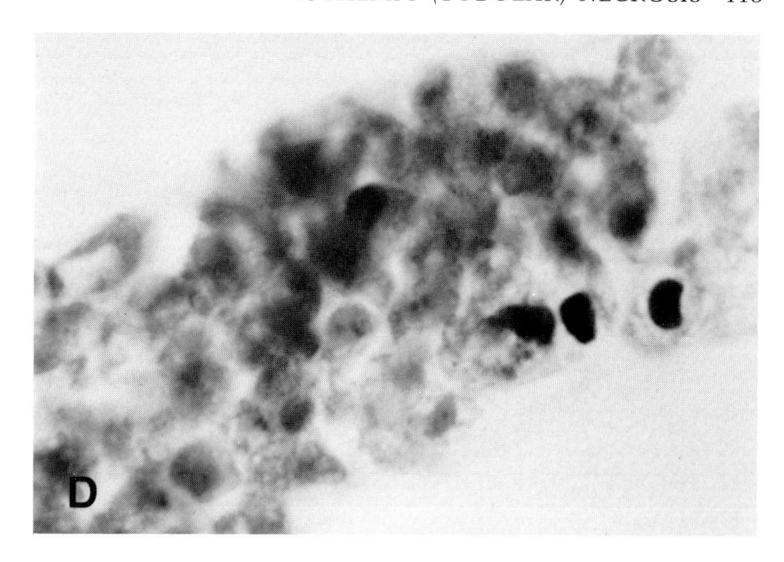

FIG. 10-1D *(continued)*
(D) The presence of numerous "ghost" cells is a characteristic finding in acute tubular necrosis. (× 1000)

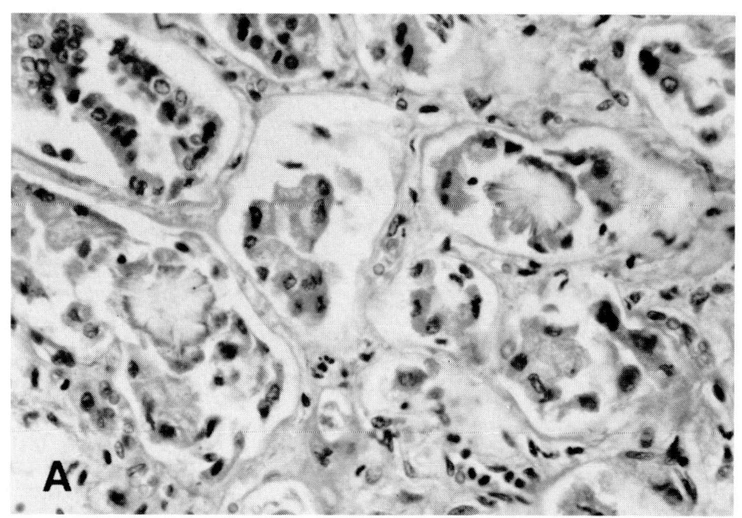

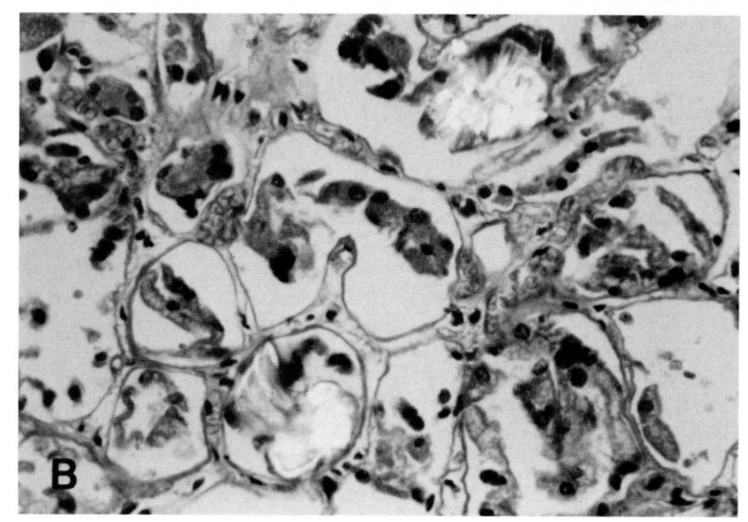

FIG. 10-2
ACUTE TUBULAR NECROSIS. **(A)** Extensive epithelial exfoliation with nuclear pyknosis. Note intratubular calcium oxalate crystals. (× 250) **(B)** Note the lack of regenerating epithelium. (Polarized × 250)

(continued)

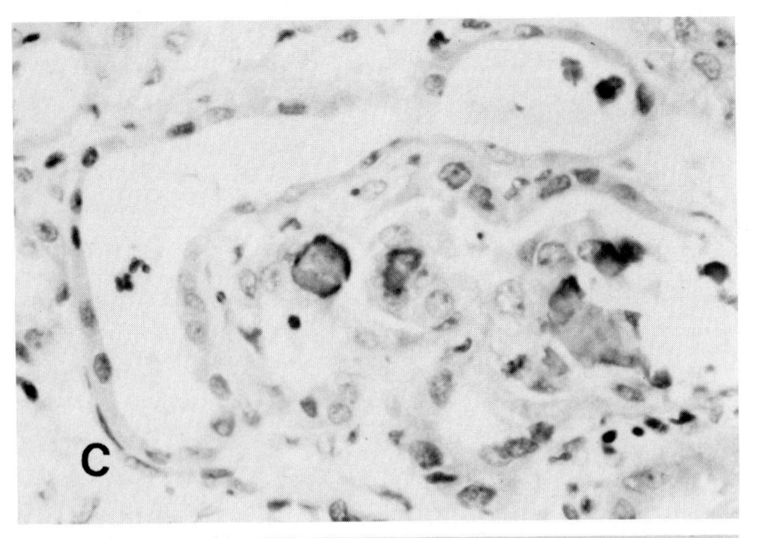

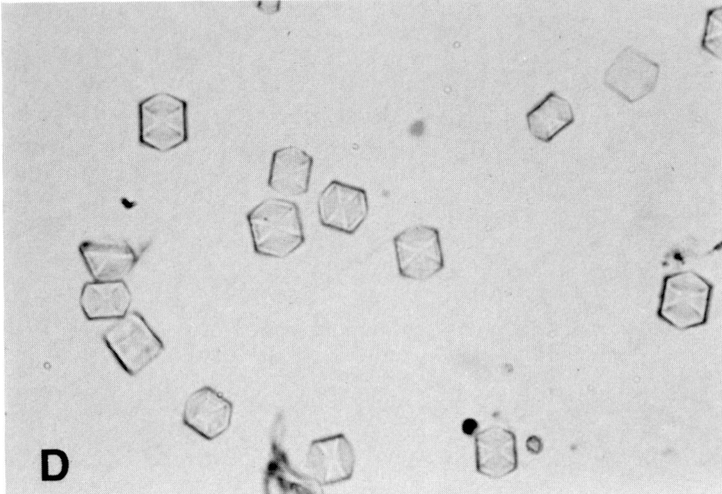

FIG. 10-2C, D *(continued)*
(C) Regenerating tubular epithelium surrounds calcified intratubular crystals. An altered layer of regenerating epithelium is also present. (× 400) **(D)** Calcium oxalate crystals in urine sediment. Note characteristic refractile cross. (× 1000)

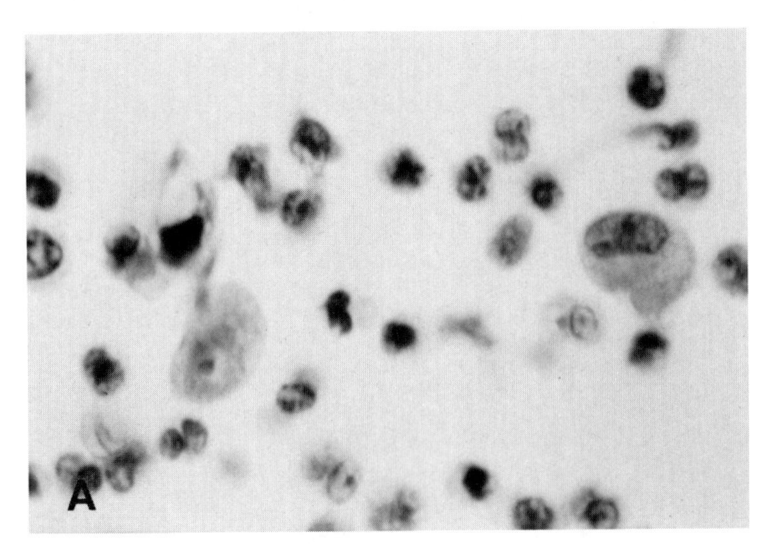

FIG. 10-3
URINE SEDIMENT IN PATIENT WITH PAPILLARY NECROSIS. **(A)** Reactive epithelial cells and neutrophilic background. (× 400)

(continued)

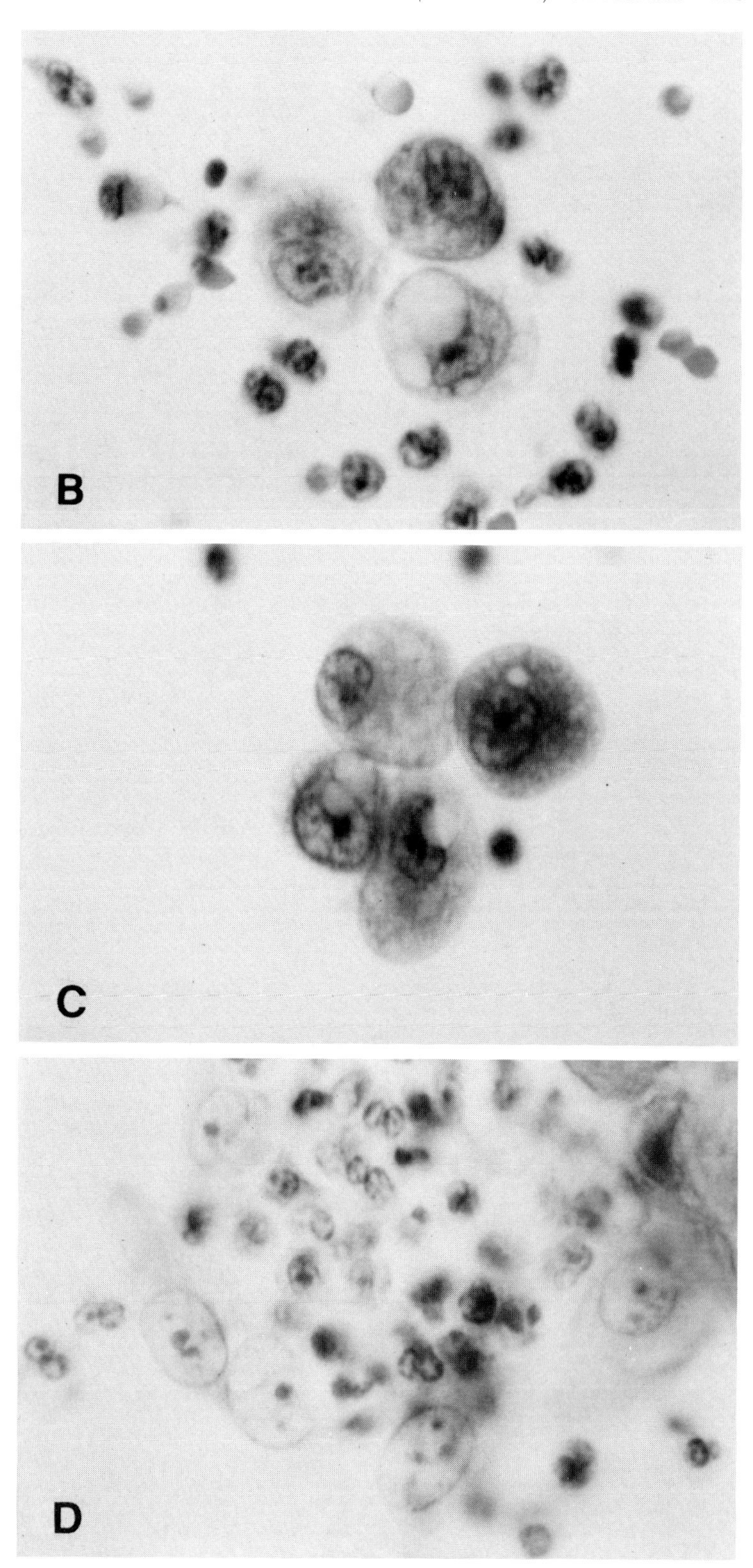

FIG. 10-3B-D *(continued)*
(B) Note epithelial cells with prominent nucleoli and vacuolated cytoplasm (× 1000) **(C)** (× 1000) **(D)** Reactive epithelial fragments and neutrophilic exudate. (× 1000)

(continued)

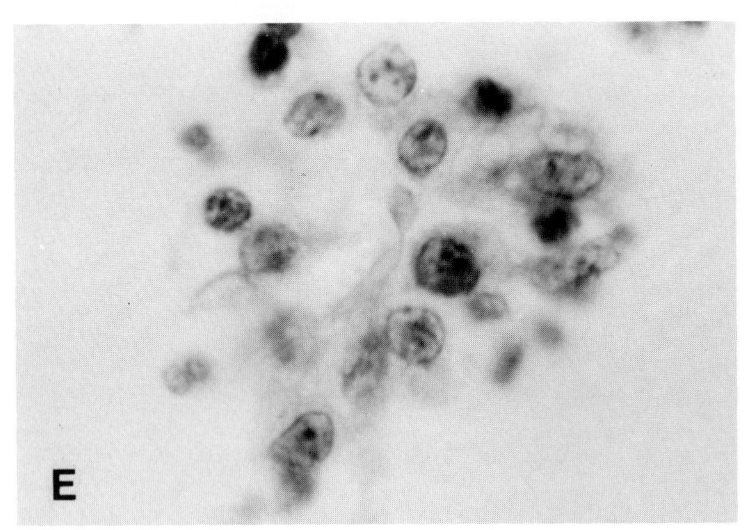

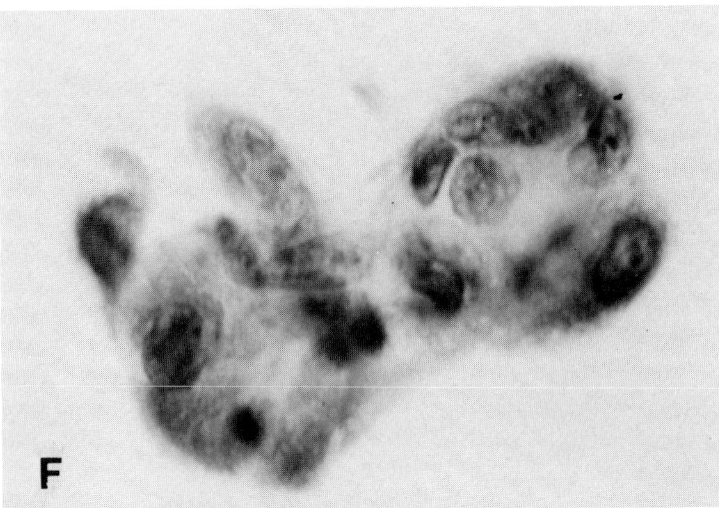

FIG. 10-3E, F *(continued)*
(E) Exfoliated reactive epithelial fragment. Note configuration suggesting portion of acinus. (× 1000) **(F)** Cylindrical epithelial fragment of probable collecting duct origin. (× 1000)

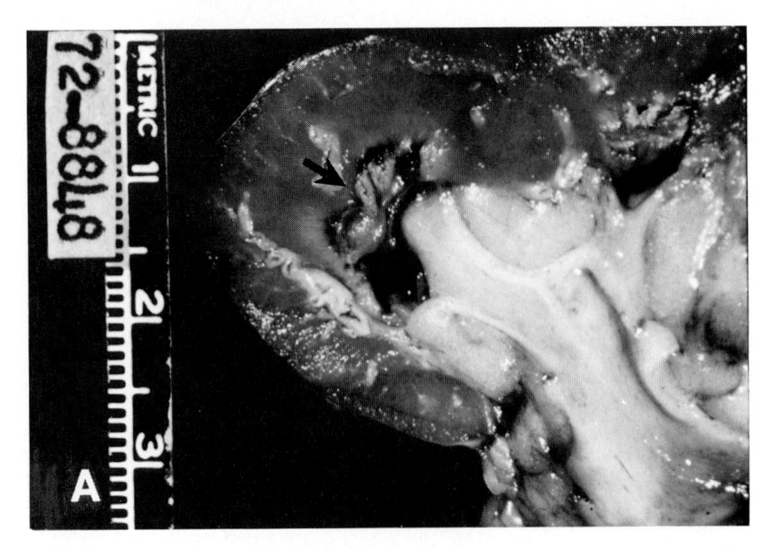

FIG. 10-4
PAPILLARY NECROSIS. **(A)** Two retracted necrotic papillae are present. Note hyperemic border (arrow). Gross.

(continued)

FIG. 10-4B-D *(continued)*
(B) The partially detached necrotic papilla protrudes into the urothelial lined calyx. Note the lack of inflammatory margin, although chronic pyelitis is present. (× 40) **(C)** Overlying cortex has chronic inflammatory infiltrate with tubular atrophy. (× 100) Patient had long history of phenacetin abuse. **(D)** Papillary necrosis with distinct inflammatory margin and linear infiltrates extending into cortex. (Tissue section whole mount.) Papillary necrosis with acute pyelonephritis.

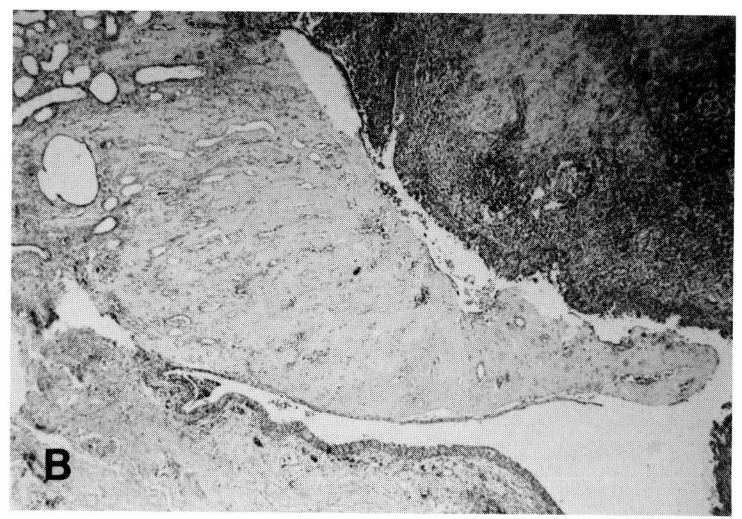

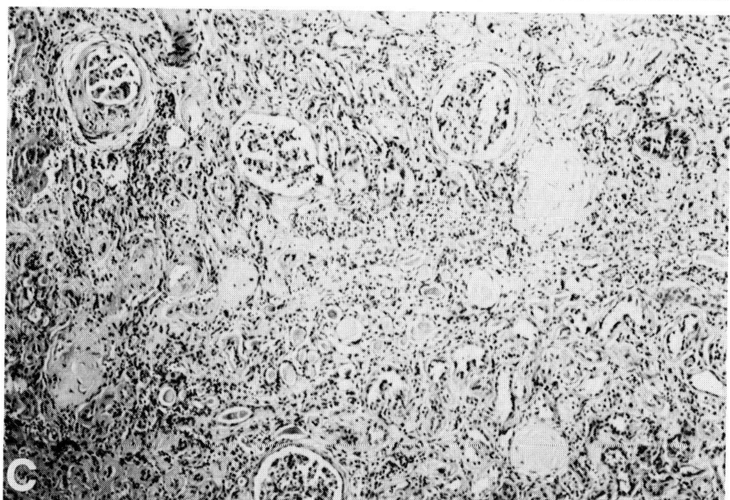

11
TUBULOINTERSTITIAL DISEASES

INTERSTITIAL NEPHRITIS

Interstitial nephritis is an inflammatory renal parenchymal process that results in tubular damage. Classification depends upon the characterization of the inflammatory infiltrate as well as the degree of interstitial scarring and nephron loss.

INFECTIOUS (PYELONEPHRITIS) TYPE

A common type of interstitial nephritis is related to ascending urinary tract infections with renal parenchymal involvement and is referred to as pyelonephritis. In acute pyelonephritis (Fig. 11-1), the renal interstitium and tubules are infiltrated by neutrophilic leukocytes, and there is variable destruction of tubules. Acute pyelonephritis is most often secondary to urinary tract infections; however, organisms are seldom identified in histologic material. In these patients, the urine sediment is characterized by bacterial organisms, neutrophils (often in clumps), epithelial cells, and occasionally casts. The identification of exfoliated renal tubular epithelial cells is helpful in localizing the disease. The identification of leukocytic casts is diagnostic of tubulointerstitial disease. In immunosuppressed and diabetic patients, fungi, particularly *Candida* species, are a common cause of renal infection (Fig. 11-2).

Chronic pyelonephritis, recognized most often at autopsy by calyceal deformity with an overlying cortical scar, may be a cause of chronic renal failure (Fig. 11-3). Zones of parenchymal scarring contain atrophic tubules with waxy casts. Interstitial inflammation, although variable, is often mild and consists of lymphocytes and plasma cells. The urine findings may be dominated by the presence of waxy casts, including broad forms, indicating chronic parenchymal disease.

Persistent bacterial urinary tract infections, particularly with *Escherichia coli,* occasionally lead to malakoplakia. Isolated renal involvement, although rare, can cause inflammatory destruction of renal parenchyma by a granulomatous infiltrate rich in foamy histiocytes (Fig. 11-4). The presence of Michaelis-Gutmann bodies distinguishes this lesion from xanthogranulomatous pyelonephritis. Michaelis-Gutmann bodies can be identified in the urine sediment, but localization is difficult.

ALLERGIC TYPE

Interstitial inflammatory lesions may be caused by a variety of drugs, for example, antibiotics and diuretics, and are termed **allergic interstitial nephritis.** In the acute phase, the histologic appearance is dominated by an inflammatory infiltrate rich in eosinophils (Fig. 11-5). The presence of eosinophiluria accompanied by exfoliated renal epithelium and pathologic casts is strong evidence of allergic interstitial nephritis. The predominance of plasma cells in the interstitial infiltrate is characteristic of chronic allergic interstitial nephritis (Fig. 11-6).

OTHER TYPES

Interstitial inflammation with tubular damage may be a significant component associated with a glomerulonephritis. The tubulointerstitial disease may dominate the urine sediment findings and obscure evidence of the glomerular lesion. In addition, specific pathogenetic mechanisms, for example, immune complex injury and antitubular basement membrane antibodies, may be identified by immunofluorescence, with or without an associated glomerulonephritis. A discussion of urine sediment findings in glomerular diseases is given in Chapter 12.

INFILTRATIVE PROCESSES

Renal parenchymal damage may be related to the accumulation or deposition of noncellular material or to the infiltration by exogenous cells. The infiltrative process may involve tubules and interstitium to varying degrees.

NONCELLULAR TYPE

Noncellular infiltrative processes include intratubular accumulation of crystals and casts with secondary damage or destruction of tubules. Nephrocalcinosis, characterized by intratubular accumulation of calcium phosphate crystals with variable calcification of tubular basement membranes, may complicate primary hyperparathyroidism. Urine sediment evidence of active parenchymal disease, for example, inflammation and hematuria, is more likely to be caused by nephrolithiasis than nephrocalcinosis.

Gouty nephropathy (Fig. 11-7) may present in the urine sediment with uric acid crystalluria and other evidence of renal parenchymal disease. The progressive accumulation of urates within tubules, predominantly within the medulla, results in eventual tubular destruction and marked inflammation.

Cast nephropathy is best typified by myeloma kidney. In this condition, tubular dysfunction is caused by the accumulation of dense casts with an associated tubular epithelial reaction. An example of myeloma casts is given in Chapter 14. Similar casts are found in light-chain disease.

CELLULAR TYPE

Parenchymal infiltration by exogenous cells generally represents secondary renal involvement by malignancy. The common types are lymphoreticular and include lymphoma, leukemia, and myeloma. Examples are given in Chapter 14. Malignant cellular infiltrates may be a cause of tubular epithelial exfoliation, but more commonly chemotherapeutic agents are responsible for the tubular injury.

BIBLIOGRAPHY

Heptinstall RH: Pathology of the Kidney, 2nd ed. Boston, Little, Brown, 1974

Schumann GB: The Urine Sediment Examination. Baltimore, Williams & Wilkins, 1980.

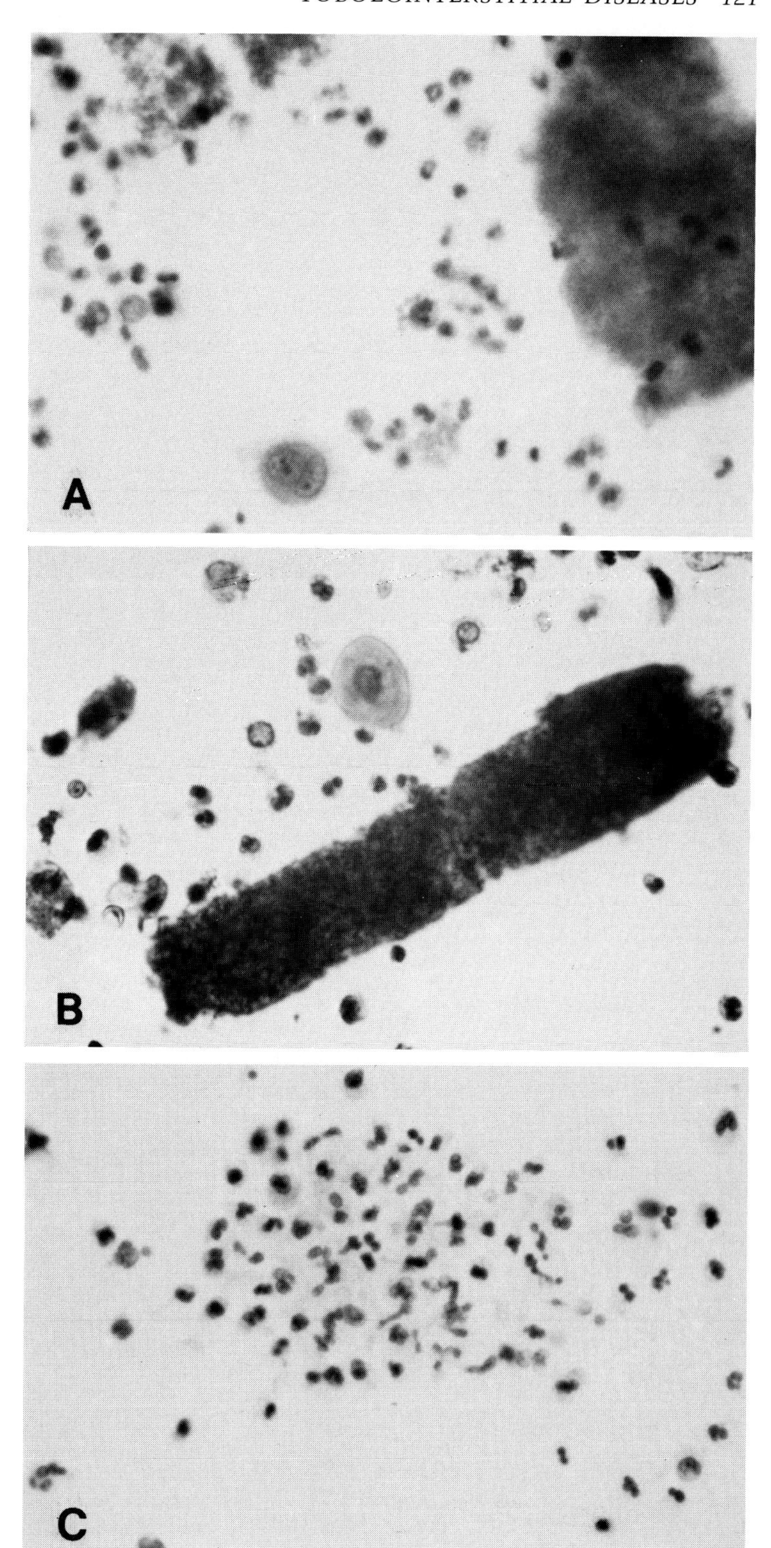

FIG. 11-1
ACUTE PYELONEPHRITIS. **(A)** Urine sediment with acute inflammatory background, bacterial colonization, and reactive epithelial cell. Note bacterial pseudocast. The lack of increased renal exfoliation and pathologic casts suggests lower urinary tract infectrion. (× 400) **(B)** Urine sediment with acute inflammatory background, transitional cell, coarse granular cast. (× 400) **(C)** Loose cluster of neutrophilic leukocytes gathered in amorphous debris. This loose cluster would not be considered a cast. (× 400)

(continued)

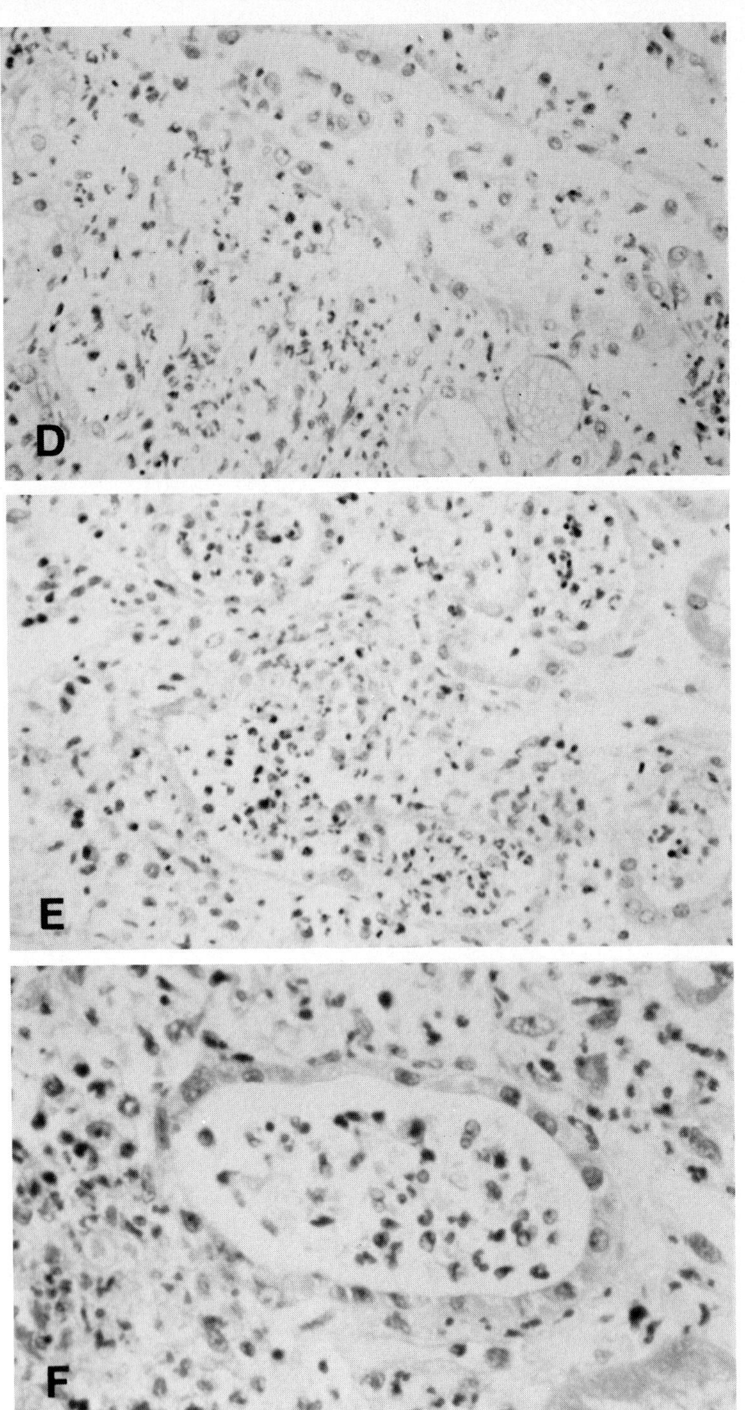

FIG. 11-1D-F *(continued)*
(D) Parenchymal infiltration by neutrophilic leukocytes is associated with tubular disruption. Regenerating epithelial cells have exfoliated into the tubular lumen. (× 250) **(E)** Tubules contain leukocytic casts and are disrupted by an acute inflammatory infiltrate. (× 250) **(F)** A tubule contains a loose aggregate of neutrophils and few degenerated epithelial cells, without definite cast formation. The edematous interstitium is infiltrated by neutrophilic leukocytes. (× 400)

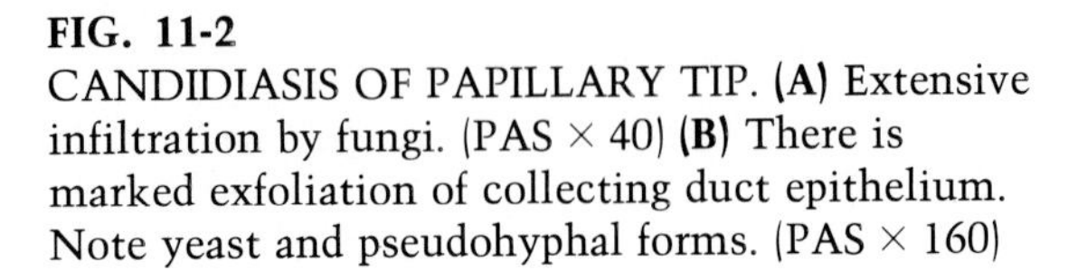

FIG. 11-2
CANDIDIASIS OF PAPILLARY TIP. **(A)** Extensive infiltration by fungi. (PAS × 40) **(B)** There is marked exfoliation of collecting duct epithelium. Note yeast and pseudohyphal forms. (PAS × 160) **(C)** Numerous fungal spores and transitional cells in urine sediment. (× 1000) **(D)** Pseudohyphal form with neutrophils and probable histiocyte. (× 1000) **(E)** (PAS × 1000) **(F)** (Grocott × 400).

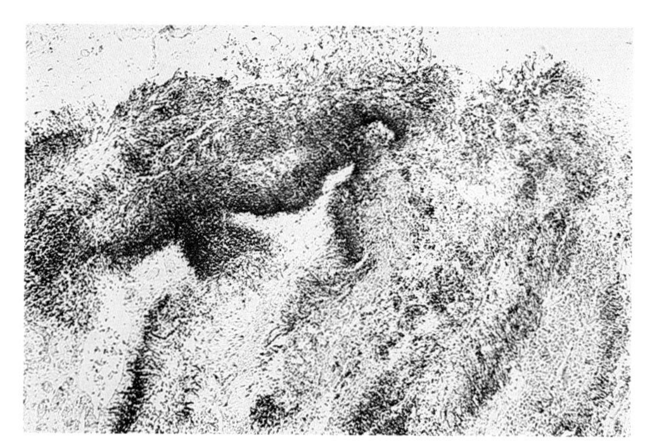

FIG. 11-2A

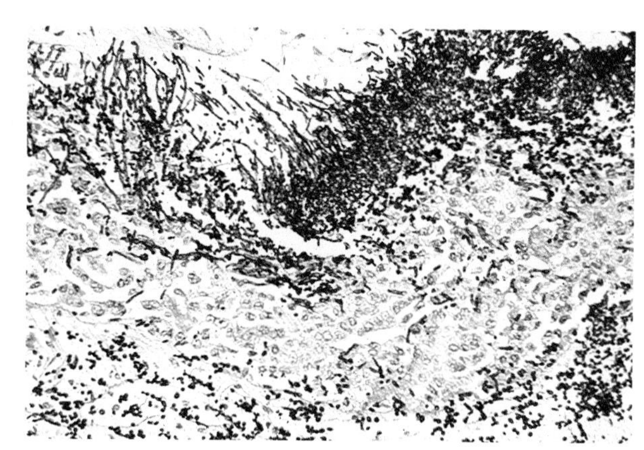

FIG. 11-2B

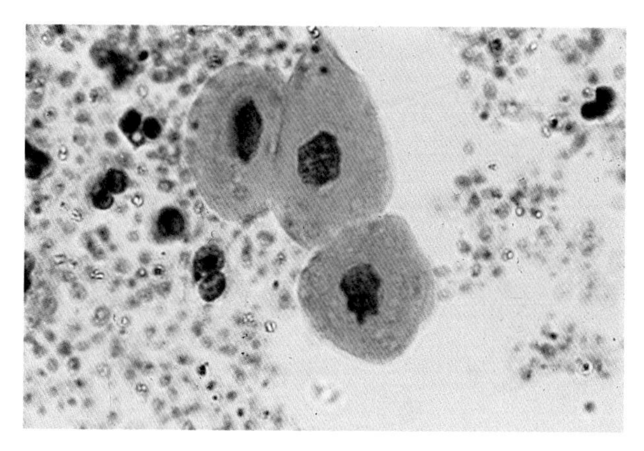

FIG. 11-2C

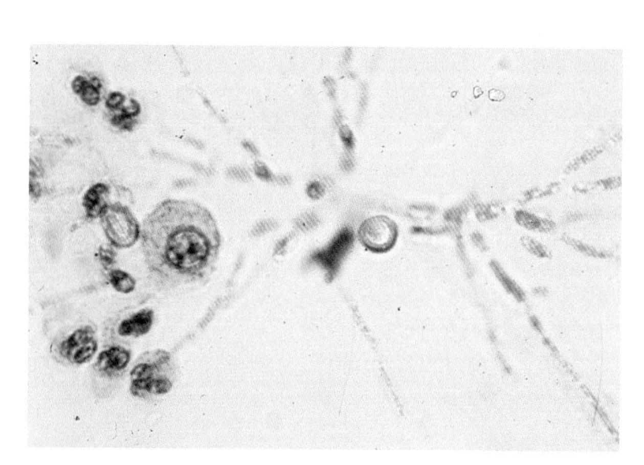

FIG. 11-2D

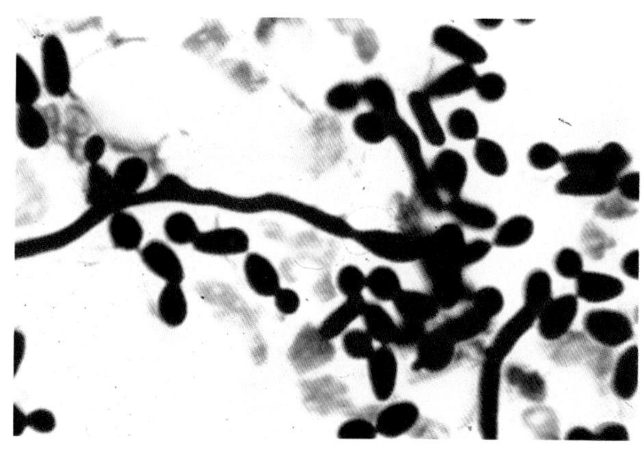

FIG. 11-2E

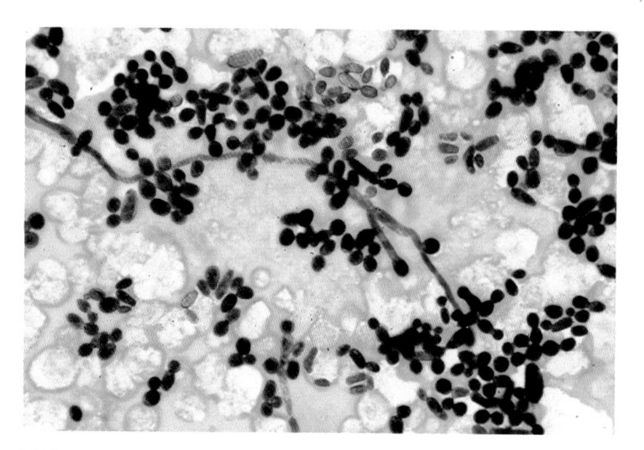

FIG. 11-2F

FIG. 11-3
CHRONIC PYELONEPHRITIS. **(A)** Cortical surface with multiple large, flat scars. (Gross) **(B)** Hemisectioned kidney. Cortical scars overlie a dilated calyx (arrow) in association with a deformed, attenuated papilla. (Gross) **(C)** Classic cortical scar with underlying blunted papilla and dilated calyx. Both medulla and cortex contain a chronic inflammatory infiltrate. Note dilated, atrophic cortical tubules with waxy casts. (× 40)

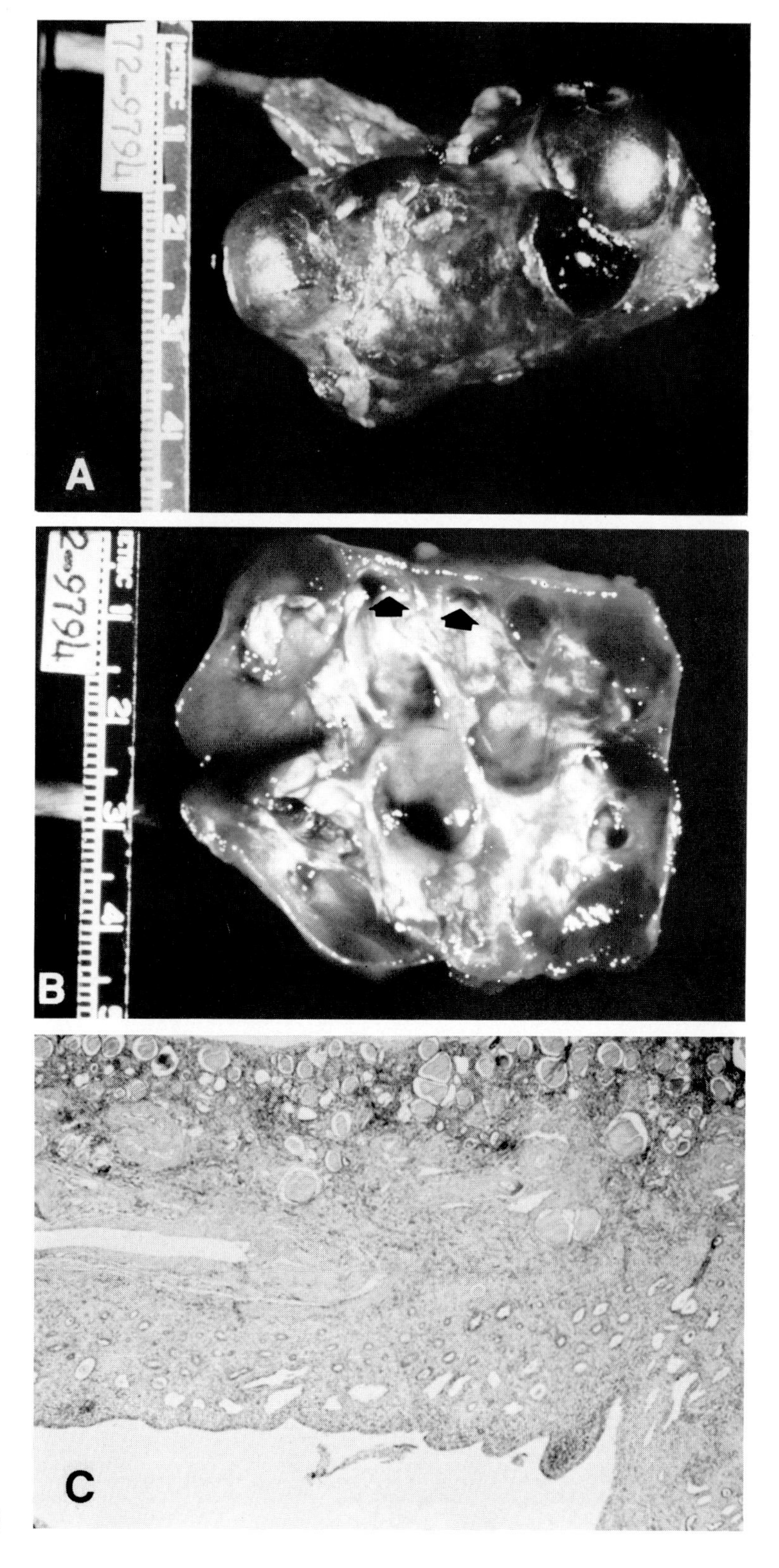

(continued)

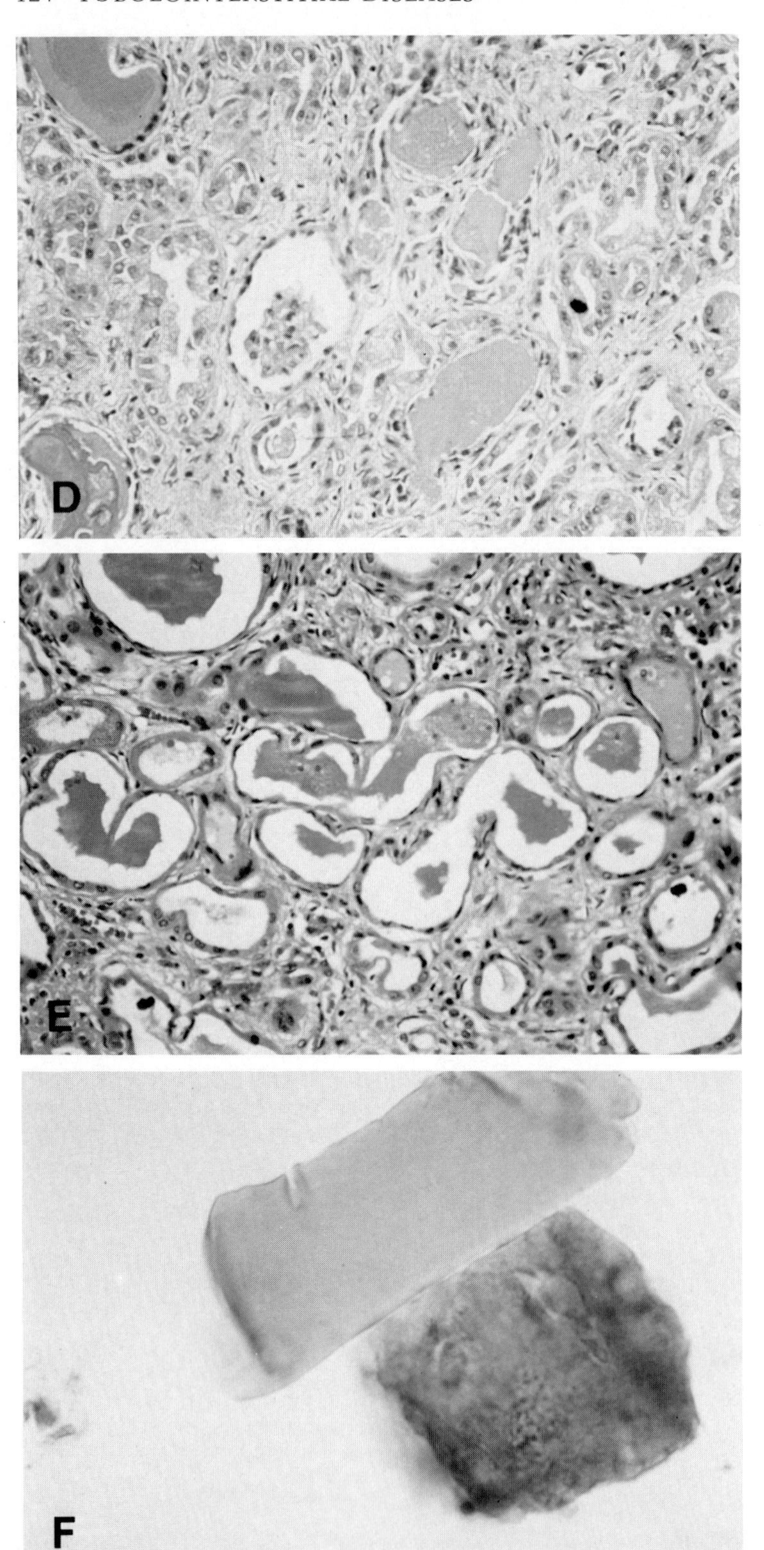

FIG. 11-3D-F *(continued)*
(D) Brittle waxy casts in tubules. (× 160) **(E)** Cystic tubular atrophy with broad waxy casts. (× 160) **(F)** Broad waxy casts characteristic of chronic parenchymal disease. Note cracks along cast borders. (× 1000)

FIG. 11-4
MALAKOPLAKIA OF RENAL CORTEX. Urine sediment with histiocytes containing a basophilic intracytoplasmic inclusion characteristic of a Michaelis-Gutmann body. **(A)** (× 1000) **(B)** (× 100) **(C)** Bowman's capsule surrounding a glomerulus is focally disrupted by a dense, inflammatory infiltrate composed predominantly of histiocytes. Tubules appear to be destroyed or replaced, and an attenuated arteriole shows mural infiltration. (× 160)

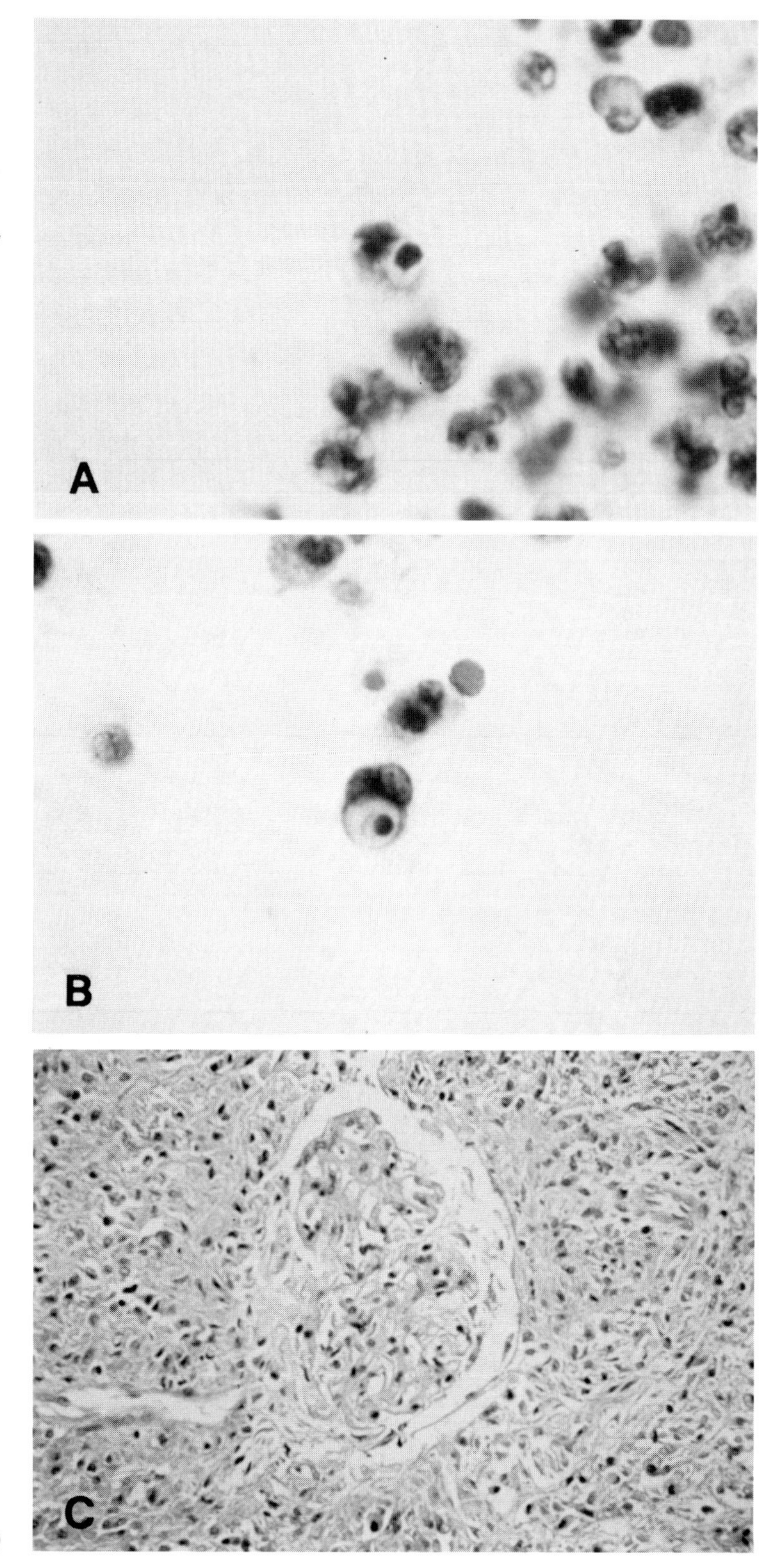

(continued)

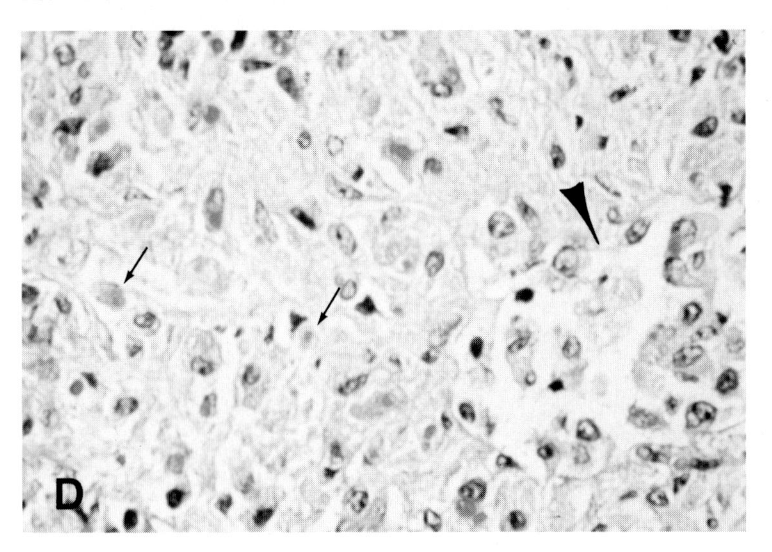

FIG. 11-4D *(continued)*
(D) Numerous PAS-positive Michaelis-Gutmann bodies (arrows) are present within the abundant foamy to slightly granular cytoplasm of histiocytes. Note histiocytes in the apparent lumen of a destroyed tubule (arrowhead). (PAS × 400)

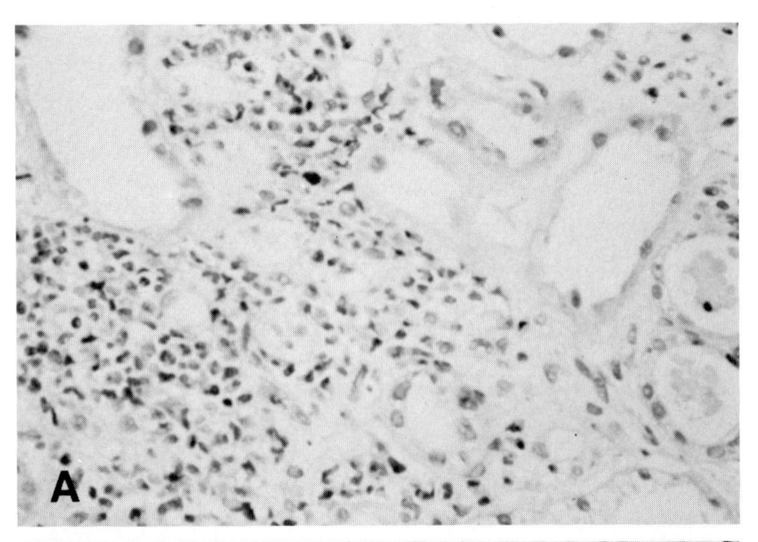

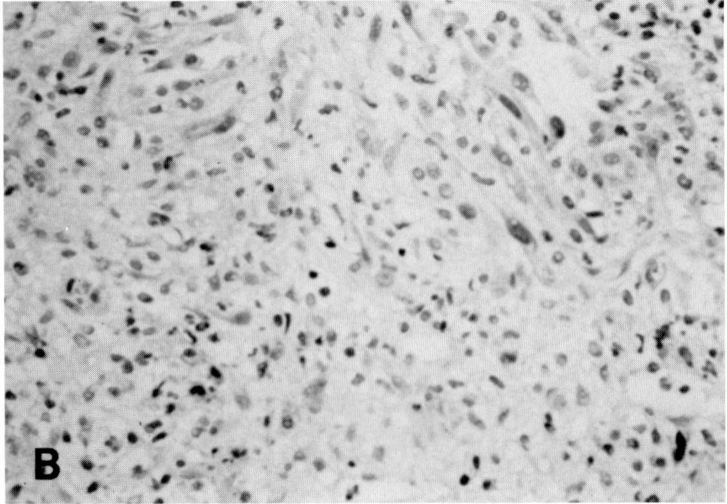

FIG. 11-5
ACUTE ALLERGIC INTERSTITIAL NEPHRITIS. **(A)** The edematous interstitium contains an inflammatory infiltrate rich in eosinophils. Tubules focally contain "ghost cells" and have an attenuated, regenerating epithelium. (× 250) **(B)** Focus of tubular disruption with fibroblastic proliferation. In addition to eosinophils, neutrophils and histiocytes are present. (× 250)

FIG. 11-6
CHRONIC INTERSTITIAL NEPHRITIS, PROBABLY DRUG-INDUCED. **(A)** The markedly edematous interstitium contains a heavy plasma cell infiltrate. Tubules have a regenerating epithelium. (× 250) **(B)** Tubules contain degenerating and necrotic epithelial cells as well as casts. (× 250) Plasma cells are often prominent in the later phases of an allergic interstitial nephritis.

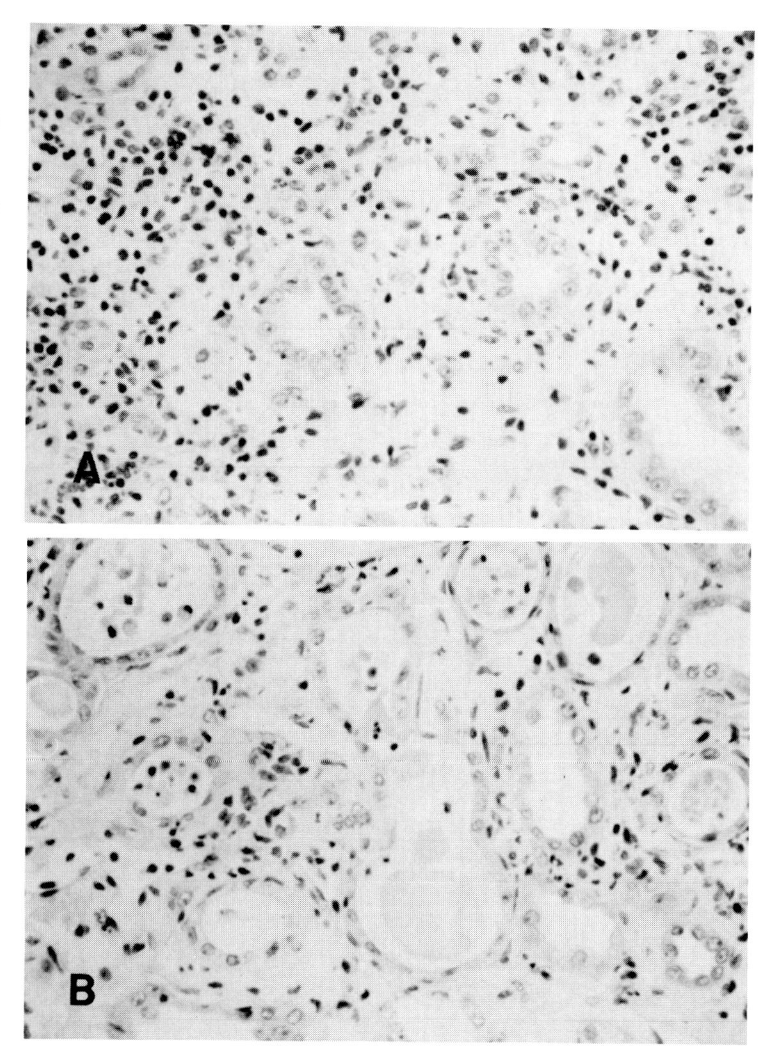

FIG. 11-7
GOUTY NEPHROPATHY. **(A)** A severely damaged collecting duct contains amorphous and few elongated crystals with a marked acute and chronic inflammatory response. (Formalin fixation × 160)

(continued)

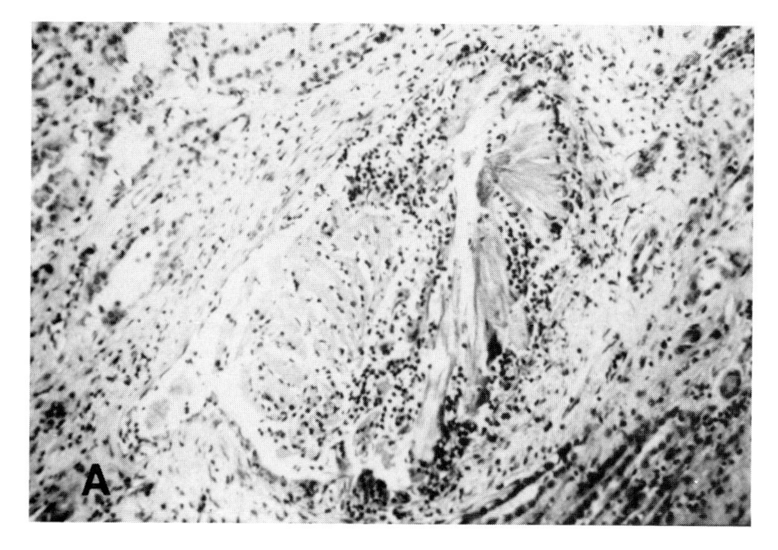

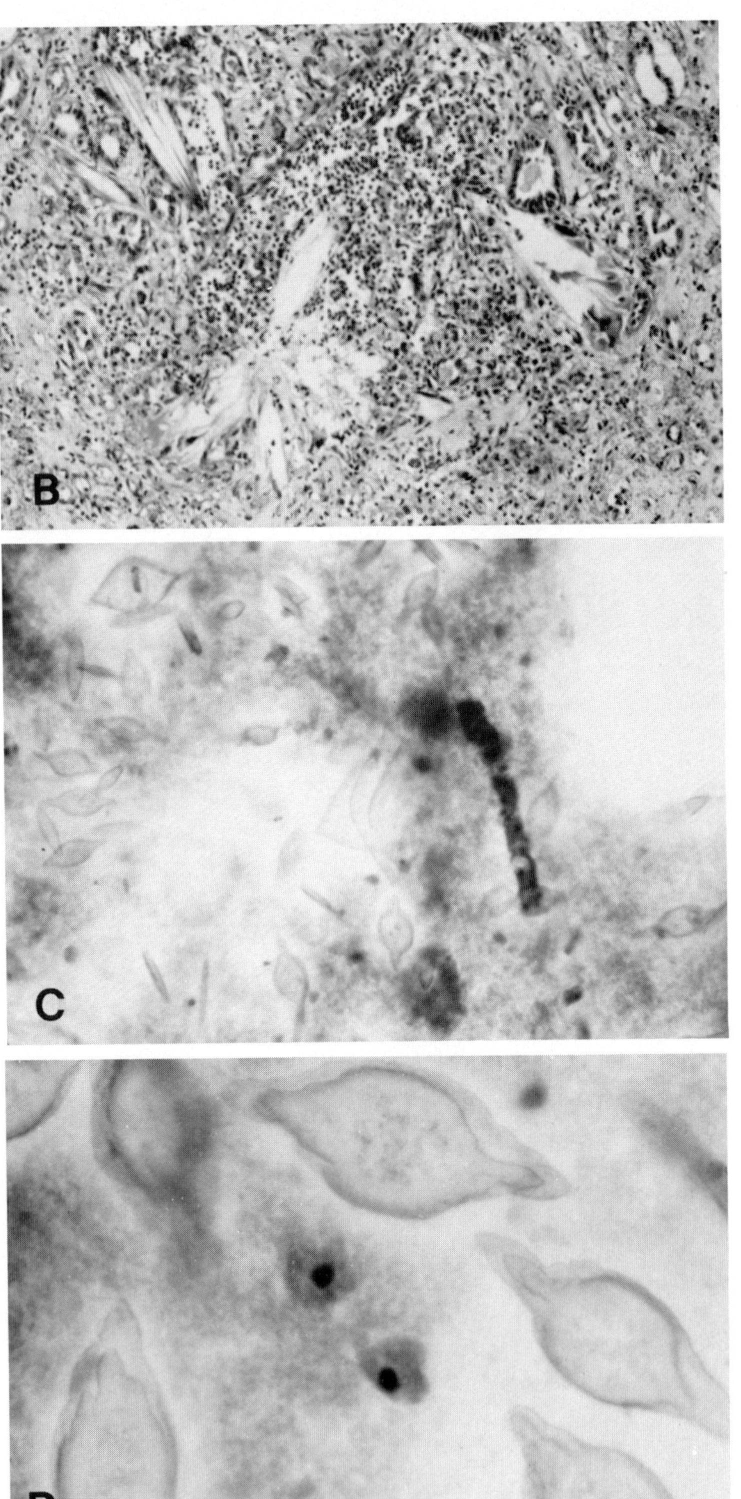

FIG. 11-7B-D *(continued)*
(B) A medullary focus of destroyed collecting ducts contains interstitial collections of elongated crystals that have evoked a giant cell reaction and intense chronic inflammation. (Formalin fixation × 160) **(C)** Amorphous urates and uric acid crystals in urine sediment. Note mixed cellular cast indicating active renal parenchymal disease. (× 160) **(D)** Note exfoliated renal tubular epithelial cells of collecting duct origin, suggesting tubular injury. (× 400)

12
GLOMERULAR DISEASES

CHANGES IN GLOMERULAR PERMEABILITY (NEPHROTIC SYNDROME)
PROLIFERATIVE, INFLAMMATORY, AND/OR NECROTIZING GLOMERULAR LESIONS
SCLEROSING (CHRONIC) GLOMERULAR LESIONS

In a normal glomerulus, plasma is filtered across the capillary wall into Bowman's space, which is continuous with the lumen of the proximal convoluted tubule. The capillary wall ultrastructurally is composed of a fenestrated endothelium, a basement membrane, and overlying epithelial foot processes joined by slit diaphragms. The semipermeable basement membrane is a major component of this filtration barrier, which normally allows passage of low molecular weight substances only. Therefore, albumin (molecular weight 69,000) is not present in the filtrate, and normally, there is less than 150 mg of protein excreted in the urine over a 24-hr period. Homeostatic tubular functions, including secretion and absorption, will modify the composition and concentration of the urine.

Abnormalities in the macroscopic urinalysis usually provide the first laboratory evidence of glomerular disease, particularly proteinuria (e.g., albuminuria). An accurate urine sediment examination in patients with an abnormal screening urinalysis or with known or suspected renal disease will often provide morphological evidence of glomerular disease. The presence of erythrocytic and blood casts is an important finding in documenting the presence of a glomerulonephritis. The activity and progression of the glomerular lesion can be followed, particularly if Papanicolaou-stained sediments are used and if results are given semiquantitatively. On the basis of urine chemistries and from the sediment findings, glomerular diseases can be grouped into three overlapping categories: (1) lesions with altered glomerular permeability (nephrotic syndrome); (2) proliferative, inflammatory, and/or necrotizing lesions; and (3) sclerosing (chronic) lesions.

CHANGES IN GLOMERULAR PERMEABILITY (NEPHROTIC SYNDROME)

Various forms of glomerular disease are associated with proteinuria. Alterations of the glomerular basement membrane related to differing mechanisms of injury, for example, immune complex or coagulation, produce an abnormal permeability to plasma proteins, particularly albumin. Ultrastructurally, basement membranes are usually intact; epithelial foot processes become fused in response to the proteinuria.

The nephrotic syndrome is defined as massive proteinuria (greater than 4 g/24 hr), hypoalbuminemia (less than 3 g/100 ml), hypercholesterolemia with lipiduria, and generalized edema. Minimal change disease, a common cause of nephrotic syndrome in children, is characterized by essentially normal-appearing glomeruli in histologic sections. Changes secondary to proteinuria can be idenfified, such as visceral epithelial swelling and hyaline droplet change in tubular epithelium. The accumulation of lipid in tubular epithelial cells led to the early descriptive diagnostic term of lipoid nephrosis.

Membranous glomerulopathy, an immune complex form of glomerular disease, is a major cause of nephrotic syndrome in adults. In the early stages (Fig. 12-1**A**), glomeruli may appear unremarkable; however, with progression, capillary wall thickening becomes apparent (Fig. 12-1**B**). The diagnostic epimembranous deposits and basement membrane spikes can be demonstrated with special stains or ultrastructurally, and with progression deposits become incorporated in the thickened glomerular basement membranes. In early stages of membranous glomerulopathy, tubules are generally well preserved, although epithelium will show hyaline droplet change and lipid accumulation (Fig. 12-1**C**). With superimposed renal vein thrombosis (Fig. 12-1**D** and **E**), tubular degeneration with epithelial exfoliation and regeneration may be prominent. In the urine sediment, this complication might be reflected by the presence of increased numbers of renal epithelial cells, erythrocytes, and leukocytes.

Another common cause of nephrotic syndrome, which may be complicated by renal vein thrombosis, is amyloidosis (Fig. 12-1**F**). In amyloidosis complicating multiple myeloma, myeloma casts in histologic sections are commonly seen (see Chap. 14, Fig. 14-9). In this clinical setting, light chains can be detected by urine chemistry methods, and abnormal casts might be observed in the urine sediment.

The characteristic urine sediment findings in patients with nephrotic syndrome will primarily reflect the lipiduria (Fig. 12-2). Therefore, fat droplets, oval fat bodies, renal epithelial foam cells, and, most importantly, fatty casts are diagnostic. Pathologic casts, that is, waxy, coarse granular, and epithelial, will also be present.

PROLIFERATIVE, INFLAMMATORY, AND/OR NECROTIZING GLOMERULAR LESIONS

Many forms of glomerulonephritis are associated with glomerular hypercellularity (i.e., mesangial and endocapillary proliferation) and infiltration by inflammatory cells (i.e., "exudation"; Fig. 12-3). The detection of hematuria and greater than 2+ proteinuria by macroscopic urinalysis will reflect leakage through the damaged capillary wall, which ultrastructurally may be evident as glomerular basement membrane (GBM) breaks. Tubular degenerative changes (Fig. 12-3**B**) and epithelial exfoliation with regeneration (Fig. 12-3**C**) are commonly observed. Typical urine sediment findings in "acute" glomerulonephritis include erythrocytic and blood casts, as well as other pathologic casts, increased renal epithelial cells, erythrocytes, and leukocytes.

Necrotizing forms of glomerulonephritis may be due to antiglomerular basement membrane antibodies (e.g., Goodpasture's syndrome), immune complex injury, or neither mechanism. With disruption of GBMs, leakage of fibrin and coagulation products into Bowman's space causes extracapillary proliferation and the formation of epithelial crescents, which, if present in greater than 70% of glomeruli, is termed **rapidly progressive glomerulonephritis** (Fig. 12-4**A**). Endocapillary proliferation and neutrophilic exudation are variable. GBM fragmentation may be identified in necrotic segments (Fig. 12-4**B**). Tubular damage is often striking and is frequently associated with exfoliation of degenerated or necrotic epithelial cells and numerous casts (Fig. 12-4**C, D, E,** and **F**). The urine sediment findings in necrotizing forms of glomerulonephritis or in rapidly progressive glomerulonephritis would be similar to acute glomerulonephritis; however, one might expect greater numbers of pathologic casts, including erythrocytic and blood, renal epithelial, coarse granular, and waxy forms, as well as the appearance of fibrillar (fibrin) casts and increased numbers of both degenerated and necrotic renal epithelial cells. Erythrocytes and leukocytes will also be present. In addition, leukocytic casts, commonly associated with tubulointerstitial disease (i.e., pyelonephritis), may also be present (Fig. 12-5).

Interstitial inflammation is a variable component in severe forms of glomerulonephritis. In addition, prominence of interstitial eosinophils or plasma cells may indicate superimposed allergic interstitial nephritis related to drug therapy (e.g., antibiotics, diuretics), and eosinophiliuria may provide a clue to this complication.

Glomerular lesions associated with systemic vasculitis may also be drug-related, and are typically necrotizing. In classical polyarteritis, however, renal involvement is usually that of parenchymal infarction. In addition, thrombotic microangiopathies, such as the hemolytic uremic syndrome (Fig. 12-6), malignant hypertension, and scleroderma, may produce acute ischemic parenchymal damage. The urinc sediment findings reflecting ischemic necrosis in these diseases would be renal epithelial fragments, a marked increase in degenerated and nccrotic ("ghost") renal epithelial cells, necrotic renal epithelial and fibrillar (fibrin) casts, as well as other pathologic casts, erythrocytes, and leukocytes.

Figure 12-7 shows the overlapping urine sediment patterns observed in renal parenchymal diseases. Diagnostic patterns include tubular injury (see Chap. 9), ischemic necrosis (see Chap. 10), interstitial nephritis (see Chap. 11), and glomerular diseases, especially the nephrotic syndrome. Glomerulonephritis may produce a broad spectrum of urine sediment changes. Indeed, the association of an interstitial nephritis or vasculitis may obscure the urine sediment findings of the glomerular lesion and result in a pattern of tubulointerstitial disease or ischemic necrosis.

Renal involvement in systemic lupus erythematosus may produce glomerular, tubulointerstitial, or vascular disease, resulting in a wide variety of urine sediment changes. Historically, using brightfield microscopy, the urine sediment findings in patients with systemic lupus erythematosus with renal involvement were described as "telescopic." This term implies that findings suggestive of acute glomerulonephritis, nephrotic syndrome, and chronic renal parenchymal disease are present in a single urine specimen. This pattern would not be unexpected in the urine from a patient with the diffuse proliferative form of lupus nephritis. However, lupus nephritis encompasses a wide variety of glomerular lesions with associated, varied prognoses. Therefore, an accurate, semiquantitative permanent urine sediment examination in lupus patients may provide a good indication of the severity of the glomerular lesion and a useful method of following the activity of the renal disease.

SCLEROSING ("CHRONIC") GLOMERULAR LESIONS

Sclerosis is a glomerular alteration characterized by an increase and thickening of the mesangial matrix, typically with thickening and contraction of GBMs, which results in collapse or compromise of glomerular capillary patency. This reaction pattern may involve part of a glomerulus, that is, segmental, or the whole glomerulus, that is, global. Glomerular sclerosis may be the apparent primary morphological alteration, as in focal segmental hyalinosis (FSH), which is frequently clinically manifested as the nephrotic syndrome, or it may represent "healing" of glomerular injury related to a proliferative or necrotizing glomerulonephritis. Glomerular sclerosis may also be secondary to chronic vascular or tubulointerstitial disease. The urine sediment, therefore, will often reflect the associated activity of the underlying glomerular lesion. With advanced glomerulonephritis, often termed **"chronic"**, when sclerosis involves greater than 60 to 70% of glomeruli and there is clinical evidence of renal insufficiency, the urine sediment will typically be less active. Pathologic casts may dominate the sediment. Progressive glomerular sclerosis is associated with tubular atrophy (Fig. 12-8**A** and **B**) and waxy casts become prominent (Fig. 12-8**C**). The appearance of numerous waxy casts, particularly broad forms, in the urine sediment is a good indication of chronic renal parenchymal disease (Fig. 12-8**D**).

BIBLIOGRAPHY

Bradley M, Schumann GB, Ward PCJ: Examination of urine. In Henry, JB (ed): Todd, Sanford, Davidsohn:

Clinical Diagnosis and Management by Laboratory Methods, 16th ed. Philadelphia, WB Saunders, 1979
Heptinstall RH: Pathology of the Kidney, 2nd ed. Boston, Little, Brown, 1974
Schumann GB, Harris S, Henry JB: An improved technique for examining urinary casts and a review of their significance. Am J Clin Pathol 69:18–23, 1978
Smith RD, Weiss M: Renal Biopsy: Technical Aspects of Light Microscopic Interpretation. Chicago, Educational Products Division, American Society of Clinical Pathologists, 1979

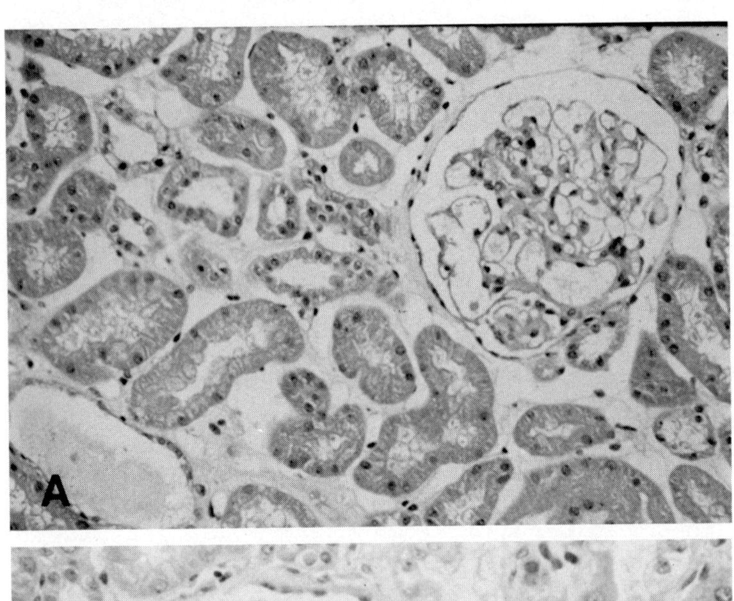

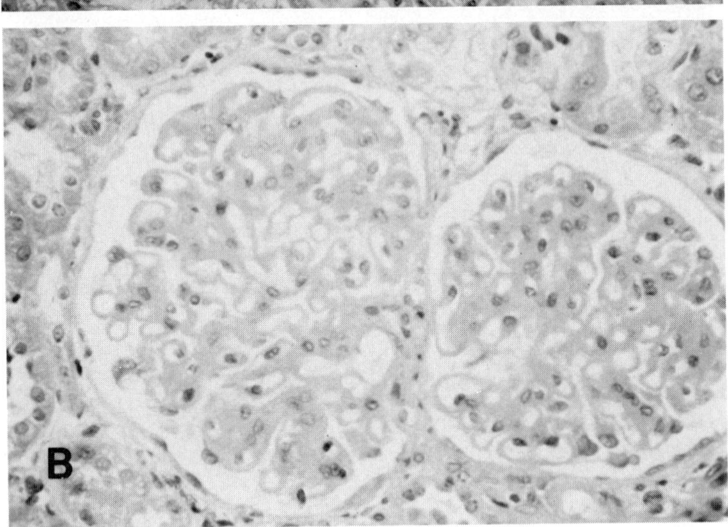

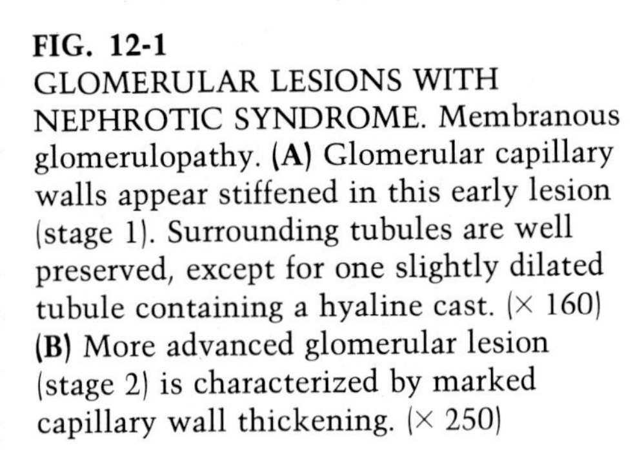

FIG. 12-1
GLOMERULAR LESIONS WITH NEPHROTIC SYNDROME. Membranous glomerulopathy. **(A)** Glomerular capillary walls appear stiffened in this early lesion (stage 1). Surrounding tubules are well preserved, except for one slightly dilated tubule containing a hyaline cast. (× 160) **(B)** More advanced glomerular lesion (stage 2) is characterized by marked capillary wall thickening. (× 250)

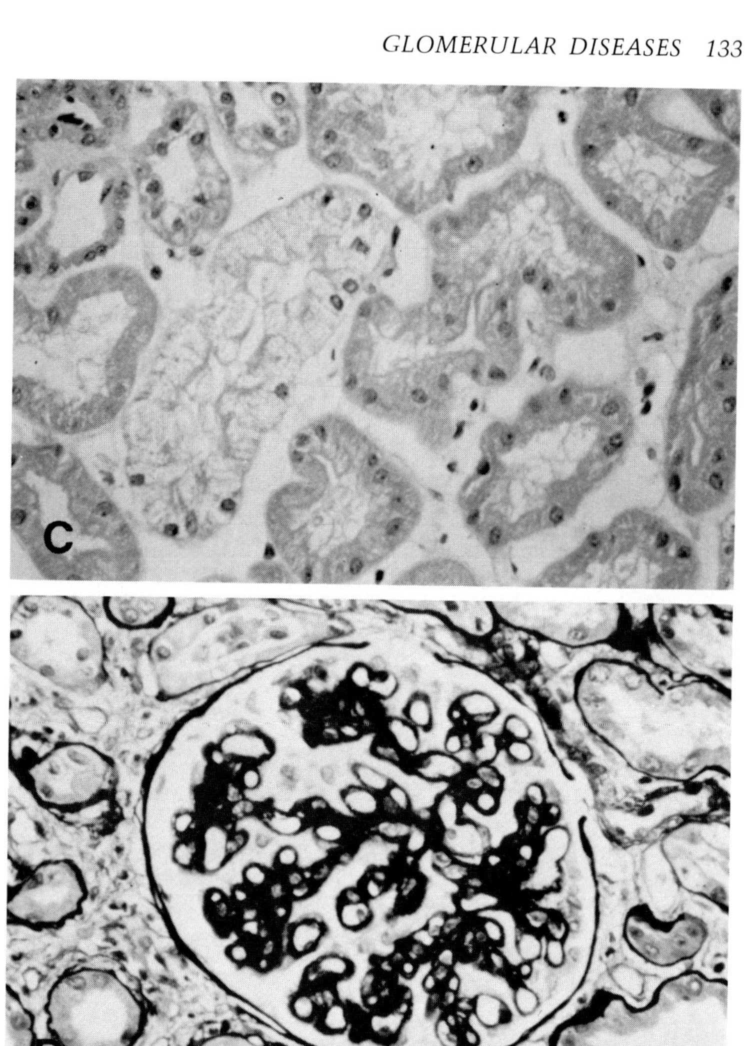

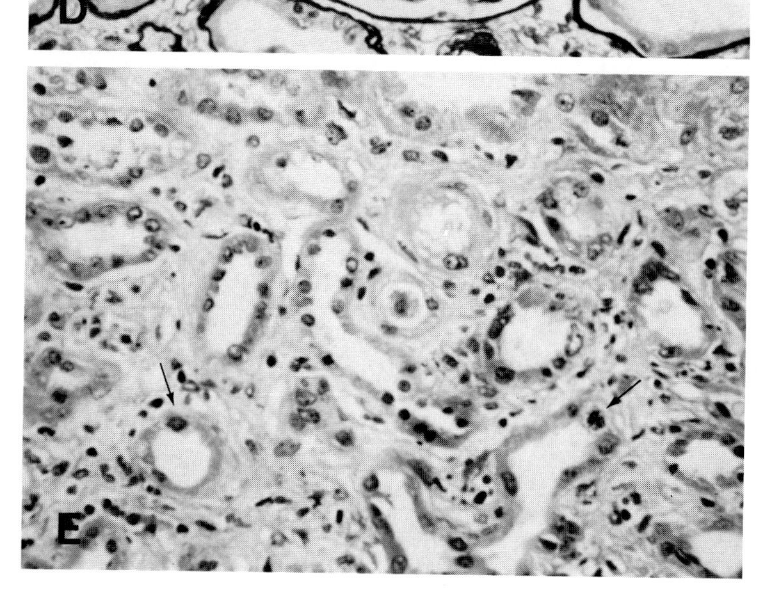

FIG. 12-1C-E *(continued)*
(C) Tubular epithelium shows hyaline droplet and fatty degenerative changes, reflecting the glomerular proteinuria. (× 250) Membranous glomerulopathy *with* renal vein thrombosis. **(D)** Glomerular basement membrane are diffusely thickened and show spiking and beading. Visceral epithelial cell swelling is secondary to the proteinuria. (Jones × 250) **(E)** Note tubular damage with epithelial flattening and occasional mitoses (arrows) indicating regenerative activity. The edematous interstitium contains a few chronic inflammatory cells, and neutrophils are present within peritubular capillaries. (Trichrome × 250)

(continued)

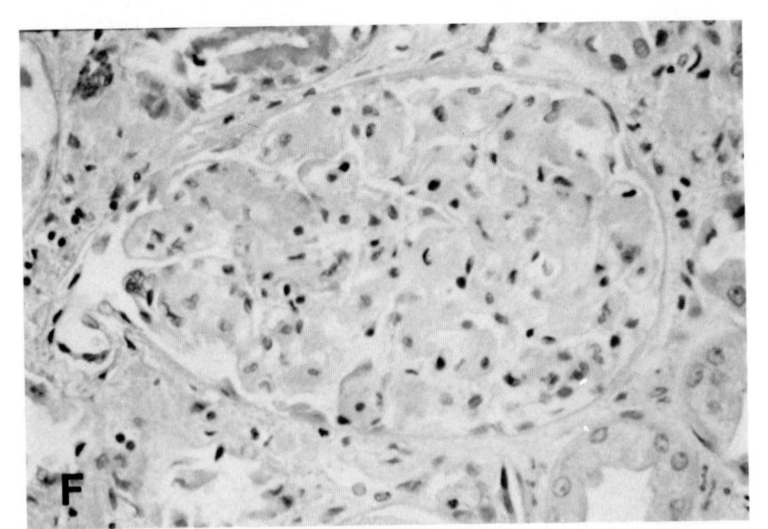

FIG. 12-1F *(continued)*
(F) Secondary renal amyloidosis. Smudgy deposits of amorphous hyaline material are present within the mesangium and along glomerular capillary walls, as well as in the interstitium. (× 250)

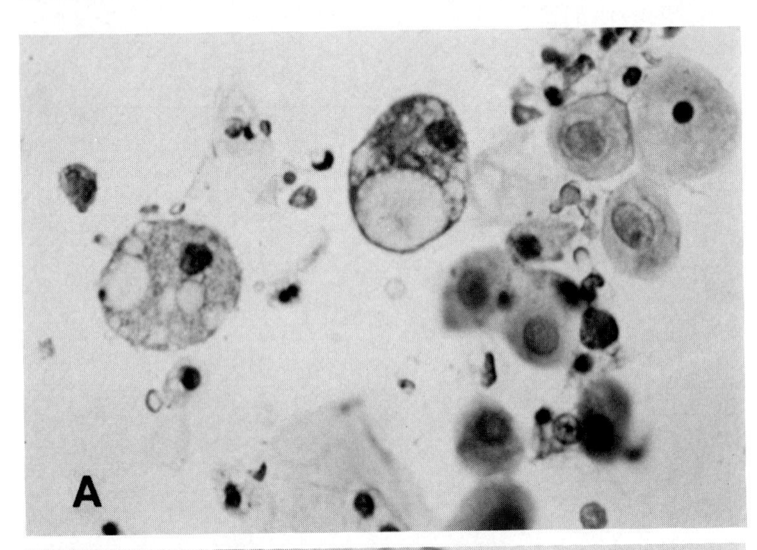

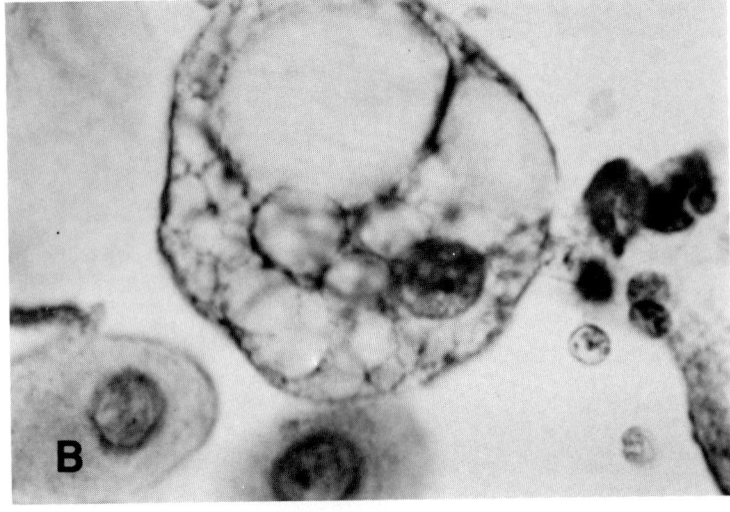

FIG. 12-2
URINE SEDIMENT IN PATIENT WITH NEPHROTIC SYNDROME RESULTING FROM PRIMARY AMYLOIDOSIS. **(A)** Vacuolated epithelial cells, urothelial cells, and mild inflammation. (× 400) **(B)** Vacuolated "foam" cell characteristic of lipiduria. (× 1000)

(continued)

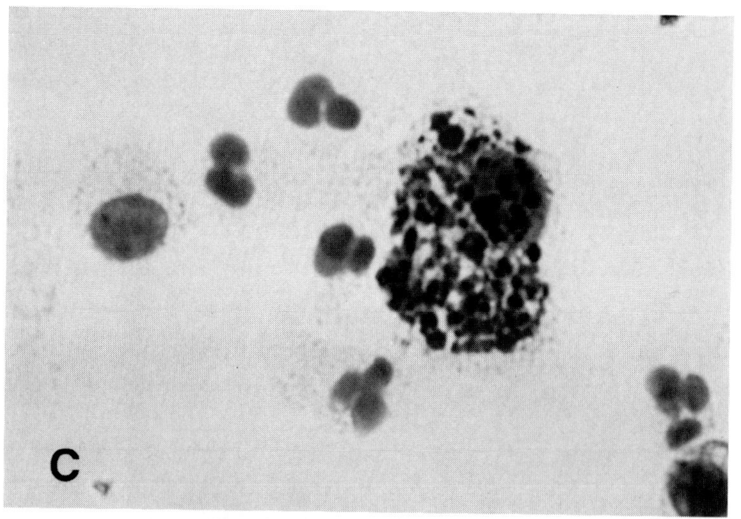

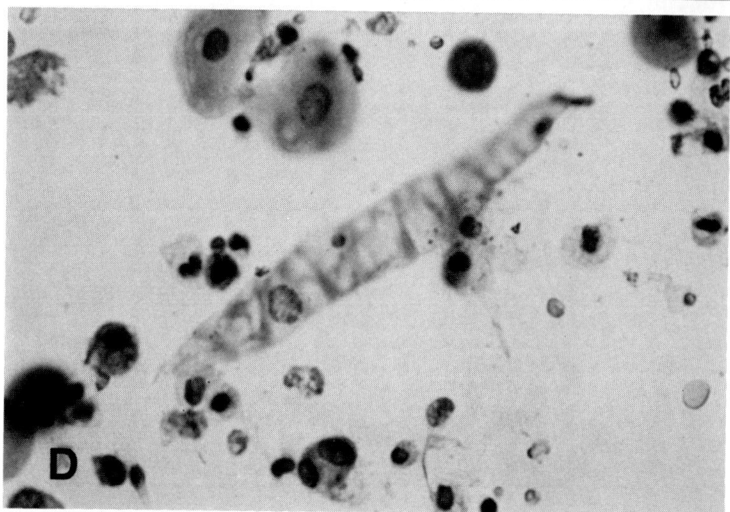

FIG. 12-2C, D *(continued)*
(C) Oil red O-positive vacuolated cells indicating "lipiduria." (Oil red O × 1000)
(D) Fatty cast. (× 400)

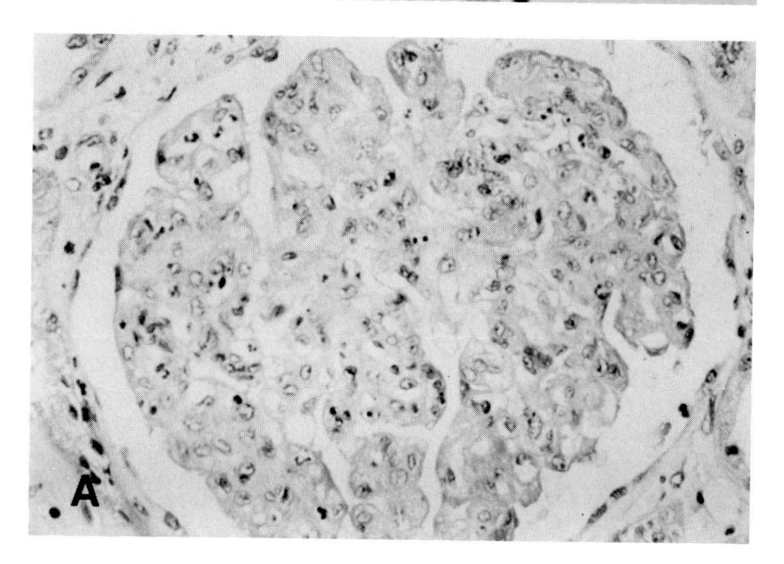

FIG. 12-3
Diffuse proliferative (and exudative) glomerulonephritis, immune complex type. **(A)** A representative glomerulus has marked proliferation of swollen mesangial and endothelial cells, as well as numerous neutrophils. (× 250)

(continued)

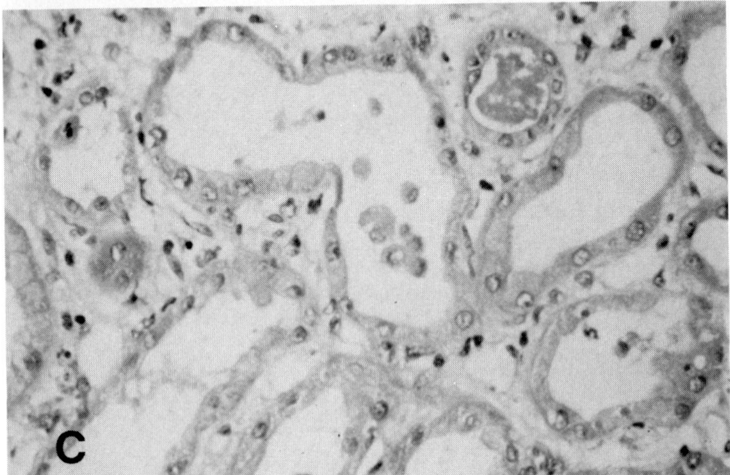

FIG. 12-3B, C *(continued)*
(B) The epithelium of proximal convoluted tubules shows hyaline droplet change and fatty degeneration, indicating glomerular proteinuria. (× 400) **(C)** Tubules focally have a flattened regenerating epithelium and contain exfoliated tubular cells. (× 250)

FIG. 12-4
RAPIDLY PROGRESSIVE GLOMERULONEPHRITIS (RPGN), IMMUNE COMPLEX TYPE. **(A)** A representative field from this needle biopsy contains a glomerulus compressed by a large circumferential epithelial crescent. Tubules are separated by interstitial edema, are dilated, have flattened epithelium, and contain numerous casts. Tubular damage is often conspicuous in severe forms of glomerulonephritis. (× 100) **(B)** A compressed glomerulus has endocapillary proliferation, neutrophilic exudation, and segmental necrosis with glomerular basement membrane fragmentation (arrow). The circumferential crescent contains neutrophils, blood, and fibrin. (Jones × 160) **(C)** Damaged tubules contain waxy and finely granular casts. (Jones × 250)

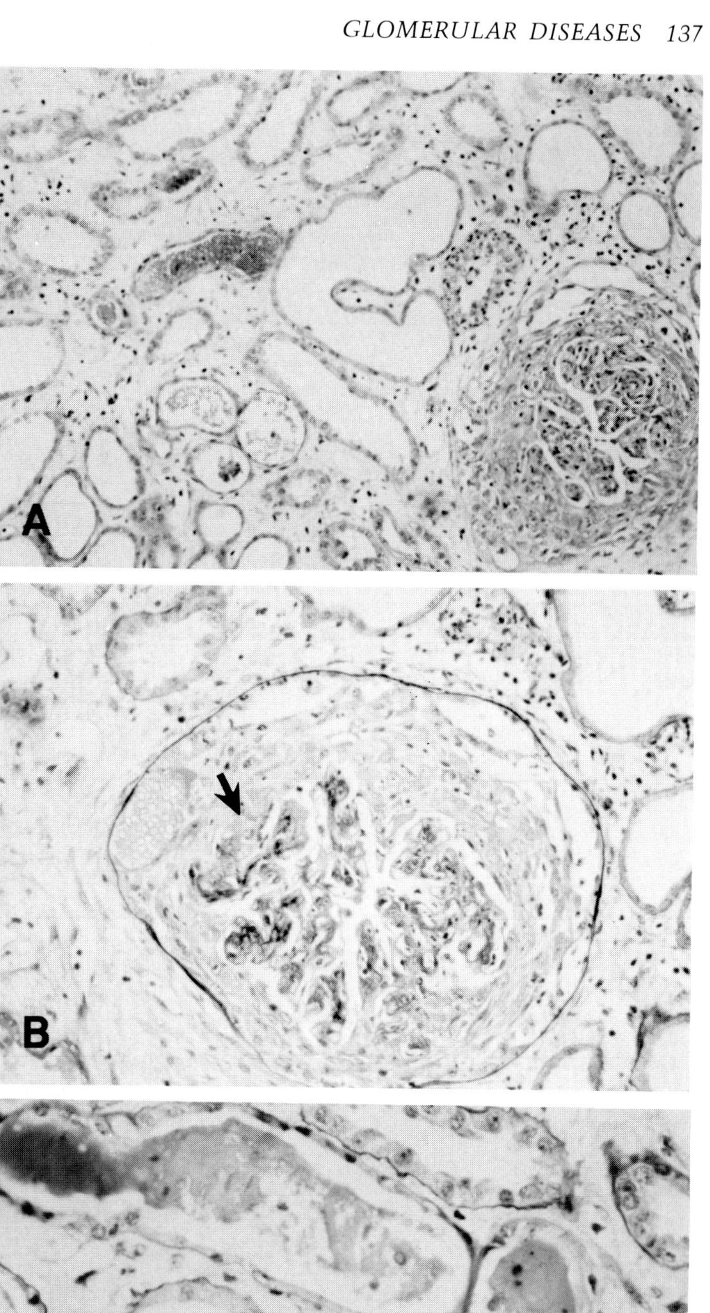

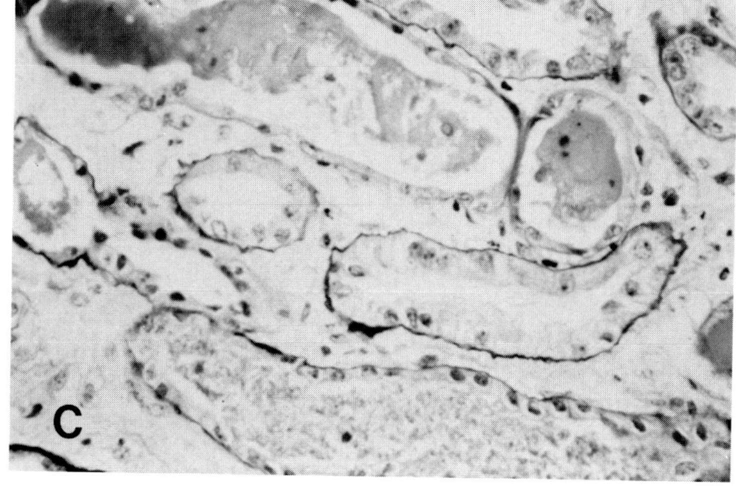

(continued)

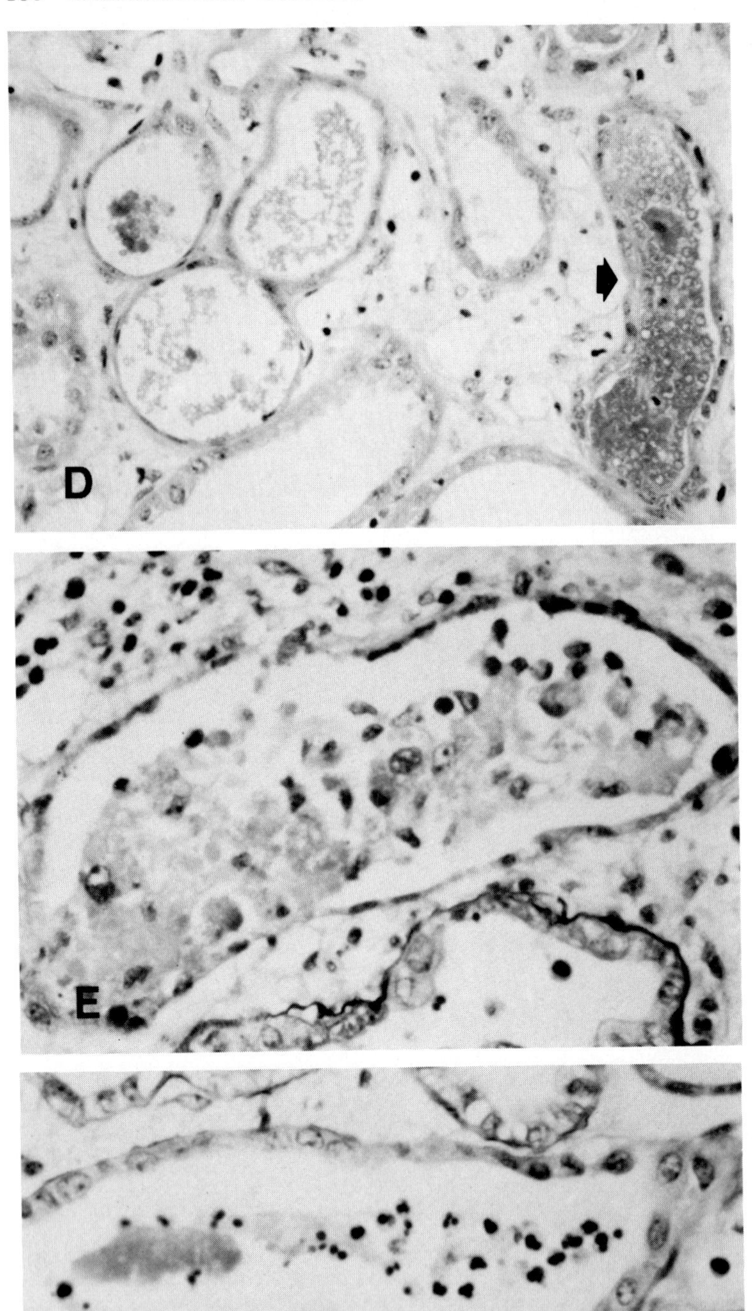

FIG. 12-4D-F *(continued)*
(D) In addition to granular casts, note an erythrocytic cast (arrow). (Jones × 250) **(E)** Desquamated renal tubular cells, many of which are degenerated or necrotic, are forming a coarsely granular cast. (Jones × 400) **(F)** Leukocytic cast. Loosely aggregated neutrophils are present within a hyaline cast matrix. The contiguous cast has a granular to waxy appearance. (Jones × 400)

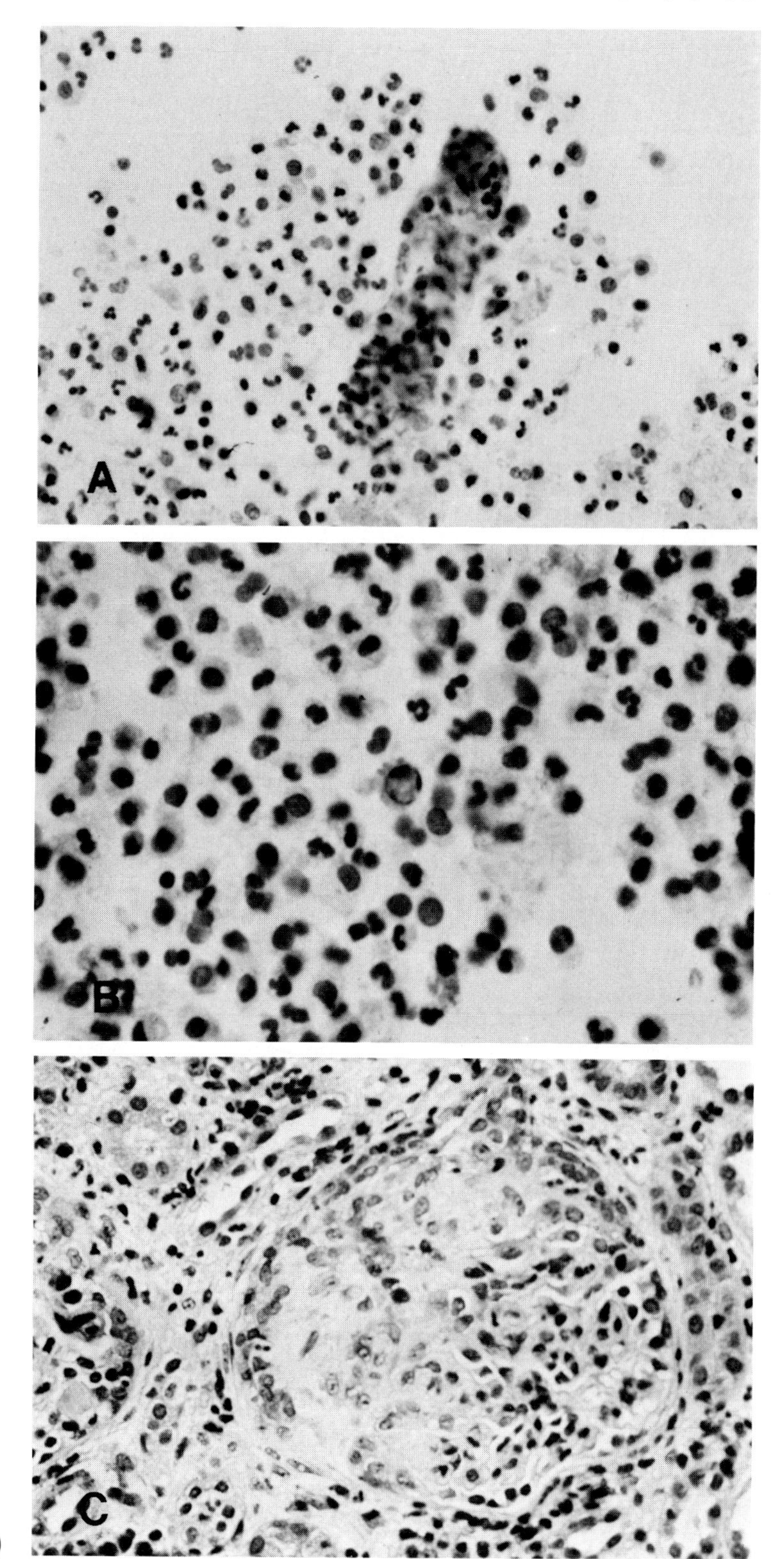

FIG. 12-5
RECURRENT NECROTIZING GLOMERULONEPHRITIS IN RENAL TRANSPLANT PATIENT. **(A)** Inflammatory sediment with leukocytic cast. Although this type of cast is more commonly seen with acute tubulointerstitial disease, it is frequently seen with glomerular lesions. (× 400) **(B)** Marked neutrophilic inflammation and scattered renal epithelial cells. Accurate identification of renal epithelium is difficult with obscuring inflammation. (× 400) **(C)** Original disease in native kidney. A segmentally necrotic glomerulus containing neutrophilic leukocytes is compressed by a eccentric epithelial crescent. Note neutrophils and mononuclear inflammatory cells in the interstitium and tubular damage with epithelial exfoliation. (× 250)

(continued)

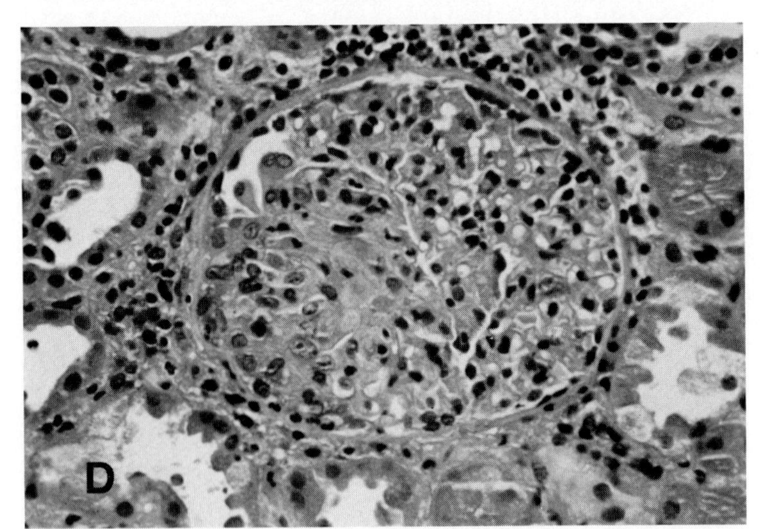

FIG. 12-5D *(continued)*
(D) Renal allograft biopsy 20 days post-transplant, demonstrating recurrence of necrotizing glomerulonephritis. Neutrophils, not a common inflammatory component of transplant rejection, are again noted in the glomerulus and interstitium. (× 250)

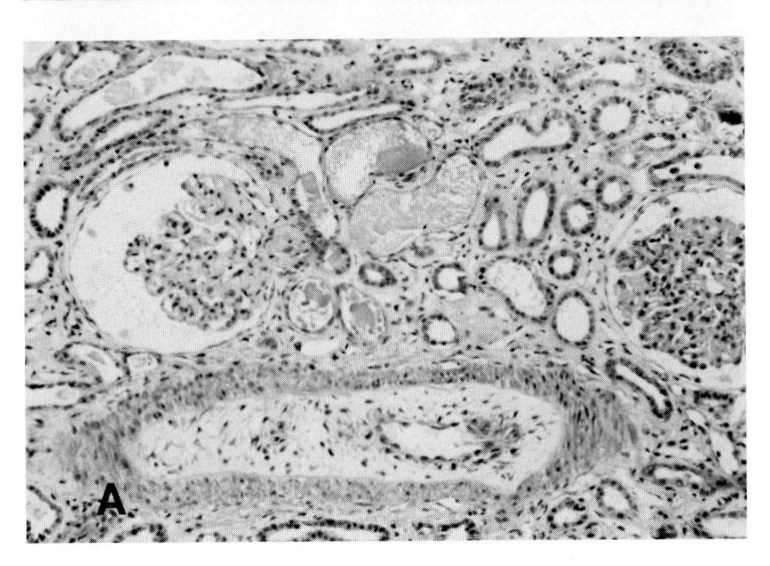

FIG. 12-6
POSTPARTUM HEMOLYTIC UREMIC SYNDROME. **(A)** An interlobular artery has a markedly narrowed lumen secondary to myxomatous-appearing intimal fibroplasia. Glomeruli show ischemic contraction. Tubular damage is prominent, and there are numerous casts. (× 100) **(B)** An afferent arteriole is obliterated and contains fibrin, neutrophilic leukocytes, and nuclear debris. (Lendrum × 250) **(C)** The sudden onset and rapid progression of the thrombotic microangiopathy has resulted in extensive tubular necrosis. Note the massive exfoliation of degenerating and necrotic renal tubular cells. (× 250)

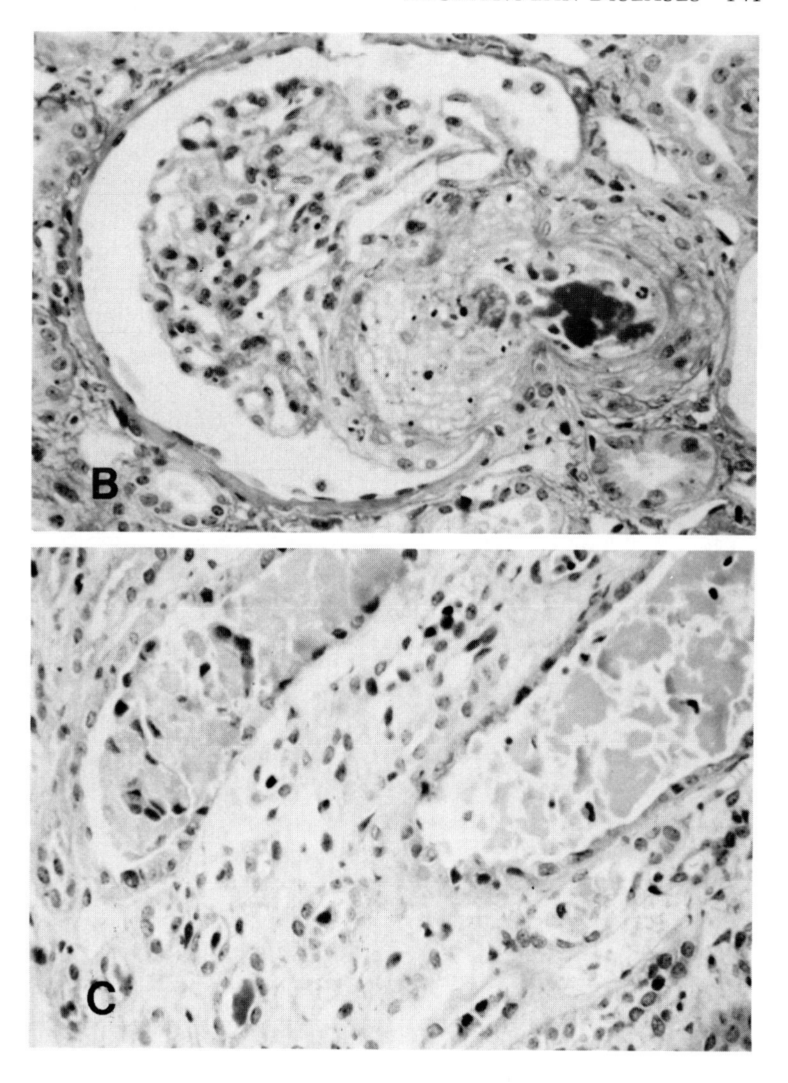
B
C

FIG. 12-7
URINE SEDIMENT PATTERNS AND DIFFERENTIAL DIAGNOSIS IN RENAL PARENCHYMAL DISEASES.

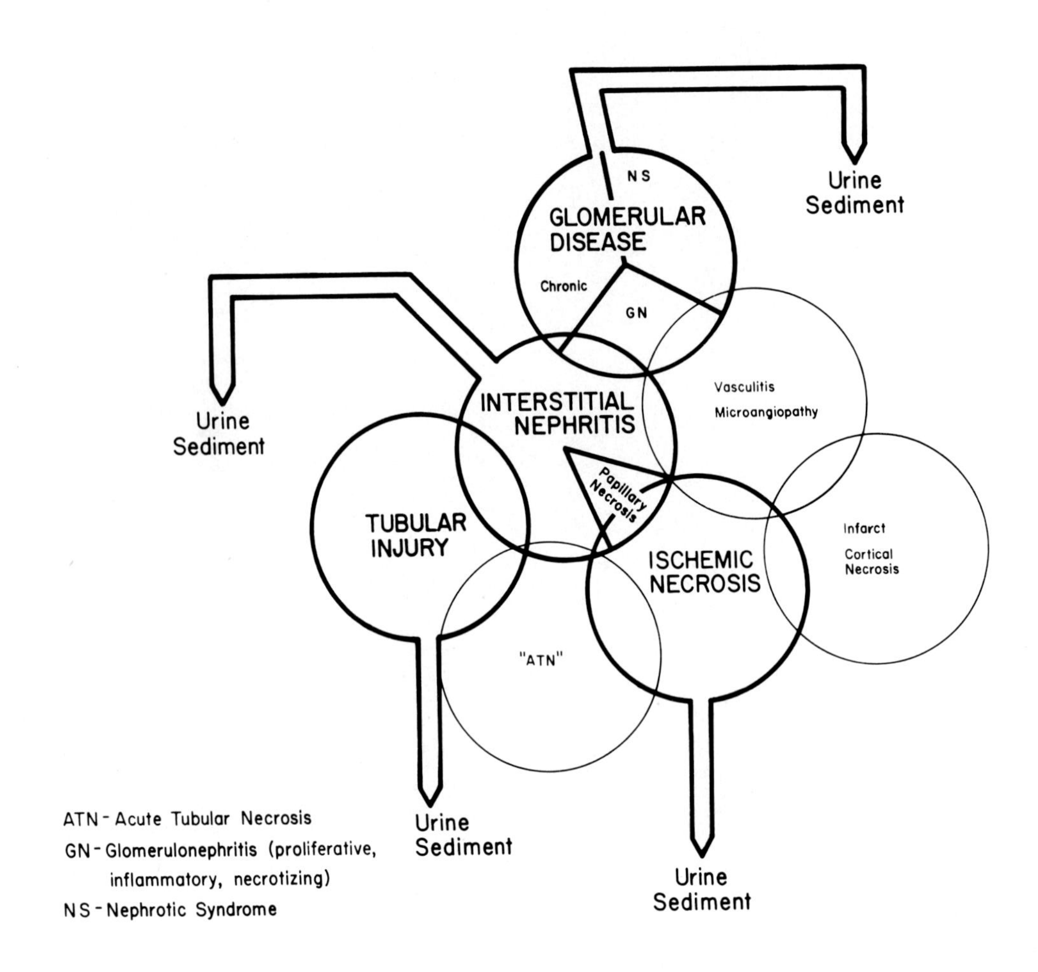

ATN - Acute Tubular Necrosis
GN - Glomerulonephritis (proliferative, inflammatory, necrotizing)
NS - Nephrotic Syndrome

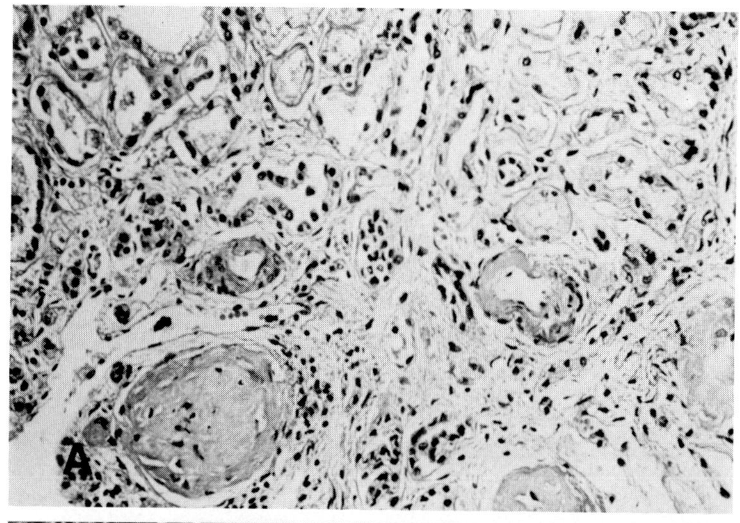

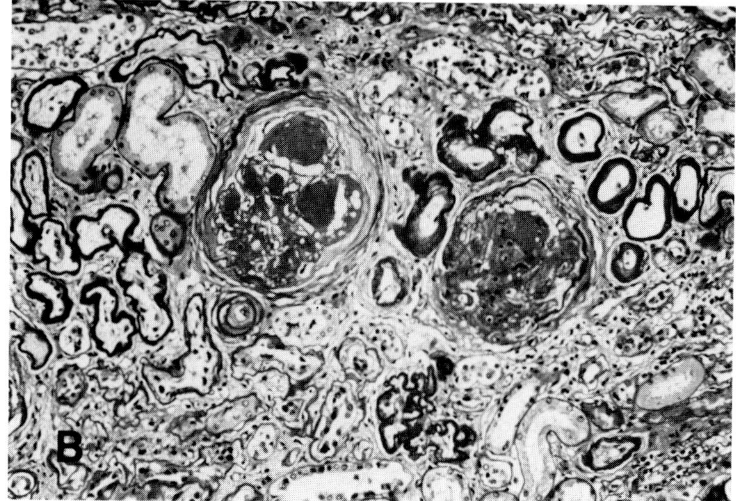

FIG. 12-8
"CHRONIC" SCLEROSING GLOMERULAR LESIONS. **(A)** A globally sclerotic glomerulus is surrounded by a zone of atrophic, shrunken tubules with intervening interstitial fibrosis and mild chronic inflammation. Arterioles have thickened, hyalinized walls (× 160) **(B)** Two glomeruli have prominent intercapillary mesangial nodules with associated loop sclerosis, basement membrane thickening, and adhesions. Note atrophic tubules with thickened, wrinkled basement membrane. (PAS × 160) Advanced diabetic glomerulosclerosis.

(continued)

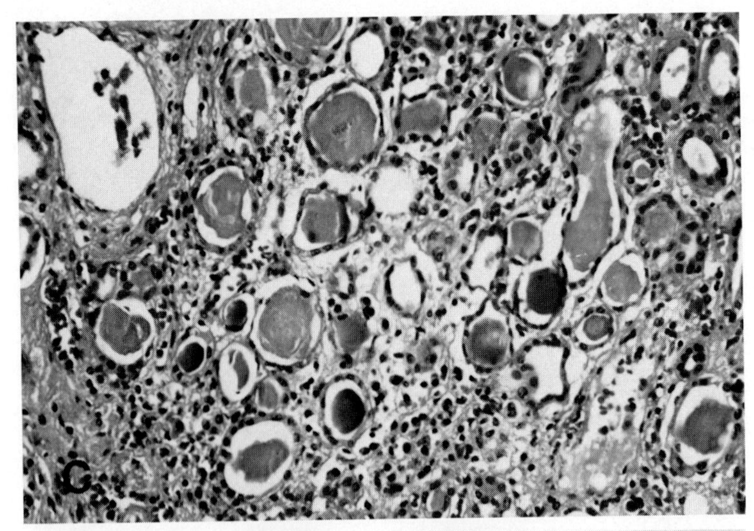

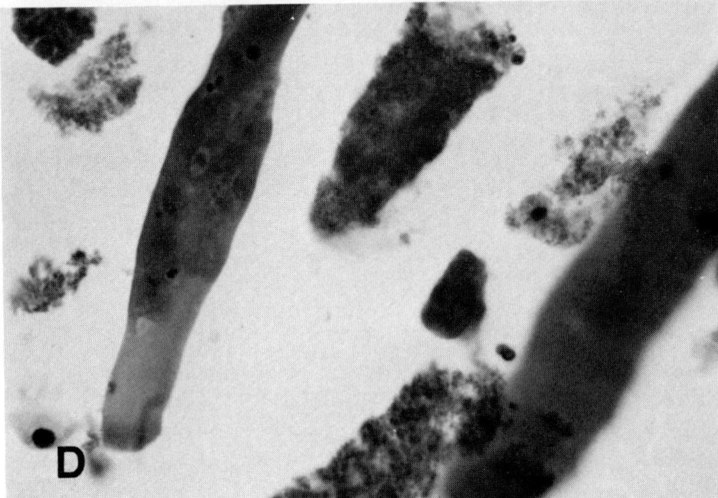

FIG. 12-8C,D *(continued)*
(C) Broad waxy casts in tubules. With progressive glomerular sclerosis and consequent devascularization, functional nephron loss is accompanied by tubular atrophy. Waxy casts are often prominent in shrunken or cystic atrophic tubules. (PAS × 160) **(D)** Waxy and granular casts in urine sediment consistent with chronic renal parenchymal disease. (× 400)

13
RENAL TRANSPLANTATION

Host immunologic responses to allograft foreign antigens are cell-mediated and humoral. Renal allograft rejection represents the cumulative effects of these mechanisms of injury, which are modified by various immunosuppressive regimens. Without therapeutic modalities, cadaveric and living-related donor allografts would be rapidly rejected.

Renal biopsies have been useful for the diagnosis of rejection reactions and have permitted the appropriate therapy to be rapidly instituted for potential reversal of the rejection reaction. They allow one to appreciate changes that carry a poor prognosis, to define a clinical endpoint in therapy, and to perform an allograft nephrectomy at a time early enough to prevent complications of immunosuppressive therapy, for example, infections.

Recently, the cytologic approach to the examination of urine sediment from renal allograft recipients has been useful in the prediction of impending rejection episodes. Historically, lymphocyturia had been used as the most important urine sediment parameter for the diagnosis of a rejection reaction, but in our experience, this cell population is a less-sensitive indicator. The progressive increased exfoliation of renal tubular epithelial cells has proved to be of greater diagnostic value. The cytodiagnosis of renal allograft rejection and the nonimmunologic causes of allograft dysfunction (e.g., infections, acute tubular necrosis [ATN]) require permanently Papanicolaou-stained urine sediments for the accurate identification of varous renal tubular epithelial cells and fragments, pathologic casts, and hematopoietic cells. In addition, it is the staining procedure of choice for the cytodiagnosis of viral inclusion-bearing cells.

The cytohistologic differential diagnostic considerations will often depend upon the post-transplantation interval. Renal allograft rejections have been traditionally classified as hyperacute, acute (early and late), and chronic. Hyperacute rejection is typically irreversible, it occurs within minutes to hours, and it involves preformed antibodies to allograft antigens. Early acute rejection can occur within 2 to 10 d following transplantation and involves cell-mediated immunity. Immunosuppression has been most valuable in the treatment of this type of rejection. In the late acute period, typically 11 or more d, vascular rejection can become an important component and may be cell-mediated or humoral. Humoral vascular rejection generally indicates a poor prognosis for the renal allograft. Chronic rejection is the cumulative effects of cell-mediated or humoral rejection occurring over a period of months to years. The terminology used in the classification of renal allograft rejection is somewhat misleading, since "acute" rejection may occur at any time during the post-transplantation period, and pathologic changes of "chronic" rejection may be observed as early as 1 mo following transplantation. The most useful cytohistologic findings in various types of renal allograft rejection are summarized in Table 13-1.

From a practical viewpoint, and on the basis of various urine sediment findings, acute rejection has been subdivided into cellular tubulointerstitial and vascular types, both of which can occur at the same time. The interpretation and differential diagnosis of morphological sediment findings are dependent upon progressive cellular changes documented by sequential sediment evaluations during the post-transplantation interval. For differential diagnostic purposes, the authors have elected to divide the post-transplantation period as follows: (1) immediate *(< 2 days),* intermediate *(days to months), and* late *(months to years).*

IMMEDIATE POST-TRANSPLANTATION PERIOD (FIG. 13-1)

The typical post-operative urine sediment findings are hematuria, variable inflammation sec-

TABLE 13-1
CYTOHISTOLOGIC FINDINGS IN RENAL ALLOGRAFT REJECTION

TYPE OF REJECTION	SUMMARY OF HISTOPATHOLOGIC FINDINGS	SUMMARY OF URINE CYTOPATHOLOGICAL FINDINGS
"HYPERACUTE" OR IMMEDIATE	Sludging of erythrocytes; neutrophilic margination in afferent arterioles and glomerular-peritubular capillaries, leading to vascular thromboses, interstitial hemorrhage, "fibrinoid" necrosis of glomeruli, and tubular necrosis (frequently cortical necrosis)	Amorphous and necrotic cellular debris Erythrocytic, blood, fibrin, and granular casts Degenerated and necrotic renal epithelial cells from convoluted tubules and collecting ducts
"ACUTE"		
Cellular tubulointerstitial	Polymorphous interstitial and perivascular infiltrate of "immunoblasts," mature lymphocytes, monocytes, and plasma cells with damage to capillary and venous endothelium and variable edema. Ischemia and infiltration of tubules leads to degeneration, necrosis, and sloughing of epithelium	Amorphous and necrotic cellular debris Moderate to marked exfoliation of renal epithelial cells from collecting ducts
Vascular	(Typically with component of acute cellular tubulointerstitial). Arterial and arteriolar mononuclear intimal infiltrate with necrotic or hyperplastic endothelium; ± mural fibrin thrombi; venous thromboses with interstitial hemorrhage common; ischemic tubular epithelial necrosis and TBM* ruptures conspicuous. With component of humoral vascular rejection, may have fibrinoid necrosis of arteries, arterioles, and glomeruli; ± patchy cortical necrosis or infarct	Renal epithelial, granular, blood, and waxy casts Erythrocytes and neutrophils Occasional lymphocytes and plasma cells Exfoliated renal epithelial fragments
"CHRONIC"	Tubular atrophy, interstitial fibrosis with scant mononuclear infiltrate; arterial-arteriolar fibromuscular intimal proliferation, ± foam cells; glomerular ischemic collapse with devascularized tufts *or* prominent "glomerulopathy" ("membranous" or "membranoproliferative" pattern)	Granular and waxy casts Occasional broad casts Mild exfoliation of renal epithelial cells from collecting ducts

* TBM = tubular basement membrane

ondary to ureteral anastomosis, and cast formation (hyaline and granular types). These sediment findings usually resolve within a 3-day period.

HYPERACUTE REJECTION

Hyperacute rejection generally results in rapid renal shutdown and, therefore, is not a major consideration in the urine sediment examination (Fig. 13-2). Renal biopsies taken minutes after anastamosis show intravascular coagulation with glomerular and arteriolar thrombosis and tubular necrosis. Because hyperacute rejection usually results in renal allograft infarction, the urine sediment findings of tubular injury and ischemic necrosis may be only transiently present (see Table 13-1).

PERFUSION NEPHROPATHY

Glomerular and tubular damage may result from maintaining donor kidneys on a perfusion pump (Fig. 13-3). The glomerular damage may result in leakage of blood, fibrin, and serum proteins. Therefore, the presence of blood, casts, and fibrin in the urine sediment would be expected. The additional sediment findings of increased renal epithelial exfoliation and other pathologic casts (renal epithelial and waxy) are suggestive of perfusion damage. This complication is usually manifested in urine sediment as tubular injury. A more complete discussion of urine sediment findings and differential diagnosis is given in Chapter 9.

ACUTE TUBULAR NECROSIS–ISCHEMIC NECROSIS

More severe tubular damage may result from prolonged warm-ischemia time to the donor kidney prior to or during transplantation or may reflect pre-existing ATN in the donor (Fig. 13-4). The urine sediment diagnosis of ATN is characterized by the increased exfoliation of degenerated and necrotic ("ghost") renal epithelial cells (Table 13-2). In addition, renal epithelial, coarsely granular, and waxy casts, amorphous and cellular debris, and mild inflammation may be present.

TABLE 13-2
DIFFERENTIAL URINE SEDIMENT CHARACTERISTICS DURING THE *IMMEDIATE* POST-TRANSPLANTATION PERIOD

URINE SEDIMENT PARAMETERS	TUBULAR INJURY ("ATN")*	ISCHEMIC NECROSIS
BACKGROUND	Amorphous and cellular debris	↑ Amorphous and cellular debris, occasionally fibrin
CELLULARITY	Mild to marked	Moderate to marked
ORGANISMS	None	None
HEMATOPOIETIC-INFLAMMATORY	Hematuria and mild acute (neutrophilic) inflammation	Hematuria and acute-chronic inflammation
EPITHELIUM	↑ Exfoliation of degenerated and necrotic *cells* of renal convoluted tubules and collecting ducts	↑ Exfoliation of degenerated and necrotic renal epithelial *cells* from collecting ducts. ↑ Exfoliation of renal epithelial *fragments*
CASTS	Granular, occasionally waxy, and rarely renal epithelial	Renal epithelial, blood, fibrin, granular and waxy, occasionally broad
INCLUSION-BEARING CELLS	None	Scattered degenerative-regenerative renal epithelial cells with intranuclear inclusions

* ATN = acute tubular necrosis

Rare renal epithelial fragments may be observed. The persistence of these fragments indicates severe renal parenchymal ischemia; for example, cortical necrosis or allograft infarct (see Table 13-2). Increased numbers of pathologic casts, especially the renal epithelial type, are also found. Ischemic necrosis in the immediate post-transplantation period often indicates surgical complications (see Fig. 13-1). Therefore, in the immediate post-transplantation period, one may see a spectrum of renal tubular epithelial exfoliation reflecting degrees of tubular damage that bridge ATN. Tubular injury and ischemic necrosis can be clearly distinguished by their morphological patterns in the urine sediment.

INTERMEDIATE POST-TRANSPLANTATION PERIOD (FIG.13-5)

PRINCIPLES OF CYTODIAGNOSIS OF ACUTE REJECTION

The urine sediment diagnosis of acute rejection requires the following: (1) identification of morphological entities, (2) a method for semiquantitative assessment, and (3) serial urine sediment examinations. By employing a Papanicolaou staining technique, urine sediment entities can be accurately identified. In combination with cytocentrifugation, this technique provides a means for semiquantitative assessment. Renal epithelial cells, lymphocytes, and casts are reported as number / 10 high-power fields. Random sediment fields should be counted with a minimum of two high-power fields at the edge of the sediment button.

A major criterion for the cytodiagnosis of acute rejection is the documentation of a progressive rise of exfoliated renal epithelial cells. Between 10 and 15 renal tubular epithelial cells per ten high-power fields with an associated dirty background and pathologic casts are sediment findings suspicious for acute rejection. With low counts or dilute urine specimens, only renal epithelial cells characteristic of collecting duct epithelium should be counted to ensure consistency of results (see Chap. 6). Urine sediments with greater than 20 renal epithelial cells per ten high-power fields with an associated dirty background, variable lymphocyturia, and pathologic casts constitute a diagnostic pattern of acute rejection (Table 13-3). In typical episodes of acute rejection, exfoliation of renal epithelial cells may exceed 100 cells per ten high-power fields. These high counts are classically associated with markedly degenerated or necrotic ("ghost") renal epithelial cells and renal fragments and are diagnostic of ischemic necrosis (see Table 13-3).

CYTOHISTOLOGIC CORRELATION OF ACUTE REJECTION

Acute renal allograft rejection may result in a complex histologic pattern that reflects pathogenetic mechanisms of injury. Acute cellular rejection is characterized by an interstitial mononuclear infiltrate and edema with tubular damage and exfoliation of renal epithelium. With severe acute cellular rejection, the diffuse dense infiltrate leads to a compromised peritubular capillary and venular circulation resulting in ischemic tubular necrosis (Fig. 13-6). With prompt therapy, the cellular infiltrate rapidly diminishes, indicating the generally favorable prognosis associated with this form of rejection. Within a 24 to 48 hr period of time, the urine sediment shows a corresponding decrease in exfoliated renal epithelial cells and a few residual casts. It should be emphasized that when ischemic necrosis is diagnosed in the urine sediment, it often represents a poor prognostic sign. It may indicate the presence of parenchymal renal vein thrombosis with hemorrhagic infarction or acute vascular rejection with patchy allograft infarction or cortical necrosis (see Fig. 13-5).

Acute vascular rejection is characterized by arterial and arteriolar damage. It usually results in more severe parenchymal ischemia with disruption of tubular basement membranes and exfoliation of renal tubular epithelium. Vascular rejection may be predominantly cell-mediated (Fig. 13-7) and, therefore, may be more responsive to immunosuppressive therapy. The presence of mural fibrin thrombi often correlates with humoral rejection and carries a worse prognosis (Fig. 13-8). In both instances, numerous exfoliated renal epithelial fragments will be present in the urine sediment. Glomerulopathy resulting from the vascular rejection with active coagulation cannot be detected in the urine sediment in the presence of the severe tubular disease (Fig. 13-9).

TABLE 13-3
DIFFERENTIAL URINE SEDIMENT CHARACTERISTICS DURING THE *INTERMEDIATE* POST-TRANSPLANTATION PERIOD

URINE SEDIMENT PARAMETERS	ACUTE CELLULAR REJECTION	ISCHEMIC NECROSIS
BACKGROUND	Amorphous and cellular debris	↑ Amorphous and cellular debris, occasionally fibrin
CELLULARITY	Mild to marked	Moderate to marked
ORGANISMS	None	None
HEMATOPOIETIC-INFLAMMATORY	Hematuria and acute-chronic inflammation, Occasional ↑ in lymphocytes and rare plasma cells	Hematuria and acute-chronic inflammation
EPITHELIUM	↑ Exfoliation of intact or degenerated renal epithelial *cells* from collecting ducts	↑ Exfoliation of degenerated and necrotic renal epithelial *cells* from collecting ducts. ↑ Exfoliation of renal epithelial *fragments*
CASTS	Renal epithelial, blood, granular, and waxy	Renal epithelial, blood, fibrin, granular, and waxy, occasionally broad
INCLUSION-BEARING CELLS	None	Scattered degenerative-regenerative renal epithelial cells with intranuclear inclusions

The identification of exfoliated renal epithelial fragments in urine sediment is diagnostic of ischemic necrosis (see Fig. 13-5). The differential cytodiagnosis of ischemic necrosis includes severe tubular injury (ATN) (Fig. 13-10), severe acute cellular rejection, vascular rejection with ischemic infarction (Fig. 13-11), and hemorrhagic infarction secondary to renal vein thrombosis (Fig. 13-12). The presence of exfoliated collecting duct epithelium along with large urothelial fragments suggests the possibility of renal papillary infarction (Fig. 13-13). Exfoliated deep urothelial cells and small fragments often reflect active rejection in the donor ureteral wall with damage to ureteral and periureteral vessels.

The persistence of ischemic necrosis in urine sediment examinations represents a poor prognostic sign and implies continued activity, progressive major parenchymal damage, and unresponsiveness to therapy (Fig. 13-14). By far, the most common cause of allograft failure and eventual nephrectomy is uncontrolled vascular rejection.

ADDITIONAL COMPLICATIONS

Major complications that involve renal allograft recipients during the intermediate post-transplantation period usually relate to drugs and infections. Drugs may cause tubular injury, toxic tubular necrosis, or allergic interstitial nephritis. These complications in the absence of rejection may give a characteristic sediment pattern (see Chap. 5, and Chap. 9). In immunosuppressed allograft recipients, the potential for bacterial, fungal, and especially viral infections, should be remembered.

LATE POST-TRANSPLANTATION PERIOD (FIG. 13-15)

CHRONIC REJECTION

Chronic rejection may present clinically months to years following renal transplantation as insidious progressive allograft dysfunction while on maintenance immunosuppressive therapy *or* as renal insufficiency secondary to one or more acute rejection episodes that have been variably modified by therapeutic modalities, including methylprednisone pulsing, antilymphocyte globulin (ALG), azothioprine, and radiation. Parenchymal damage results from the cumulative effects of cell-mediated and humoral rejection. Histologically, chronic vascular rejection is typically found, and is characterized by arterial myointimal proliferation and fibrosis (Fig. 13-16). Vascular compromise results in glomerular ischemic collapse and sclerosis, interstitial fibrosis, and tubular atrophy. A mononuclear interstitial inflammatory infiltrate is typically present, but it is often scant and located around veins at the corticomedullary junction. As a reflection of chronic parenchymal disease, the urine sediment will often contain granular and waxy casts, including occasional broad forms, with mild renal epithelial exfoliation (see Table 13-1).

The histological changes of chronic vascular rejection, which may be noted in allograft biopsies as early as 3 to 4 weeks post-transplanation, are often accompanied by active ("acute") vascular, as well as variable cellular tubulointerstitial, rejection (see Fig. 13-11). Transplant glomerulopathy (see Table 13-1) is occasionally a prominent feature and may be related to immune complex injury or renal intravascular coagulation. The urine sediment pattern is often masked by the findings of ischemic necrosis.

Acute rejection episodes may occur during the late post-transplantation period and may cause sudden allograft dysfunction with increased urine sediment activity, suggesting acute rejection (Fig. 13-17). Sudden cessation of immunosuppression therapy, or a rapid tapering due to concurrent infection, may result in a severe acute rejection reaction. In this clinical setting, there is frequently a component of vascular rejection, and the urine sediment is likely to show evidence of ischemic necrosis (Table 13-4).

ADDITIONAL COMPLICATIONS

Drug-related complications, metabolic conditions—such as nephrocalcinosis (Fig. 13-18), and infections may also occur during the late post-transplantation period. Unexplained urine sedi-

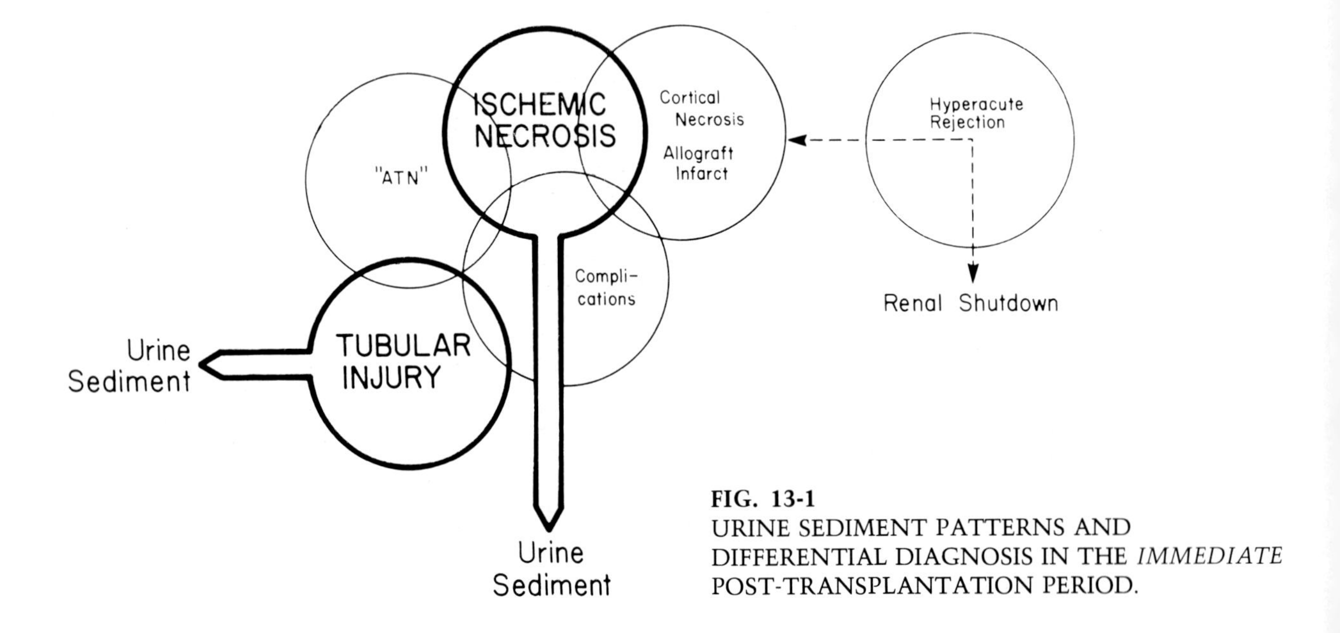

FIG. 13-1
URINE SEDIMENT PATTERNS AND DIFFERENTIAL DIAGNOSIS IN THE *IMMEDIATE* POST-TRANSPLANTATION PERIOD.

TABLE 13-4
DIFFERENTIAL URINE SEDIMENT CHARACTERISTICS DURING THE *LATE* POST-TRANSPLANTATION PERIOD

URINE SEDIMENT PARAMETERS	ACUTE CELLULAR REJECTION	CHRONIC REJECTION	ISCHEMIC NECROSIS
BACKGROUND	Amorphous and cellular debris	Clean to slightly amorphous	↑ Amorphous and cellular debris, occasionally fibrin
CELLULARITY	Mild to marked	Mild to moderate	Moderate to marked
ORGANISMS	None	None	None
HEMATOPOIETIC-INFLAMMATORY	Hematuria and acute-chronic inflammation Occasional ↑ in lymphocytes and rare plasma cells	Hematuria and mild Acute-chronic inflammation	Hematuria and acute-chronic inflammation
EPITHELIUM	↑ Exfoliation of intact or degenerated renal epithelial *cells* from collecting ducts	Slight ↑ in degenerated renal epithelial cells from the collecting duct	↑ Exfoliation of degenerated and necrotic renal epithelial *cells* from collecting ducts. ↑ Exfoliation of renal epithelial *fragments*
CASTS	Renal epithelial, blood, granular, and waxy	Granular and waxy, occasionally broad	Renal epithelial, blood, fibrin, granular, and waxy, occasionally broad
INCLUSION-BEARING CELLS	None	None	Scattered degenerative-regenerative renal epithelial cells with intranuclear inclusions

ment findings suggesting a "glomerulopathy" invokes a new differential diagnosis—transplant glomerulopathy versus recurrent or *de novo* glomerulonephritis. Some glomerular lesions, such as focal segmental hyalinosis, may recur in renal allografts with alarming rapidity and present as massive proteinuria early (days to weeks) in the post-transplantation period. Others, such as dense-deposit disease, may be pathologically documented by renal biopsy prior to the appearance of clinical manifestations. In the late post-transplantation period, unexplained marked proteinuria or hematuria with erythrocytic or blood casts suggests a complicating glomerular lesion. Renal biopsy, with immunofluorescence and ultrastructural studies, is often required to establish a diagnosis of recurrent or *de novo* glomerulonephritis (Fig. 13-19). A discussion of the morphological features of glomerulonephritis is given in Chapter 12 (Fig. 12-5).

BIBLIOGRAPHY

Bossen EH, Johnston WW, Amatulli J, et al: Exfoliative cytopathologic studies in organ transplantation. III. The cytologic profile of urine during acute renal allograft rejection. Acta Cytol 14:176–181, 1970

Herbertson BH, Evans DB, Calne RY, et al: Percutaneous needle biopsies of renal allografts: The relationship between morphological changes present in biopsies and subsequent allograft function. Histopathy 1:161–178, 1977

Porter KA: Rejection in treated renal allografts. Symposium on Tissue Organ Transplant. J Clin Pathol (Suppl) 20:518, 1967

Schumann GB, Burleson RL, Henry JB, et al: Urinary cytodiagnosis of acute renal allograft rejection using the cytocentrifuge. Am J Clin Pathol 67:134–140, 1977

Taft PD, Flax MH: Urine cytology in renal transplantation: Association of renal tubular cells and allograft rejection. Transplant 4:194–204, 1966

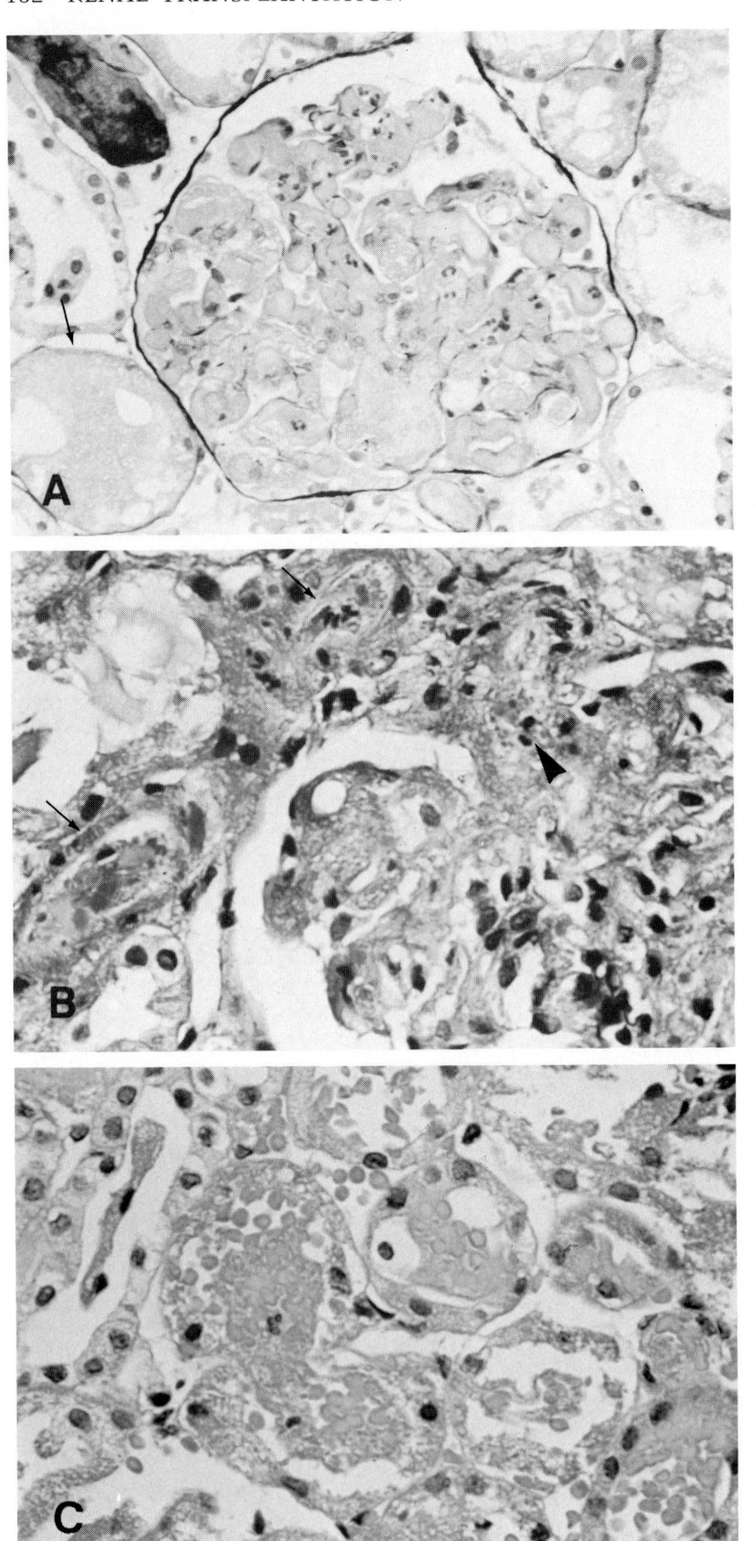

FIG. 13-2
HYPERACUTE REJECTION. **(A)** Glomerular capillaries are occluded by thrombotic material and contain numerous marginating neutrophilic leukocytes. Mesangial and endothelial cellularity appear diminished, and karyorrhexis is present. A tubule that has partially lost its epithelium contains vacuolated mucoprotein substance (arrow) (Jones × 250) **(B)** This hypocellular glomerular segment contains prominent subendothelial and intracapillary fuchsinophilic thrombi. Similar thrombotic material is present in the axial pole (arrowhead) and the afferent arteriole, which has swollen or necrotic endothelium and medial necrosis (arrows). (Trichrome × 400) **(C)** Tubules contain erythrocytes and forming erythrocytic casts. Cytoplasmic degenerative changes, nuclear pyknosis, and focal loss of epithelium are present. Peritubular capillaries are dilated but contain no neutrophilic leukocytes in this field. (× 250)

(continued on p. 155)

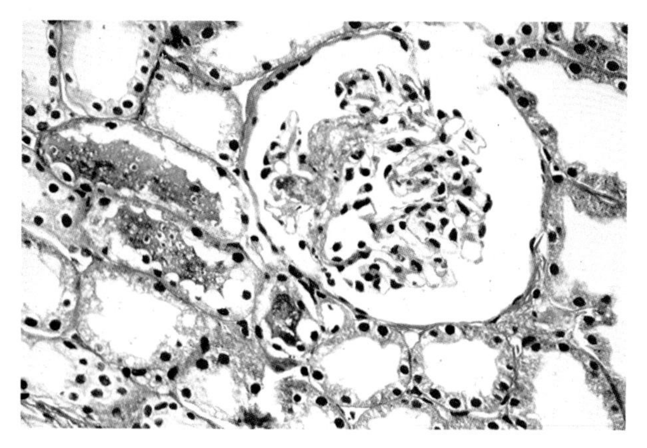

FIG. 13-3A

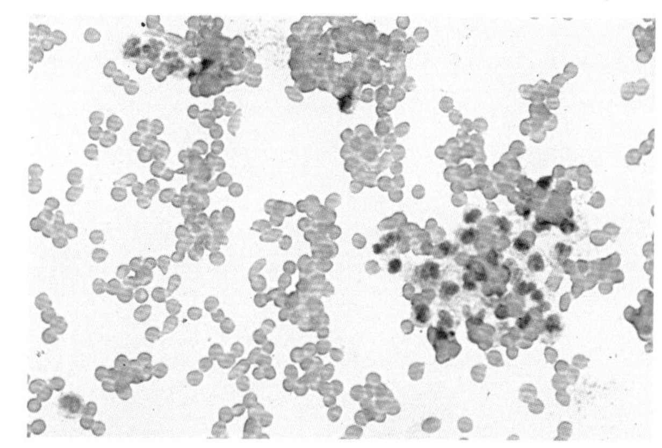

FIG. 13-3B

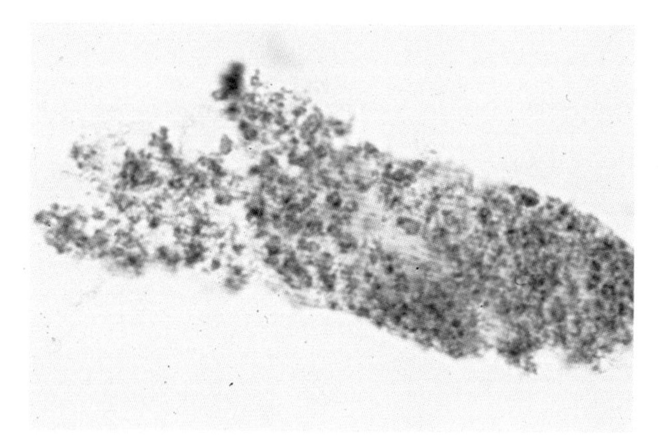

FIG. 13-3C

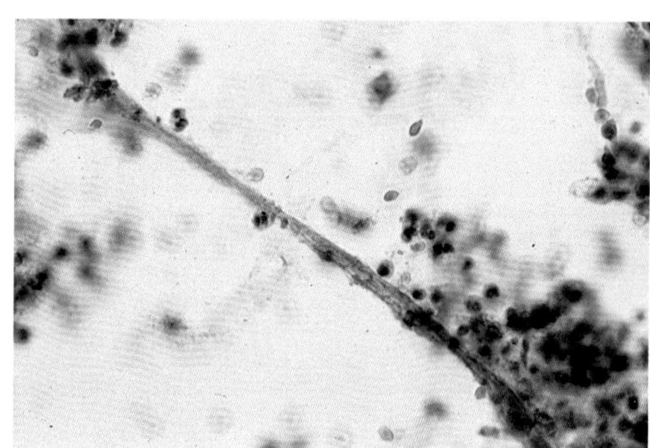

FIG. 13-3D

FIG. 13-3
PERFUSION NEPHROPATHY. **(A)** Glomerulus with a large granular-fibrillar Lendrum-positive thrombus has slight collapse of capillary loops, which appear bloodless. Glomerular and peritubular capillaries do not contain neutrophilic leukocytes. Granular casts are present within proximal convoluted tubules, which have a vacuolated epithelium. (Lendrum × 250) **(B)** Marked hematuria and neutrophilic clump. (× 400) **(C)** Granular cast with probable blood pigment. (× 1000) **(D)** Fibrin strand. (× 1000) The presence of blood and coagulation products with renal epithelial exfoliation indicate a glomerular leak.

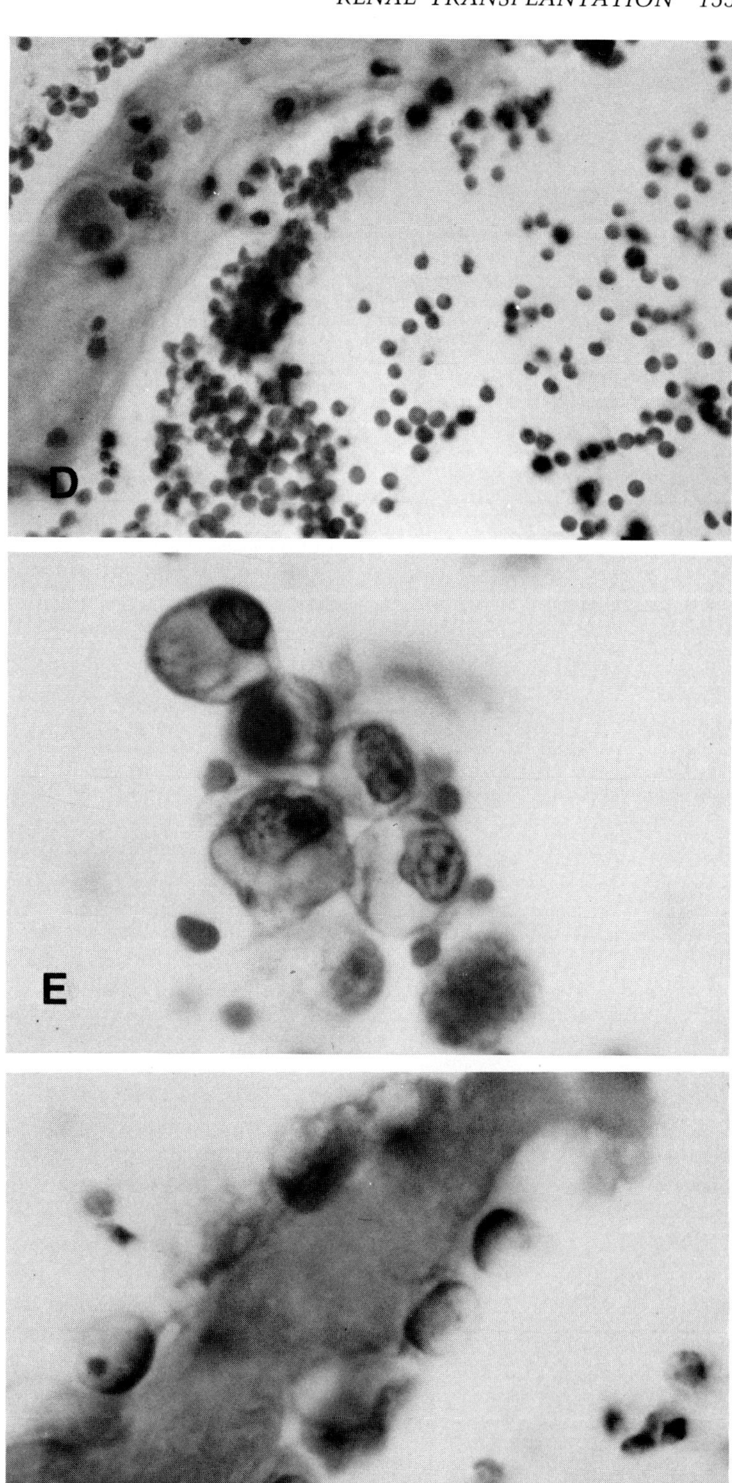

FIG 13-2D-F *,continued.*
(D) Background containing numerous erythrocytes and fibrillar (fibrin) cast in urine sediment. (× 400) **(E)** Loose cluster of vacuolated epithelial cells in urine sediment. These reactive cells with degenerative vacuoles are probably of ductal origin. (× 400) **(F)** Epithelial fragment encasing a waxy cast. This structure is commonly seen in ischemic necrosis. (× 400)

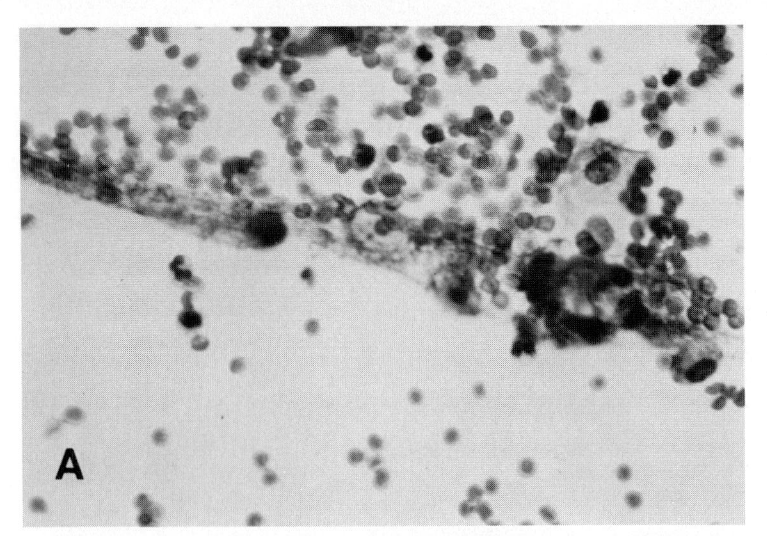

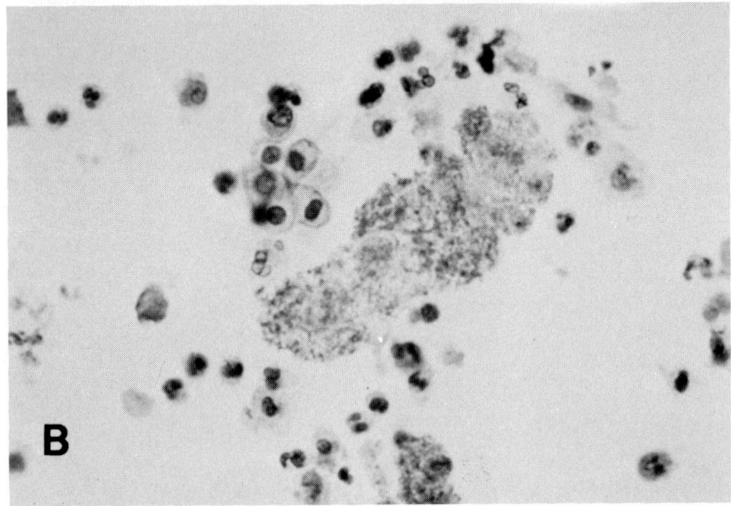

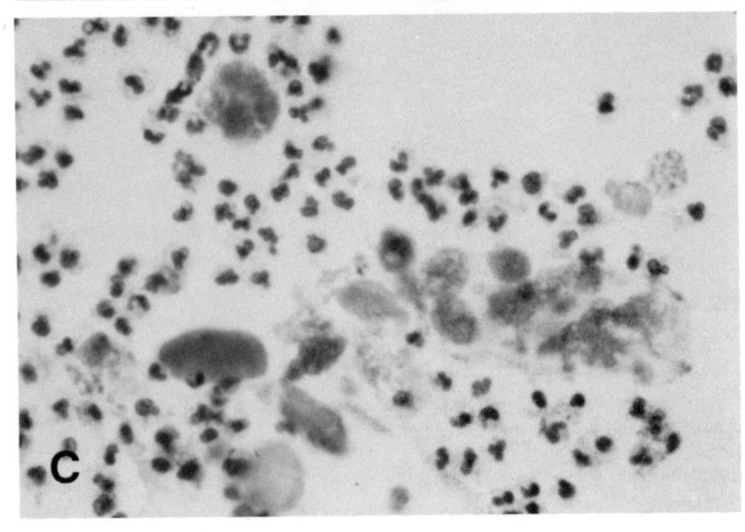

FIG. 13-4
ACUTE TUBULAR NECROSIS. (**A**) Numerous erythrocytes and fibrillar material (fibrin) in urinary sediment. (× 400) (**B**) Urine sediment with slightly dirty background containing acute inflammatory cells, renal epithelial clusters, and granular casts. (× 400) (**C**) Urine sediment with marked acute inflammation and a necrotic renal epithelial cast. This cast is characteristic of tubular necrosis (ischemic necrosis). (× 400)

(continued)

FIG. 13-4D-F *(continued)*
(D) Cylindrical renal epithelial fragment. Renal fragments present in sediment indicate ischemic necrosis. (× 1000) **(E)** Urine sediment with dirty background and numerous broad granular to waxy casts. (× 400) **(F)** Broad degenerating renal epithelial cast undergoing transition to a waxy cast. This pathologic cast indicates tubular injury and stasis in dilated tubules or terminal collecting duct. (× 1000)

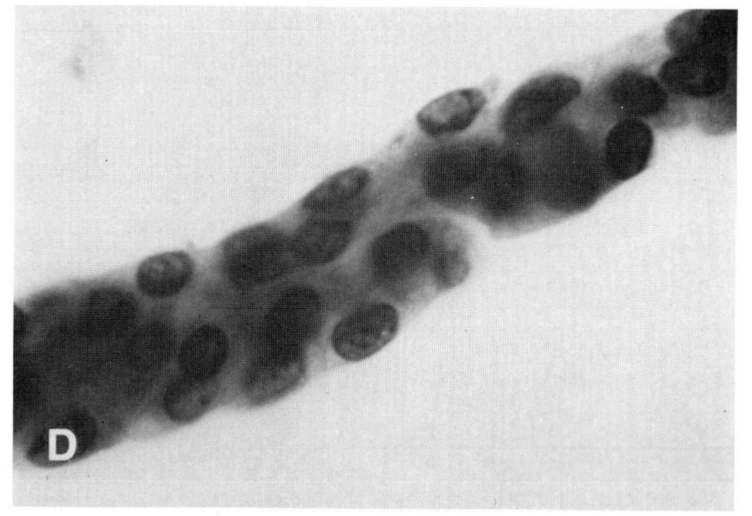

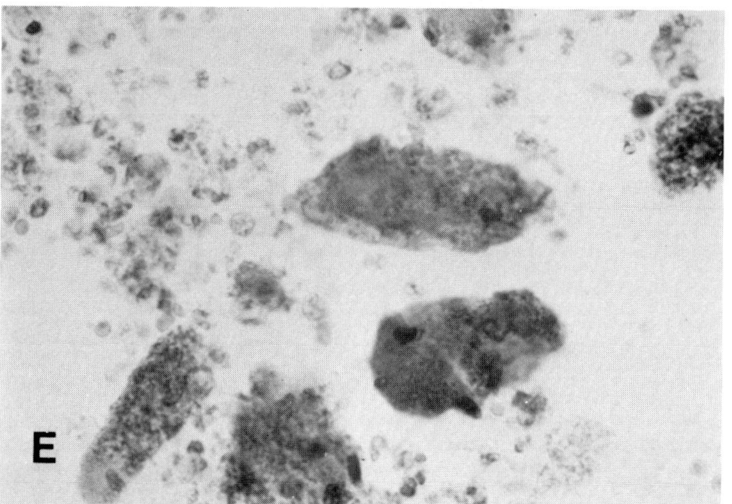

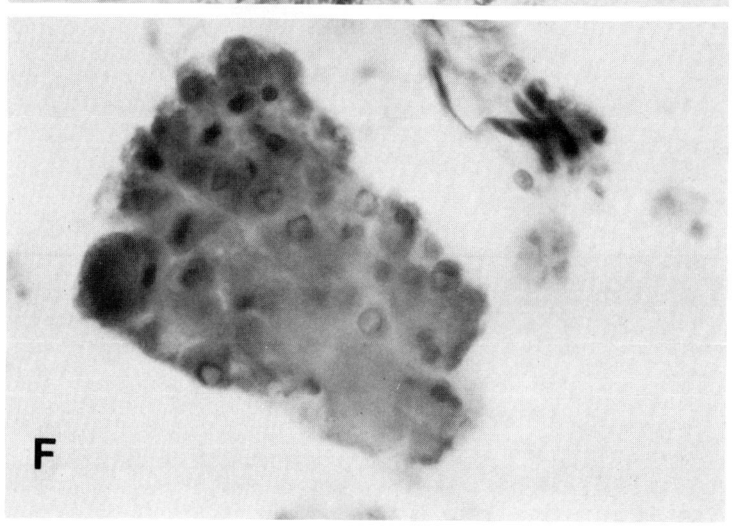

(continued)

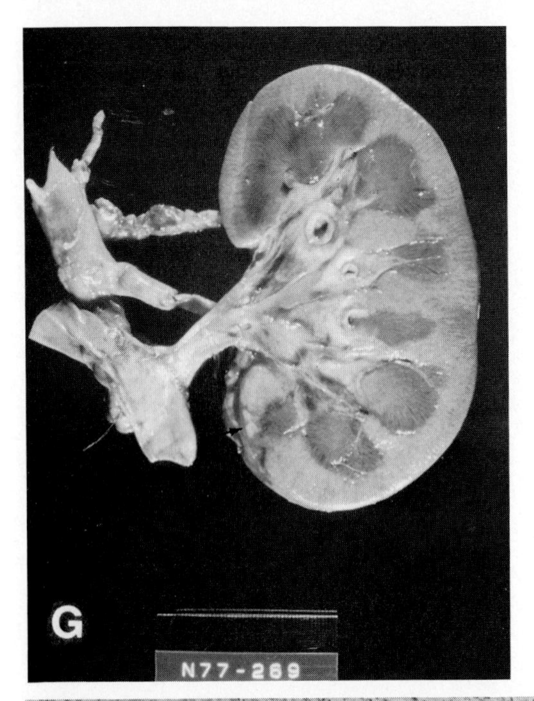

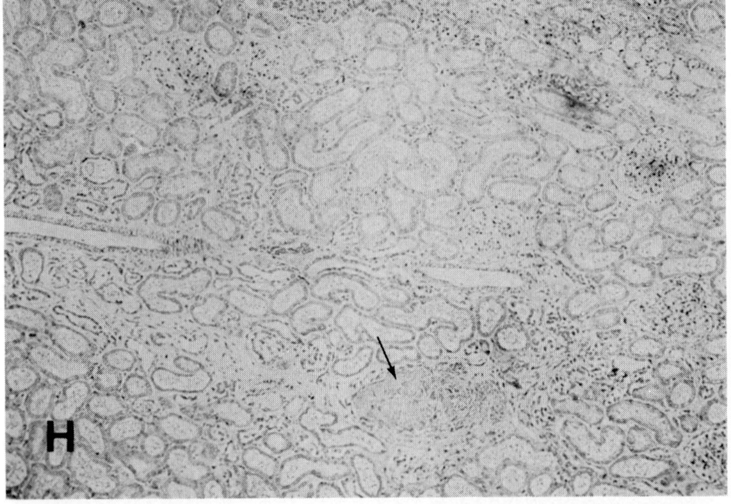

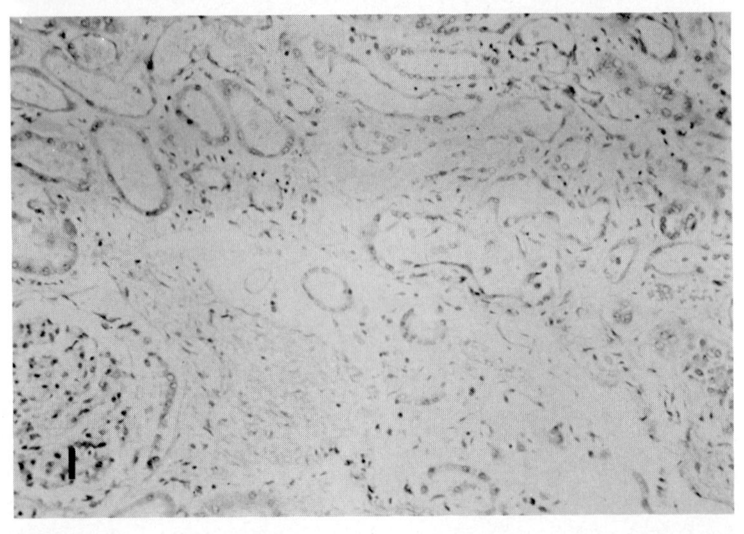

FIG. 13-4G-I *(continued)*
(G) Transplant kidney at autopsy. The pale edematous bulging cortices are sharply demarcated from the congested medulla. A small discrete infarct is present in the lower pole (arrow; gross). **(H)** Tubules are dilated, angulated, and lined by an attenuated epithelium. There is focal rupture of tubules with interstitial deposits of intraluminal contents (Tamm-Horsfall protein; arrow). (× 400) **(I)** Juxtamedullary region with tubular damage and interstitial edema. A perivascular cellular infiltrate typical of rejection is absent (× 160)

(continued)

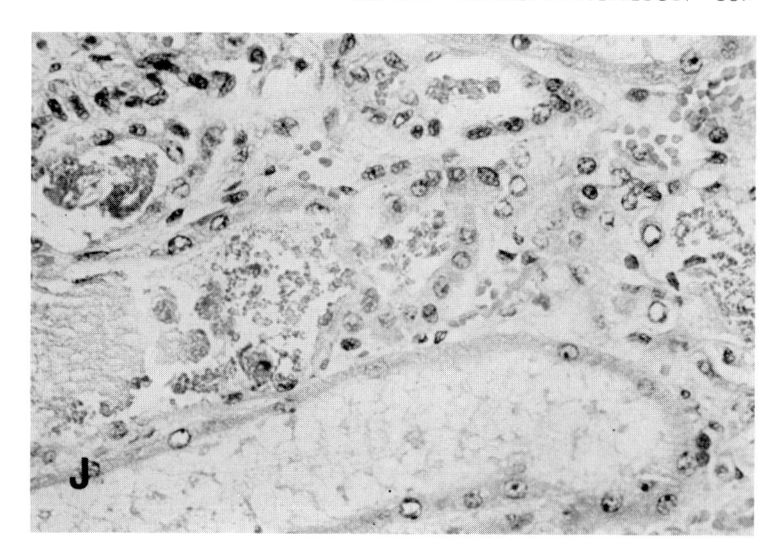

FIG. 13-4J *(continued)*
(J) Exfoliated renal epithelium is degenerated and necrotic; granular casts are forming. The intact, attenuated epithelium is regenerative. (× 250)

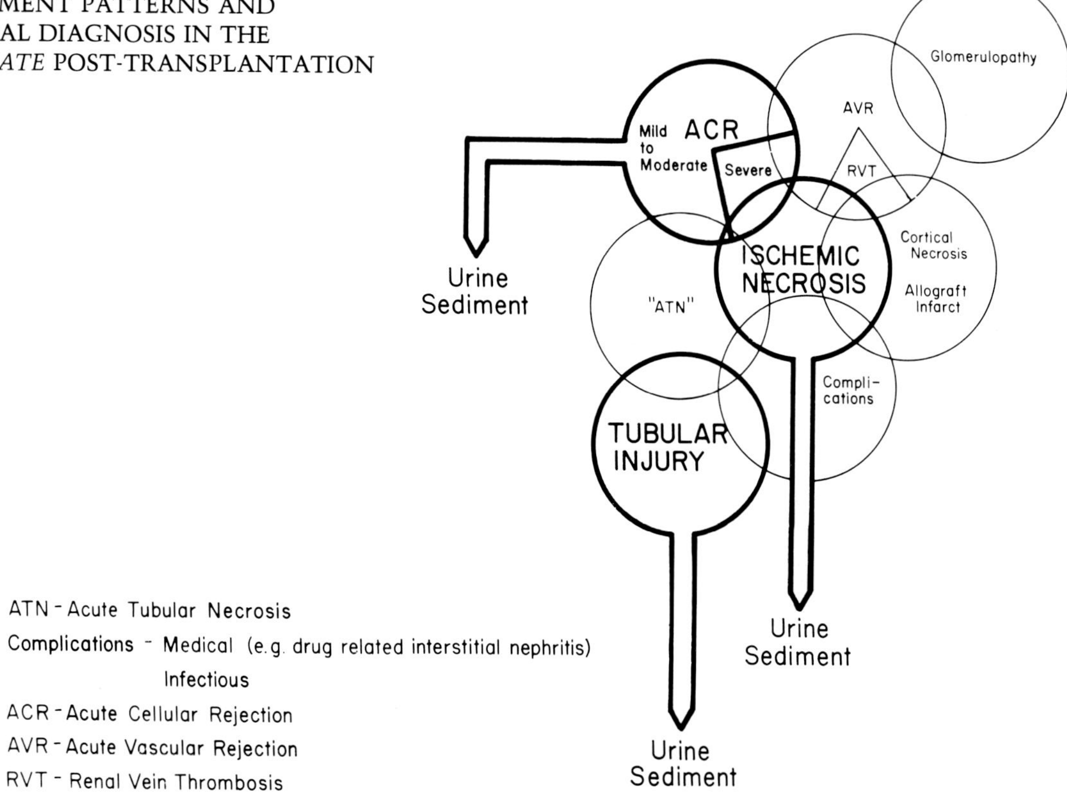

FIG. 13-5
URINE SEDIMENT PATTERNS AND DIFFERENTIAL DIAGNOSIS IN THE *INTERMEDIATE* POST-TRANSPLANTATION PERIOD.

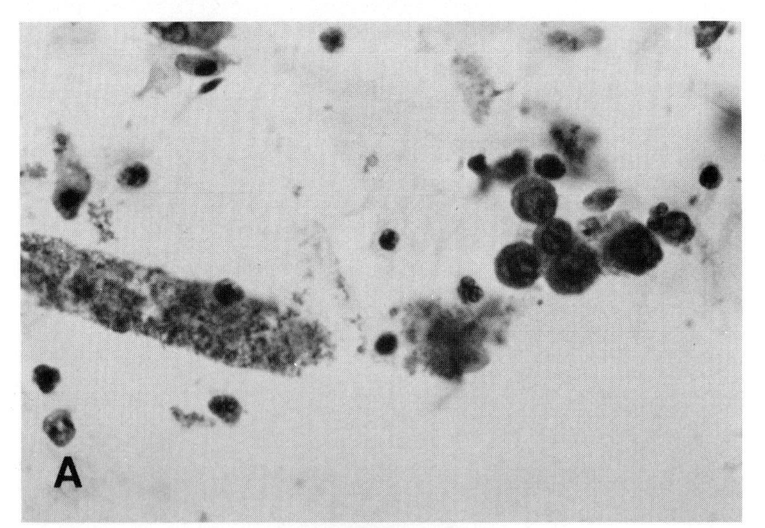

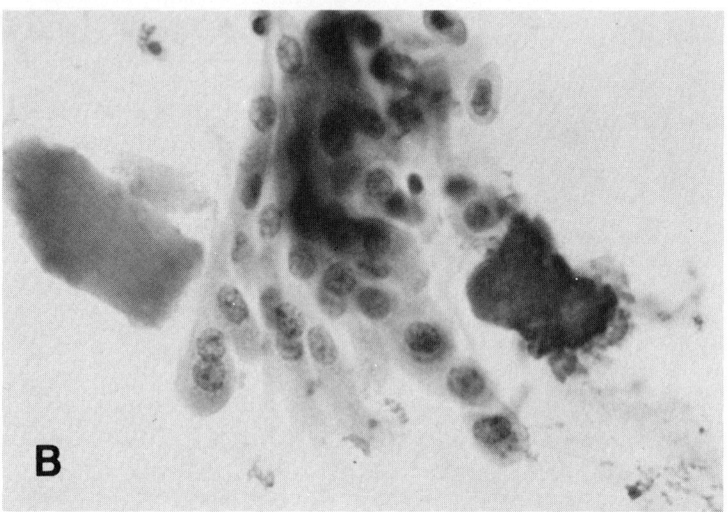

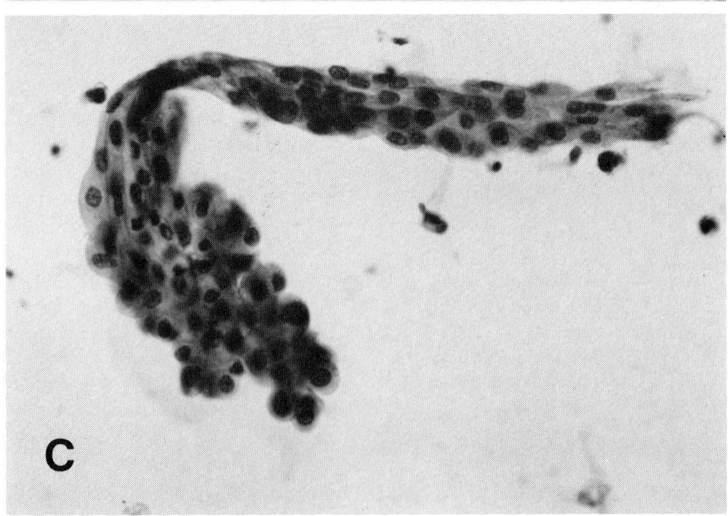

FIG. 13-6
ACUTE CELLULAR REJECTION. **(A)** Urine sediment with dirty background, increased numbers of renal tubular epithelial cells, and a granular cast. (× 400) **(B)** Waxy casts and large epithelial fragments. (× 400) **(C)** Exfoliated cylindrical renal epithelial fragment from terminal collecting duct in urine sediment. (× 400)

(continued)

FIG. 13-6D-F *(continued)*
(D) Renal fragments in urine sediment following acute rejection indicate ischemic necrosis. (× 1000) **(E)** The edematous interstitium contains a moderate inflammatory infiltrate. Tubules are slightly dilated, frequently have flattened epithelium, and contain exfoliated cells and casts. (× 100) **(F)** Interstitial inflammation with infiltration of tubules and basement membrane disruption. Note intraluminal dense hyaline casts. (× 250)

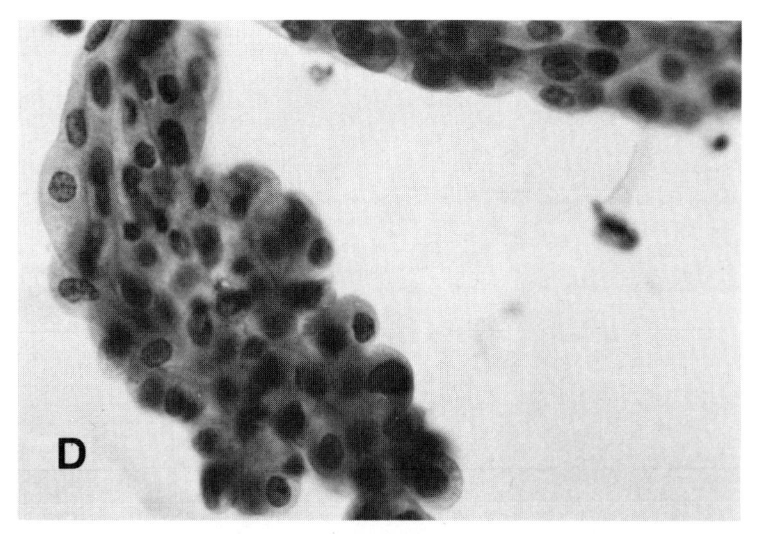

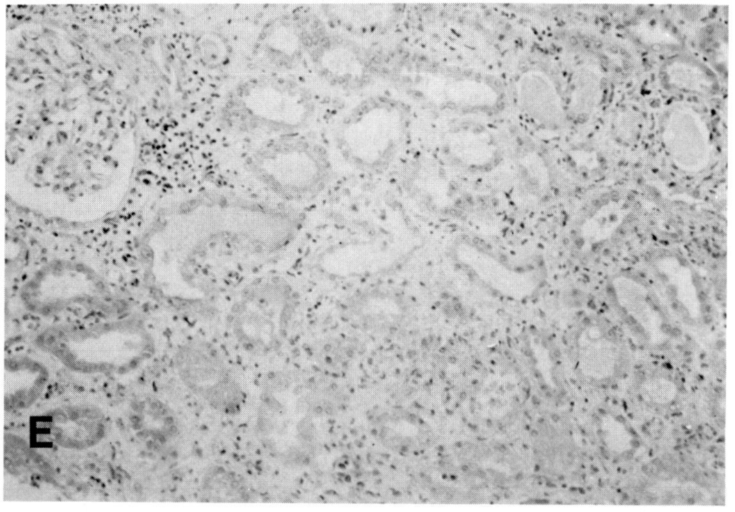

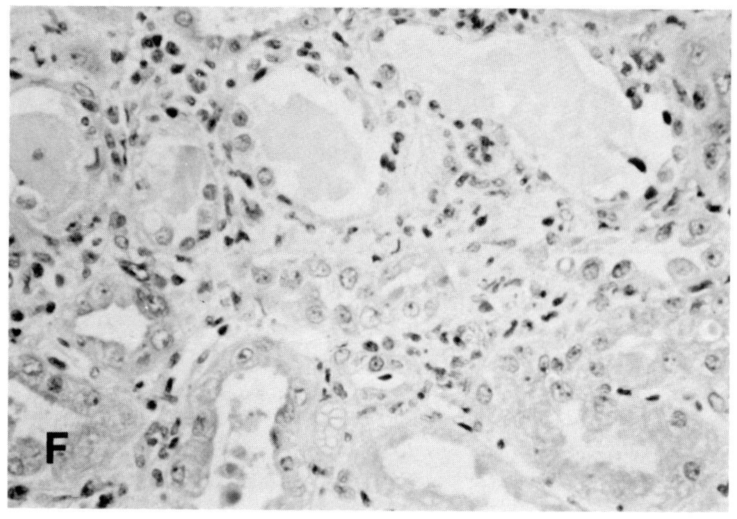

(continued)

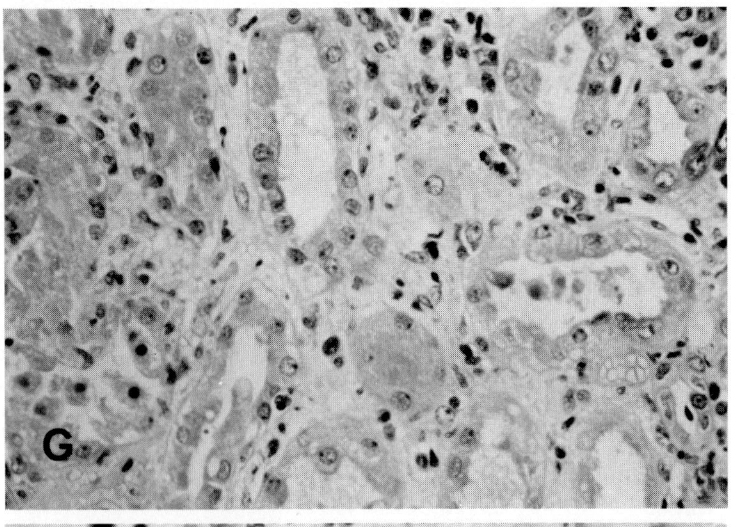

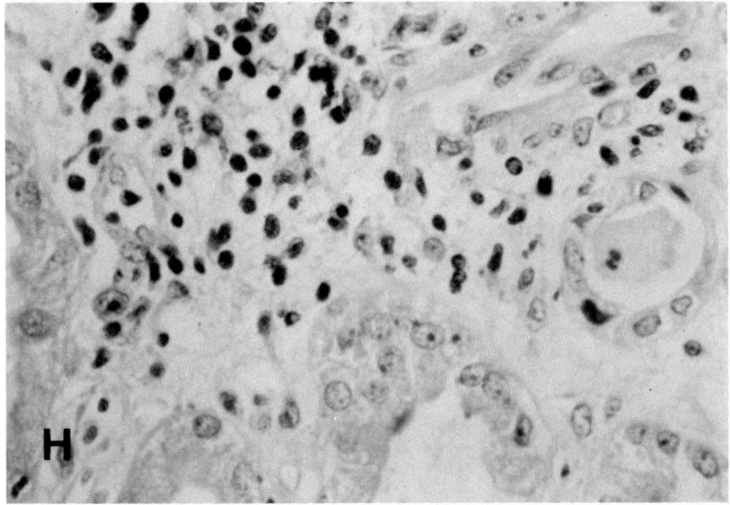

FIG. 13-6G,H *(continued)*
(**G**) Exfoliation of renal epithelium with nuclear pyknosis, karyolysis, and numerous "ghost" cells. Intact epithelium appears regenerative. (× 250) (**H**) The interstitial cellular infiltrates consists of lymphoblasts, lymphocytes, and few histiocytes. (× 400)

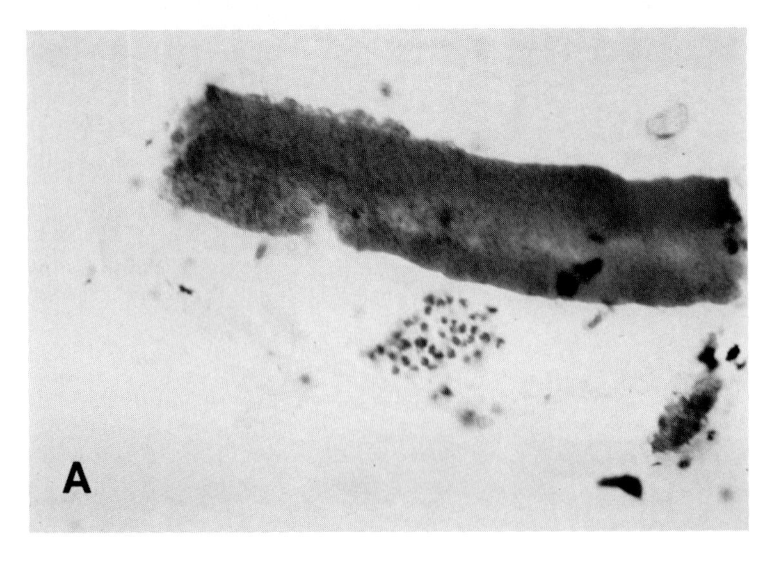

FIG. 13-7
SEVERE ACUTE REJECTION WITH ISCHEMIC NECROSIS. (**A**) Blood cast in urine sediment. (× 400)

(continued)

FIG. 13-7B-D *(continued)*
(B) Urine sediment with renal epithelial cast and several individual degenerated renal epithelial cells. (× 400) **(C)** Degenerating convoluted renal epithelial cast disintegrating into a waxy cast. (× 400) **(D)** Partially homogenized waxy cast in urine sediment. Note brittle nature of this type of pathologic cast. (× 400)

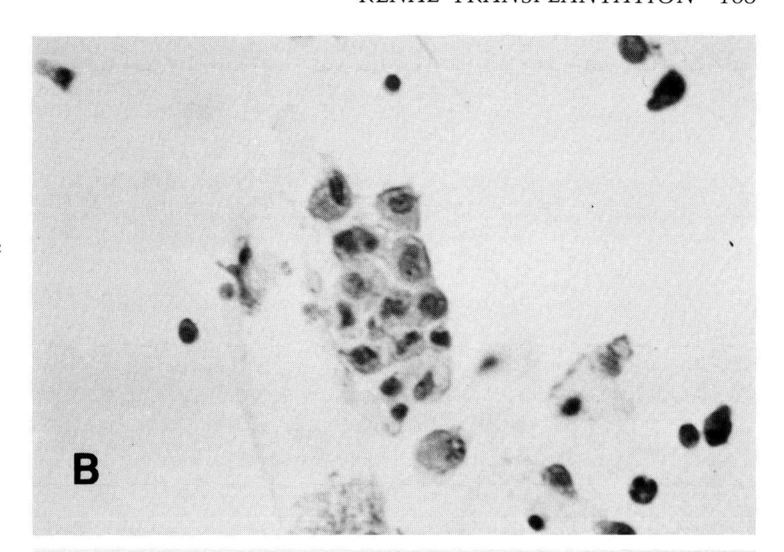

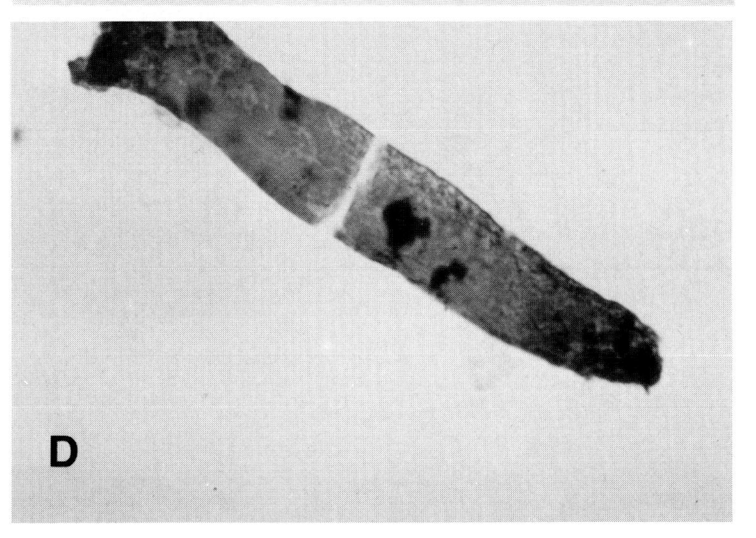

(continued)

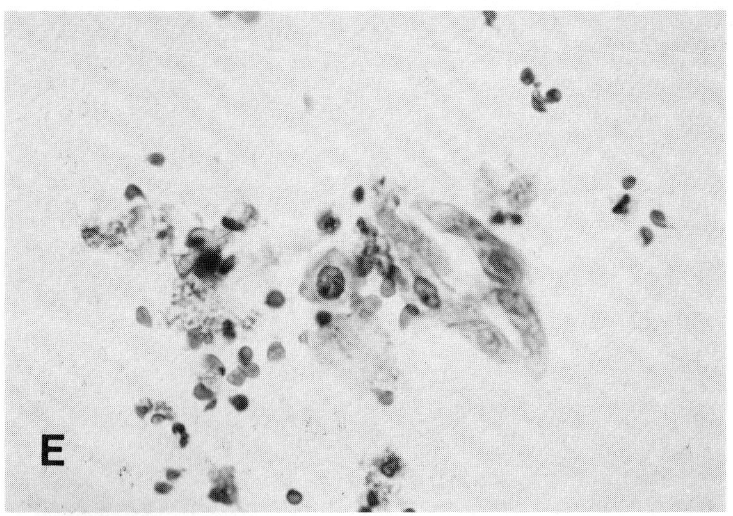

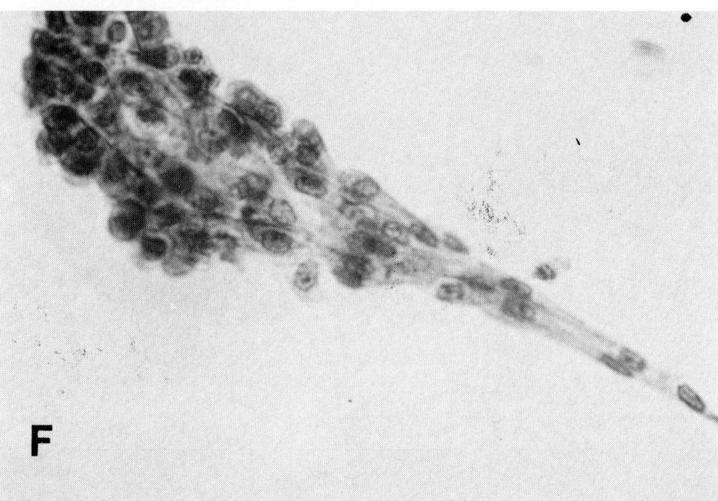

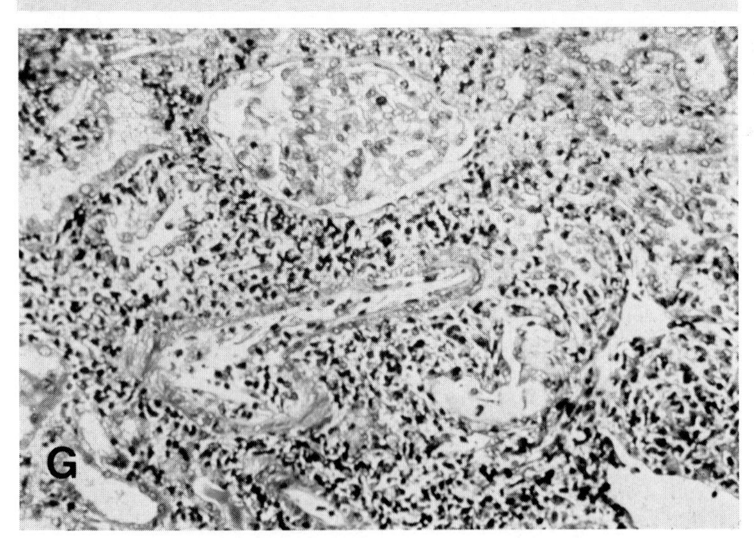

FIG. 13-7E-G *(continued)*
(E) Urine sediment with erythrocytes and degenerating and necrotic epithelial cells. (× 400) **(F)** Large sheet of epithelium with elongated cells. Numerous epithelial fragments are typically seen in transplant patients with ischemic necrosis. (× 400) **(G)** Acute cellular rejection with vascular component. A dense perivascular interstitial inflammatory infiltrate is present, and numerous tubules have disrupted basement membranes. An interlobular artery also has a mononuclear intimal infiltrate. (Trichrome × 160)

(continued)

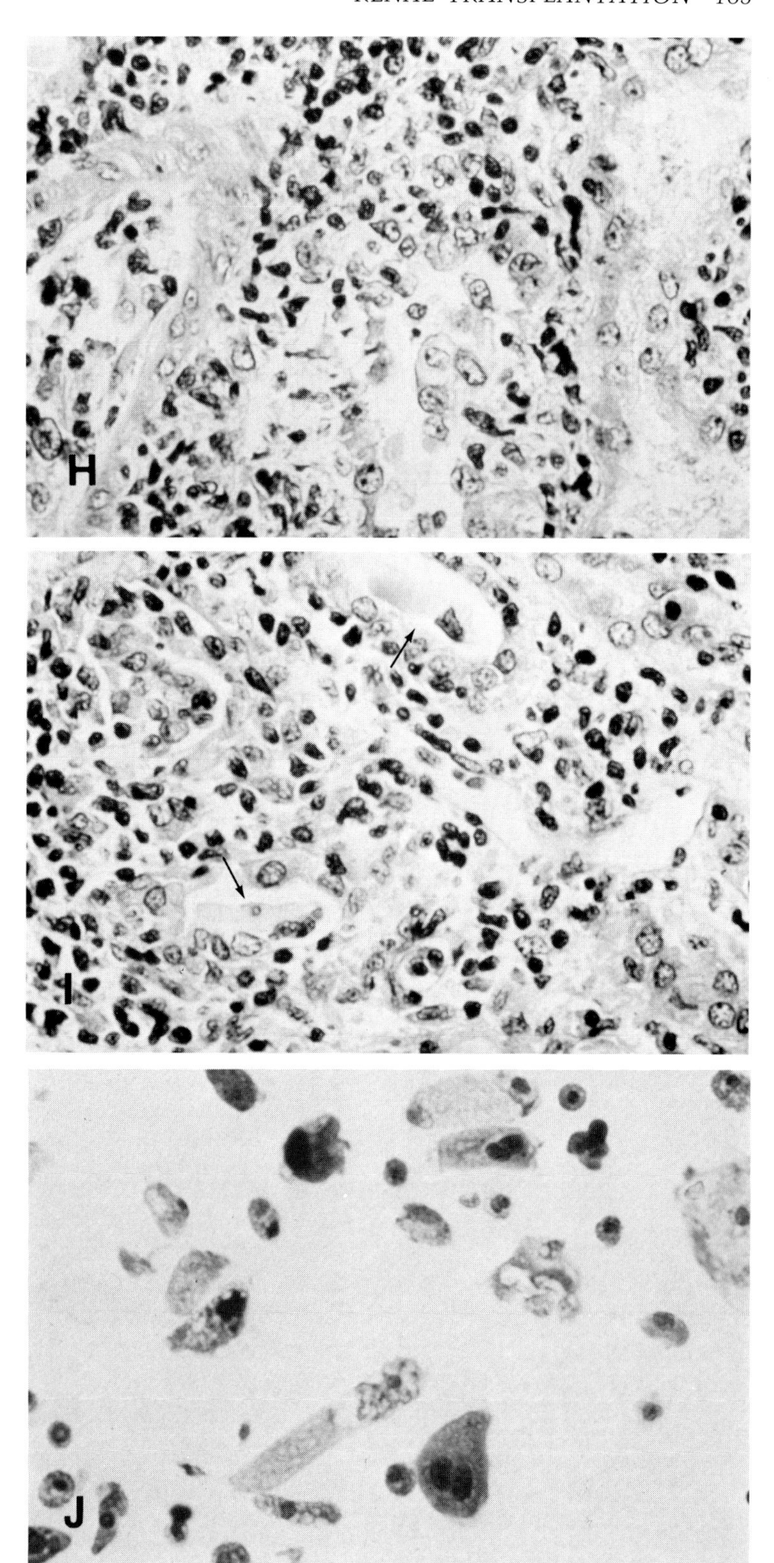

FIG. 13-7H-J *(continued)*
(H) An interlobular artery has a lymphocytic-mononuclear intimal infiltrate. Note the adjacent disrupted tubule with markedly reactive epithelium and the dense interstitial lymphocytic infiltrate. (× 400) **(I)** Tubules are disrupted by the marked inflammatory infiltrate of lymphocytes with few lymphoblasts and occasional histiocytes and plasma cells. Exfoliation of degenerating renal epithelium is striking, and cytologic changes secondary to regeneration are prominent. Occasional exfoliated cells appear to have waxy castlike material in the cytoplasm (arrows). (× 400) **(J)** Urine sediment with clean background, numerous bizarre epithelial cells with nuclear enlargement, prominent nucleoli; and abundant cytoplasm. Note the necrotic smaller cells in the background. The bizarre epithelial cells represent exfoliated regenerative epithelium. (× 400)

(continued)

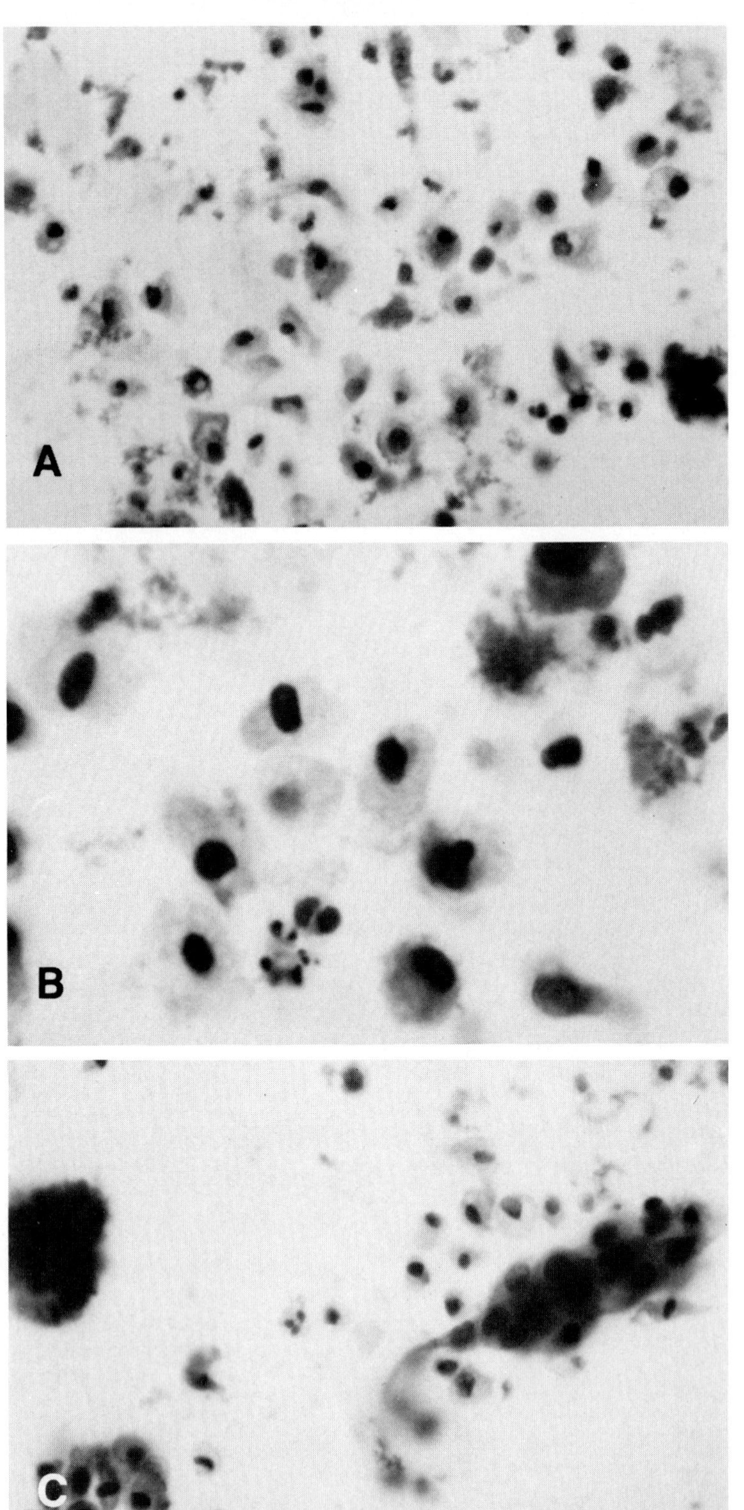

FIG. 13-8
ACUTE VASCULAR REJECTION WITH ISCHEMIC TUBULAR NECROSIS. **(A)** Urine sediment with dirty background containing cellular debris and numerous exfoliated renal tubular epithelial cells. (× 400) **(B)** Slightly degenerated renal tubular epithelial cells, probably of early collecting duct origin. Increased exfoliation of renal cells is an important sediment parameter to follow in diagnosing rejection. Renal epithelial cells should not be mistaken for lymphocytes. (× 100) **(C)** Cylindrical renal epithelial fragment in the urine sediment. The presence of fragments usually follows renal tubular cell exfoliation and pathologic cast formation. (× 400)

(continued)

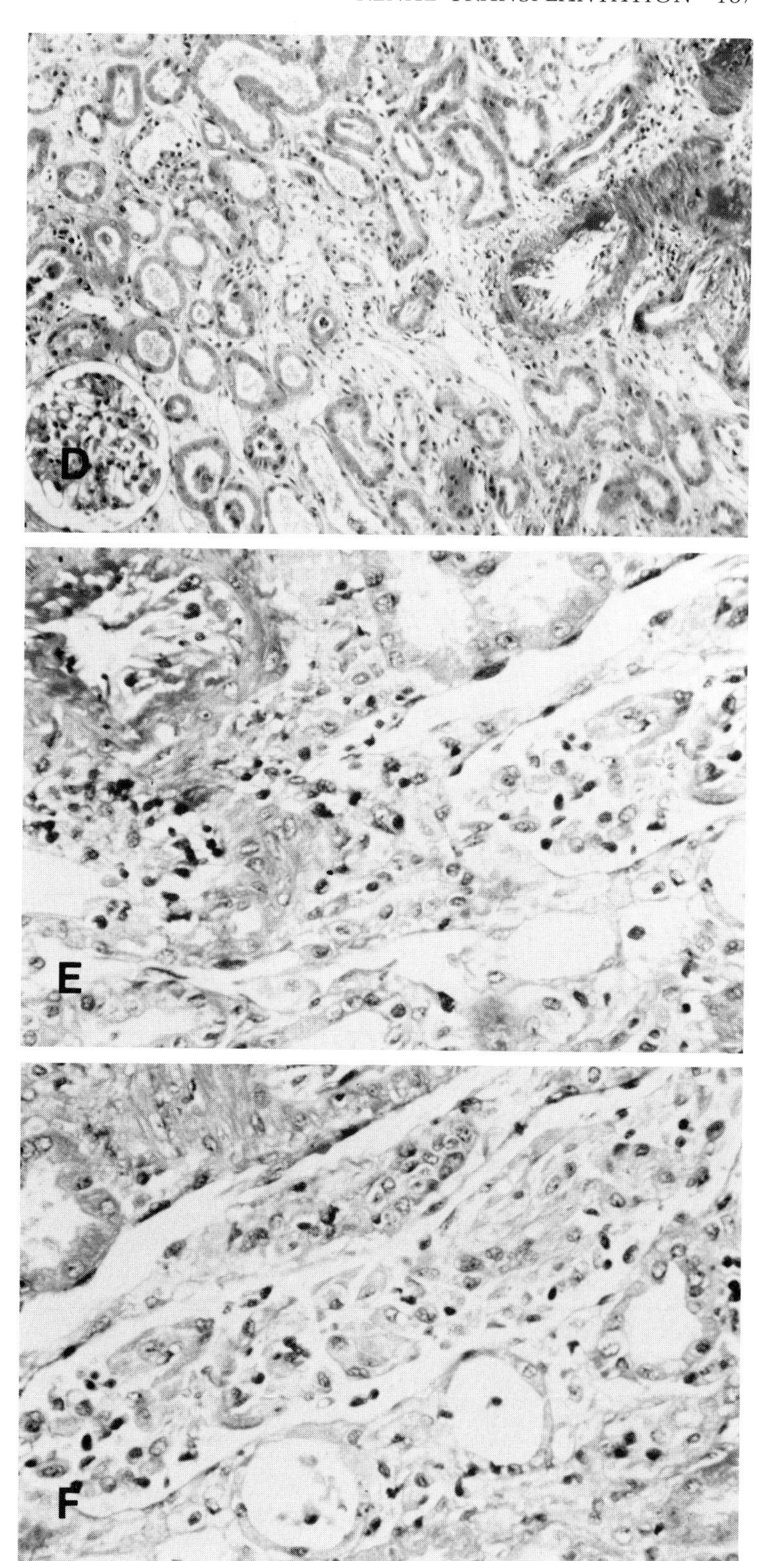

FIG. 13-8D-F *(continued)*
(D) Renal biopsy. Arterial branches have mural fibrin deposition with endothelial swelling, proliferation, and necrosis. The edematous interstitium contains a mild lymphocytic (plasmacytic) inflammatory infiltrate. (Trichrome × 100) **(E)** The presence of a scant interstitial infiltrate may reflect prebiopsy therapy of acute cellular rejection. (Trichrome × 250) **(F)** The presence of sheets, clusters, and individual renal epithelial cells within an apparently dilated venule may be artifactual, but demonstrates the marked degree of renal tubular cell exfoliation possible with acute vascular rejection. (Trichrome × 250)

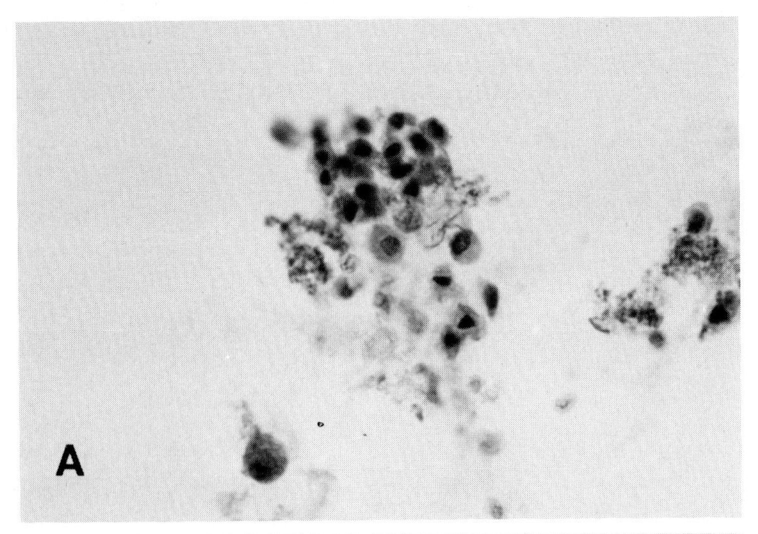

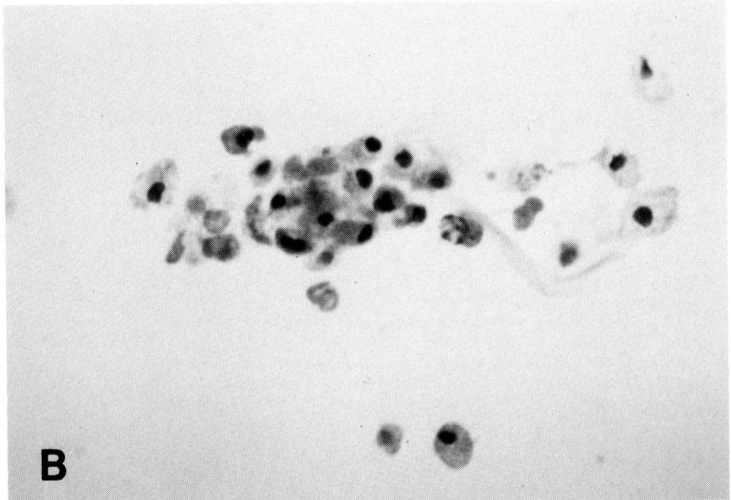

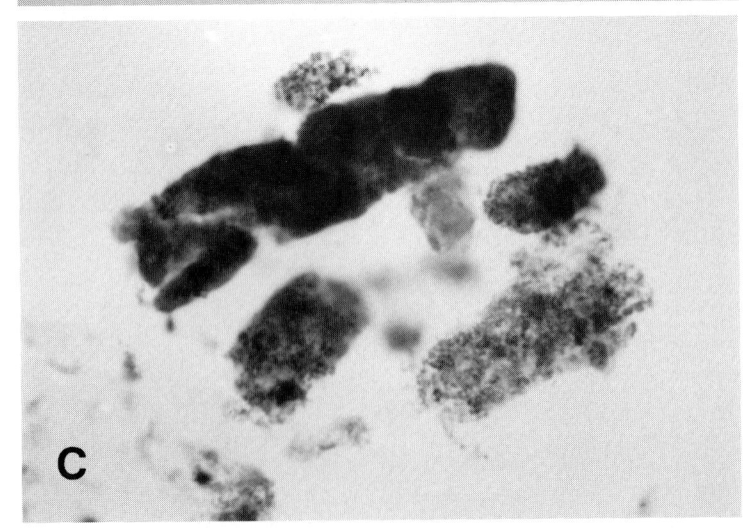

FIG. 13-9
ACUTE CELLULAR AND VASCULAR REJECTION WITH GLOMERULOPATHY AND FOCAL ISCHEMIC TUBULAR EPITHELIAL NECROSIS. **(A)** Urine sediment with loose cluster of renal tubular epithelial cells and cellular debris in background. (× 400) **(B)** Renal epithelial cast. Note granular cast matrix. **(C)** Coarse granular casts. (× 400)

(continued)

FIG. 13-9D-E *(continued)*
(D) Broad coarse granular cast indicating severe renal parenchymal disease with stasis in terminal collecting duct or tubular dilatation. (× 1000) **(E)** Allograft nephrectomy during episode of severe acute cellular rejection. The kidney is markedly enlarged and has a pale, swollen cortex. (Gross)

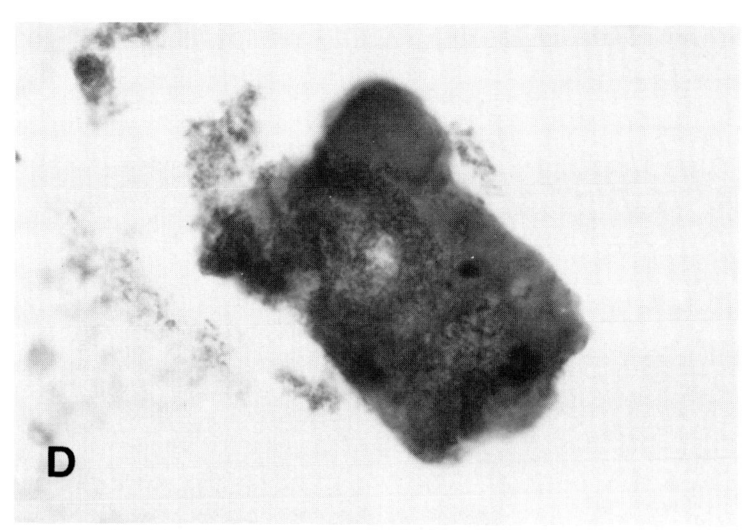

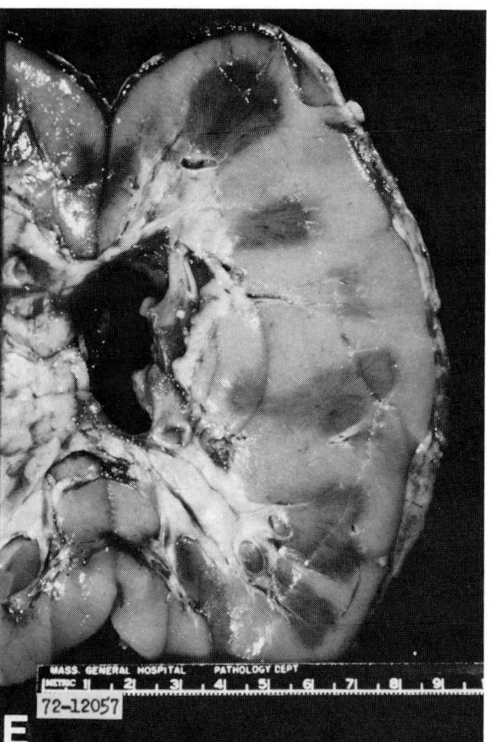

(continued)

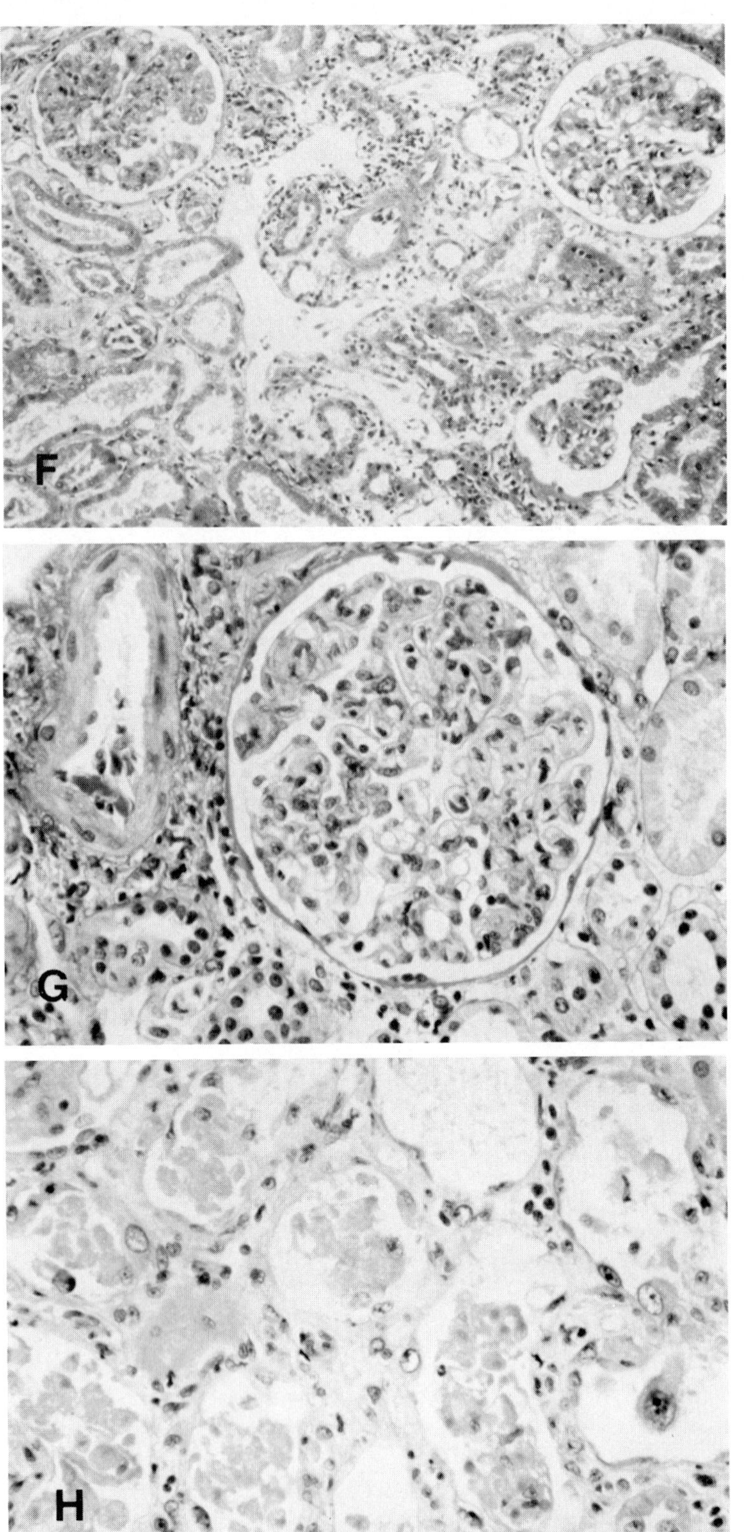

FIG. 13-9F-H *(continued)*
(F) Interstitial edema and inflammation are most prominent around a dilated venule. Tubules frequently have a flattened epithelium and contain casts. Glomeruli appear hypercellular, and there is diminished glomerular capillary patency. (Trichrome × 100) **(G)** An interlobular artery has a small intimal fibrin thrombus with associated endothelial damage. The hypercellular glomerulus contains numerous neutrophilic leukocytes and swollen cells. (Lendrum × 250) **(H)** Tubules are filled with exfoliated "ghost" cells. Occasional cells with nuclear pyknosis or karyolysis are present. Few exfoliated, as well as attached, renal tubular epithelial cells demonstrate marked regenerative cytologic changes. (× 250)

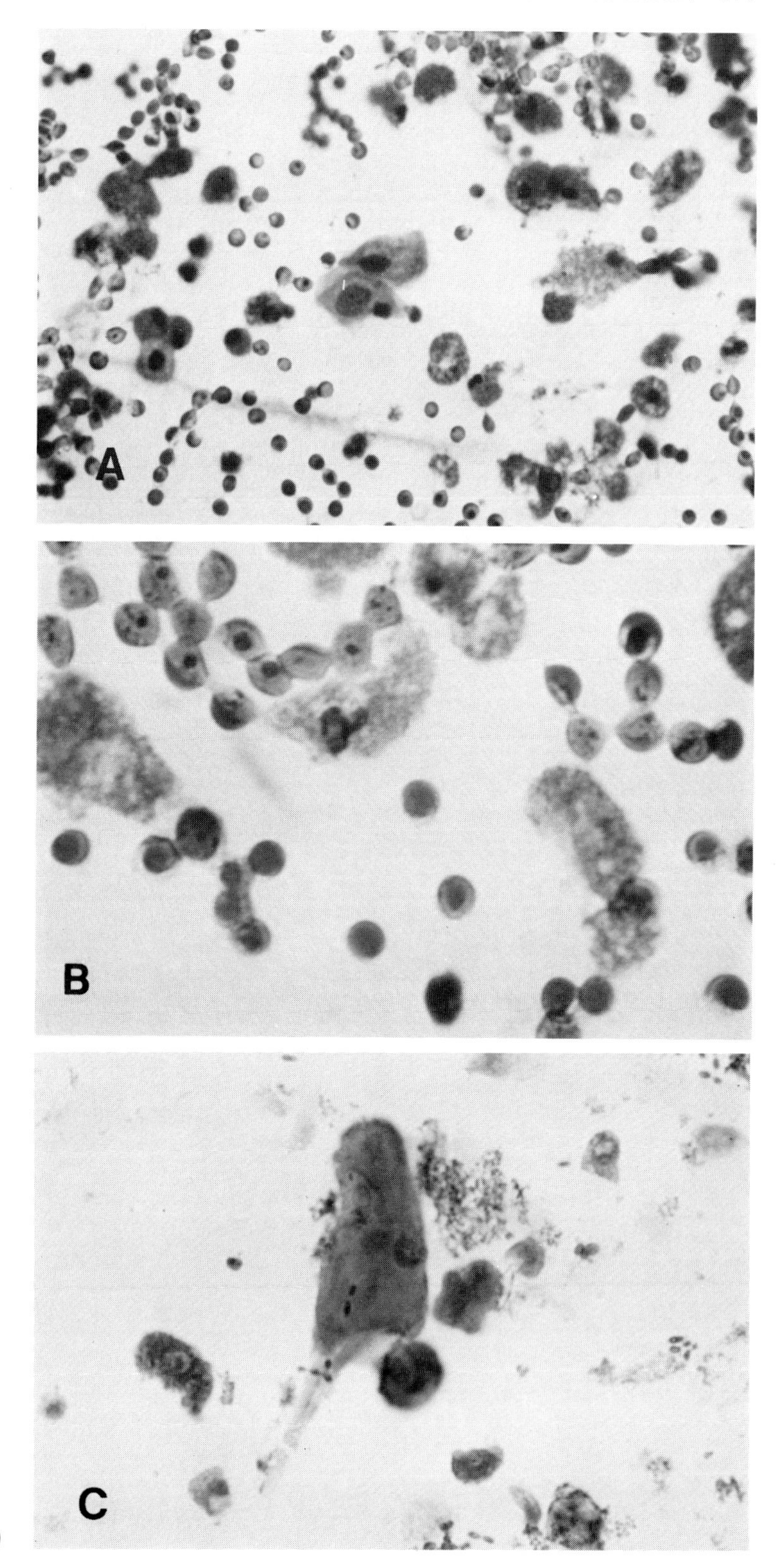

FIG. 13-10
VASCULAR REJECTION WITH ACUTE TUBULAR NECROSIS AND FOCAL INFARCT. (**A**) Urine sediment with degenerated and necrotic renal tubular epithelial cells and numerous erythrocytes. (× 400) (**B**) Note "ghost" cells. (× 1000) (**C**) Necrotic cells and waxy cast. (× 400)

(continued)

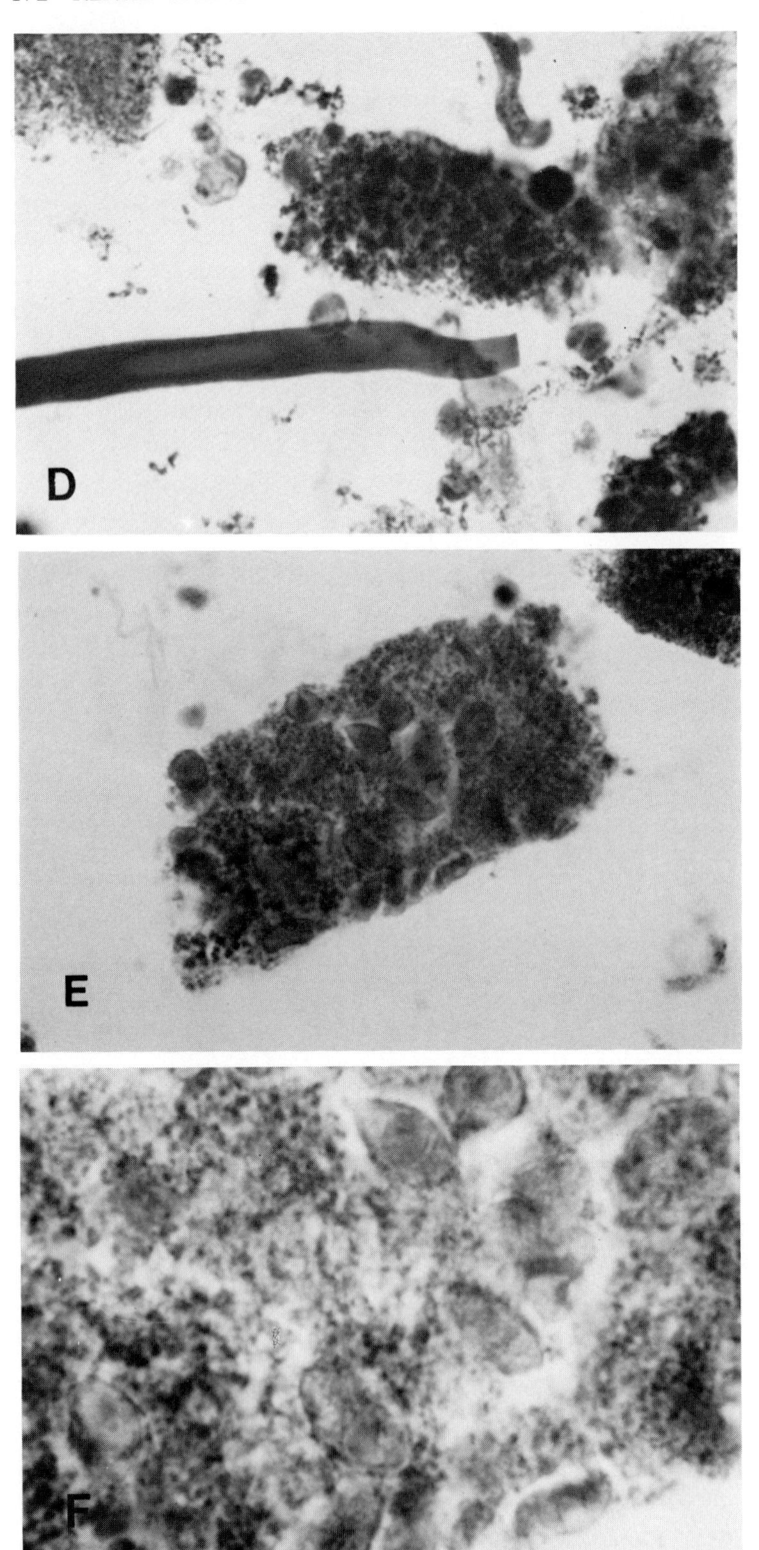

FIG. 13-10D-F *(continued)*
(D) Note variability in cast size and composition. (× 400) **(E)** Broad necrotic renal epithelial cast. (× 400) **(F)** (× 1000)

(continued)

FIG. 13-10G-H *(continued)*
(G) Transplant kidney with focal infarct surrounded by hemorrhagic zone. Cortical pallor with hemorrhage mottling is due to severe acute cellular rejection. (Gross) **(H)** A predominantly lymphocytic infiltrate is present at the corticomedullary junction adjacent to an arcuate artery having an intimal mononuclear inflammatory infiltrate. (× 160)

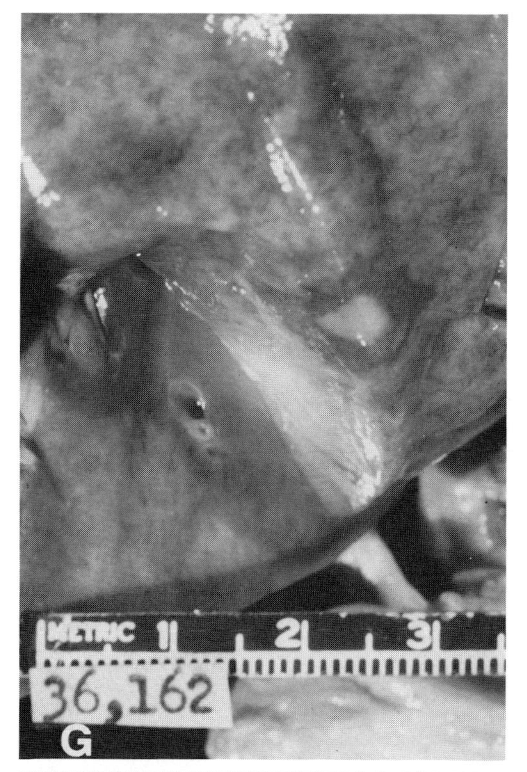

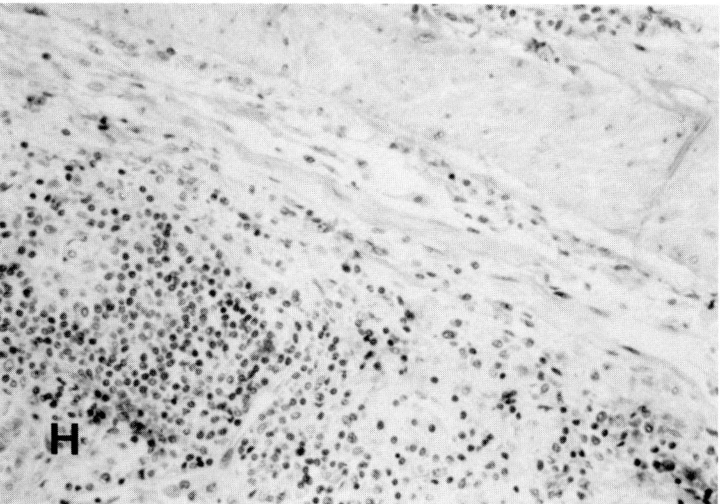

(continued)

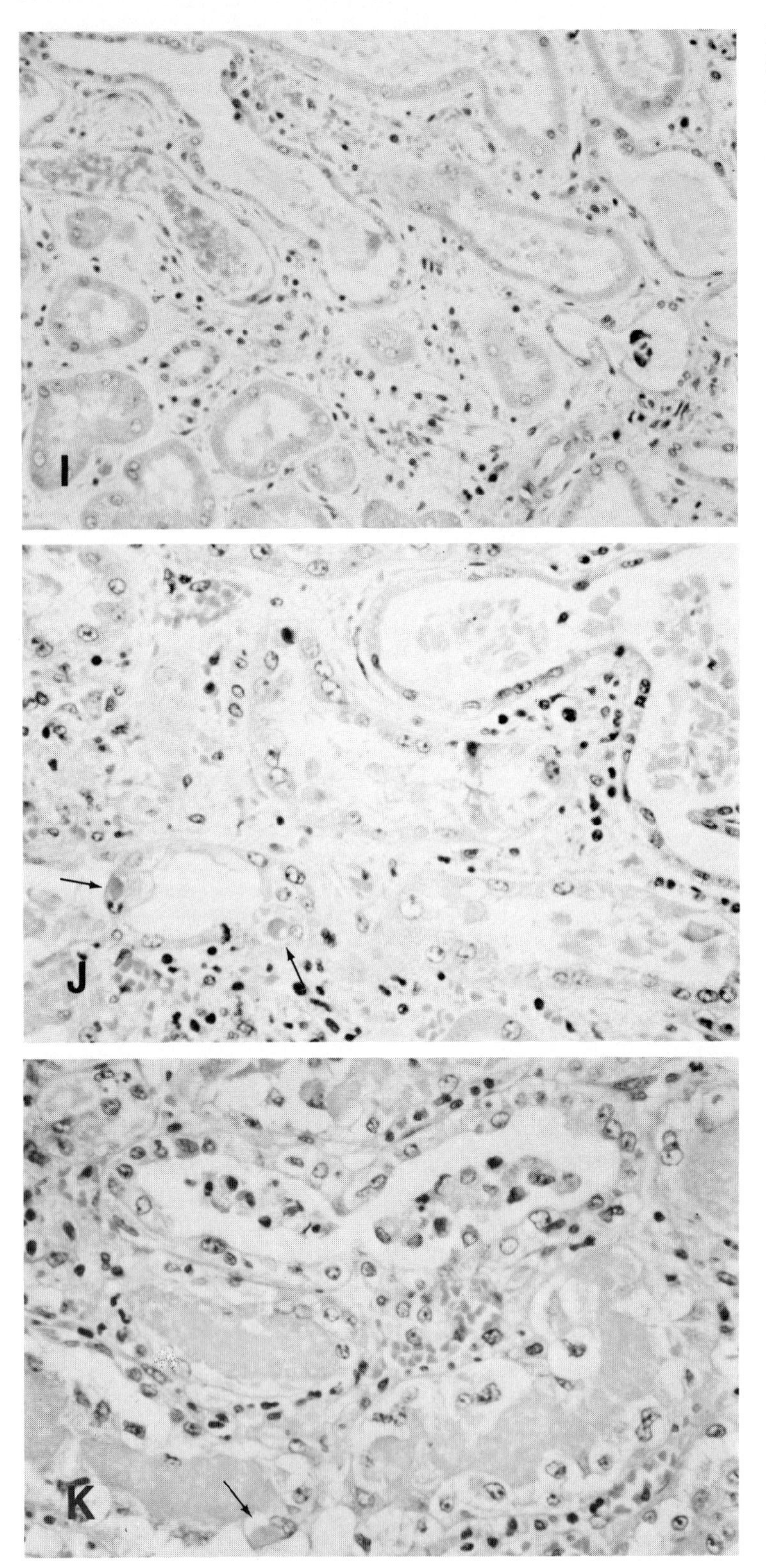

FIG) 13-10I-K *(continued)*
(I) Tubules are dilated, have a flattened, regenerating epithelium, and contain broad hyaline casts, coarse granular casts, and exfoliated renal epithelial "ghost" cells. Lymphocytes and few plasma cells are present in the edematous interstitium. (× 160) **(J)** Numerous "ghost" cells within damaged tubules. Attached epithelium appears regenerative; occasional cells exhibit nuclear pyknosis and several contain large eosinophilic intracytoplasmic bodies (arrows). (× 250) **(K)** Tubules contain abundant waxy cast material. Regenerating proximal convoluted tubular cells have vacuolated cytoplasm and occasionally contain globules of similar material (arrow). A distal convoluted tubule contains exfoliated degenerating renal tubule epithelial cells. (× 250)

(continued)

FIG. 13-10L *(continued)*
(L) A broad, dense hyaline cast appears attached to intact and exfoliated renal epithelium and focally appears phagocytized by renal epithelium (arrows). Similar material in an adjacent tubule is partially mineralized. (× 250)

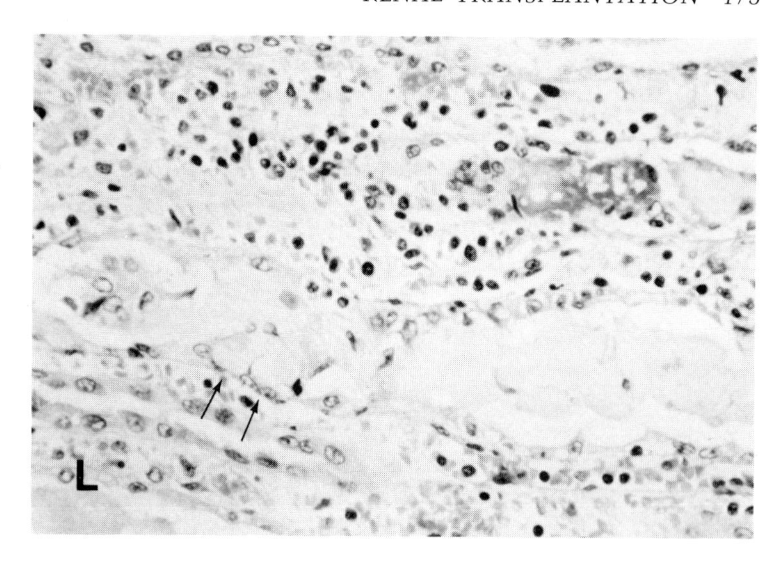

FIG. 13-11
ACUTE CELLULAR REJECTION, CHRONIC VASCULAR REJECTION WITH GLOMERULOPATHY, AND FOCAL INFARCT. **(A)** A large interlobular artery has a markedly narrowed lumen secondary to myointimal proliferation. Intraluminal and subendothelial mononuclear inflammatory cells are present. Interstitium surrounding the artery and adjacent vein contains an infiltrate of lymphocytes and plasma cells. (× 250) **(B)** A hypercellular glomerulus contains swollen cells, and capillary patency is diminished. Nonfuchsinophilic cytoplasmic hyaline droplets (arrow) in renal tubular epithelium and an erythrocytic cast are present. Congested peritubular capillaries contain neutrophilic leukocytes. (Trichrome × 400)

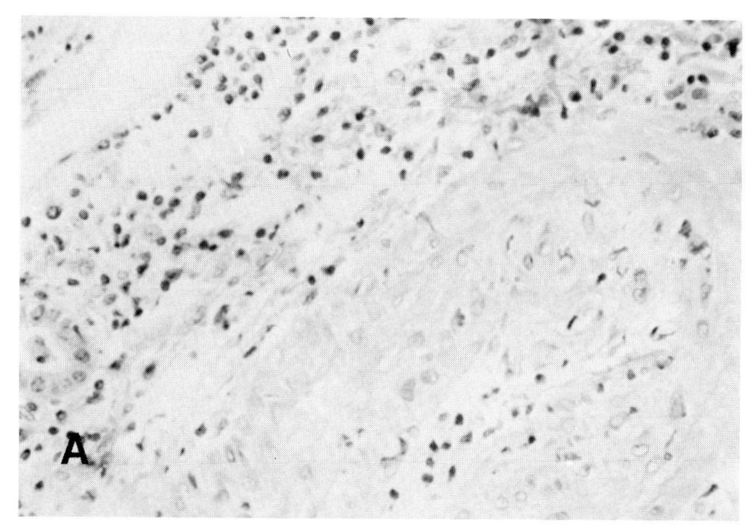

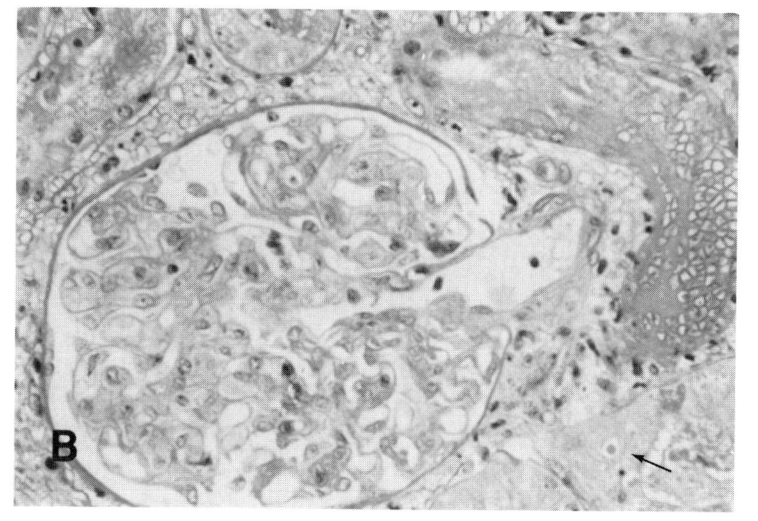

(continued)

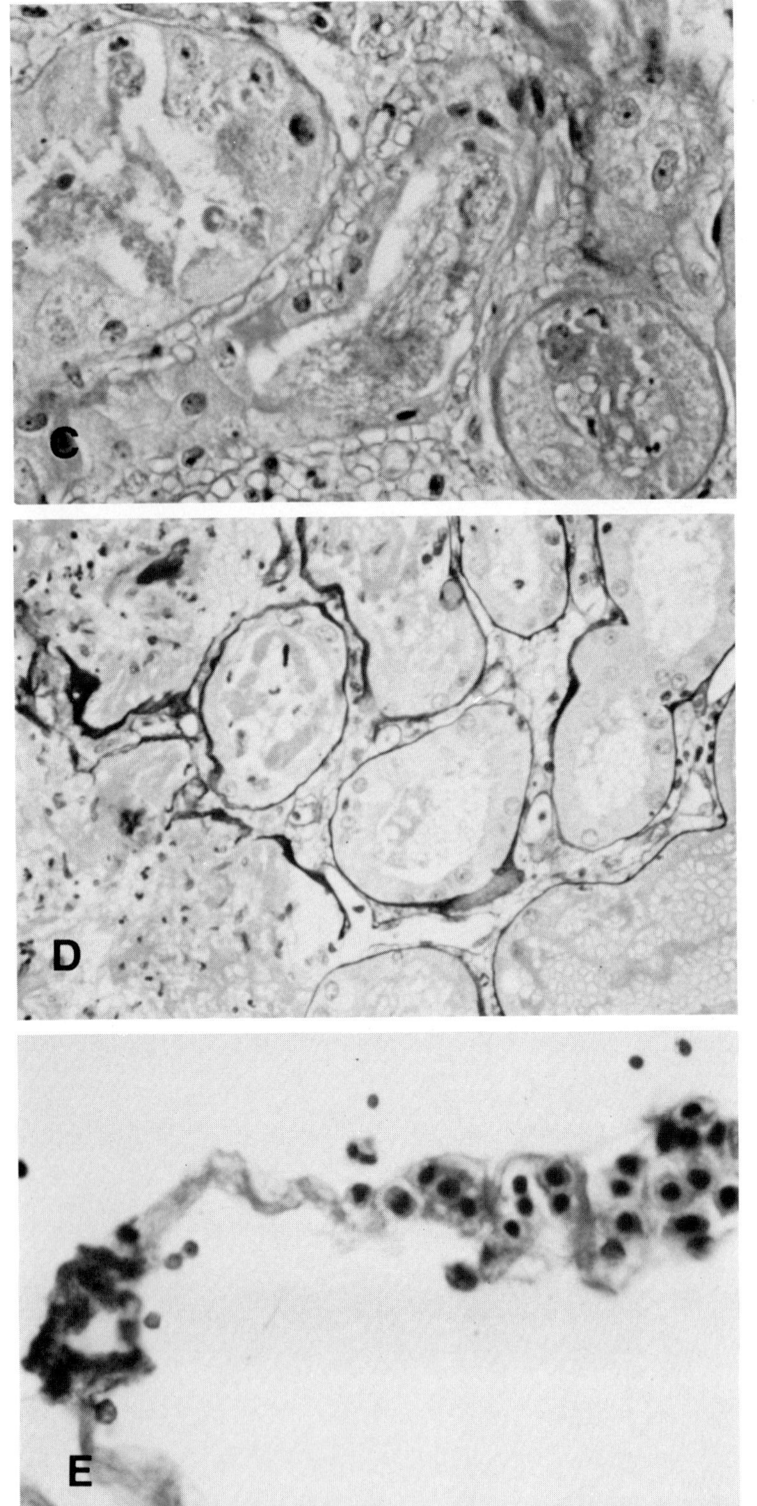

FIG. 13-11C-E *(continued)*
(C) Fibrillar cast. A tubule with necrotic epithelium contains neutrophils and blood. (Trichrome × 400) **(D)** Focal infarct. Interstitial and intratubular hemorrhage with fibrin, neutrophils, and tubular disruption are present at the margin of a focal infarct. (Jones × 250) **(E)** Renal epithelial cast in urine sediment. Note fibrillar texture of cast matrix suggesting fibrin. (× 400)

(continued)

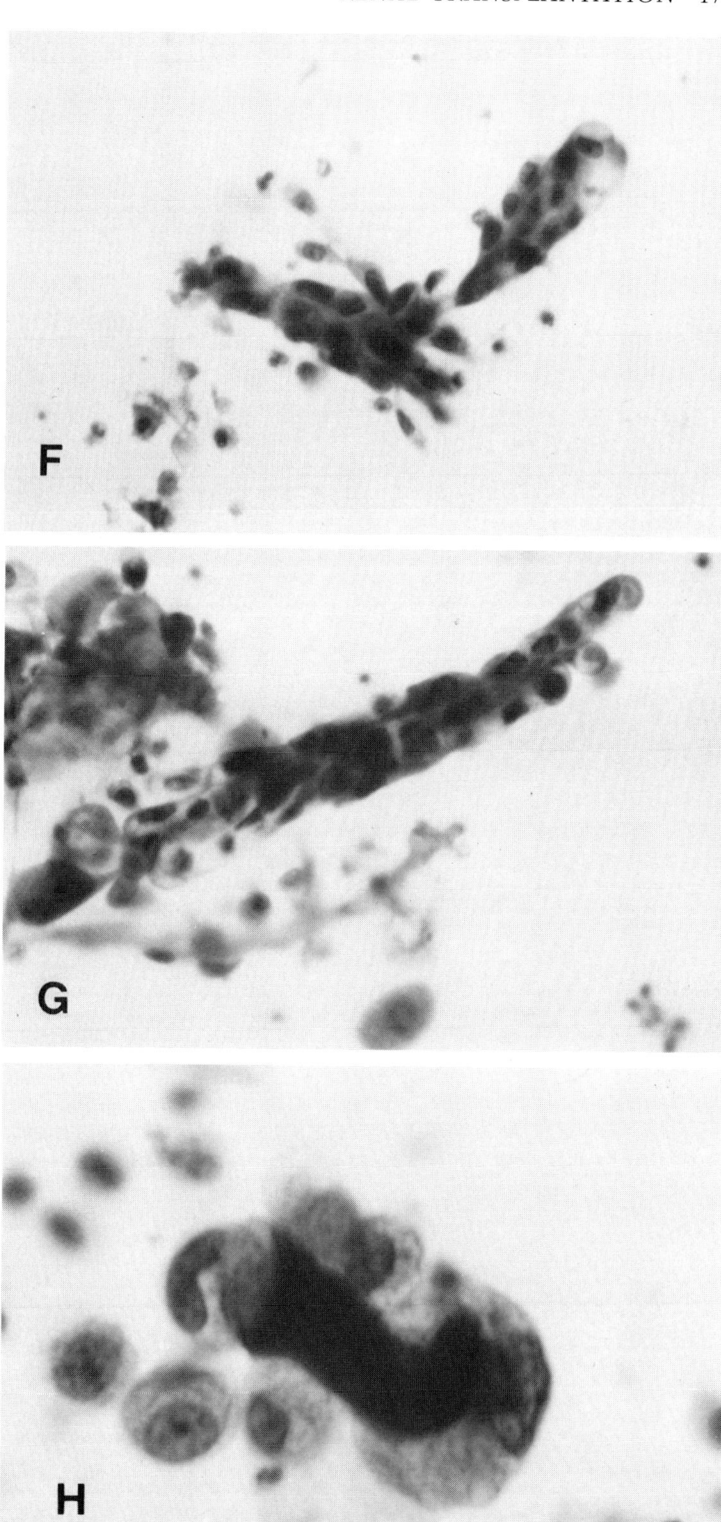

FIG. 3-11F-H *(continued)*
(F) Renal epithelial fragment in urine sediment indicating ischemic necrosis (× 400) **(G)** Exfoliated cylindrical renal epithelial fragment from collecting duct. (× 400) **(H)** Renal epithelial cast encasement. (× 1000)

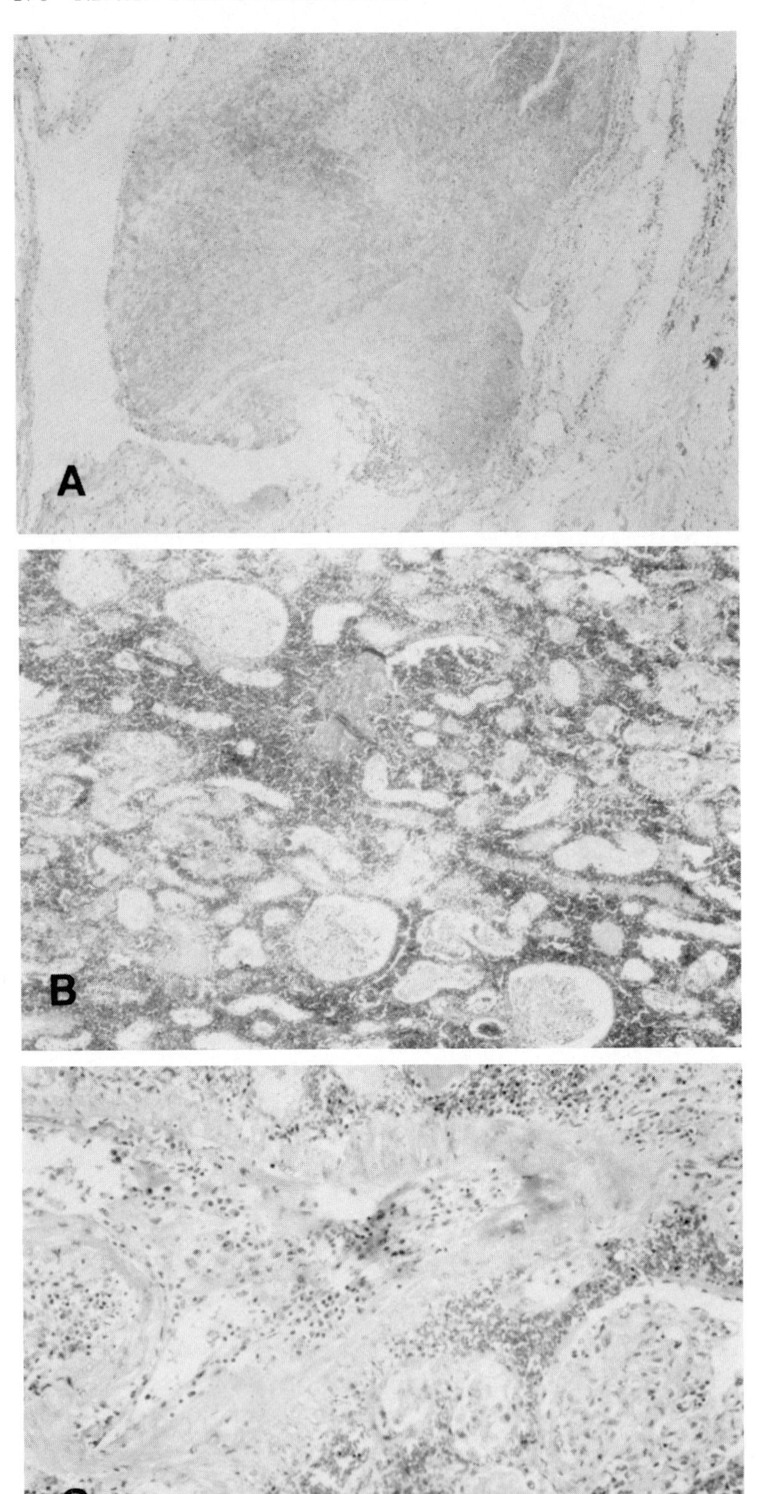

FIG. 13-12
ACUTE VASCULAR REJECTION WITH VENOUS THROMBOSIS AND HEMORRHAGIC INFARCTION. **(A)** Organizing thrombus in major renal vein branch. (×40) **(B)** Hemorrhagic infarct of renal allograft. (×40) **(C)** Acute vascular rejection with a marked intimal monoculear inflammatory infiltrate. Note the surrounding interstitial inflammation and hemorrhage with extensive ischemic necrosis of renal epithelium. (× 100)

(continued)

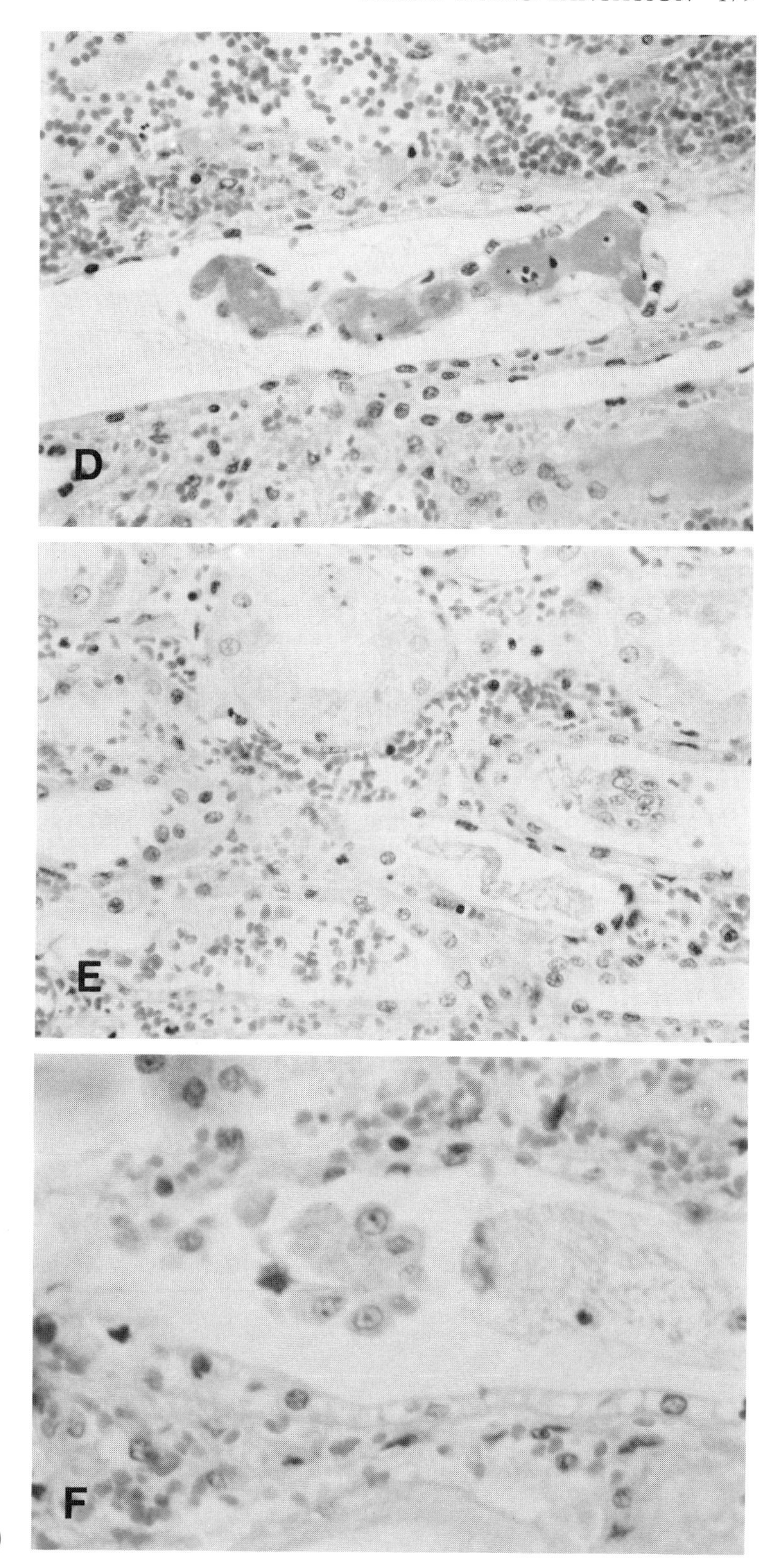

FIG. 13-12D-F *(continued)*
(D) Waxy cast-like material is encased by renal epithelial cells and is focally-attached to a collecting duct. (× 250) **(E)** Sheet-like epithelial fragment, fibrillar cast, and erythrocytes within tubules. (× 250) **(F)** Papillary renal epithelial fragment in tubule. A fibrillar cast is also present. (× 400)

(continued)

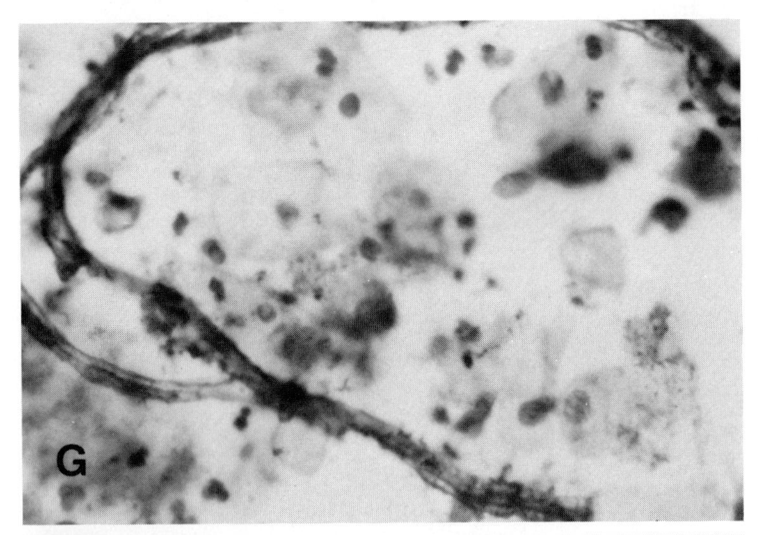

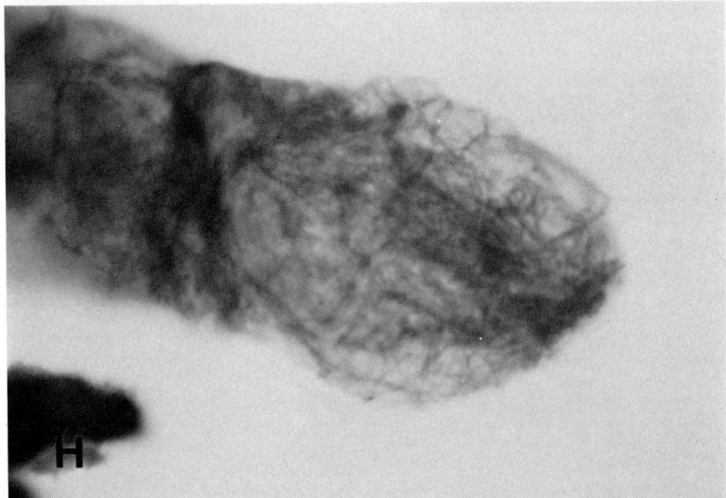

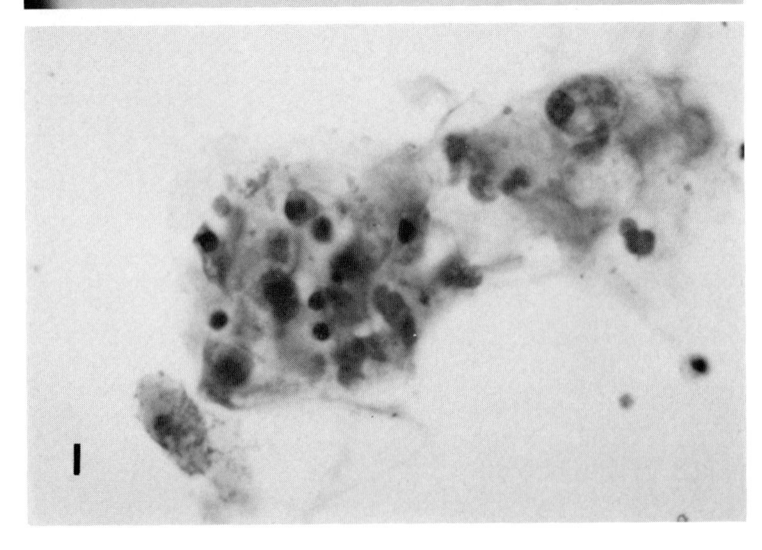

FIG. 13-12G-I *(continued)*
(G) Urine sediment with thready material composed of aggregated fibrils. (× 400) **(H)** Fibrin cast in urine sediment. (× 1000) **(I)** Renal epithelial cast containing markedly degenerated and necrotic cells. (× 400)

(continued)

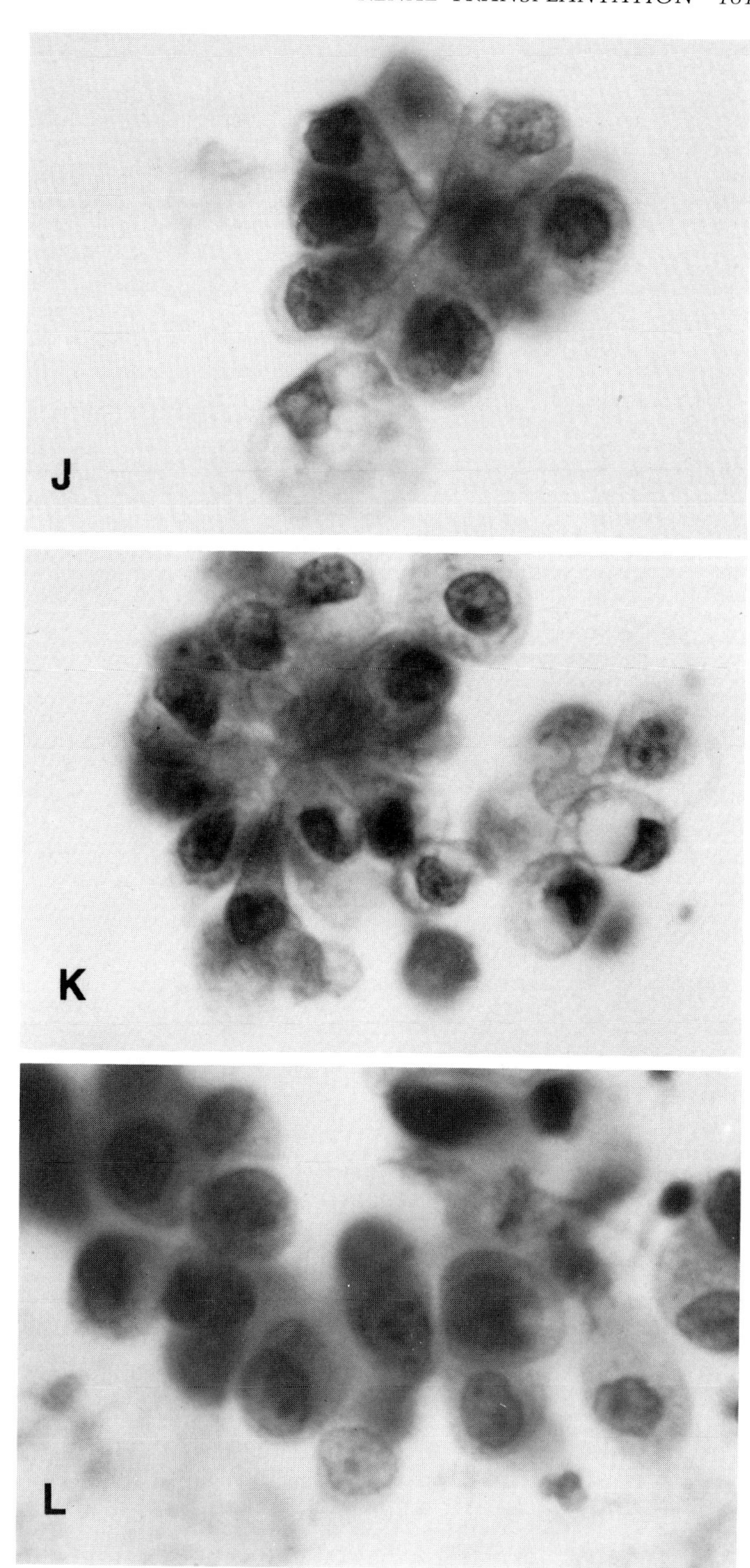

FIG. 13-12J-L *(continued)*
(J to L) Exfoliated renal epithelial fragments in urine sediment, indicating ischemic necrosis. (× 1000)

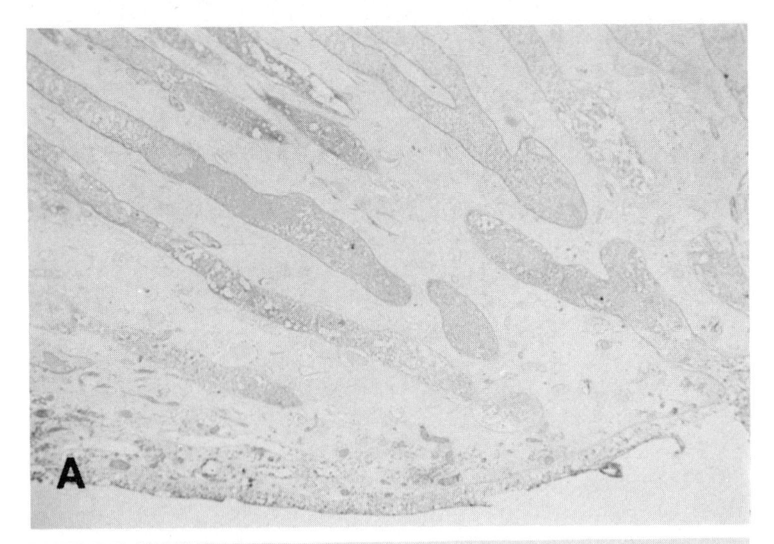

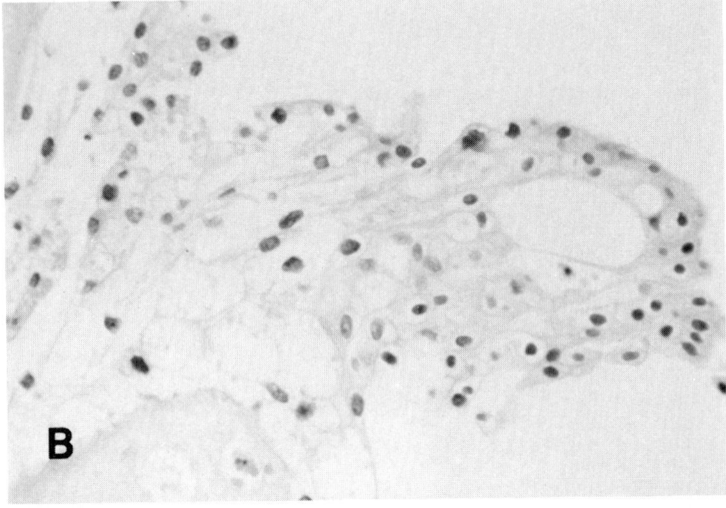

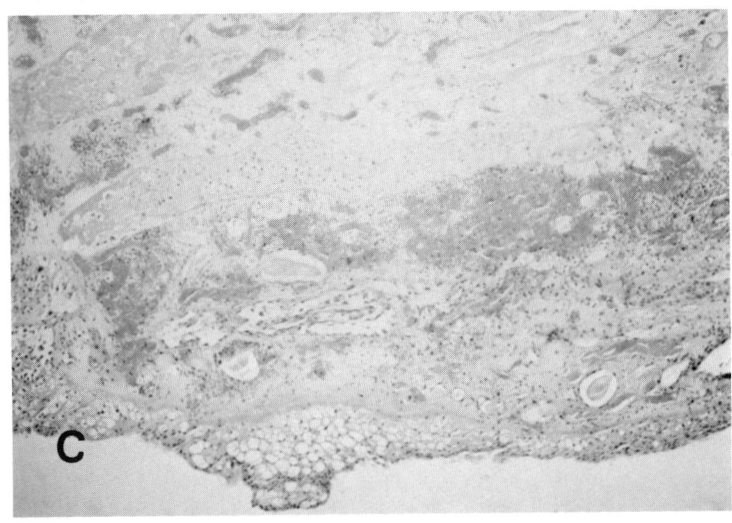

FIG. 13-13
ISCHEMIC INFARCTION OF ALLOGRAFT. Infarcted renal papilla. **(A)** Urothelium is generally necrotic. An epithelial fragment protrudes from the opening of the collecting duct. (× 25) **(B)** The sheet-like fragment has the appearance of transitional epithelium composed of intermediate and superficial cells. Note prominent cytoplasmic vacuolization and occasional eosinophilic droplets. (× 250) **(C)** Note exfoliation of urothelium, and collecting duct epithelium. (× 40)

(continued)

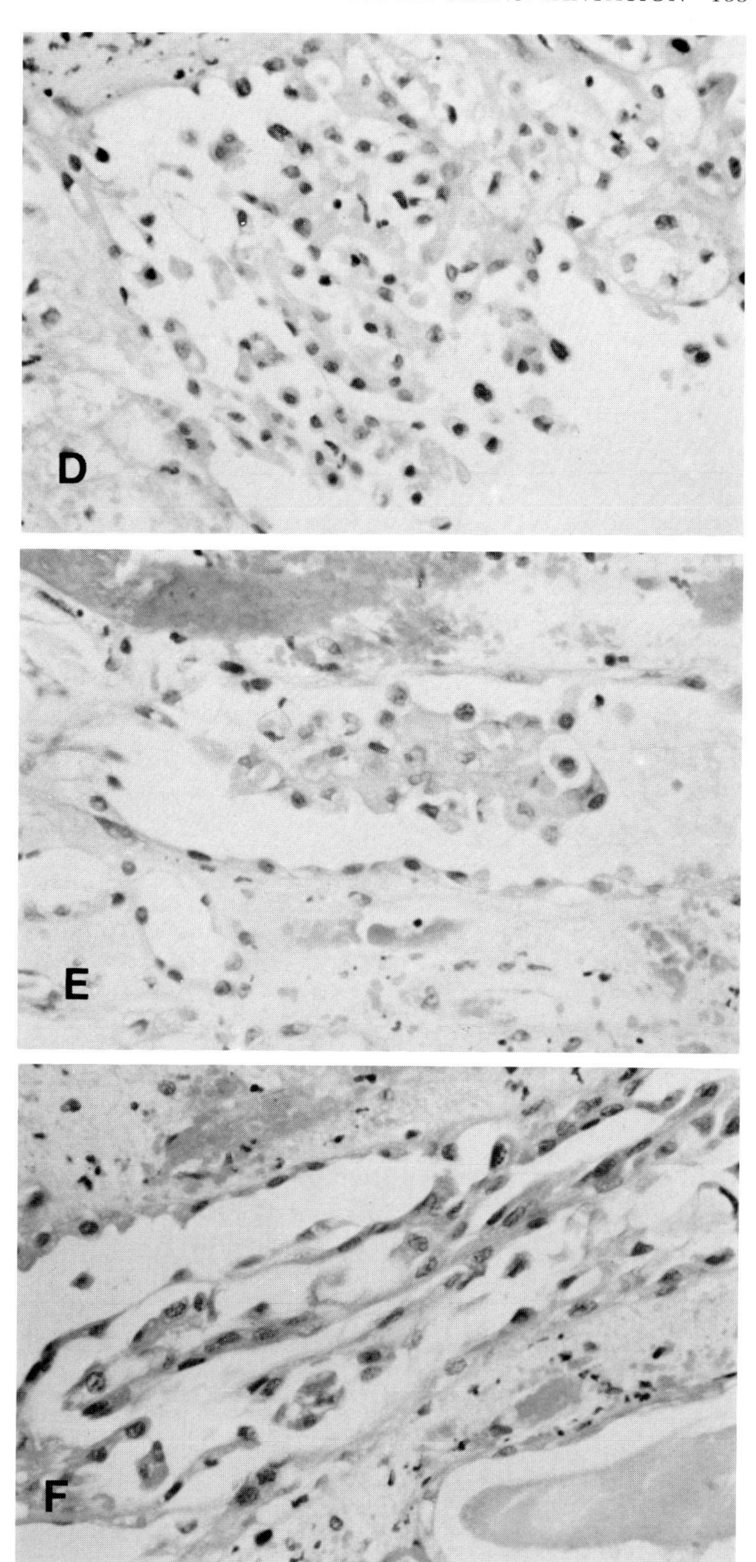

FIG. 13-13D-F *(continued)*
(D) Exfoliation of a loosely cohesive sheet of transitional cells from the papillary urothelium. Many cells are degenerated or necrotic. (× 250) **(E)** and **(F)** Spindle-cell epithelial fragments. These exfoliating fragments within collecting ducts contain many spindle-like cells with elongated nuclei and tapered cytoplasm. Attached epithelium is attenuated and has a similar appearance, suggesting derivation of spindle cell fragments from epithelium in an early phase of regeneration. (× 250)

(continued)

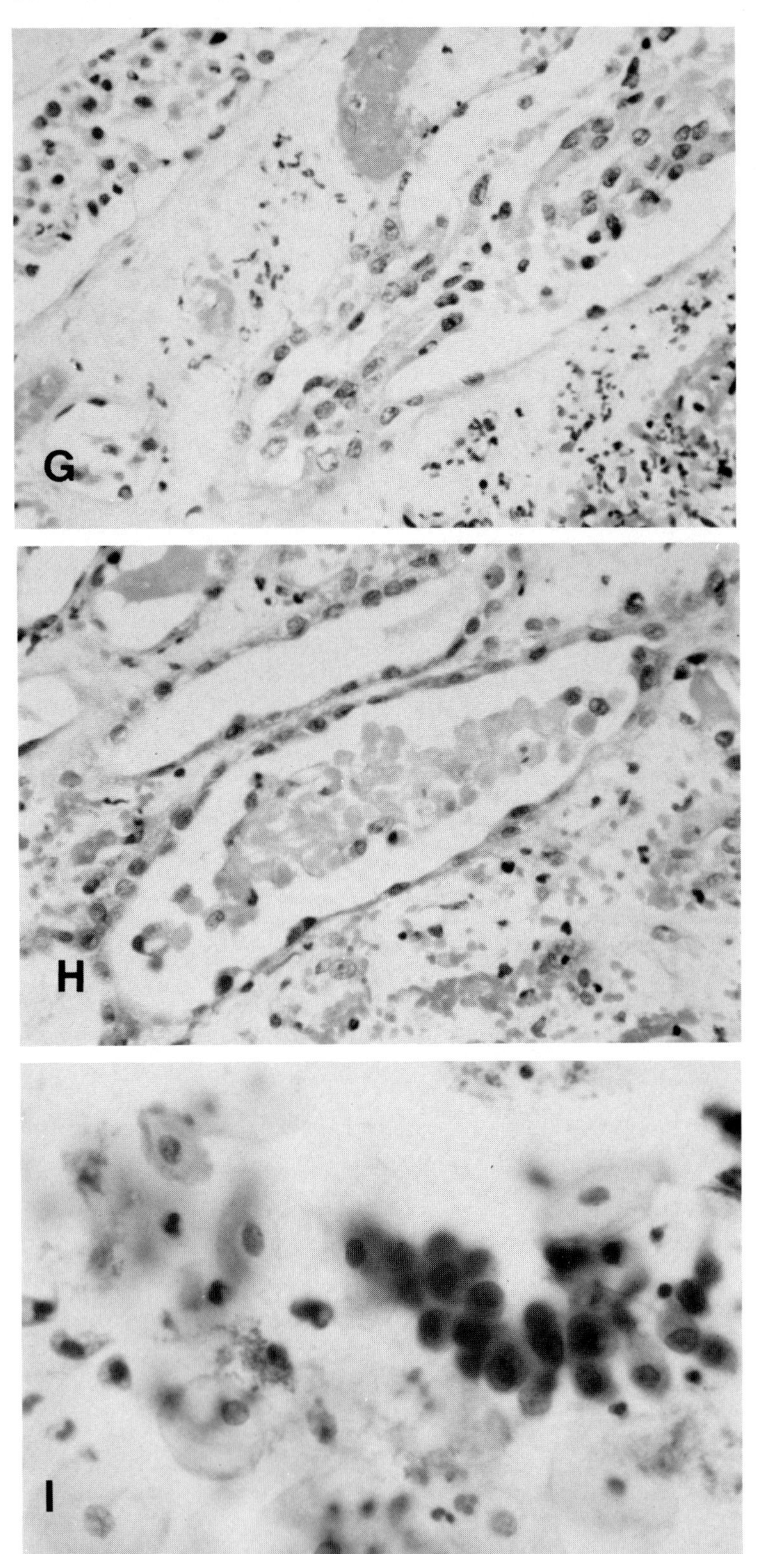

FIG. 13-13G-I *(continued)*
(G) Renal epithelial fragment with cellular degeneration and necrosis. Few elongated cells are evident at this site of loose attachment to the regenerating epithelium. (× 250) **(H)** Intratubular aggregate of necrotic renal epithelium ("ghost cells.") (× 250) **(I)** Urine sediment with dirty background, urothelial cells, and epithelial fragment. (× 400)

(continued)

FIG. 13-13J-L *(continued)*
(J) Urine sediment with cylindrical renal epithelial fragment of collecting duct origin, indicating ischemic necrosis. (× 400) **(K)** Broad necrotic renal epithelial cast and cylindrical renal epithelial fragment in urine sediment. (× 400) **(L)** Broad mixed cast in urine sediment. Note renal epithelial cells and homogeneous waxy material. (× 400)

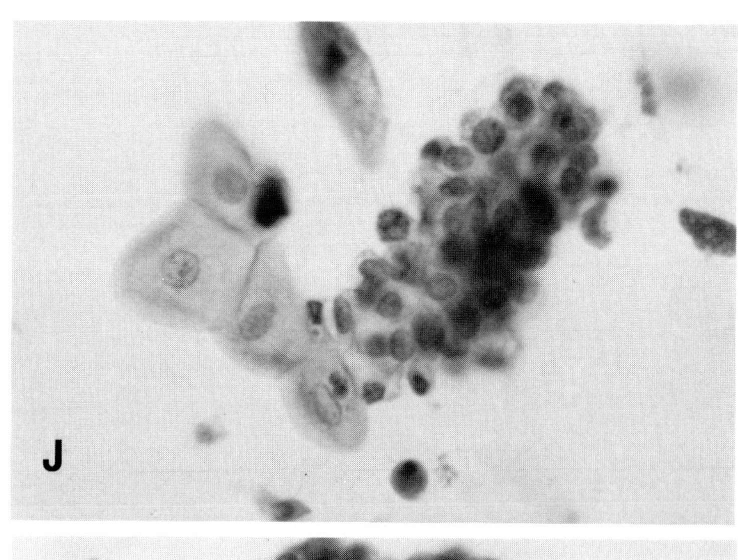

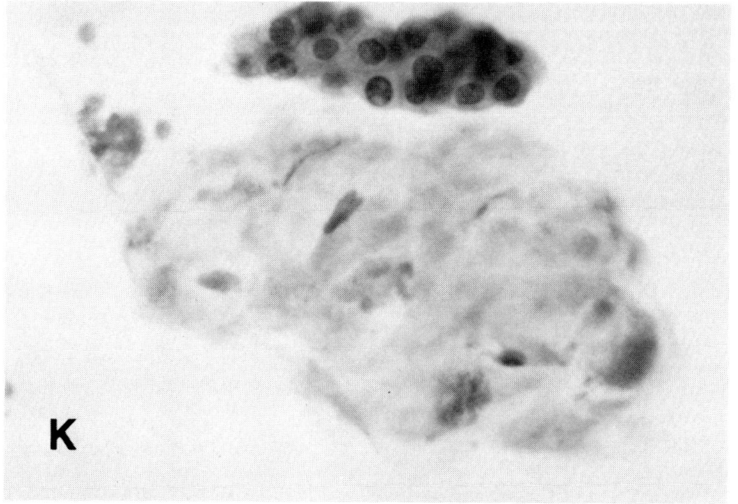

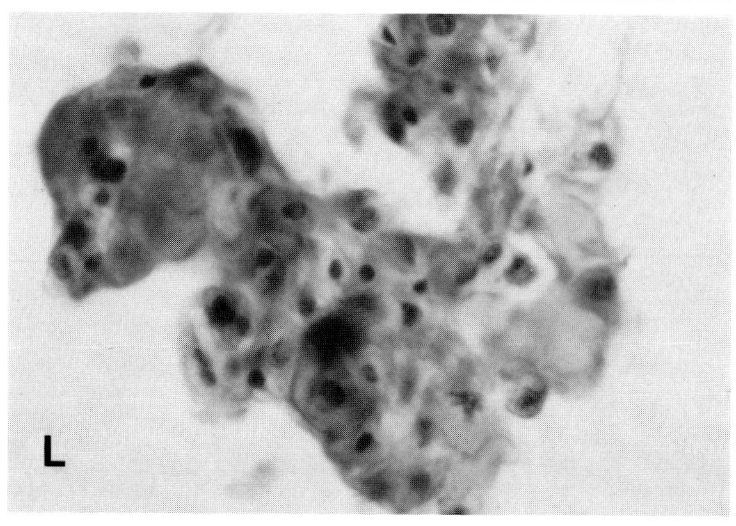

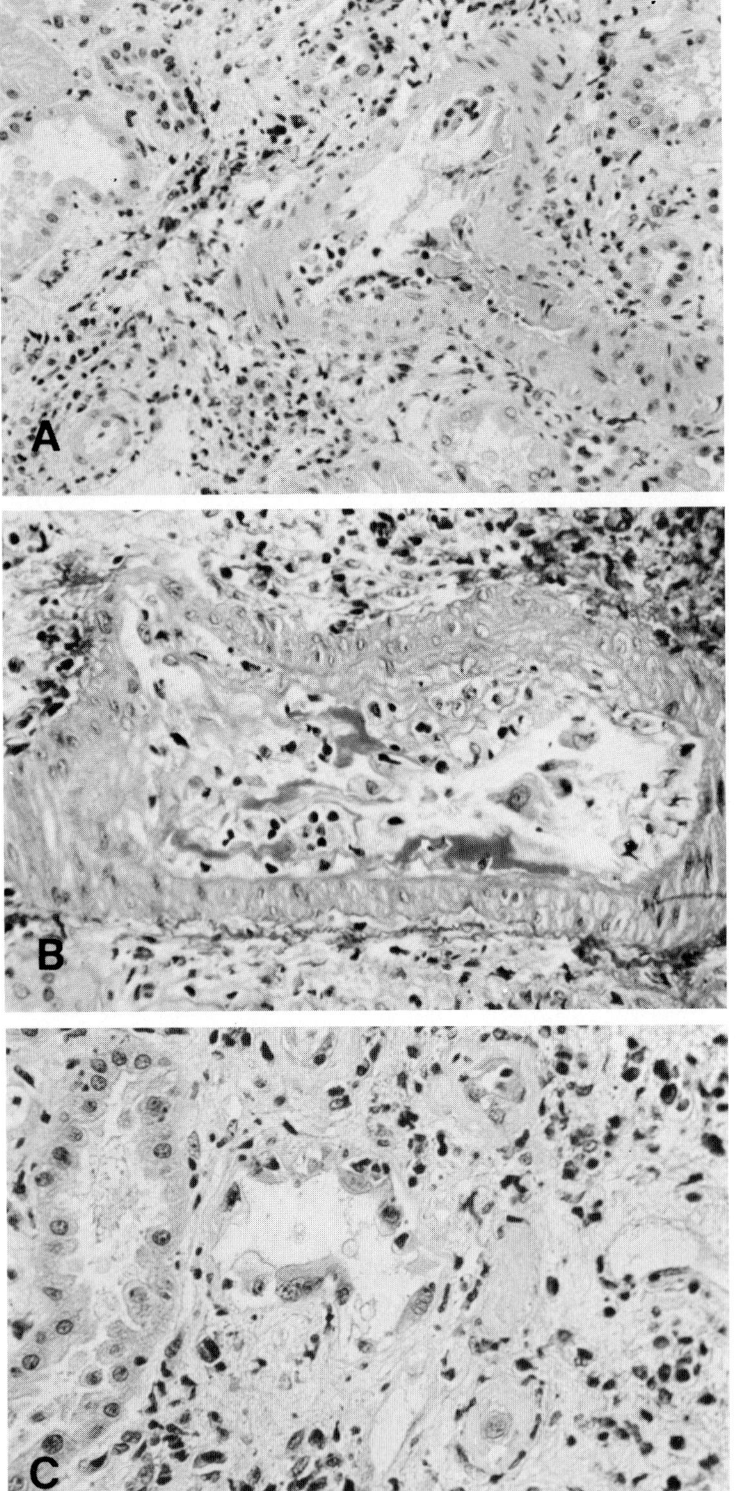

FIG. 13-14
VASCULAR REJECTION UNRESPONSIVE TO THERAPY. Allograft biopsy. **(A)** Interlobular artery with mural fibrin deposition and a mononuclear intimal infiltrate. Note the perivascular inflammatory infiltrate typical of acute cellular rejection. (× 160) **(B)** Interlobular artery. Note reactive endothelium, fibrin thrombus, and intimal thickening with infiltration by neutrophils and mononuclear cells. (× 250) **(C)** Disrupted tubule with regenerating epithelium. Note lymphocytes and plasma cells in the edematous interstitium. (× 400)

(continued)

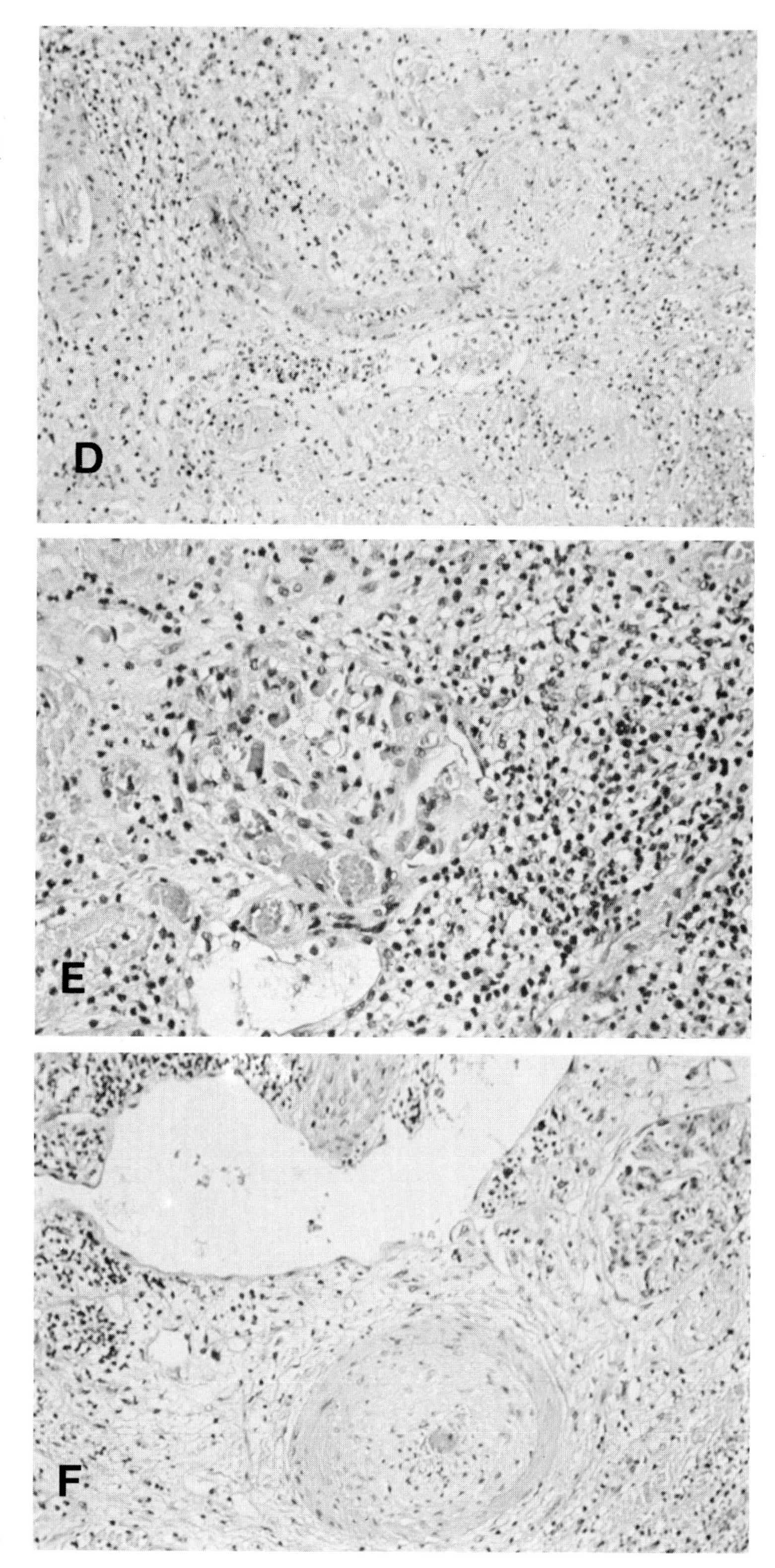

FIG. 13-14D-F *(continued)*
Allograft nephrectomy. **(D)** Ischemic infarction. (× 100) **(E)** Dense lymphocytic infiltrate adjacent to infarcted zone. (× 160) **(F)** Interlobular artery with marked luminal narrowing due to myointimal proliferation characteristic of chronic vascular rejection. Note the lymphocytic perivenular infiltrate, indicating active acute cellular rejection. (× 160)

(continued)

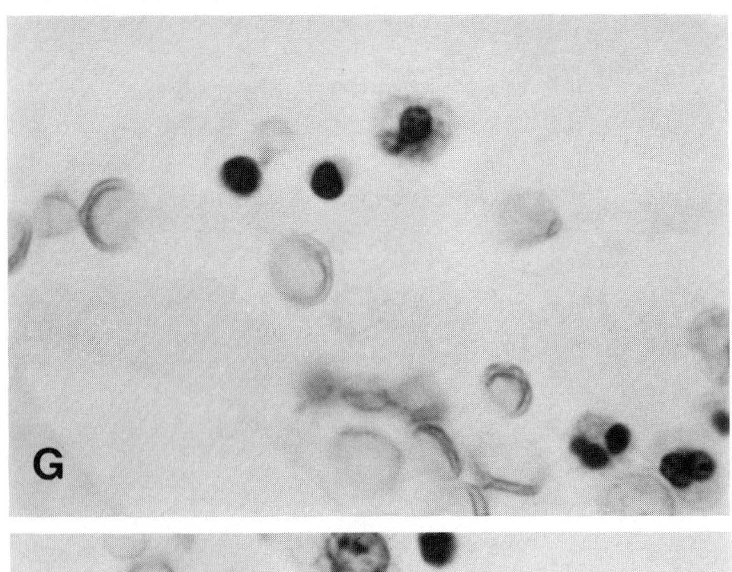

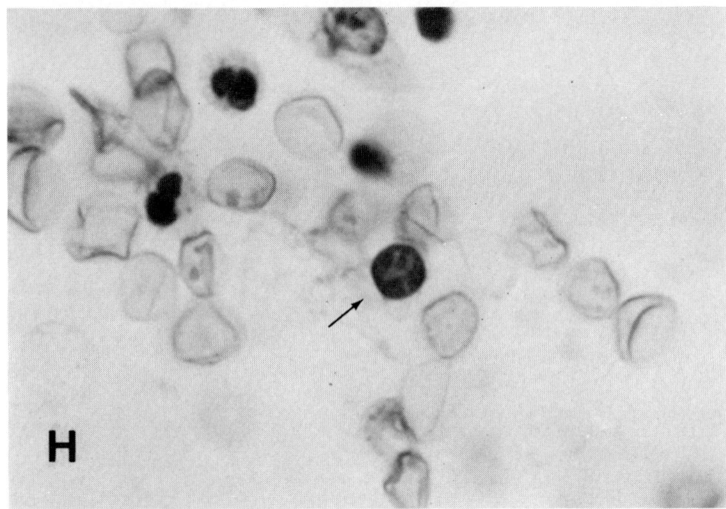

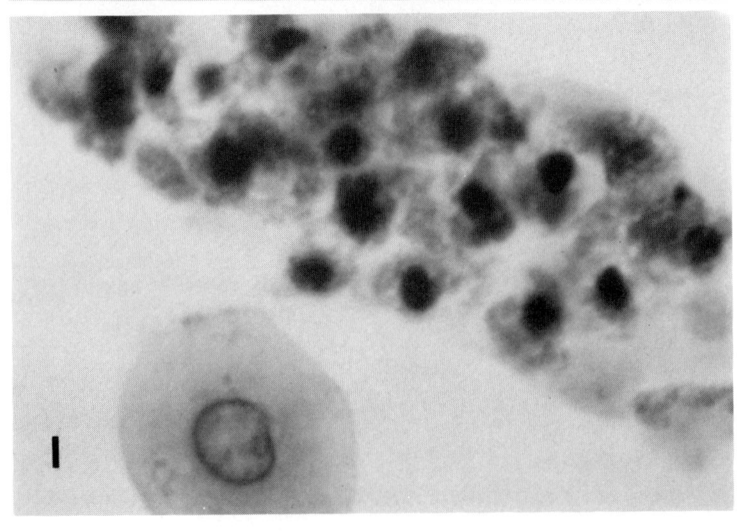

FIG. 13-14G-I *(continued)*
(G) Urine sediment with lymphocytes, neutrophils, and erythrocytes. (× 1000) **(H)** Note immature lymphocyte (arrow). (× 1000) Lymphocyturia is a variable finding dependent upon immunosuppressive therapy. (× 1000) **(I)** Renal epithelial cast in urine sediment. (× 1000)

(continued)

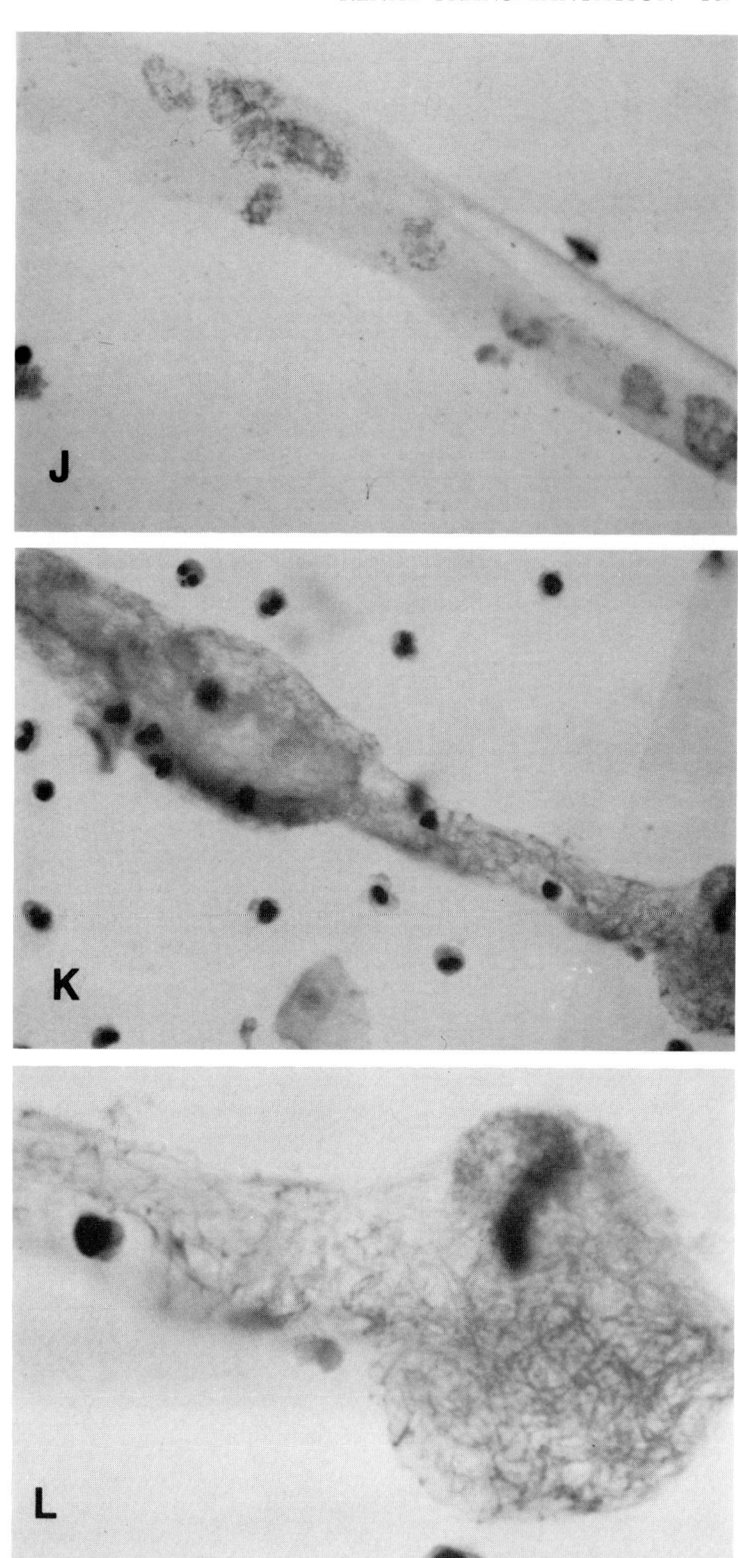

FIG. 13-14J-L *(continued)*
(J) Necrotic renal epithelial cast in urine sediment. Note "ghost" cells embedded in hyaline matrix. (× 1000) **(K)** Fibrin cast in urine sediment. (× 400) **(L)** (× 1000).

FIG. 13-15
URINE SEDIMENT PATTERNS AND DIFFERENTIAL DIAGNOSIS IN THE *LATE* POST-TRANSPLANTATION PERIOD.

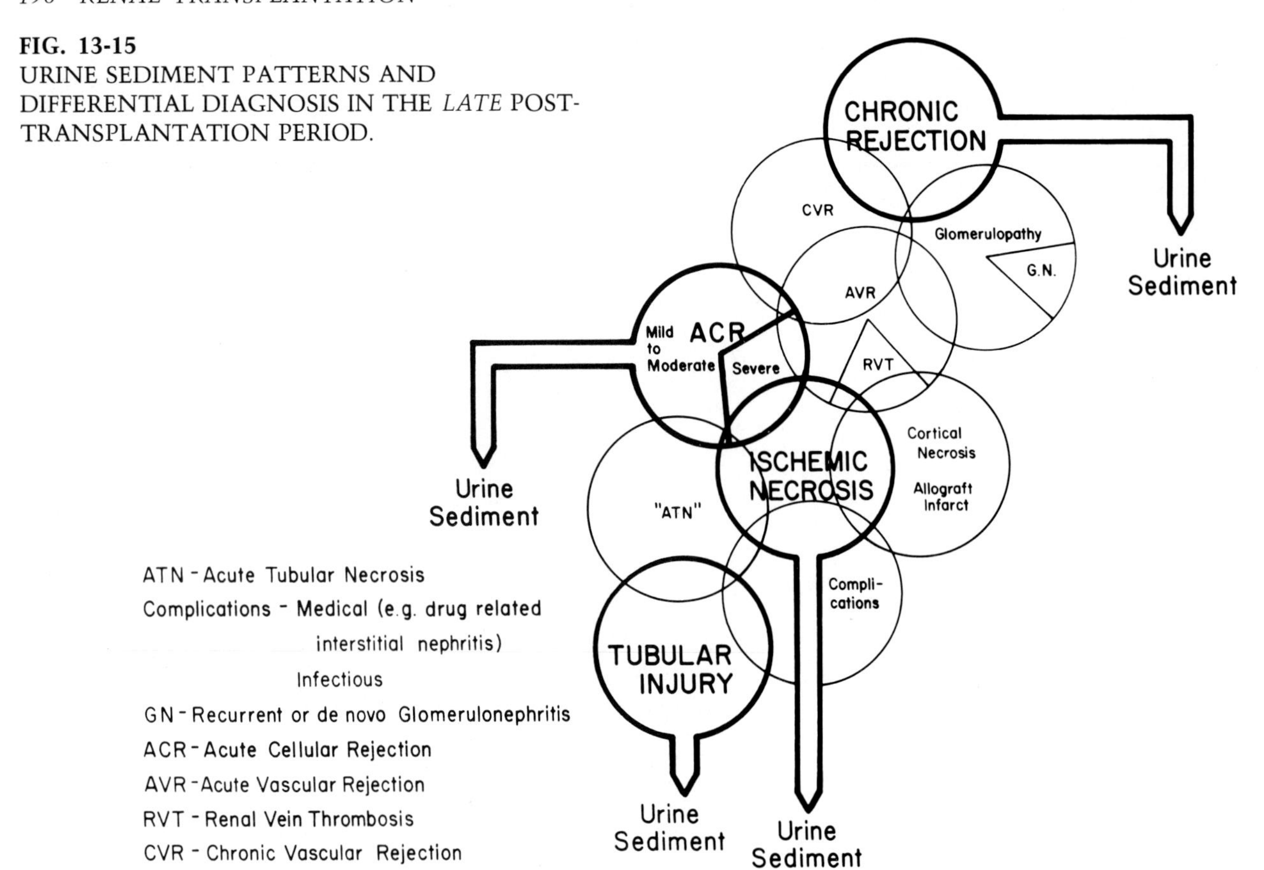

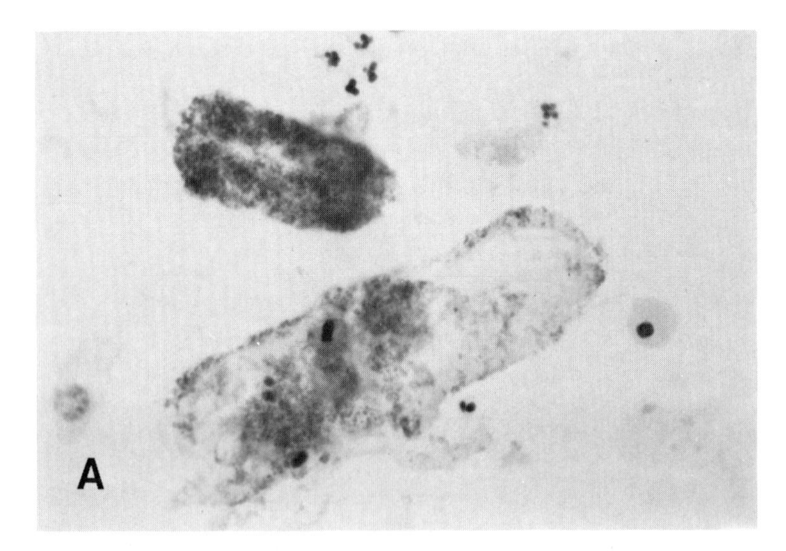

FIG. 13-16
CHRONIC REJECTION. **(A)** Urine sediment w granular casts and rare exfoliated renal tubular epithelial cells. (× 400)

(continued)

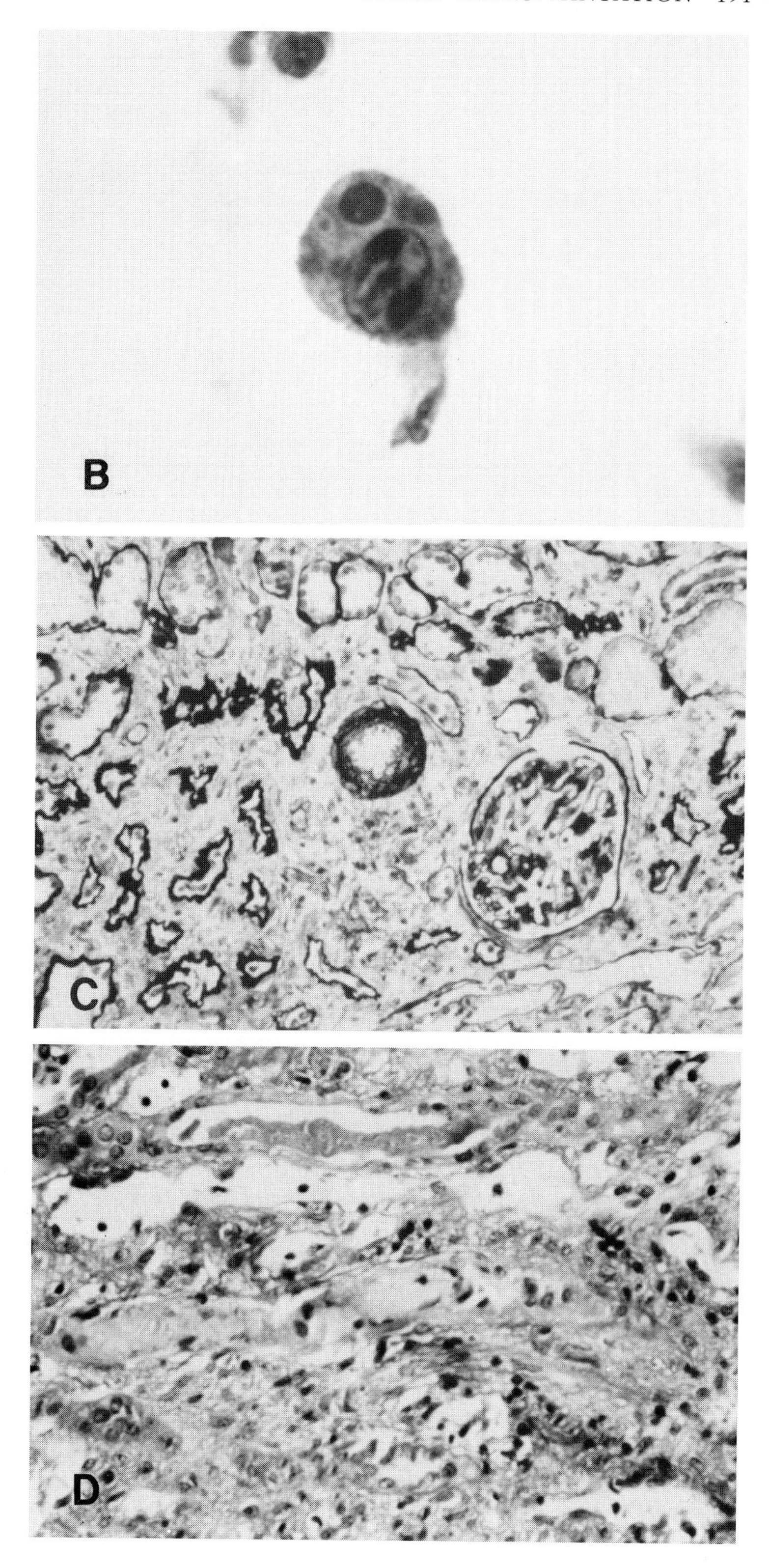

FIG. 13-16B-D *(continued)*
(B) Degenerative intranuclear and intracytoplasmic inclusions in exfoliated renal epithelial cell. (× 1000) Allograft biopsy 14 months post-transplant for progressive decline in function. **(C)** A glomerulus with slightly thickened mesangium is surrounded by atrophic tubules and a fibrotic interstitium. Note the lack of interstitial inflammation. (Jones × 160) **(D)** Medulla with waxy and granular casts. (Trichrome × 250)

(continued)

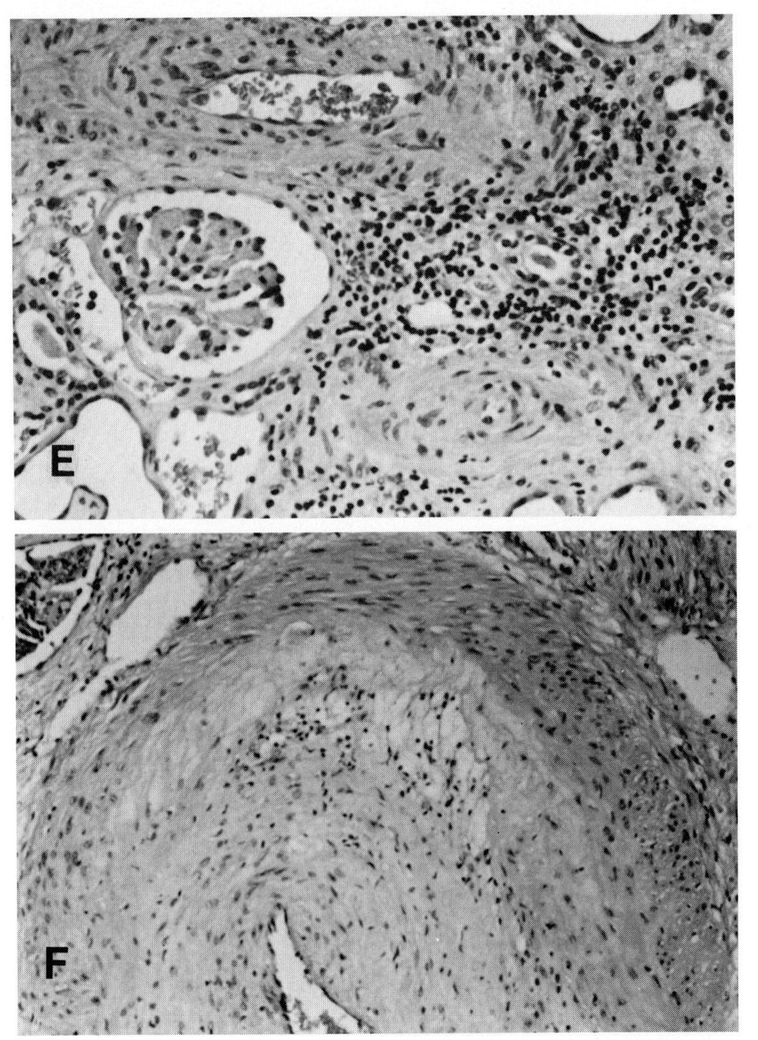

FIG. 13-16E-F *(continued)*
Chronic vascular rejection. Allograft nephrectomy 3 months posttransplant. **(E)** Interlobular arteries have markedly thickened, fibrotic walls with diminished lumens. There is a mild lymphocytic perivascular interstitial infiltrate. Note the atrophic tubules with waxy casts and the shrunken avascular glomerulus. (× 160) **(F)** Arcuate artery with marked intimal fibrosis, a scant mononuclear infiltrate, and foam cells. (× 160)

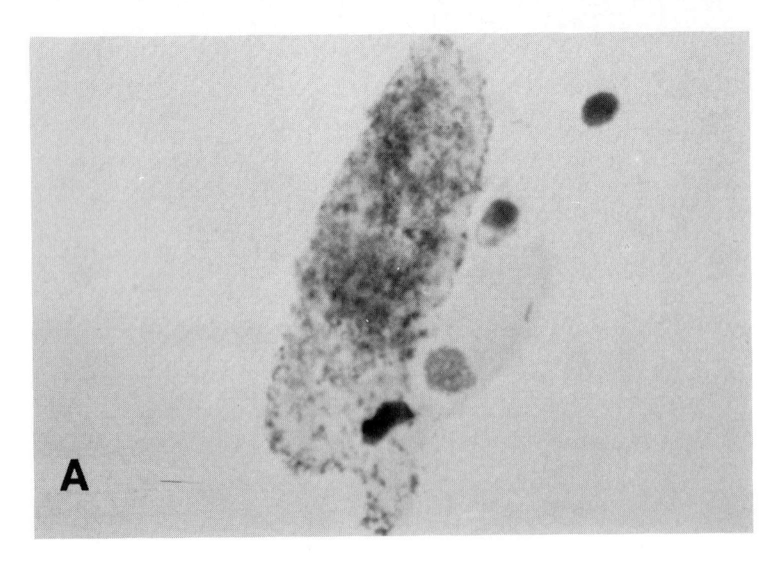

FIG. 13-17
ACUTE CELLULAR REJECTION AND VASCULOPATHY 4 YEARS POST-TRANSPLANT. **(A)** Finely granular cast and rare necrotic cell in urine sediment. (× 1000)

(continued)

FIG) 13-17B-D *(continued)*
(B) Mixed cellular cast containing several lymphocytes. (× 1000) **(C)** Urine sediment with long waxy cast. (× 400) **(D)** Urine sediment with granular and waxy casts. (× 1000)

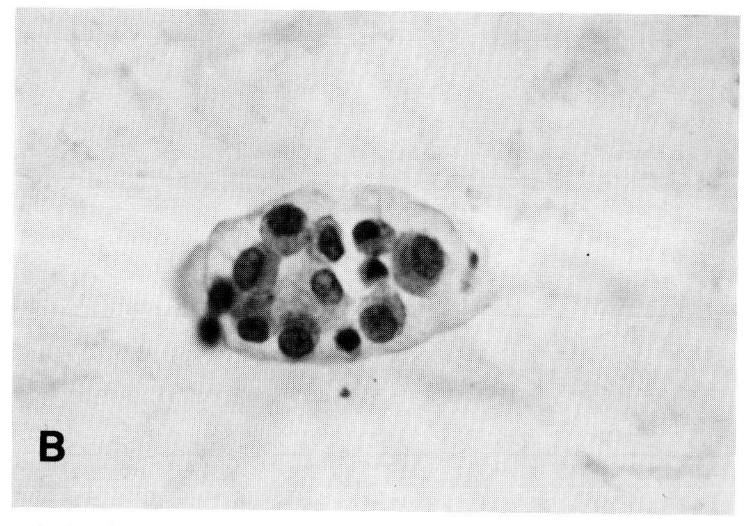

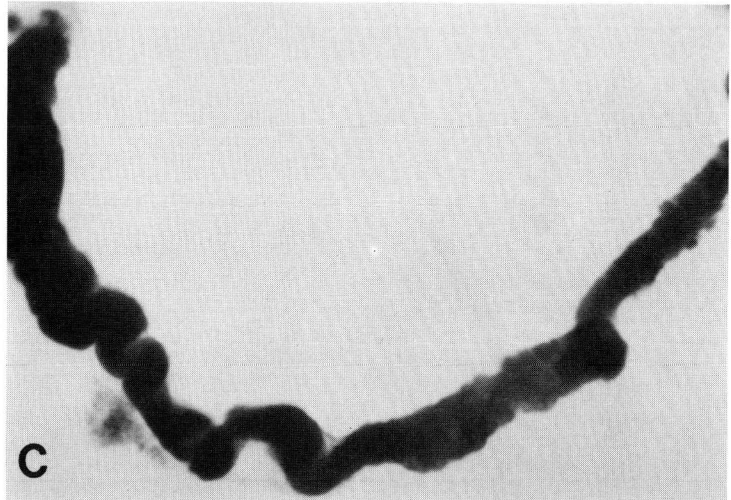

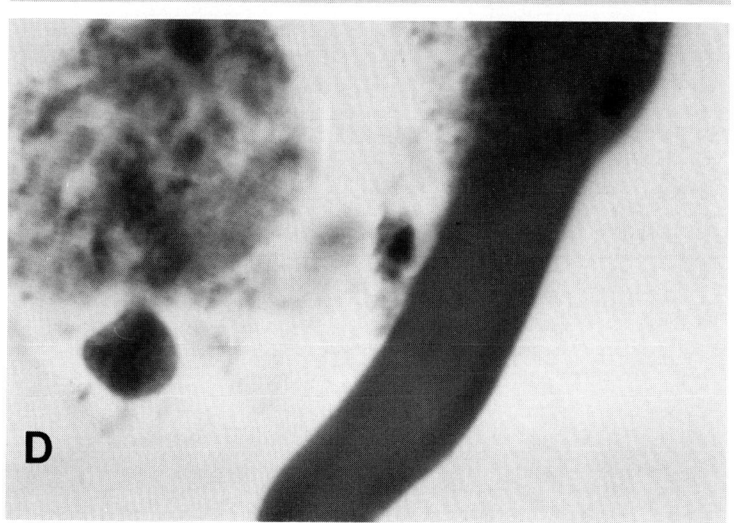

(continued)

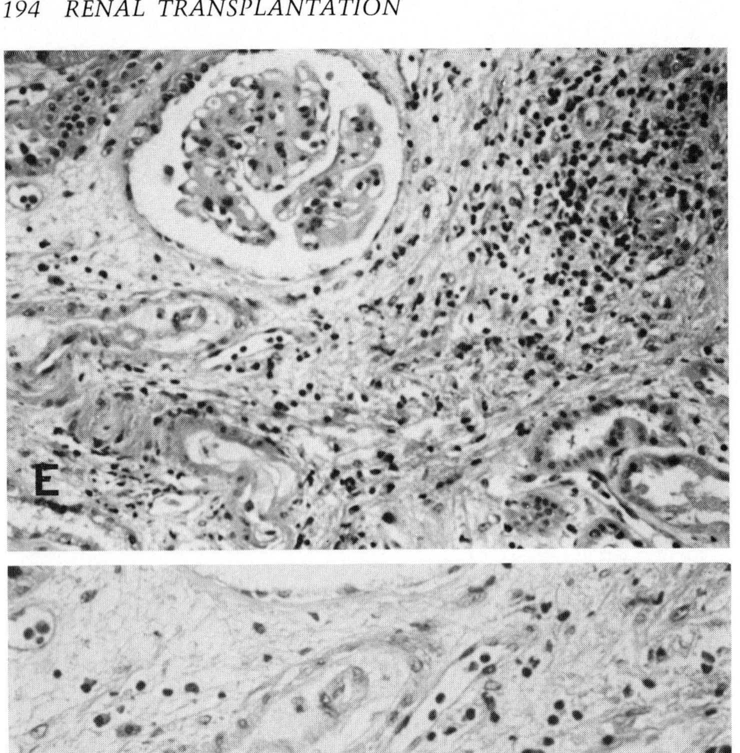

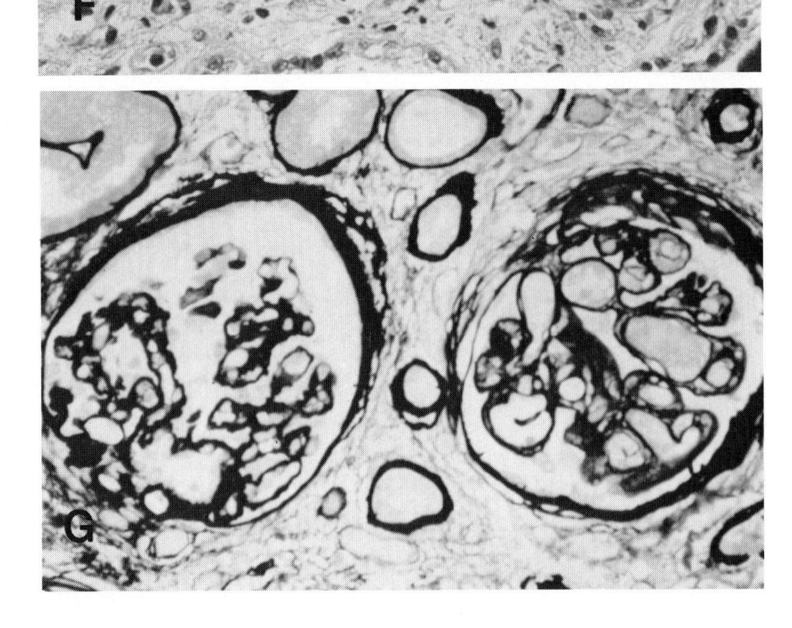

FIG. 13-17E-G *(continued)*
(E) The edematous interstitium contains an infiltrate of lymphocytes and plasma cells. The glomerulus has a widened mesangium and diminished capillary patency. Note compromise of arterial lumens. (× 160) **(F)** Small arteries have an edematous intimal zone with apparent foam cells. (× 250) **(G)** One glomerulus (right) contains fibrin thrombi and has reduplicated and reticulated capillary basement membranes. The adjacent glomerulus (left) shows ischemic collapse with thickening and wrinkling of capillary basement membranes. (Jones × 250)

(continued)

FIG. 13-17H *(continued)*
(H) Glomerulus with reduplicated capillary basement membranes, endothelial swelling, and subendothelial fibrin. Note segmental prominence of intracapillary mononuclear leukocytes (? lymphocytes). (Lendrum × 400) Stenosis of upper pole segmental artery documented by arteriogram. Biopsy performed following sudden onset of malignant hypertension (230/160).

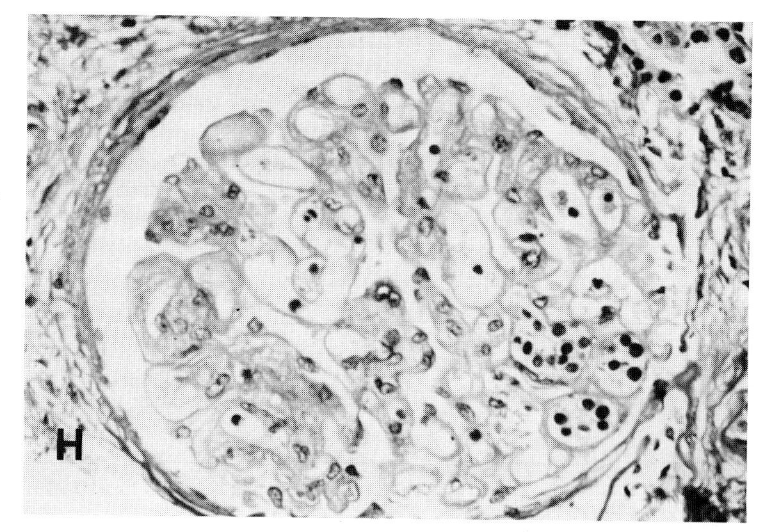

FIG. 13-18
NEPHROCALCINOSIS. **(A)** Urine sediment with inflammation, urothelial cells, pathologic waxy cast, and triple phosphate crystal (arrow). (× 400) **(B)** Triple phosphate crystals in urine sediment. (× 1000)

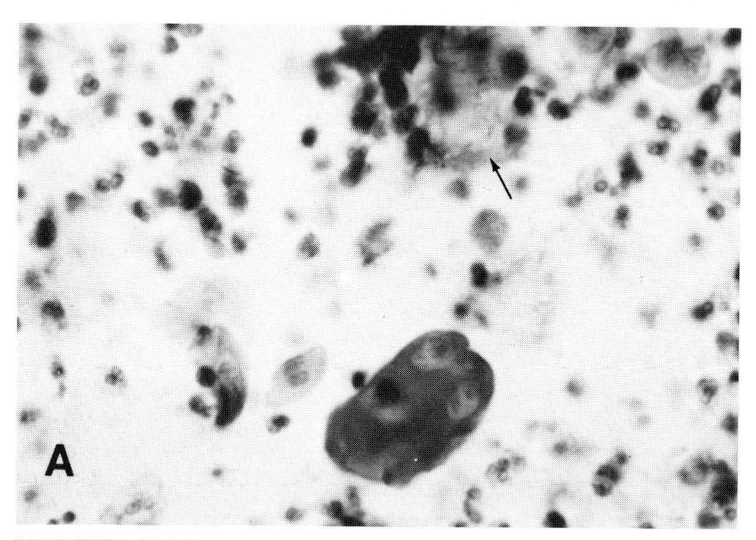

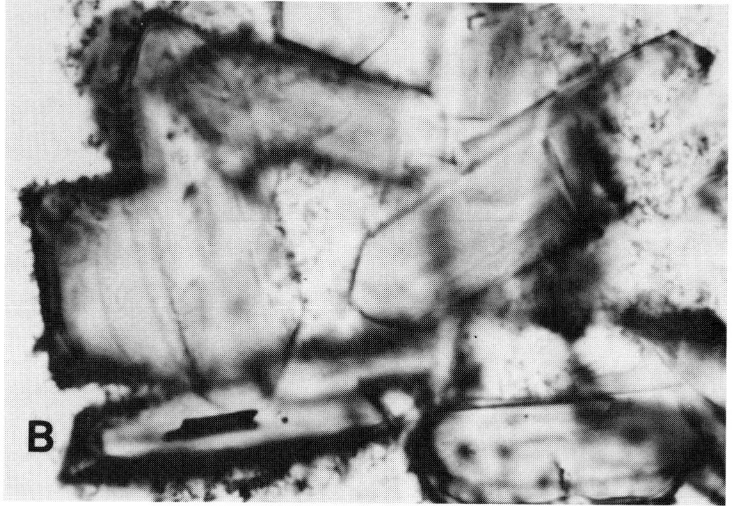

(continued)

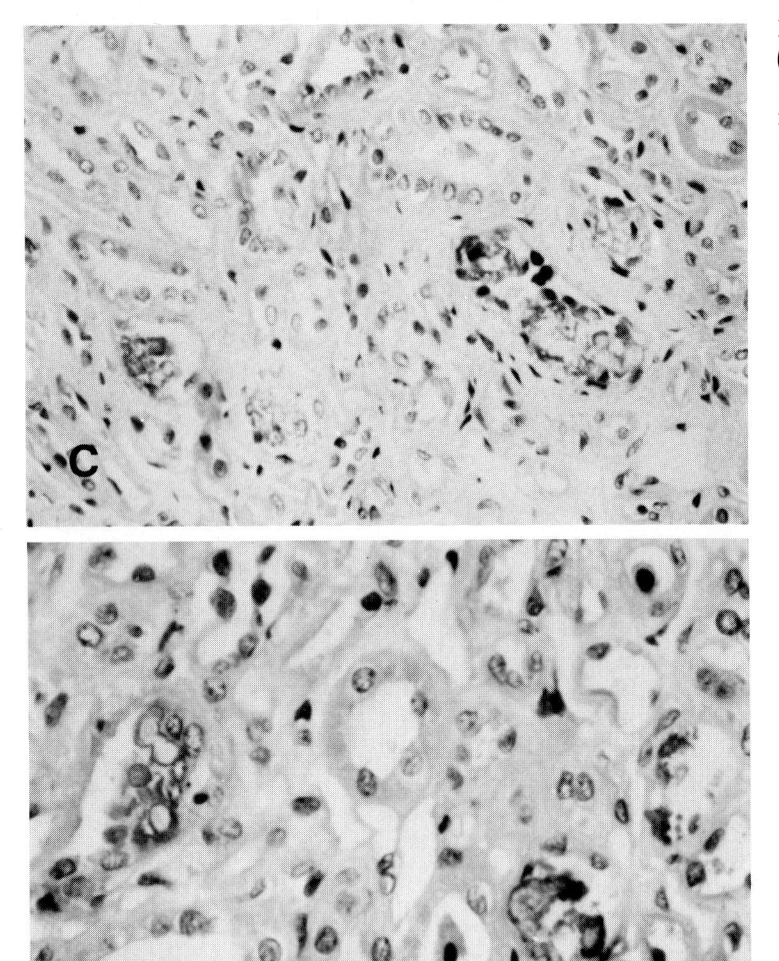

FIG. 13-18C, D *(continued)*
(C) Calcified crystals within tubules. (× 250) **(D)** Note exfoliation of individual renal tubular cells with pyknotic nuclei. (× 400)

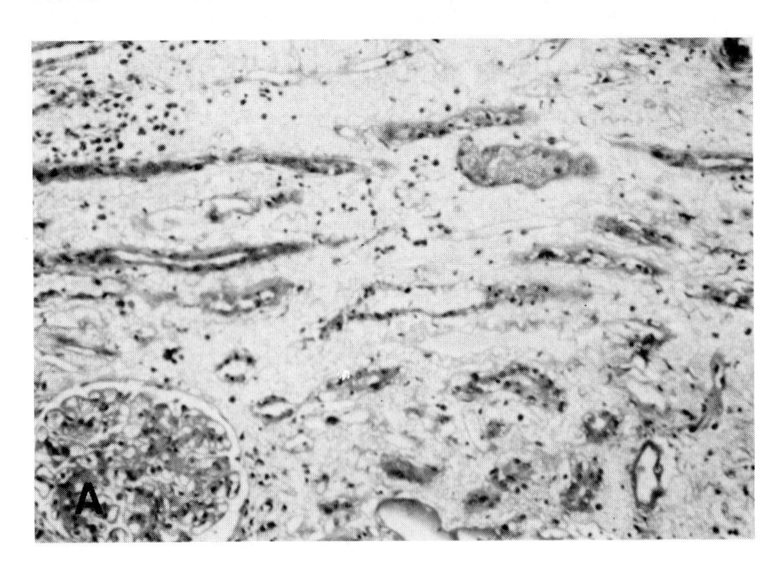

FIG. 13-19
COMPLICATIONS OF TRANSPLANTATION. Glomerulopathy with renal vein thrombosis (RVT). Allograft biopsy 13 months post-transplant for proteinuria (3 gm) and increased serum creatinine (6.5 mg/dl). **(A)** Striking interstitial edema with a mild plasma cell infiltrate. (× 100)

(continued)

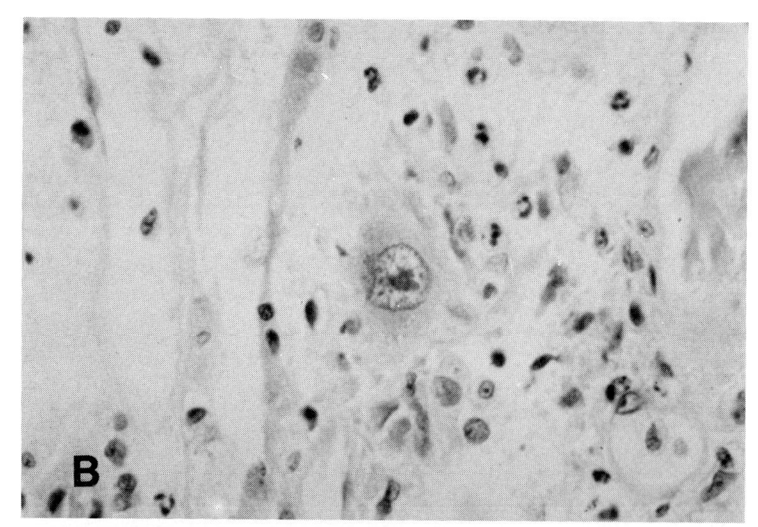

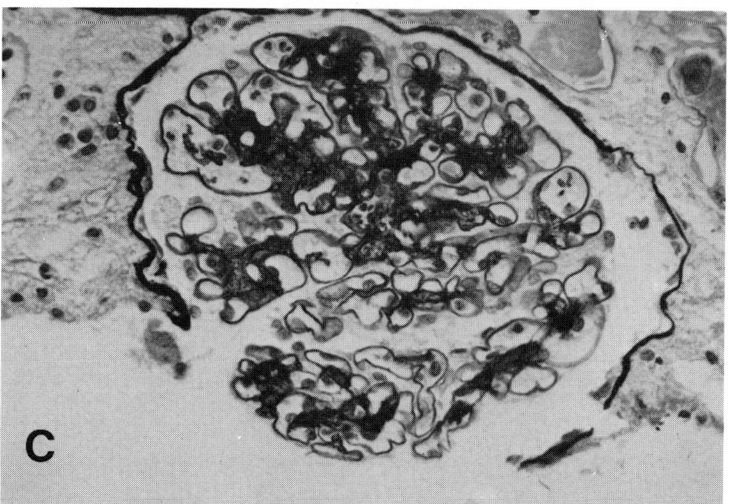

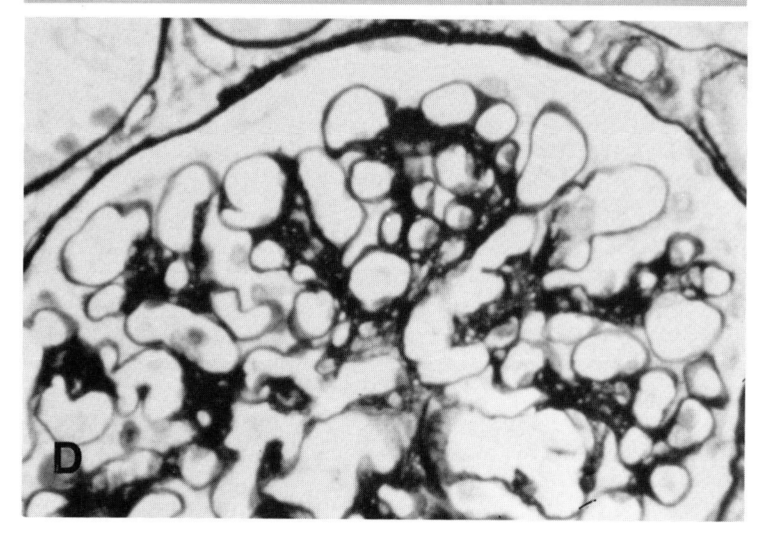

FIG. 13-19B-D *(continued)*
(B) Cytomegalic cell of probable tubular epithelial origin. The markedly enlarged nucleus has chromatin clearing and a prominent irregular nucleolus resembling a viral inclusion. Note neutrophils within peritubular capillaries. (× 400) **(C)** Glomerulus with increased mesangial matrix and stiffened capillary walls with slight irregular thickening of basement membranes. Note the prominence of intracapillary neutrophils. (Jones × 250) Repeat biopsy following heparin therapy and documentation of RVT by venogram. **(D)** The slightly thickened glomerular basemant membranes lack identifiable spikes. (Jones × 400)

(continued)

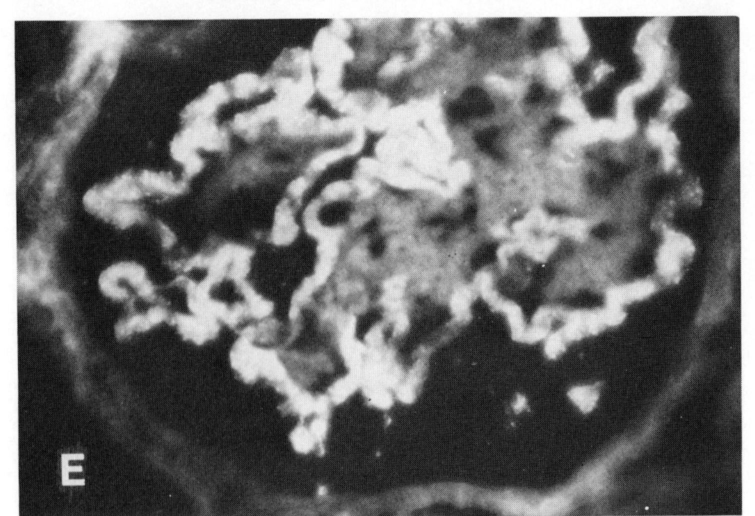

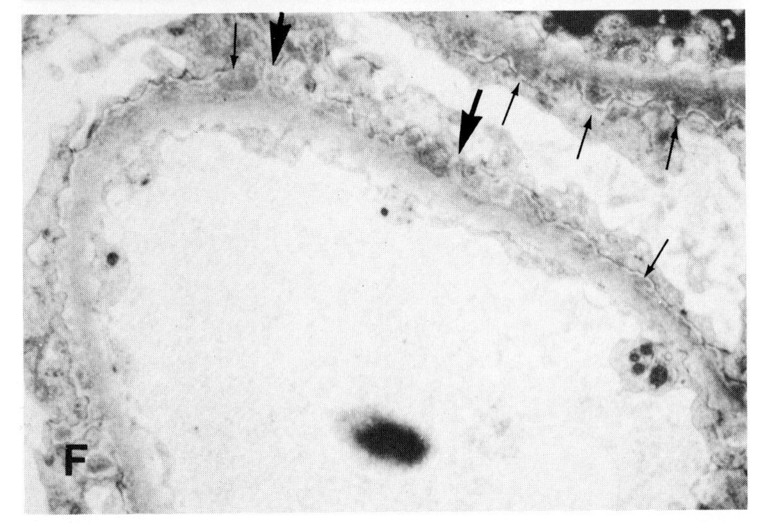

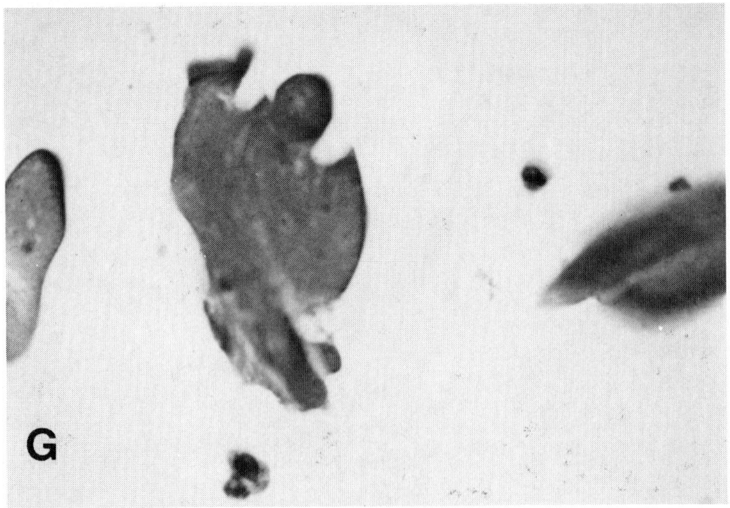

FIG. 13-19E-G *(continued)*
(E) Immunoflourescent stain for human IgG. Note finely granular deposits along glomerular capillary walls (× 400). **(F)** Electron micrograph of glomerulus, demonstrating epimembranous deposits (small arrows) with occasional basement membrane spikes (large arrows). (× 15,200) These findings suggest *de novo* membranous glomerulopathy. **(G)** Globular material in urine sediment. (× 400)

(continued)

FIG. 13-19H-J *(continued)*
(H) Renal epithelial cells with waxy cytoplasmic material (× 1000) These inclusion cells are frequently associated with waxy casts. **(I)** Renal epithelial fragment in urine sediment indicating ischemic necrosis. (× 1000) **(J)** Urine sediment with bizarre binucleated epithelial cell with marked nuclear enlargement, chromatin clumping and clearing, and prominent nucleoli. (× 1000) Cellular features correlate with **B.**

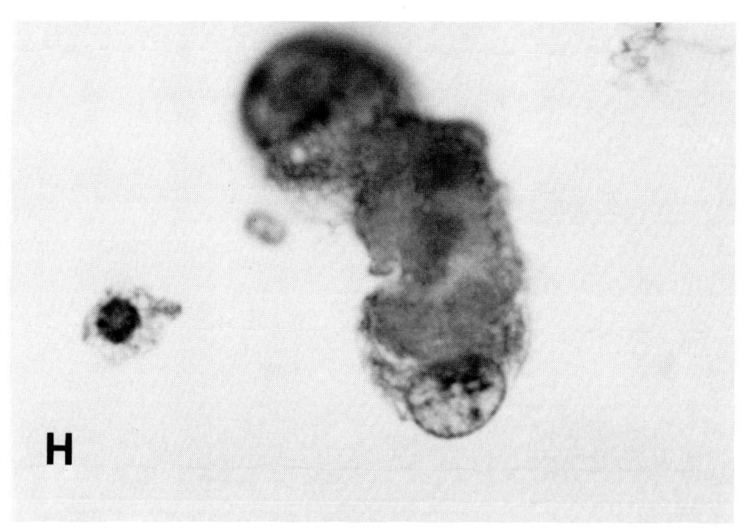

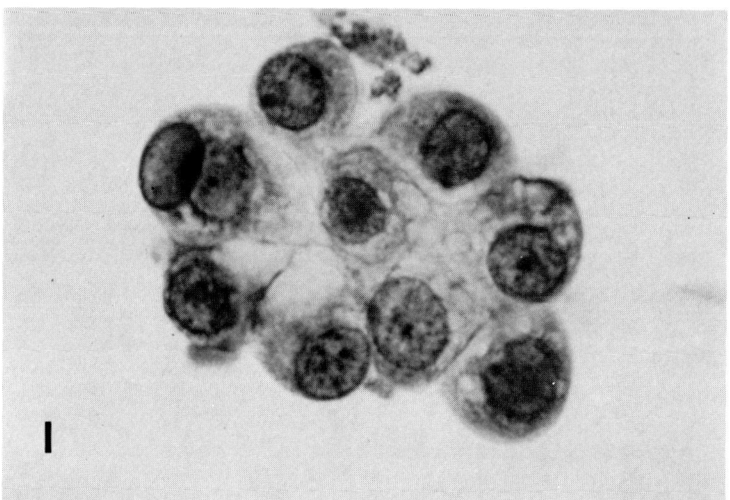

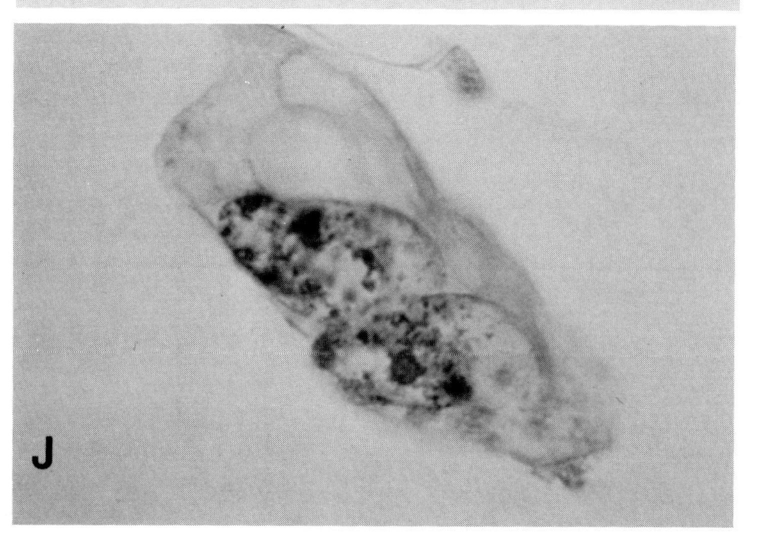

(continued)

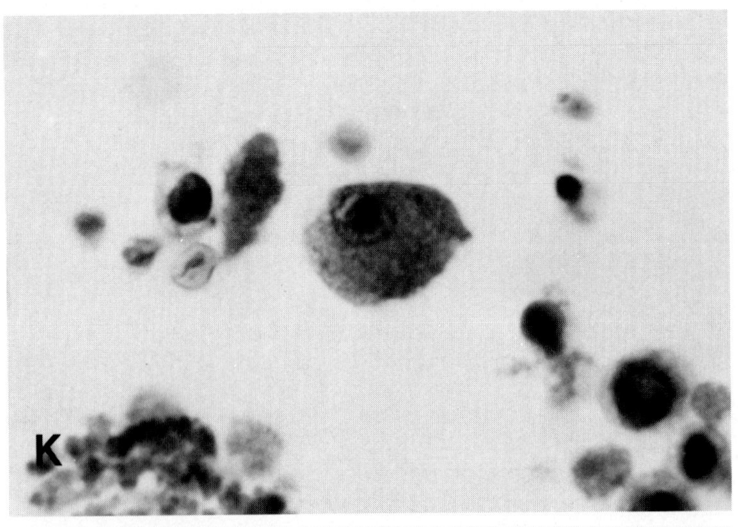

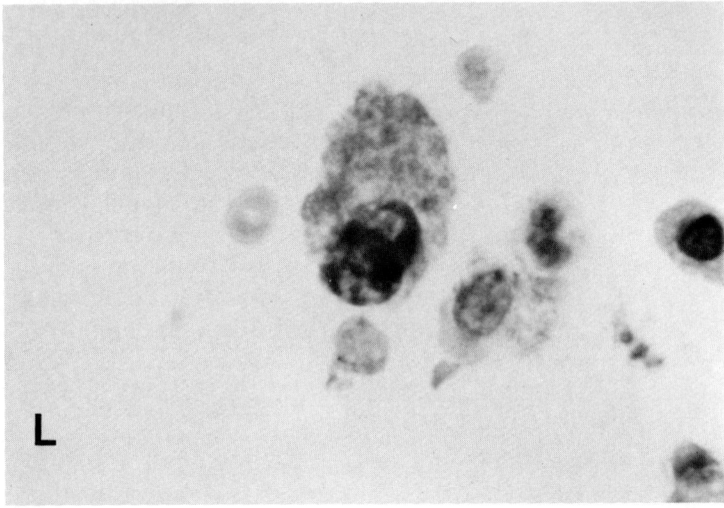

FIG. 13-19K-L *(continued)*
(K) and **(L)** Urine sediment with intranuclear inclusion-bearing cells suspicious for viral disease. (× 400)

14
RENAL NEOPLASIA

The urine sediment examination has limited value in the diagnosis of renal neoplasia. The exfoliation of malignant cells into the urine stream generally requires extension to the collecting system. Fine-needle aspiration is the most reliable cytologic technique in diagnosing renal neoplasia. Short of this technique, the only definitive means of diagnosis is by nephrectomy.

PRIMARY NEOPLASIA

RENAL CELL CARCINOMA (ADENOCARCINOMA)

This tumor, which has been shown ultrastructurally to arise from proximal convoluted tubular epithelium, is considered an adenocarcinoma. Histologically, a variety of growth patterns may be observed, including tubular, trabecular, and papillary, and there is a characteristic rich capillary vascularity (Fig. 14-1). The cytologic appearance is also quite variable, although the clear cell type is most commonly recognized. In histologic material, these cells may look deceptively benign because of the abundant clear cytoplasm and relatively small nuclei. In specimens obtained by fine-needle aspiration, nuclear enlargement, mild hyperchromasia, and macronucleoli are more obvious. The abundant finely vacuolated or foamy cytoplasm characteristically stains positive for glycogen and fat, although this feature is not diagnostic. In addition, in tumors composed predominantly or solely of granular cells (Fig. 14-2), the fat stain can be negative. Pleomorphic forms of renal adenocarcinoma, including a sarcomatoid variant, are occasionally encountered (Fig. 14-3). Grading based upon cytologic appearance is of questionable value in determining prognosis. The stage, including renal vein and capsular invasion, is a more important determinant.

WILMS' TUMOR

This malignant neoplasia arising from renal blastema occurs in young children. Histologically, there is a variable component of primitive blastema, spindle cell stroma, and tubular differentiation (Fig. 14-4). To our knowledge, a primary cytodiagnosis of Wilms' tumor has not been established from a urine sediment examination.

METASTATIC NEOPLASIA

Renal involvement by malignant tumor is commonly observed at autopsy as a component of widespread metastases. Lung carcinoma is a common metastatic source, and renal metastases may result from hematogenous spread (Fig. 14-5).

Lymphoreticular neoplasia may also involve

renal parenchyma as a part of widespread dissemination (Figs. 14-6 and 14-7). On rare occasions extensive infiltration has resulted in renal dysfunction. With lesser degrees of parenchymal replacement, urine sediment patterns of tubular injury may be related to chemotherapy.

In patients with multiple myeloma, the urine sediment findings may reflect the type of renal disease. Atypical plasma cells have been observed in urine (Fig. 14-8). The findings of nephrotic syndrome may indicate renal amyloidosis, and in myeloma kidney with cast nephropathy, urine sediment may show evidence of tubular injury with pathologic casts (Fig. 14-9).

BIBLIOGRAPHY

Bennington JL, Beckwith JB: Tumors of the Kidney, Renal Pelvis and Ureter: Atlas of Tumor Pathology, 2nd series, Fasicle 12. Washington, D.C., Armed Forces Institute of Pathology, 1975

Schumann GB: The Urine Sediment Examination. Baltimore, Williams & Wilkins, 1980.

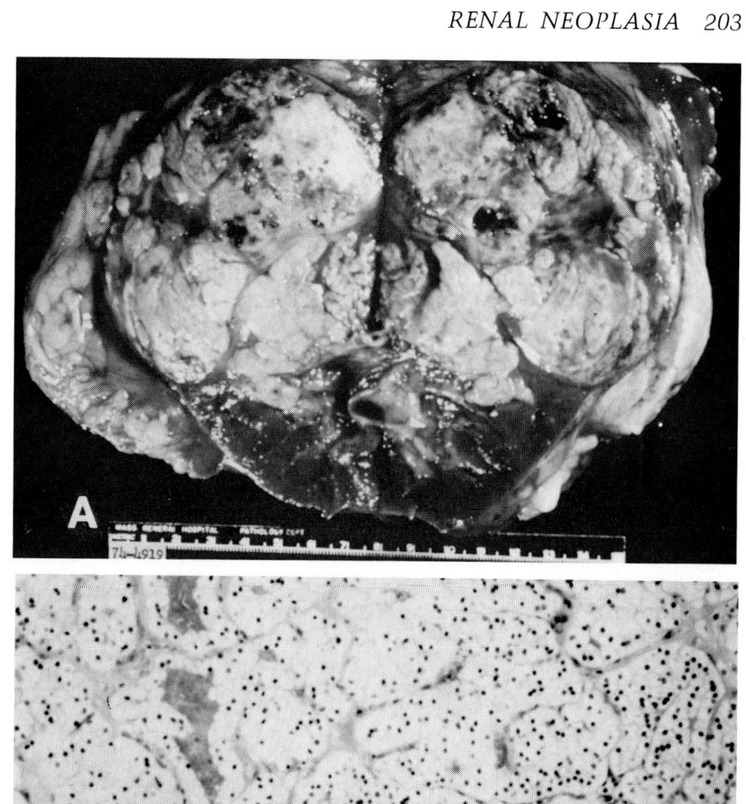

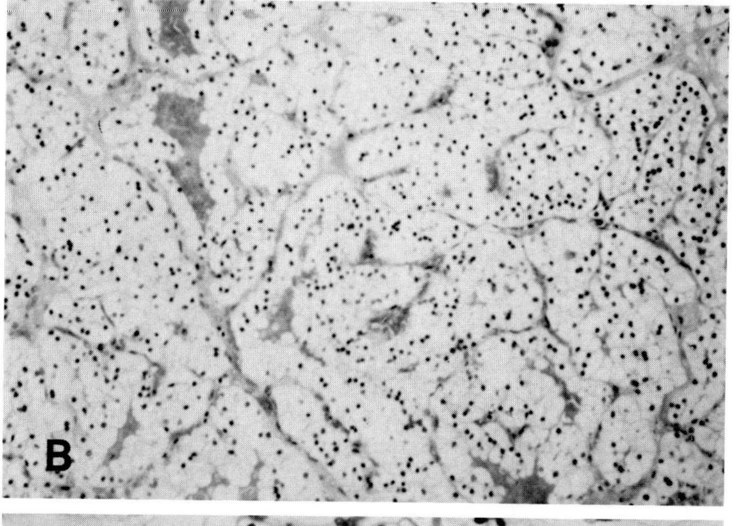

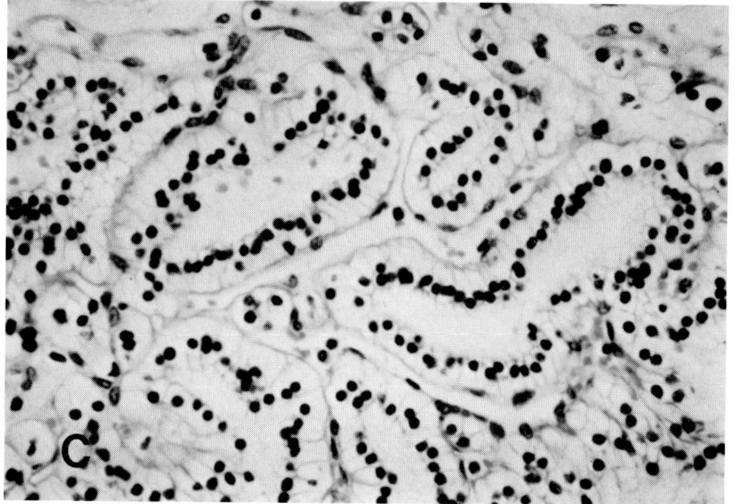

FIG. 14-1
RENAL CELL ADENOCARCINOMA, CLEAR CELL TYPE. **(A)** The upper and midportion of this bivalved kidney are replaced by a lobulated yellow-tan tumor. The mottled appearance is due to foci of fibrosis, hemorrhage, and necrosis. The renal capsule and pelvis are invaded. (Gross) **(B)** Nests and trabeculae are composed of cells with abundant clear cytoplasm. Note the characteristic rich capillary vascularity and scant intervening stroma. (× 100) **(C)** Tubular growth pattern with delicate fibrovascular stroma. Tubules are lined by columnar clear cells with uniform, hyperchromatic nuclei. (× 250)

(continued)

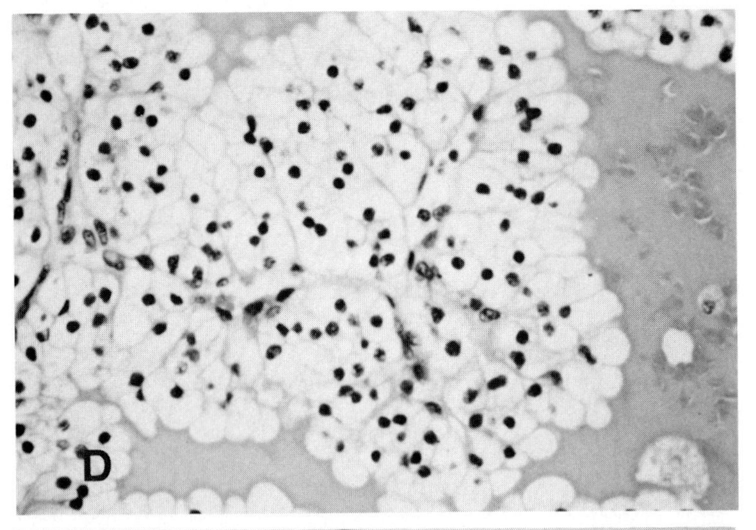

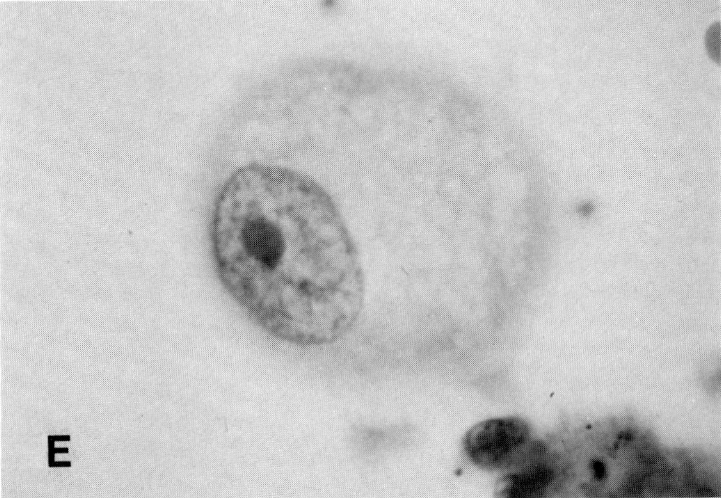

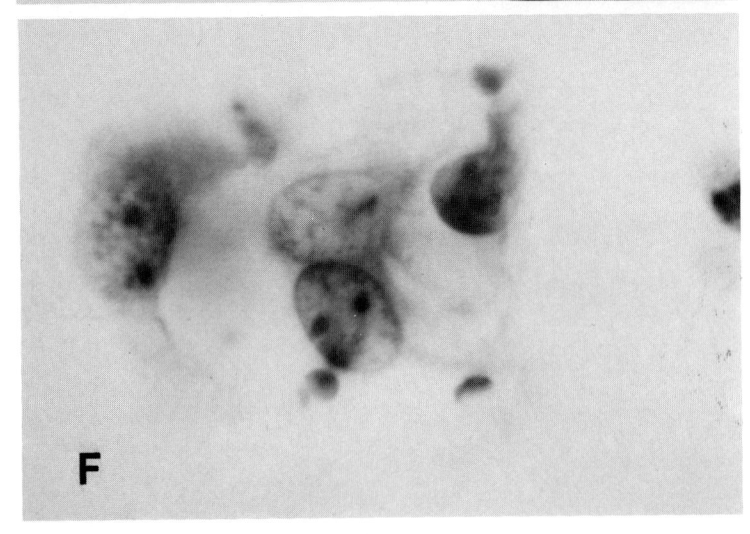

FIG. 14-1D-F *(continued)*
(D) Proliferation of cells around a delicate branching fibrovascular stalk forms a papillary projection within a tubular lumen containing erythrocytes and proteinaceous material. Note polygonal cells with abundant clear cytoplasm, distinct cell borders, and uniform hyperchromatic nuclei. (× 250) **(E)** Malignant epithelial cell derived from adenocarcinoma. Nuclear enlargement, mild hyperchromasia, macronucleolus, and abundant foamy cytoplasm are characteristic of renal cell carcinoma. Fine-needle aspiration. (× 1000) **(F)** Loose cluster of malignant cells derived from renal cell carcinoma. Hyperchromatic nuclei with prominent nucleoli and abundant foamy or vacuolated cytoplasm. Fine-needle aspiration. (× 1000)

FIG. 14-2
RENAL CELL ADENOCARCINOMA, GRANULAR CELL TYPE. **(A)** Bivalved kidney with subtotal replacement by an orange-brown lobulated tumor with oentral scarring. (Gross) **(B)** Tubules are lined by cells with abundant granular cytoplasm. Note unclear hyperchromasia and moderate pleomorphism. (× 250) Foci containing clear cells were not present in this tumor.

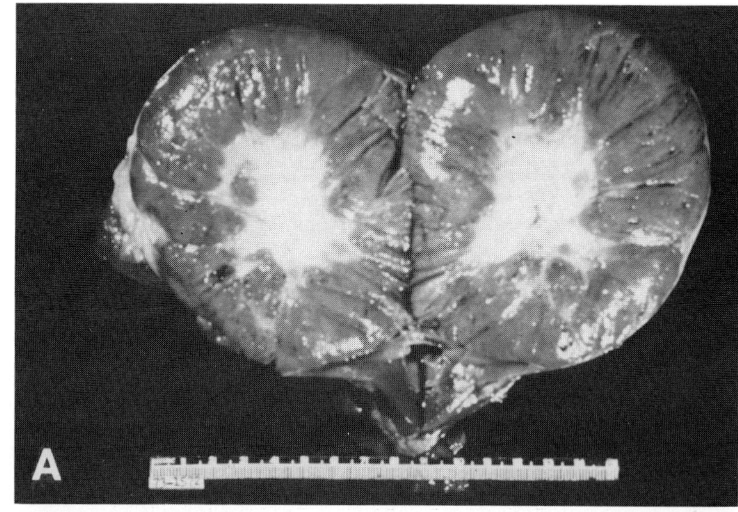

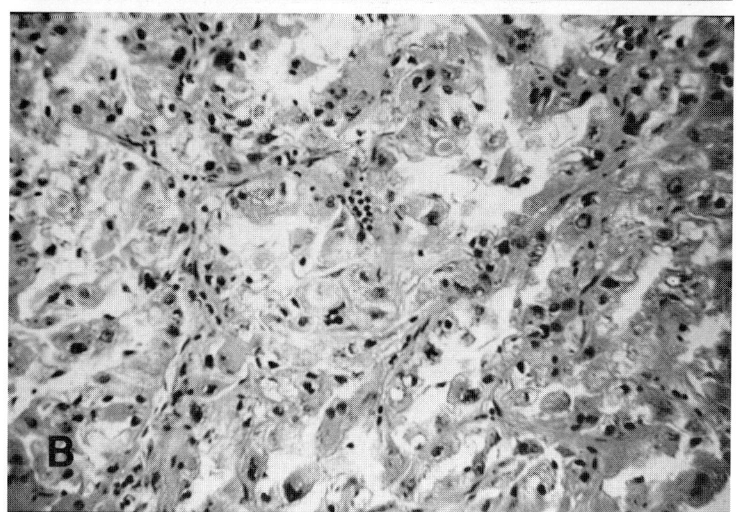

FIG. 14-3
RENAL CELL ADENOCARCINOMA, PLEOMORPHIC TYPE. **(A)** Exfoliated pleomorphic malignant cells (× 400)

(continued)

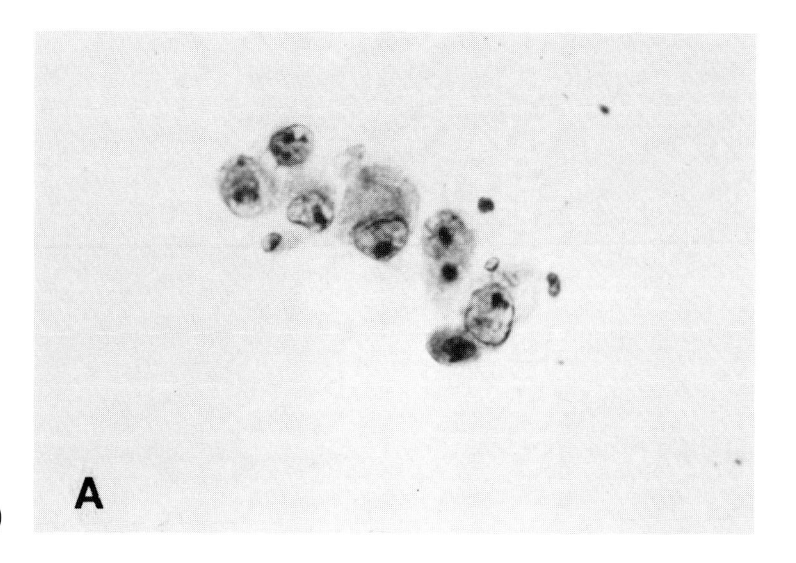

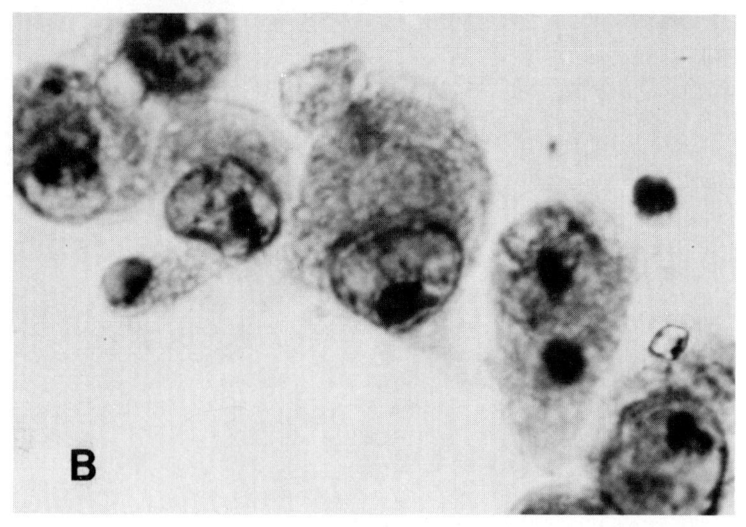

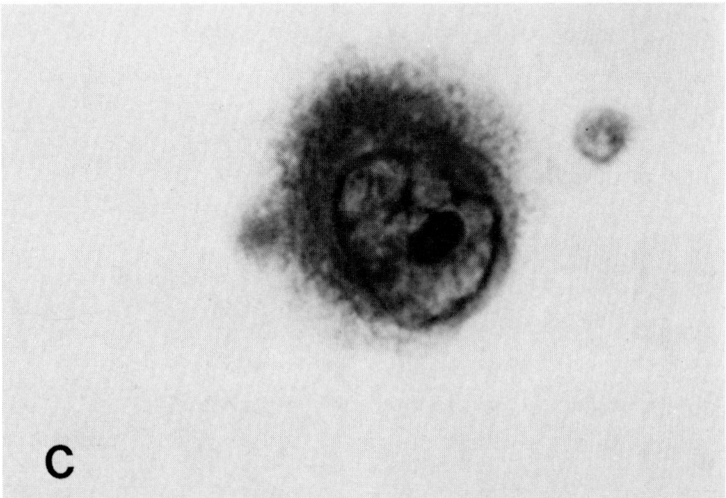

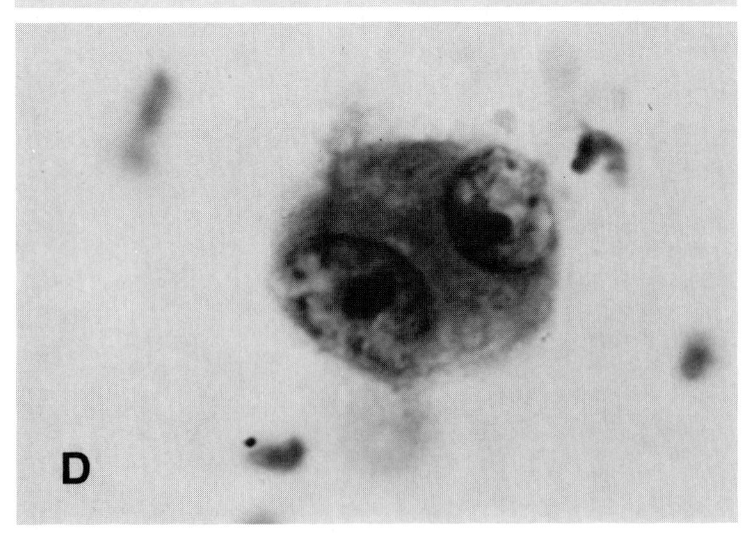

FIG. 14-3B-D *(continued)*
(B) (× 1000). **(C)** Note macronucleolus and finely vacuolated cytoplasm. (× 1000) **(D)** Binucleate malignant cell with prominent nucleoli.

(continued)

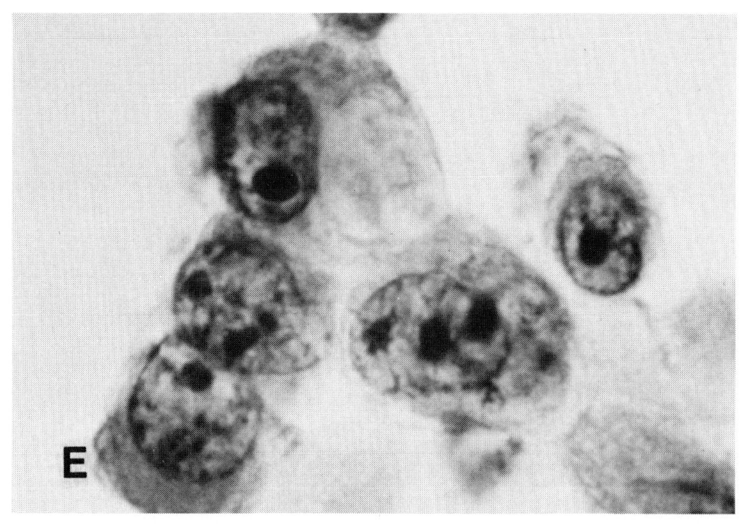

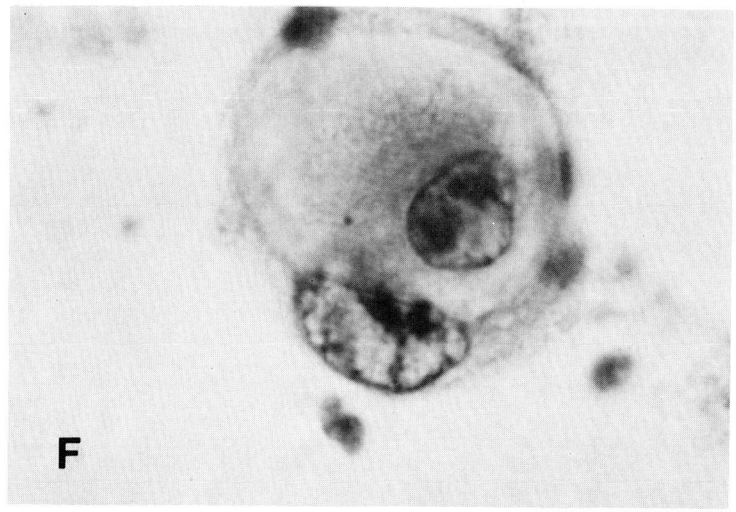

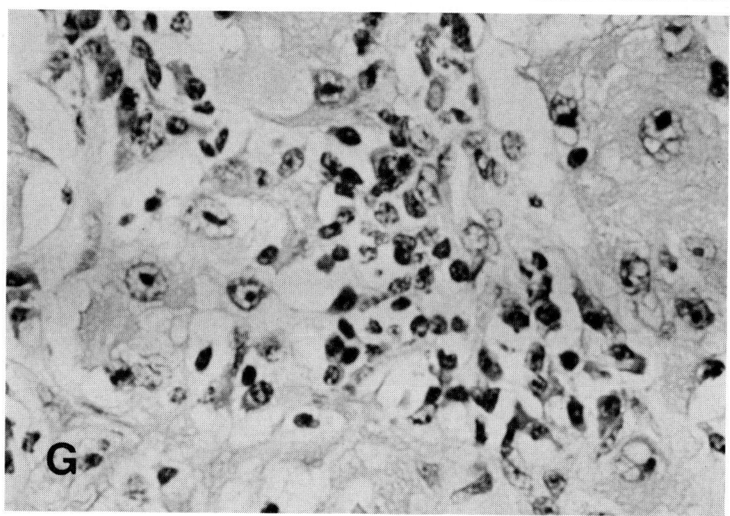

FIG. 14-3E-G *(continued)*
(E) Fragment of malignant cells. Note anisonucleosis. (× 1000) **(F)** Note vacuolization. (× 1000) **(G)** Lymph node with metastatic carcinoma. Cells have abundant granular and vacuolated cytoplasm with indistinct borders. (× 400)

(continued)

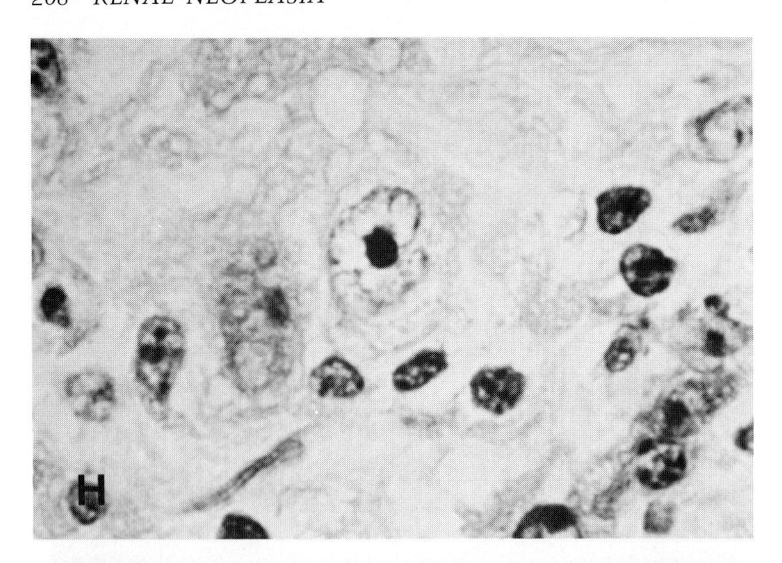

FIG. 14-3H *(continued)*
(H) Note the nuclear pleopmorphism and macronucleoli. (× 1000)

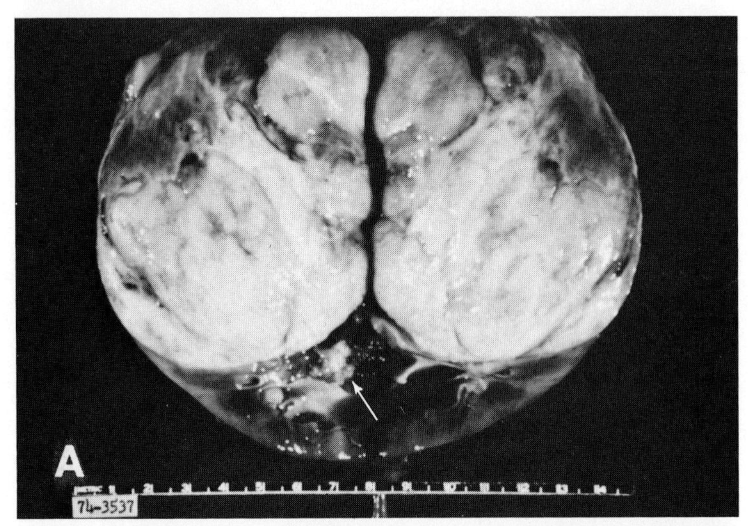

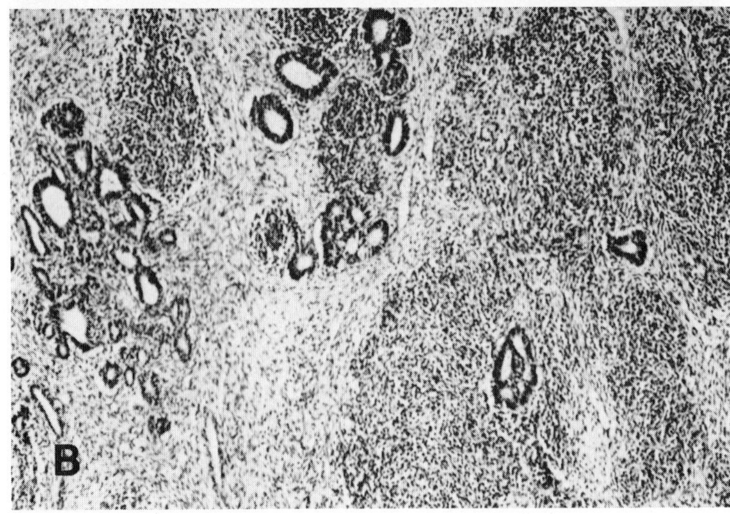

FIG. 14-4
WILMS' TUMOR (NEPHROBLASTOMA). **(A)** This psuedoencapsulated tumor has a typical pale, bulging cut surface. Slight variegation is due to recent and old hemorrhage. Renal vein invasion (arrow) is present. (Gross) **(B)** Typical mixed appearance with well-formed tubular structures, spindle cell stroma, and nodules of blastema. (× 100)

(continued)

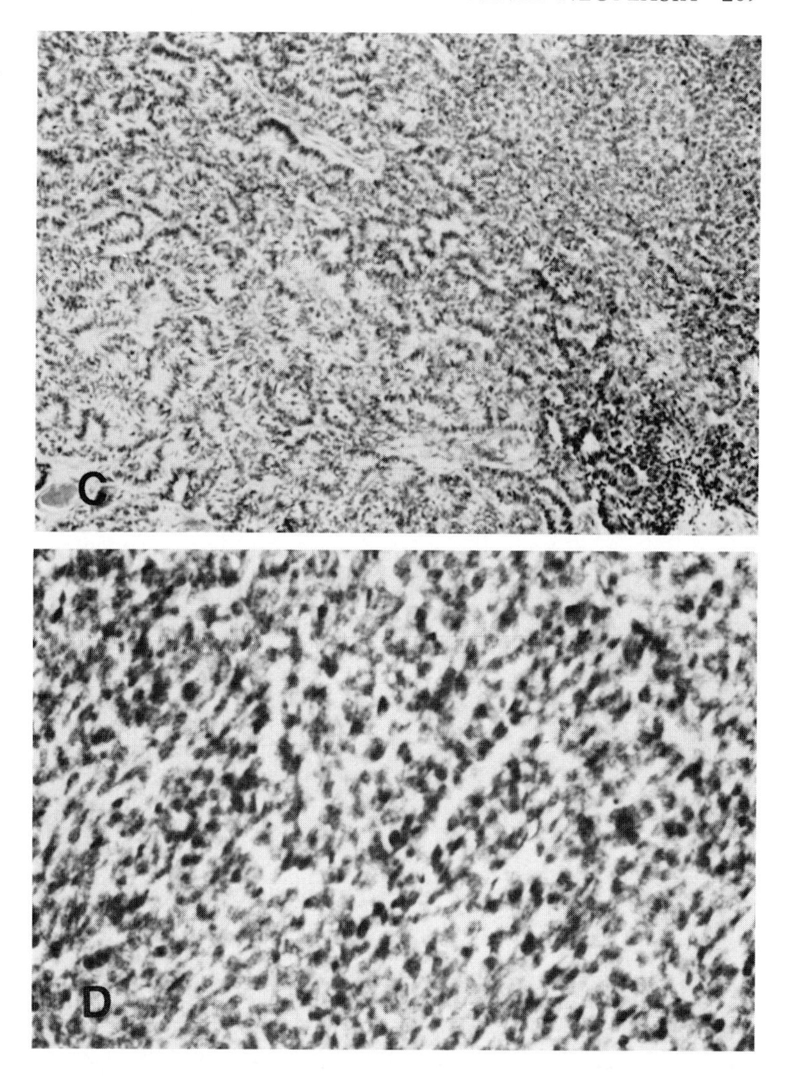

FIG. 14-4C-D *(continued)*
(C) Focus of blastema (upper right) merges with a zone of embryonal tubular differentiation. (× 100) **(D)** Blastema is composed of tightly packed small cells with scant cytoplasm and hyperchromatic nuclei. (× 400)

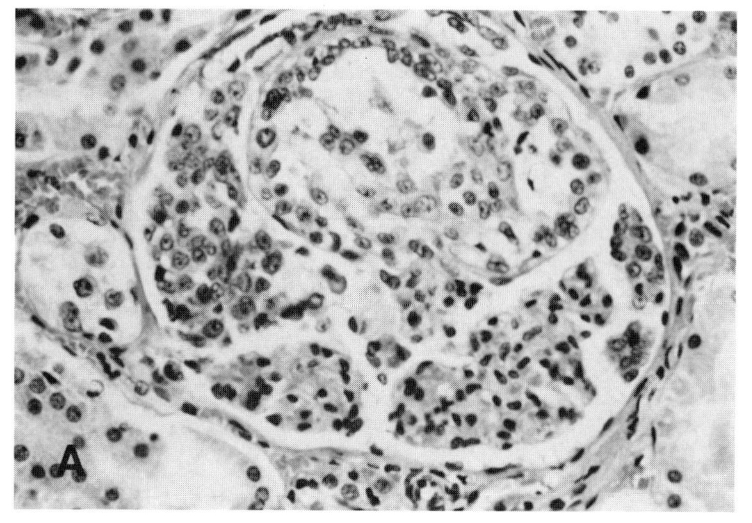

FIG. 14-5
METASTATIC CARCINOMA. **(A)** A glomerulus is segmentally replaced by malignant cells having uniform hyperchromatic nuclei with nucleoli and granular eosinophilic cytoplasm. There is a suggestion of acinar formation. Note malignant cells in Bowman's space and in an adjacent proximal convoluted tubule. (× 400)

(continued)

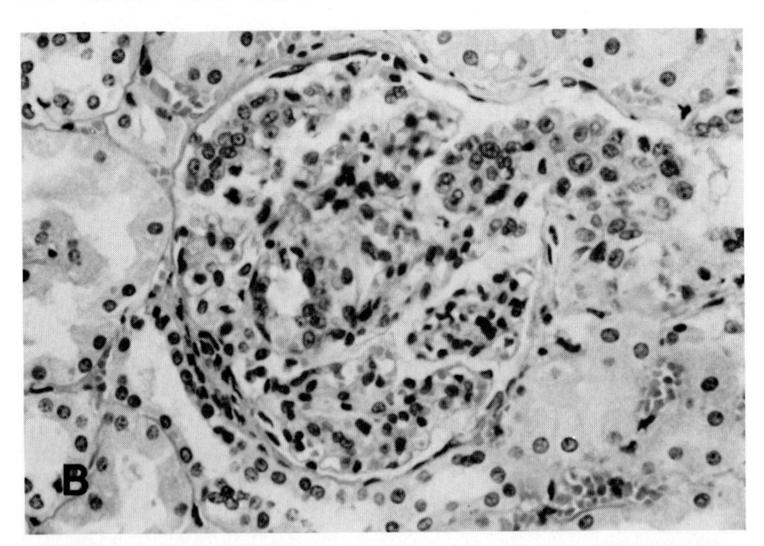

FIG. 14-5B *(continued)*
(B) Another glomerulus contains metastatic tumor. Note large clusters of malignant cells in Bowman's space and the corresponding proximal convoluted tubule. (× 400) Patient died with widespread metastatic carcinoma from a primary lung tumor.

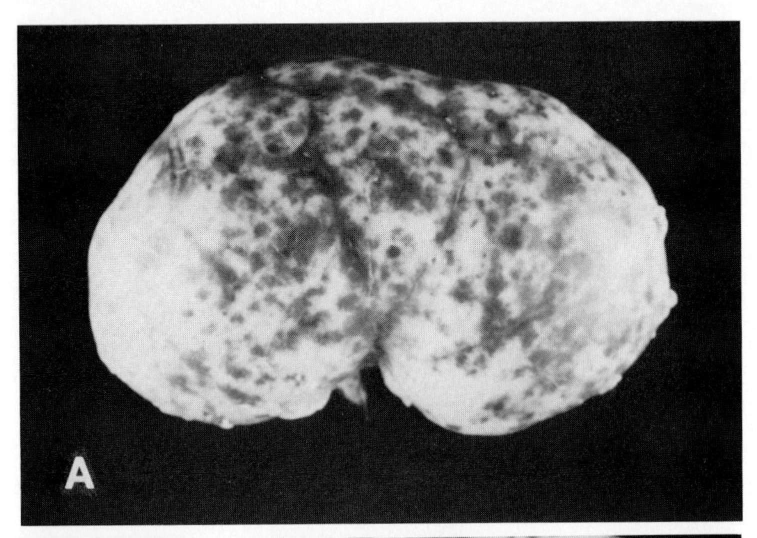

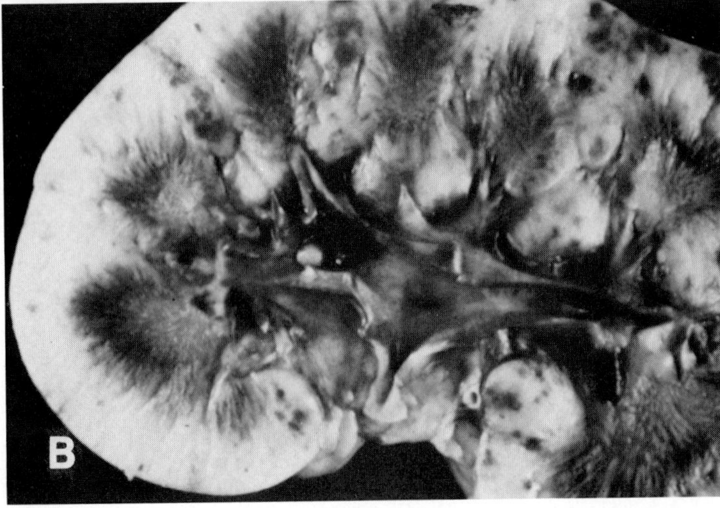

FIG. 14-6
LEUKEMIC INFILTRATE IN KIDNEY. **(A)** The cortical surface is diffusely mottled with pale zones representing leukemic infiltrates. Note extensive ecchymoses indicating parenchymal hemorrhage. (Gross) **(B)** Cut surface. Note blood clot in the renal pelvis. (Gross) Urinary tract mucosal hemorrhage is a common complication in leukemic patients with thrombocytopenia.

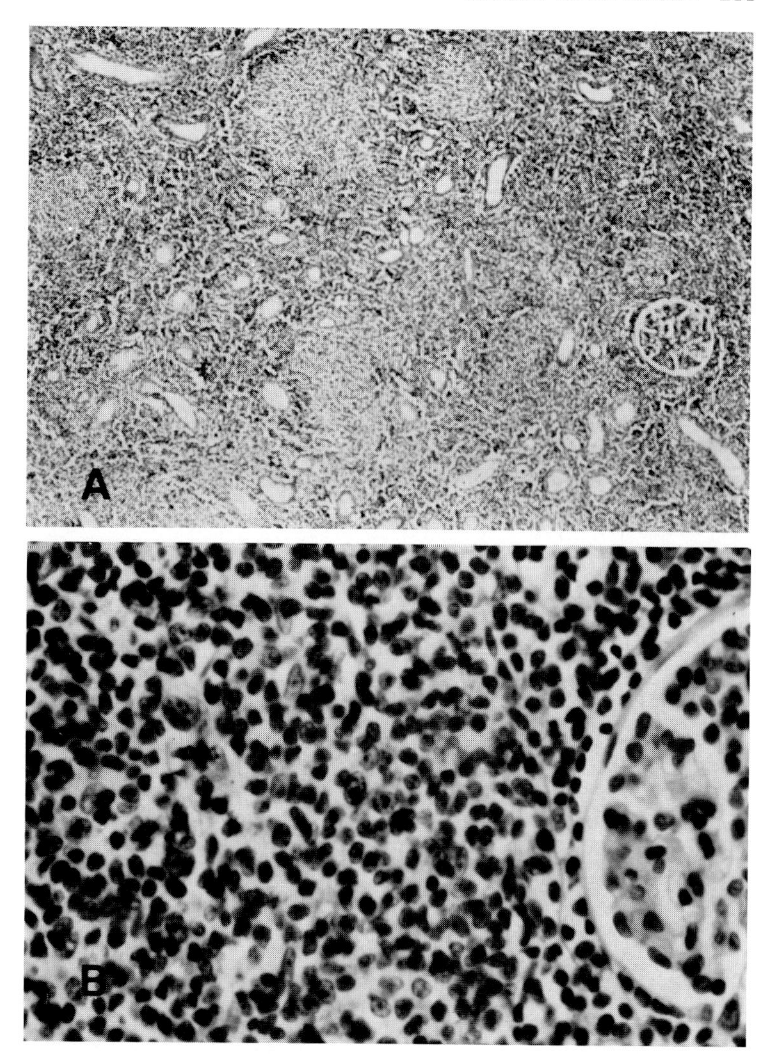

FIG. 14-7
LYMPHOMATOUS INFILTRATE IN KIDNEY. **(A)** Confluent interstitial infiltrate with distinct nodules in the cortex. The widely separated tubules appear diminished. (× 40) **(B)** The infiltrate is composed of immature lymphoid cells having nucleoli. (× 400)

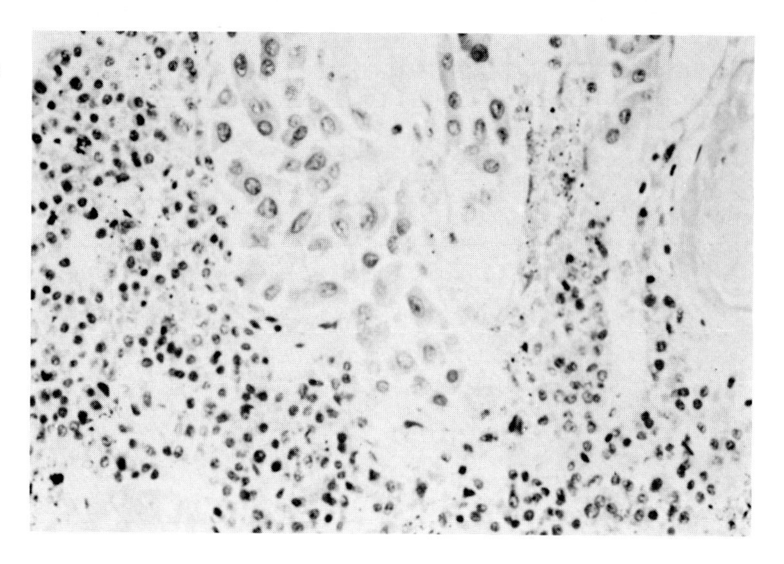

FIG. 14-8
MYELOMA KIDNEY. Interstitial infiltrate containing plasmacytoid cells. Note tubular epithelial exfoliation. (× 250)

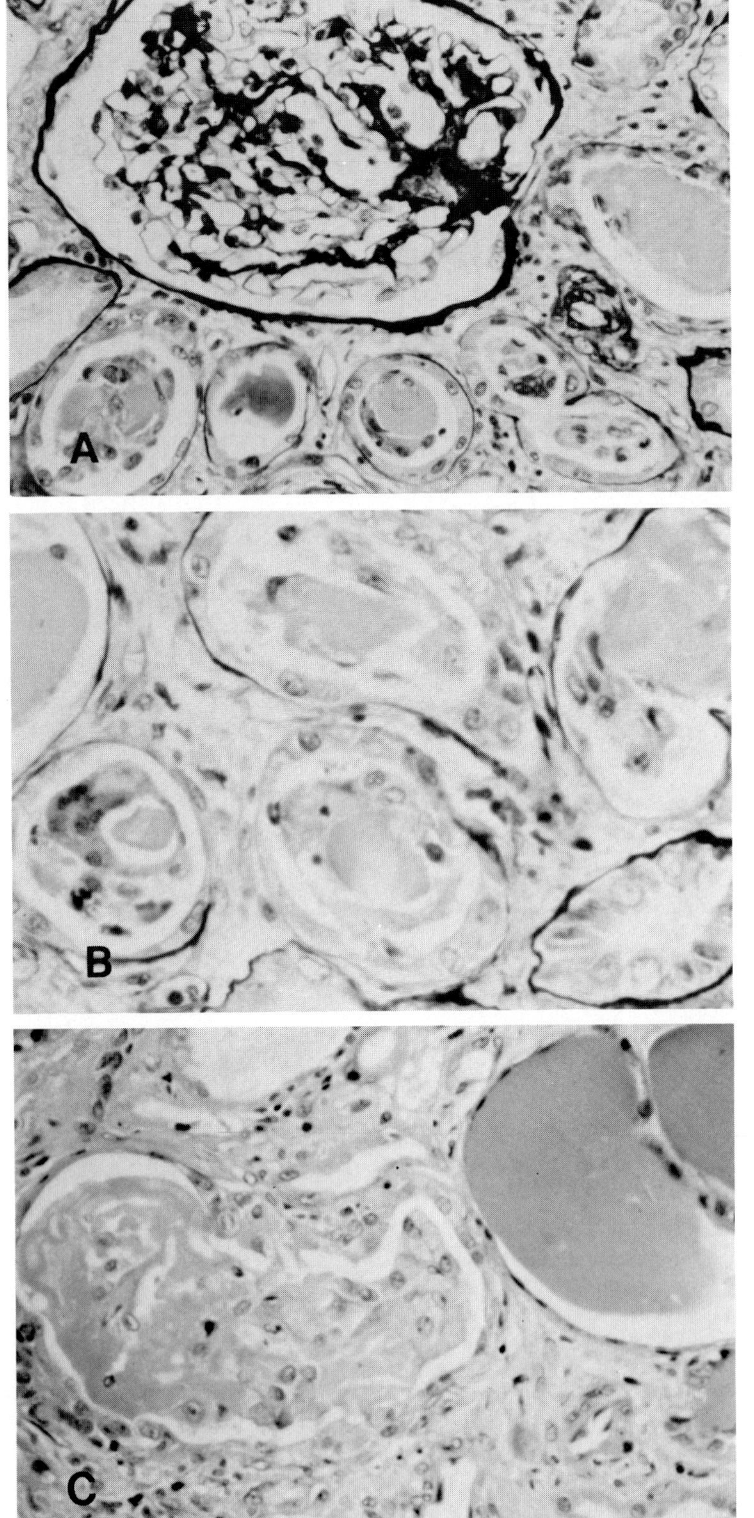

FIG. 14-9
MYELOMA KIDNEY WITH CAST NEPHROPATHY. **(A)** Tubules contain hard-appearing casts with associated epithelial reaction. Amyloid deposition is not present in glomeruli. (Jones × 250) **(B)** Myeloma cast with syncytial epithelial reaction. (Jones × 400) Myeloma cast in dilated tubule with disrupted basement membrane. Note broad waxy casts in adjacent tubules. **(C)** (× 250).

(continued)

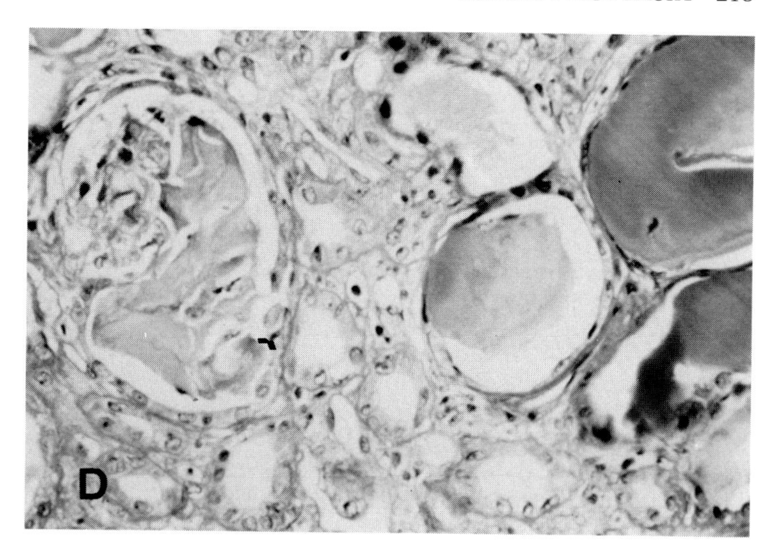

FIG. 14-9D *(continued)*
(D) Trichrome × 250)

PART IV

LOWER URINARY TRACT DISEASES

15

INFLAMMATORY LESIONS OF THE LOWER URINARY TRACT

Inflammatory lesions of the lower urinary tract may involve the urethra, bladder, and ureter. From the urine sediment standpoint, bladder lesions, termed **cystitis,** *are most often encountered. They are usually classified by the inflammatory cell component as acute, chronic, or allergic. The etiologies are varied and include infectious, for example, bacterial, fungal, viral, and parasitic; irritative, for example, calculi; allergic; chemical; or radiation-induced (see Chap. 16). Urothelial degenerative, regenerative, or proliferative changes may also influence the histologic classification and produce important urine sediment findings such as viral cellular changes and metaplasia. These inflammatory lesions cause increased exfoliation of urothelium with an inflammatory background. The lack of renal casts and renal tubular epithelium is necessary for localization to the lower urinary tract.*

ACUTE LESIONS

Acute cystitis (Fig. 15-1) is characterized by neutrophilic infiltration of the bladder lamina propria with variable edema and hemorrhage. It is reflected in the urine sediment by exfoliation of superficial and deep urothelial cells with an acute inflammatory background. The presence of hemorrhage without inflammation suggests the possibility of a bleeding diathesis (Fig. 15-2). Bacteria are not frequently observed histologically, but their presence in the urine sediment with an associated inflammatory background and exfoliated urothelium helps to establish an etiologic diagnosis (Fig. 15-3).

CHRONIC LESIONS

In chronic cystitis, the lamina propria typically contains an inflammatory infiltrate of lymphocytes and plasma cells with variable mixtures of histiocytes and eosinophils. Squamous metaplasia is frequently associated (Fig. 15-4). A variant of chronic cystitis, termed **follicular cystitis,** is histologically recognized by the presence of large lymphoid follicles in the lamina propria (Fig. 15-5). With this condition, the urine sediment examination may show lymphocyturia and exfoliated deep urothelial cells; however, specific cytodiagnosis requires identification of immature lymphocytes, plasma cells, and reticular cells.

Although bacterial infections, particularly with gram-negative organisms, are a common cause of chronic cystitis, other infectious etiologies include fungi and viruses. *Candida* species is the most frequent type of fungus, especially in immunosuppressed and diabetic patients (Fig. 15-6). Viral diseases are rapidly being recognized as a cause of lower urinary tract infections. Classic virally-induced cellular changes are given in Chapters 5 and 8. Viral infections may also cause a proliferative or metaplastic urothelial change

(Fig. 15-7).

Malakoplakia represents an unusual inflammatory response to chronic bacterial urinary tract infection. The histologic and urine sediment diagnosis requires the identification of macrophages containing Michaelis-Gutmann bodies (Fig. 15-8).

ALLERGIC LESIONS

Lower urinary tract lesions felt to be of an allergic etiology are poorly understood. The identification of eosinophils in tissue and in urine sediment is the hallmark of an acute allergic response. In eosinophilic cystitis a dense infiltrate of eosinophils is present in the lamina propria and wall (Fig. 15-9). This may result in eosinophiluria and exfoliation of urothelial cells.

PROLIFERATIVE AND/OR METAPLASTIC LESIONS

Proliferative and metaplastic urothelial lesions have a variable inflammatory component. Proliferative cystitis generally refers to cystitis cystica and cystitis glandularis. Although von Brunn's nests may be a variant of normal urothelium, they frequently increase with inflammatory lesions and can be considered the simplest form of proliferative cystitis. Cystic von Brunn's nests may be lined by urothelial cells, that is, cystitis cystica or by columnar epithelium, that is, cystitis glandularis (Fig. 15-10). Occasionally, this columnar epithelium may be found in the urine sediment. Metaplastic squamous epithelium may occur on the mucosal surface as well as line cysts within the lamina propria. These proliferative and metaplastic changes can also occur in the ureter and produce urine sediment changes (Fig. 15-11).

BIBLIOGRAPHY

Koss LG: Tumors of the Urinary Bladder. Atlas of Tumor Pathology, 2nd series, Fascicle 11. Washington, D.C., Armed Forced Institute of Pathology, 1975

Koss LG: Diagnostic Cystology and Its Histopathologic Bases. 3rd ed. Philadelphia, JB Lippincott, 1979

Voogt HJ, Rathert P, Beyer-Boon ME: Urinary Cytology. Phase Contrast Microscopy and Analysis of Stained Smears. New York, Springer-Verlag, 1977

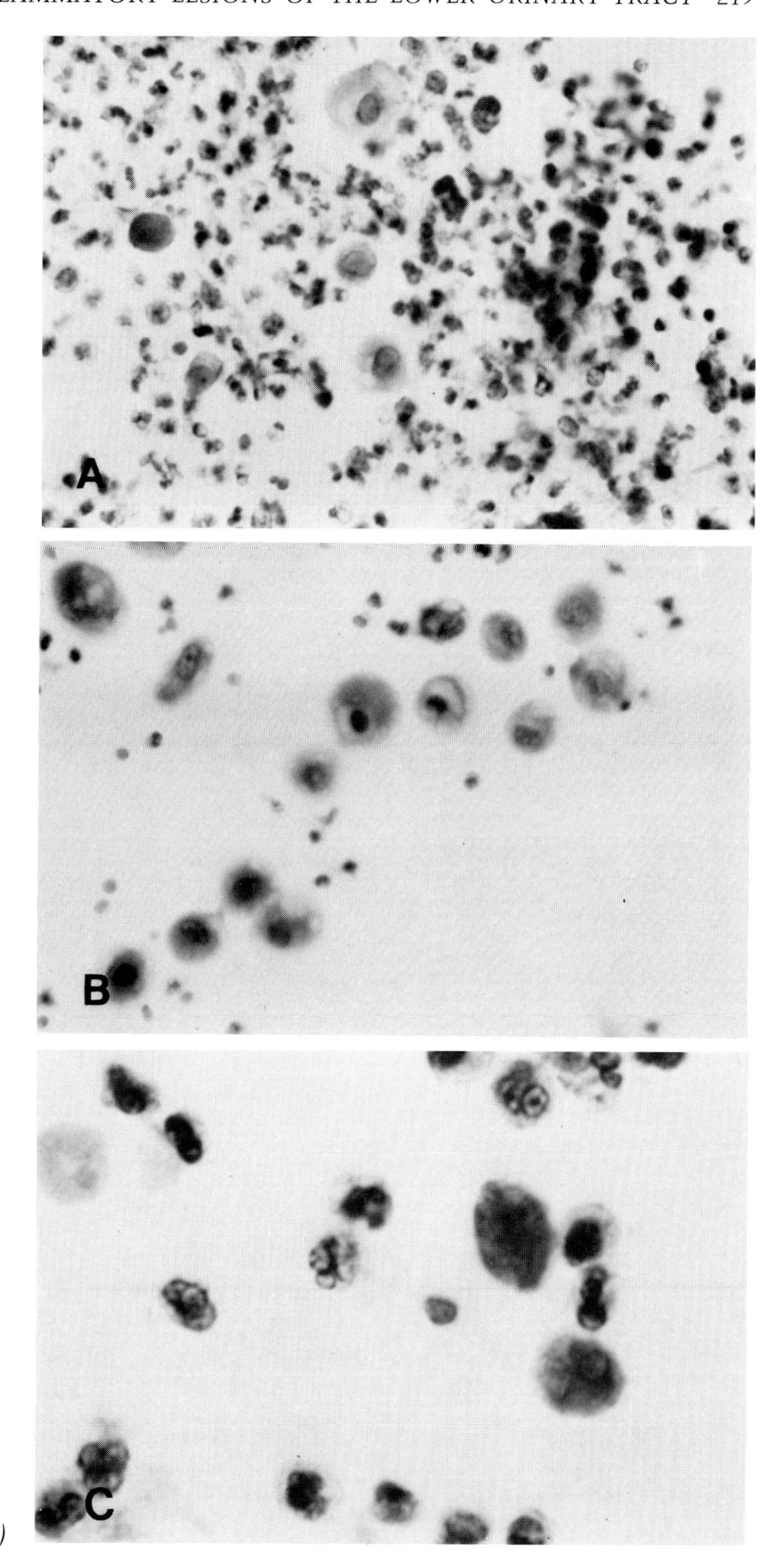

FIG. 15-1
ACUTE CYSTITIS. **(A)** Neutrophilic exudate with urothelial cells. (× 250) **(B)** Deep urothelial cells with degenerative changes. Note cytoplasmic vacuoles. (× 400) **(C)** Segmented neutrophils and degenerated deep urothelial cells with faintly stained nuclei. These degenerated cells should not be mistaken for trichomonads. (× 1000)

(continued)

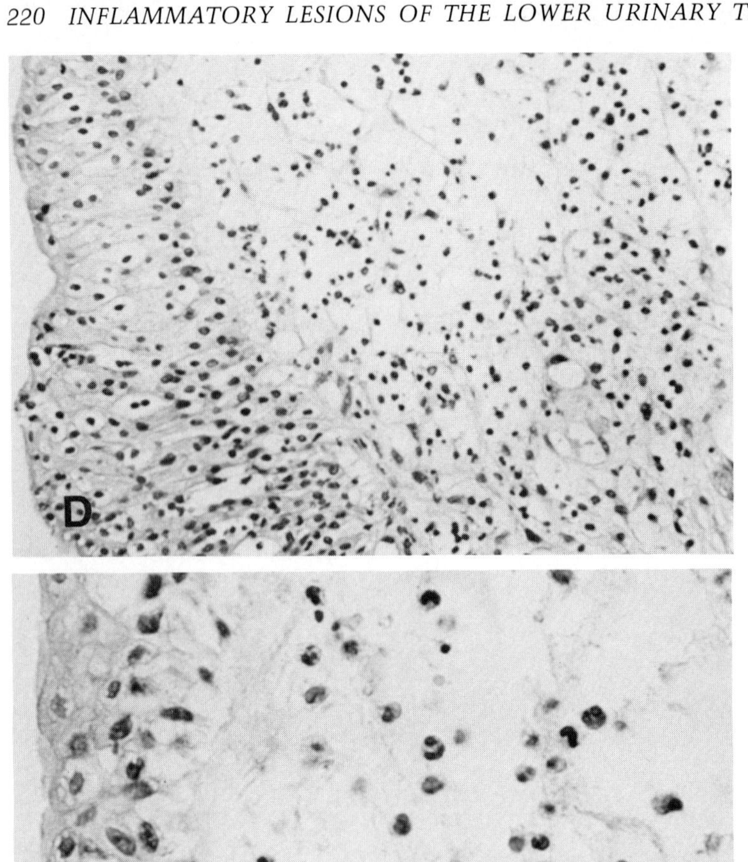

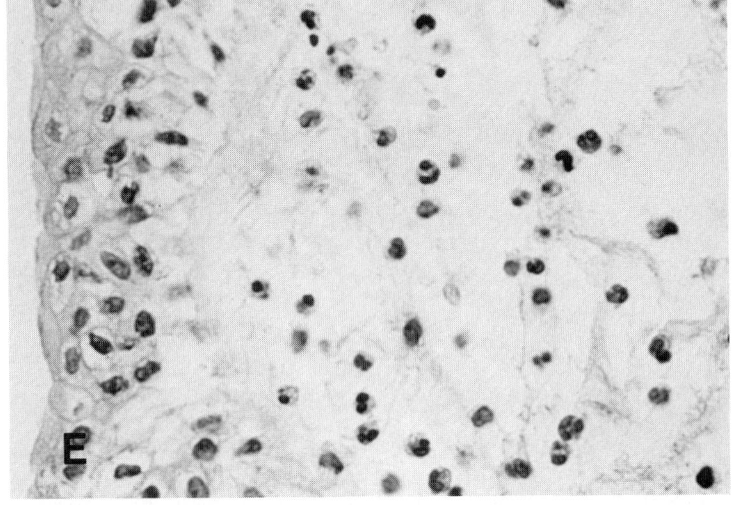

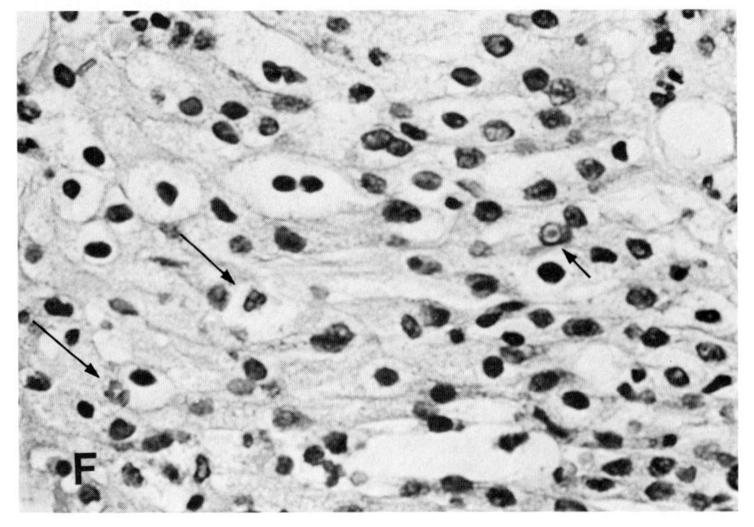

FIG. 15-1D-F *(continued)*
(D) A marked inflammatory infiltrate is present in the edematous lamina propria. Overlying urothelium is hyperplastic but shows no cytologic atypia. (× 160) **(E)** Inflammatory infiltrate composed predominantly of neutrophilic leukocytes. (× 400) **(F)** Neutrophils have migrated into the hyperplastic urothelium. Transitional cells are frequently vacuolated and occasionally contain intracytoplasmic leukocytes (long arrows). One cell (short arrow) appears to have an intranuclear inclusion, suggesting viral infection. (× 400)

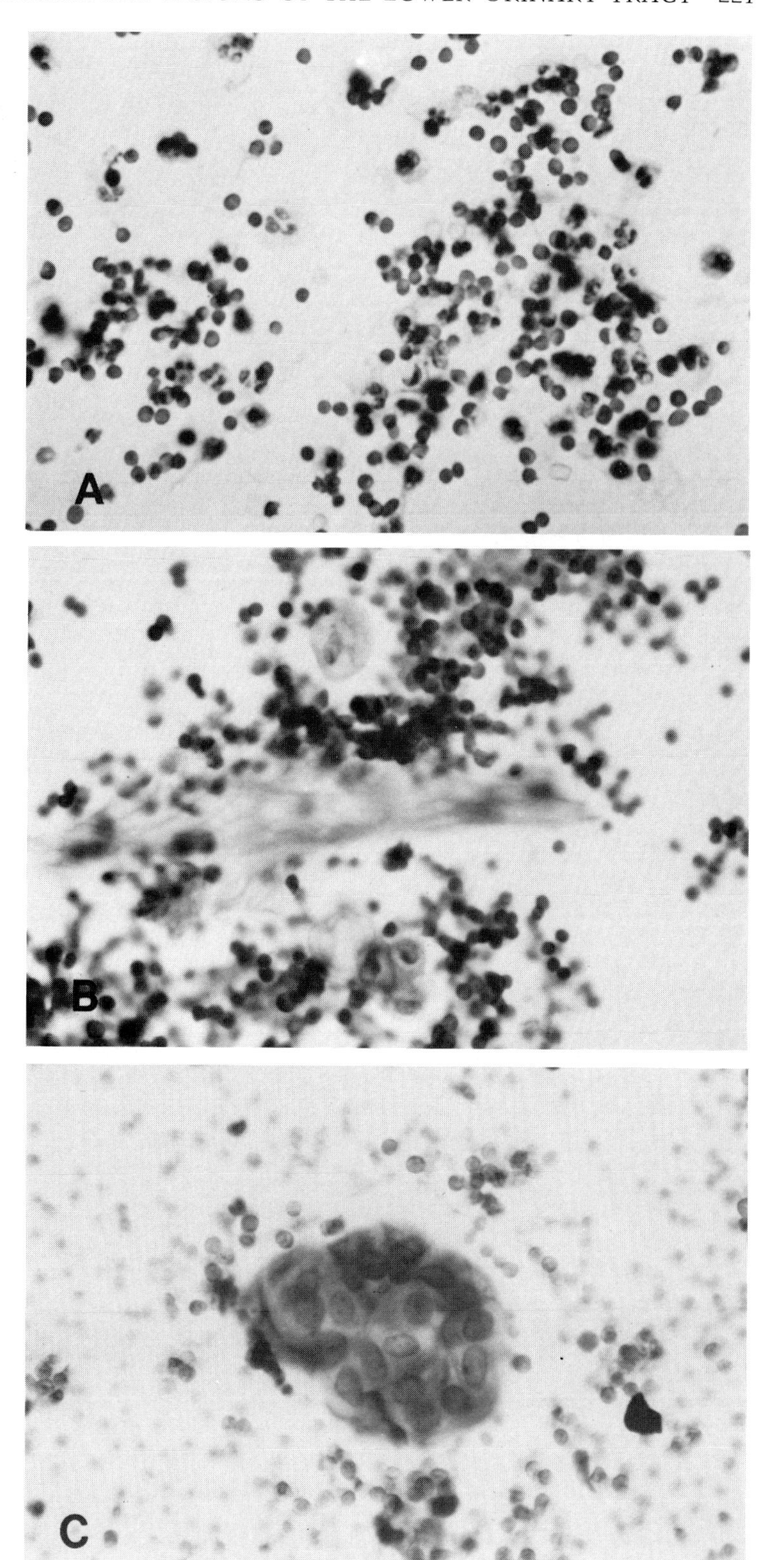

FIG. 15-2
HEMORRHAGIC CYSTITIS. **(A)** Bloody background with fibrin, neutrophilic leukocytes, and reactive transitional cells. (× 250) **(B)** (× 400) **(C)** Exfoliated urothelial cells and the lack of casts and renal cells indicate lower urinary tract disease. (× 400)

(continued)

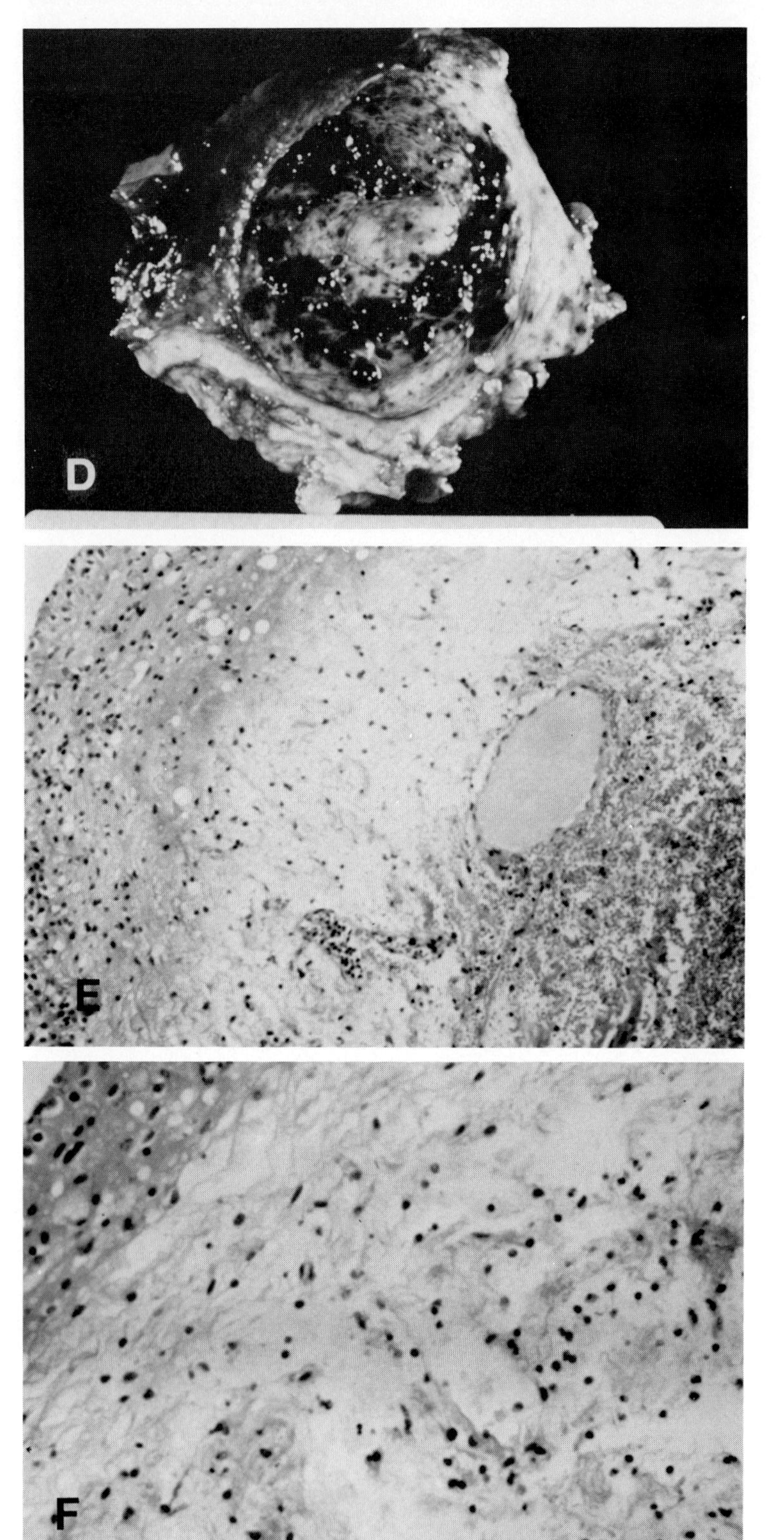

FIG. 15-2D-F *(continued)*
(D) The bladder mucosa has multiple petechiae and confluent ecchymoses. (Gross) **(E)** There is extensive recent hemorrhage into the edematous lamina propria. (× 100) **(F)** Note the relatively scant neutrophilic infiltrate, suggesting a primary hemorrhagic diathesis rather than an acute cystitis. (× 160)

FIG. 15-3
BACTERIAL URINARY TRACT INFECTION. **(A)** Inflammatory background with bacteria. (× 160) **(B)** Neutrophils, urothelial cells, and bacilli usually indicate bacterial cystitis. (× 1000)

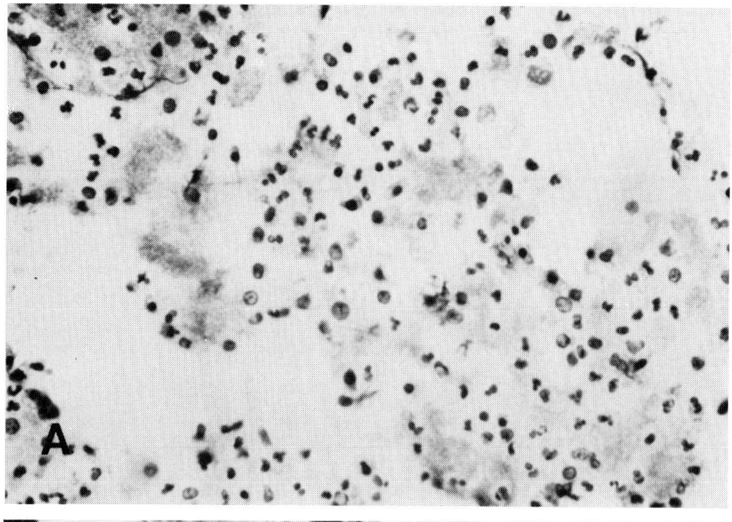

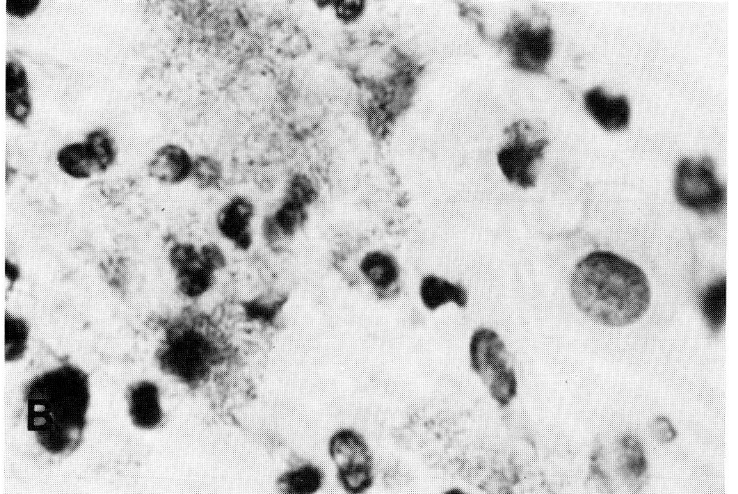

FIG. 15-4
CHRONIC CYSTITIS WITH SQUAMOUS METAPLASIA. **(A)** Inflammatory background with urothelial cells and squames. (× 250)

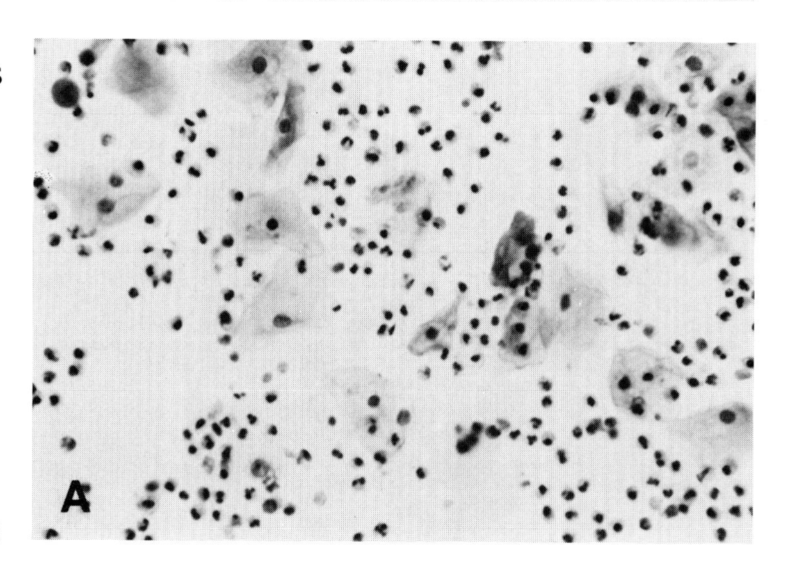

(continued)

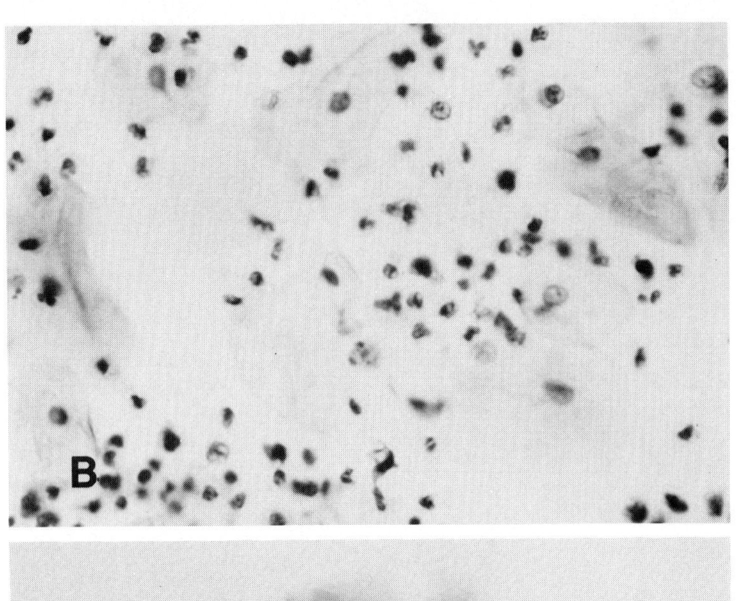

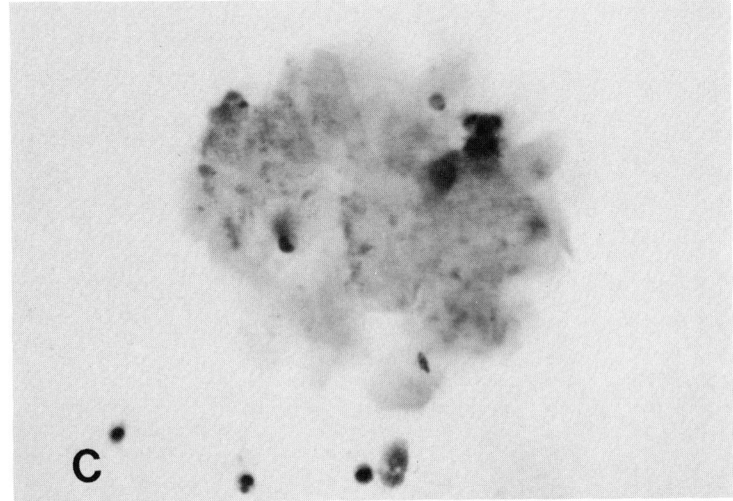

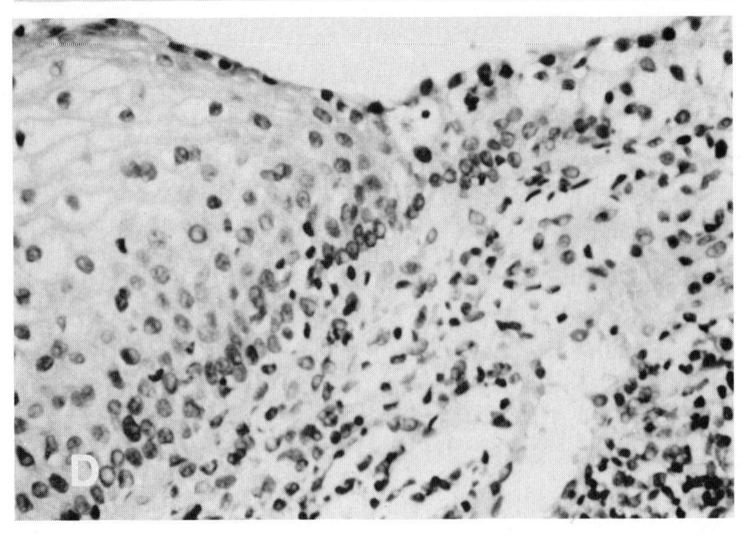

FIG. 15-4B-D *(continued)*
(B) Note squames, some of which are anucleated. (× 400) **(C)** Cluster of anucleated squames consistent with hyperkeratosis or contamination. (× 1000) **(D)** Lymphocytes and histiocytes are present in the lamina propria. The overlying urothelium (right) is vacuolated, and there is an abrupt transition to a metaplastic, glycogenated squamous epithelium (left). (× 250)

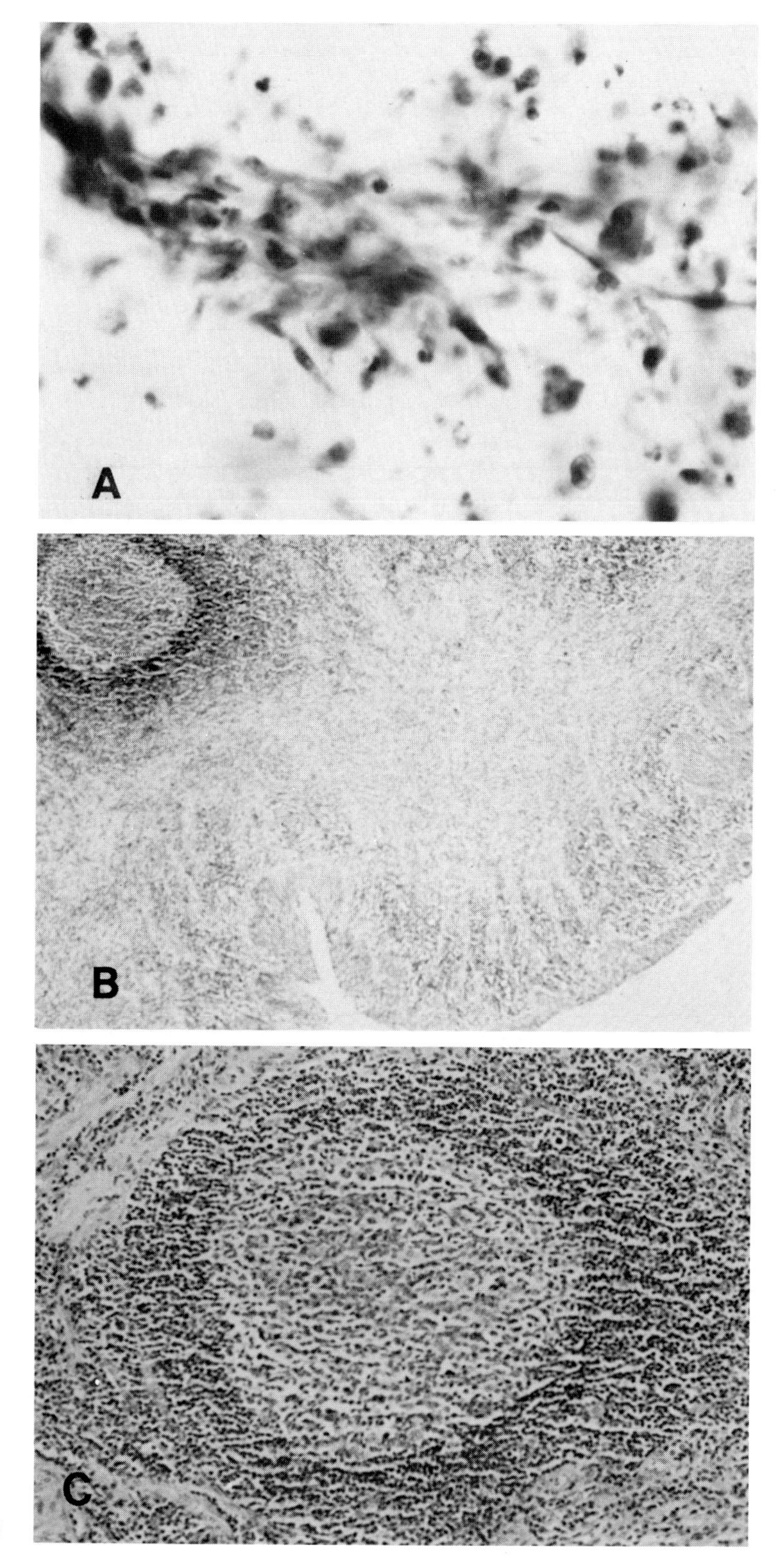

FIG. 15-5
FOLLICULAR CYSTITIS. **(A)** Lymphocyturia and deep urothelial cells. In the absence of immature lymphocytes, plasma cells, and reticular cells, a diagnosis cannot be suggested. (× 250) **(B)** The lamina propria is widened and contains a dense inflammatory infiltrate with a large lymphoid follicle. (× 40) **(C)** Lymphoid follicle with reactive germinal center. (× 100)

(continued)

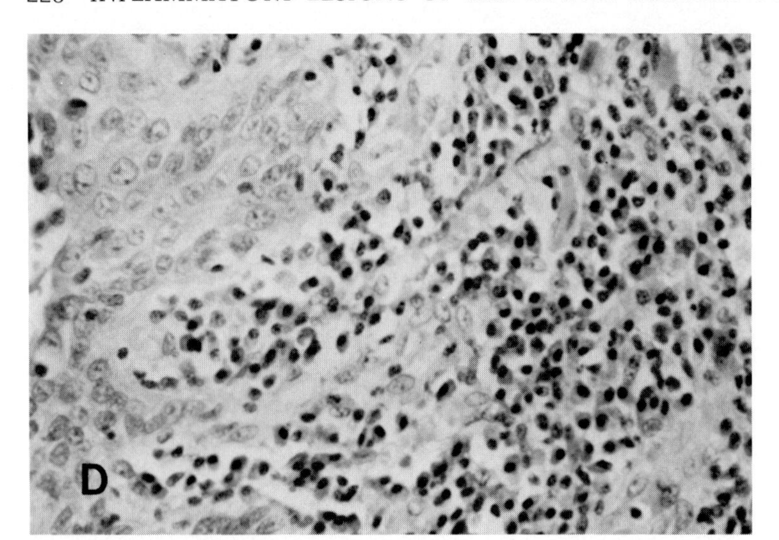

FIG. 15-5D *(continued)*
(D) Plasma cells are a major component of the inflammatory infiltrate. Urothelial budding with abnormal maturation is present. Urothelial cells have enlarged nuclei with chromatin clearing and clumping and prominent nucleoli typical of reactive or regenerating cells. (× 250)

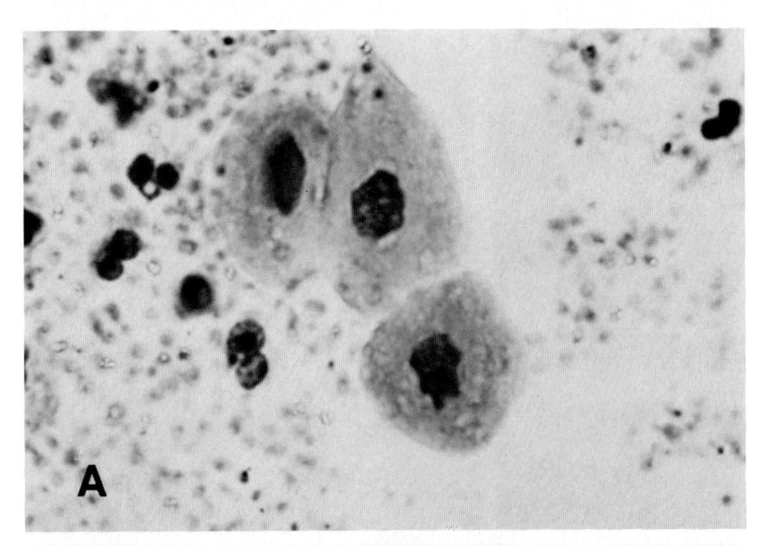

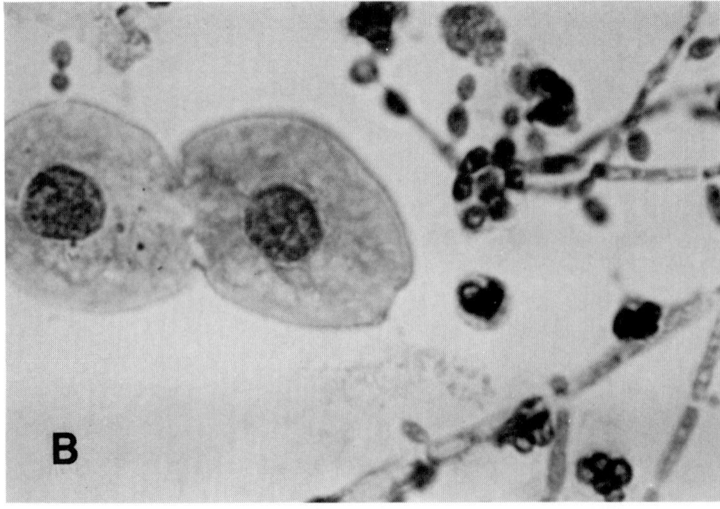

FIG. 15-6
CANDIDIASIS. **(A)** Neutrophils, urothelial cells, and numerous yeasts are present. (× 1000) **(B)** Both yeast and pseudohyphal forms of *Candida* species. Budding forms of yeast are needed to distinguish fungi from erythrocyte stroma and cellular debris. Pseudohyphal forms suggest an invasive infection. (× 1000)

(continued)

FIG. 15-6C-D *(continued)*
(C) Denuded bladder urothelium is replaced by a confluent growth of fungi. (× 100) **(D)** Yeast forms and pseudohyphae of *Candida* are present along the ureteral mucosa and invade the lamina propria without associated inflammatory response. (Grocott × 250)

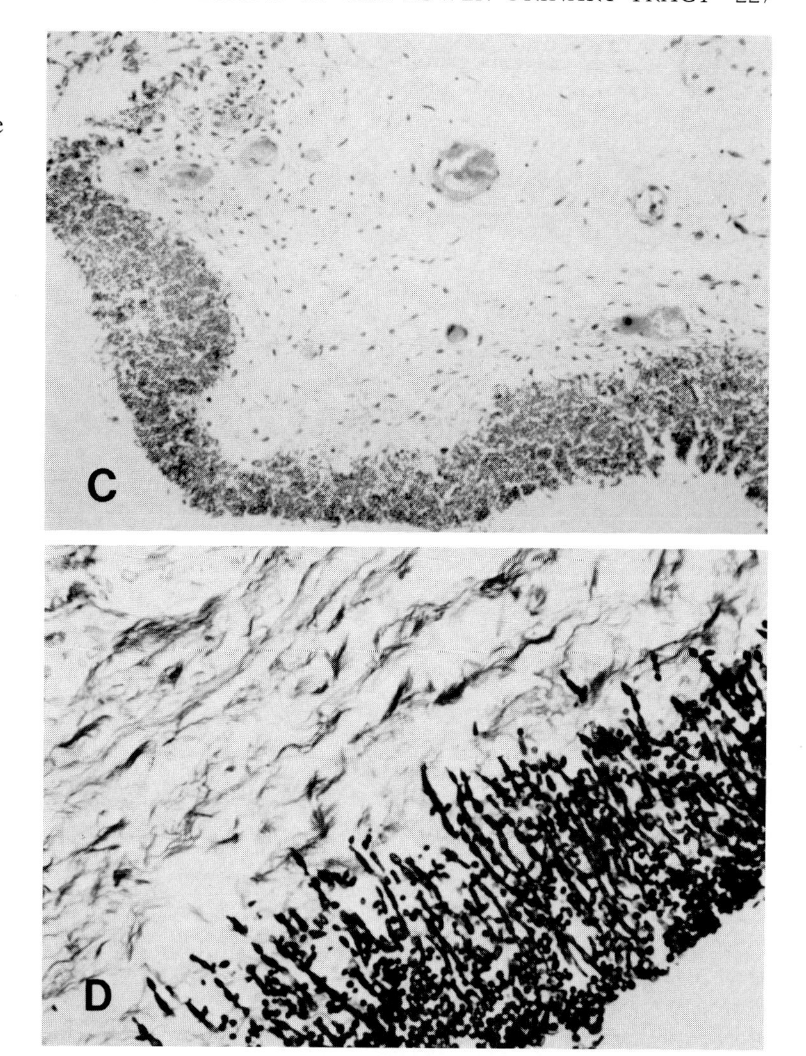

FIG. 15-7
CONDYLOMA ACUMINATUM IN BLADDER. **(A)** Mildly dysplastic epithelial cells with hyperkeratosis and parakeratosis. (× 250)

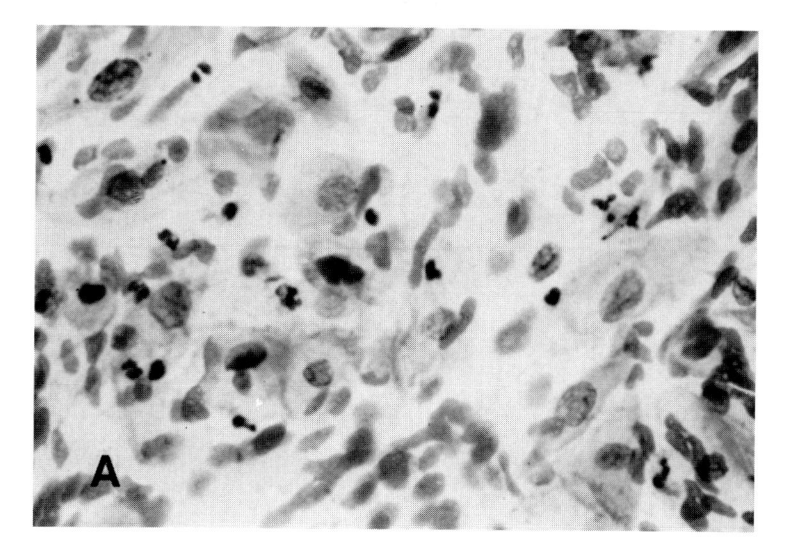

(continued)

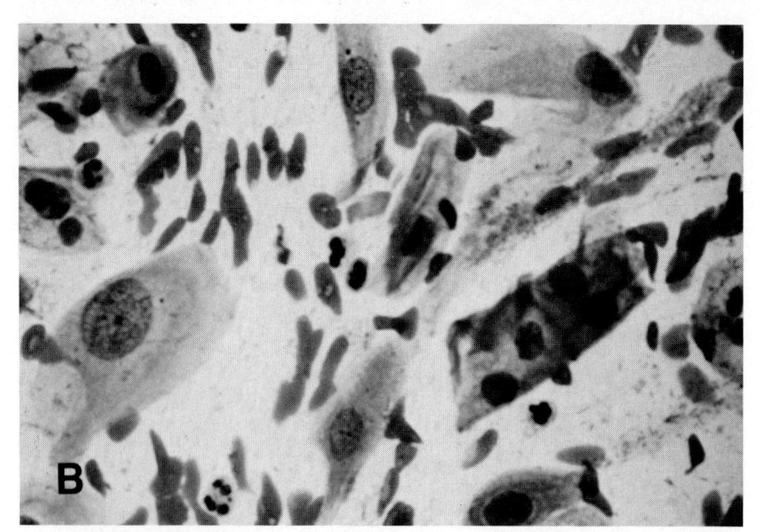

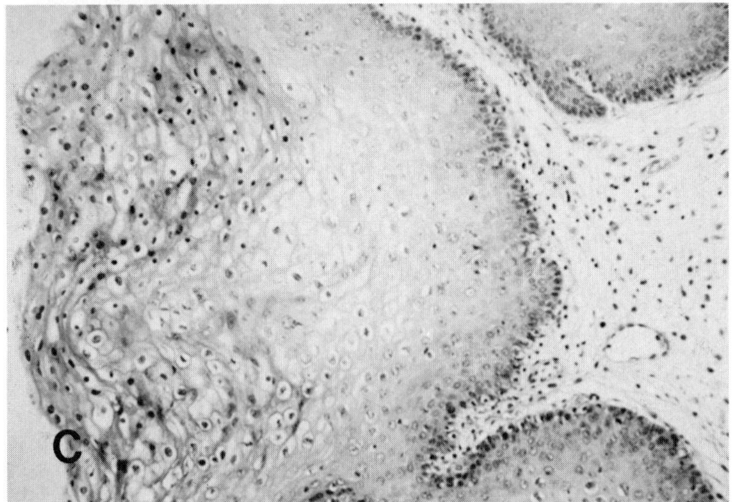

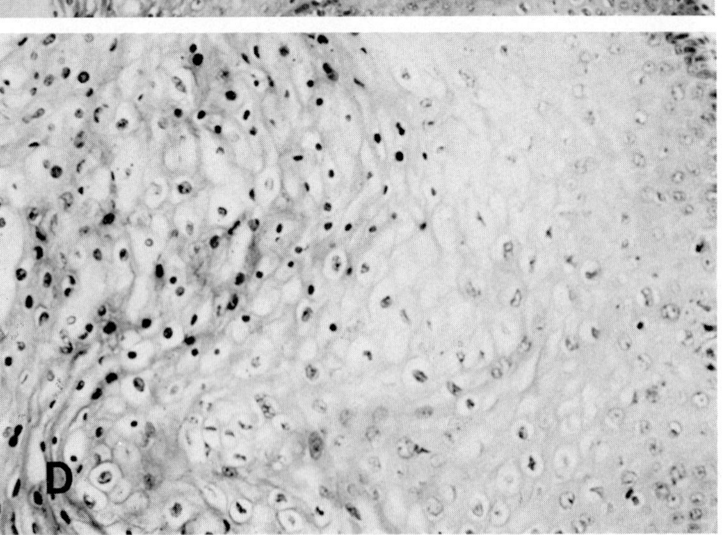

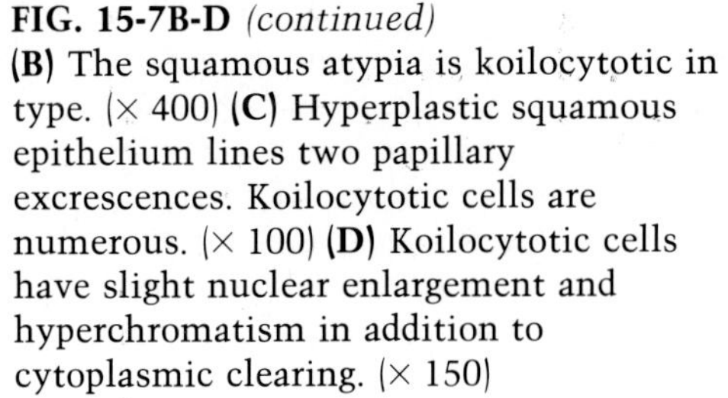
FIG. 15-7B-D *(continued)*
(B) The squamous atypia is koilocytotic in type. (× 400) **(C)** Hyperplastic squamous epithelium lines two papillary excrescences. Koilocytotic cells are numerous. (× 100) **(D)** Koilocytotic cells have slight nuclear enlargement and hyperchromatism in addition to cytoplasmic clearing. (× 150)

FIG. 15-8
MALAKOPLAKIA. **(A)** Urine sediment with inflammatory exudate. (× 250) **(B)** Neutrophils and a histiocyte. Note bilobed nucleus and single micronucleolus of histiocyte. (× 1000) **(C)** Michaelis-Gutmann body within macrophage. Note the characteristic dark basophilic intracytoplasmic inclusion. (× 1000) **(D)** The PAS stain often fails to accentuate the intracytoplasmic inclusion in urinary macrophages. (× 1000) **(E)** Gross appearance of malakoplakia in excised bladder specimen. The mucosa is thickened by confluent yellow-tan plaques. **(F)** Beneath an intact urothelium are sheets of inflammatory cells, the majority of which are histiocytes. (PAS × 150) **(G)** Histiocytes have abundant foamy to slightly granular cytoplasm and contain numerous intracytoplasmic-Michaelis-Gutmann bodies, which are round to oval and frequently laminated. (PAS × 250) **(H)** Michaelis-Gutmann bodies are variably calcified. (von Kossa stain × 400).

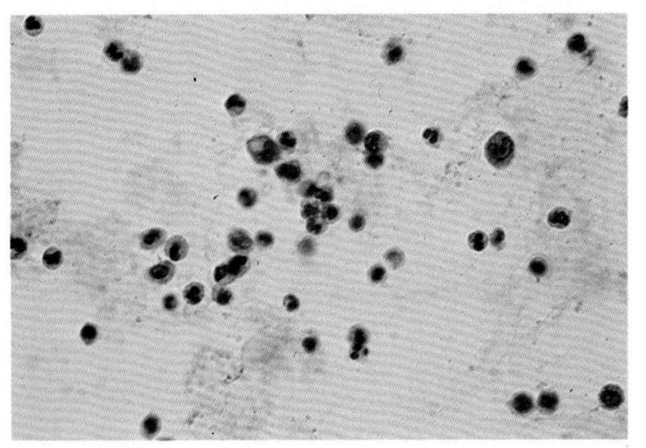

FIG. 15-8A

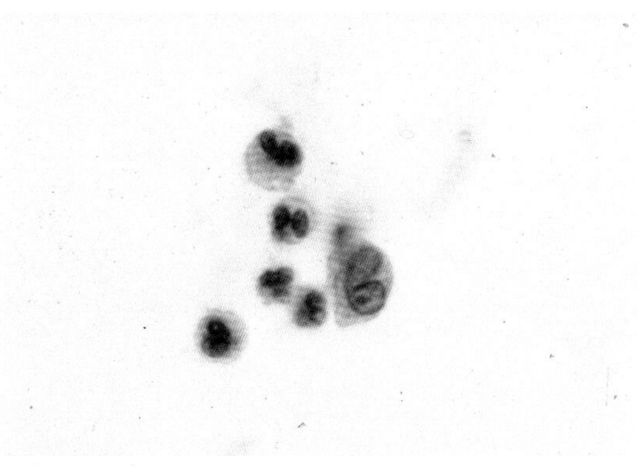

FIG. 15-8B

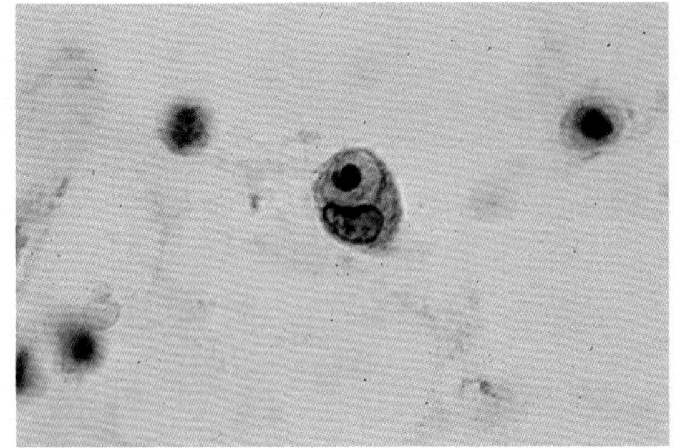

FIG. 15-8C

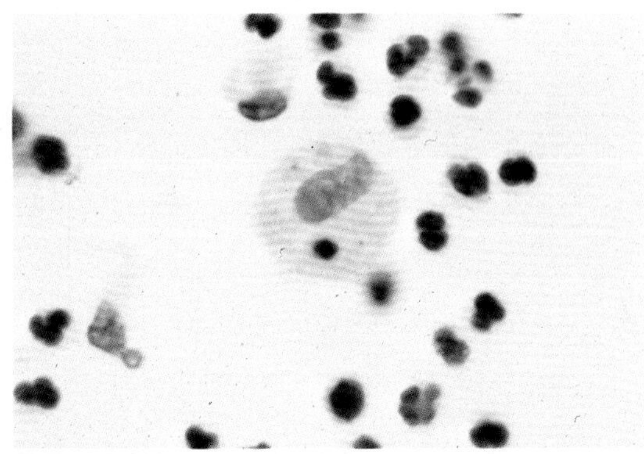

FIG. 15-8D

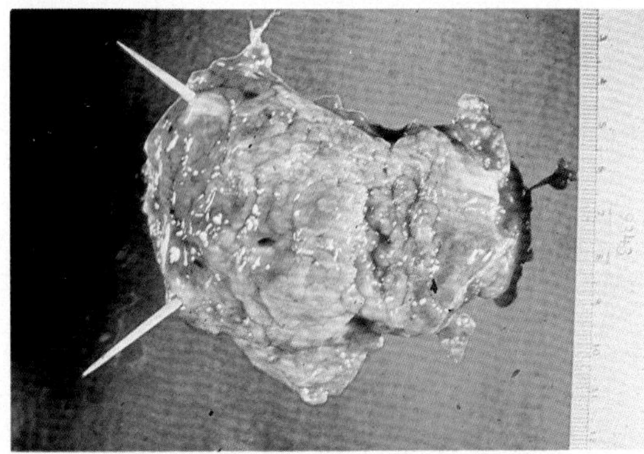

FIG. 15-8E

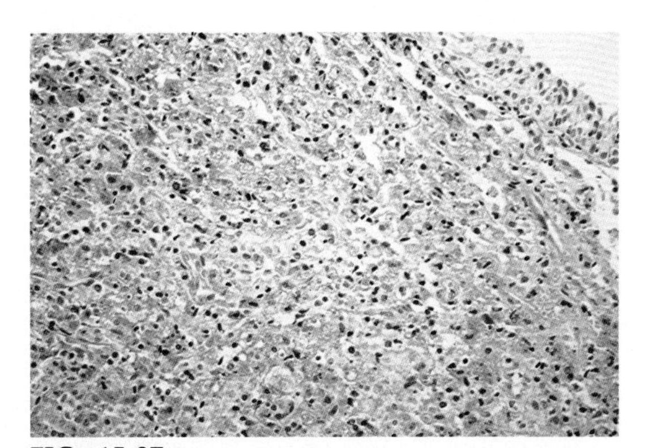

FIG. 15-8F

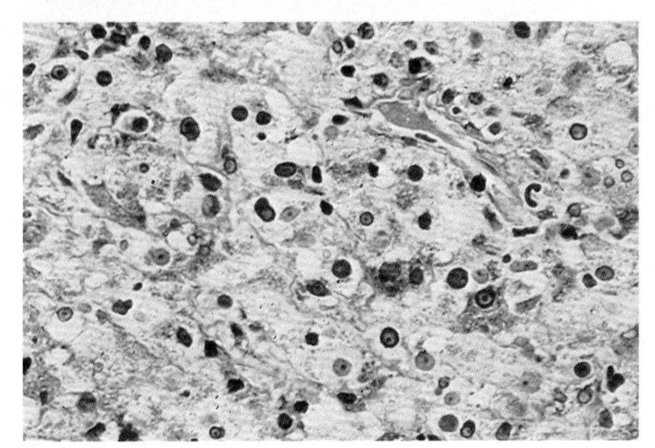

FIG. 15-8G

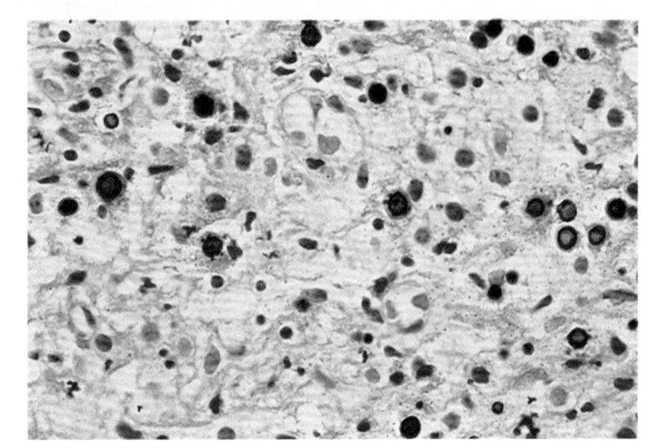

FIG. 15-8H

FIG. 15-9

EOSINOPHILIC CYSTITIS. **(A)** Urine sediment with inflammation, cellular debris, and numerous reactive urothelial cells. (× 250) **(B)** Numerous reactive superficial and deep urothelial cells with nuclear enlargement and micronucleoli. Note eosinophil (arrow). (× 1000) **(C)** Numerous eosinophils adherent to mucus. (× 1000) **(D)** The lamina propria is edematous, congested, and contains a dense infiltrate of eosinophils. The overlying urothelium appears slightly thickened. (× 100) **(E)** Focal clustering of eosinophils around dilated venule at junction of lamina propria and muscular wall. Smooth muscle bundles are separated by edema and inflammation. (× 160) **(F)** A focus containing eosinophils, histiocytes, and proliferating, reactive fibroblasts, suggests histologically the possibility of eosinophilic granuloma. (× 250)

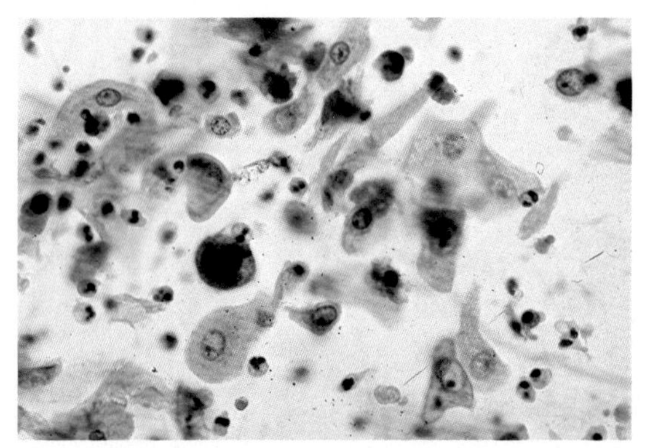
FIG. 15-9A

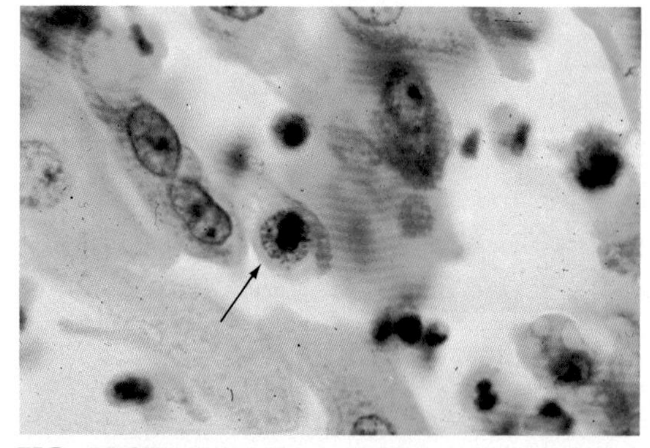
FIG. 15-9B

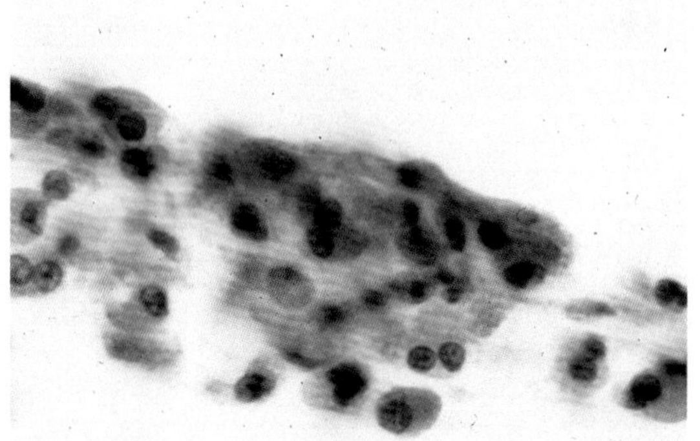
FIG. 15-9C

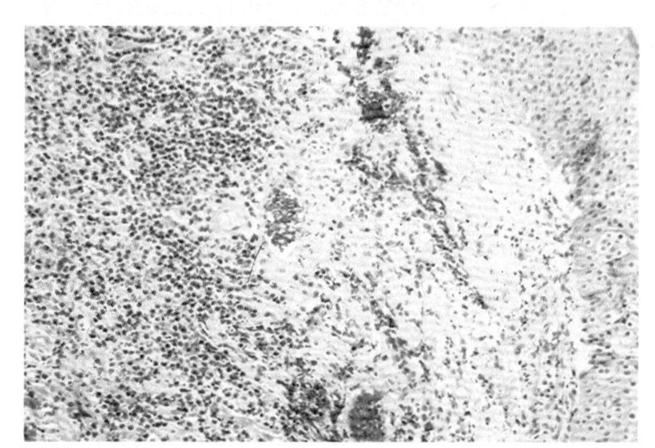
FIG. 15-9D

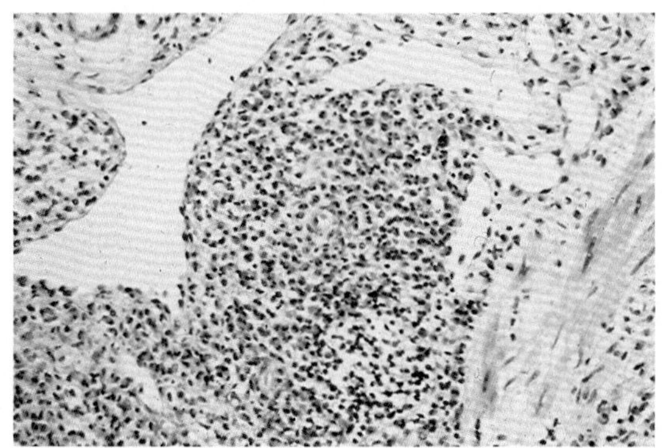
FIG. 15-9E

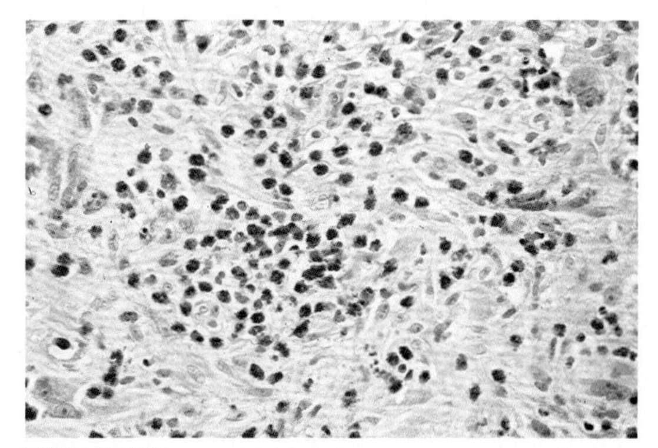
FIG. 15-9F

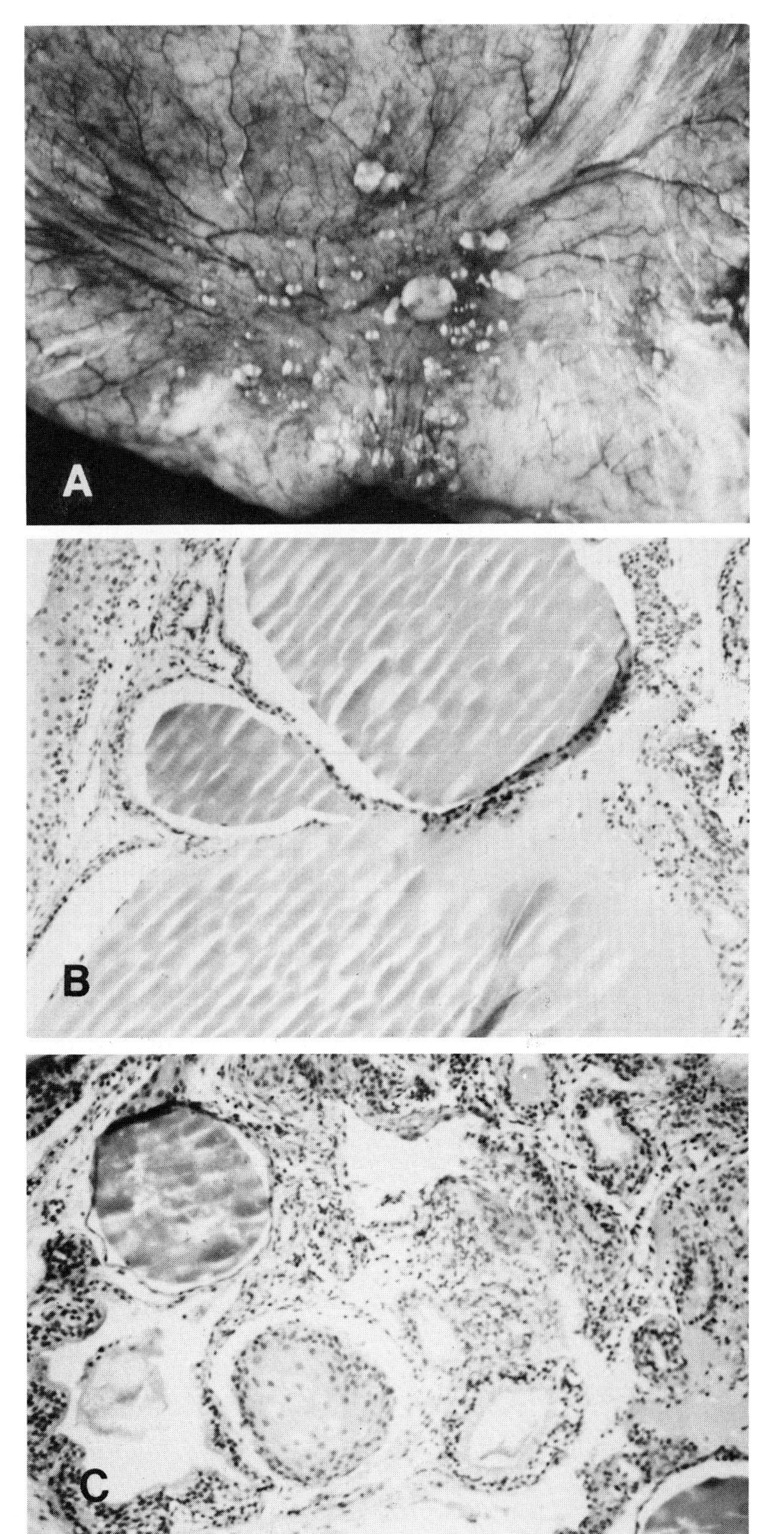

FIG. 15-10
CYSTITIS CYSTICA. **(A)** Multiple cysts of varying sizes in bladder trigone. Prominence of vasculature represents postmortem artifact. (Gross) **(B)** Cystitis cystica with squamous metaplasia. Squamous metaplasia of the surface epithelium overlying cysts within the lamina propria. (× 100) **(C)** Cystitis cystica and glandularis with squamous metaplasia. Within the lamina propria is a solid cluster of metaplastic squamous epithelium. Occasional cysts are lined by a columnar epithelium. (× 100)

(continued)

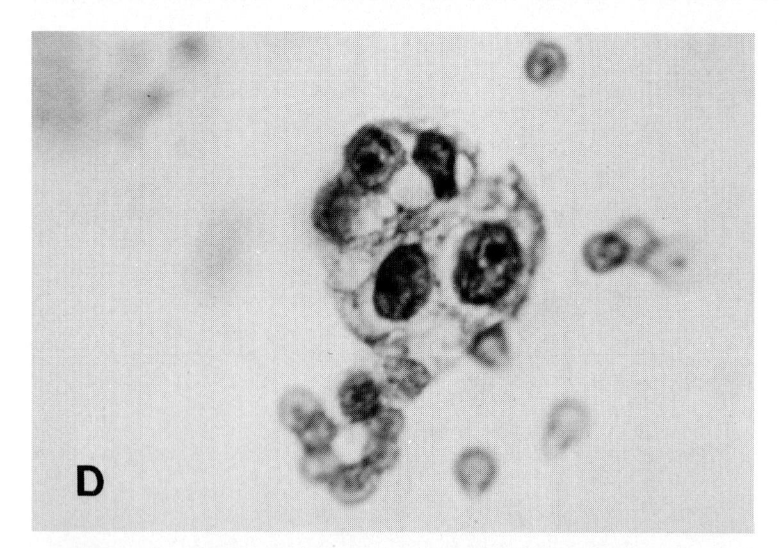

FIG. 15-10D *(continued)*
(D) Acinar epithelial clusters in urine sediment. Note eccentric nuclei containing prominent nucleoli and abundant foamy cytoplasm. Urine specimen obtained from patient with cystitis glandularis. (× 1000)

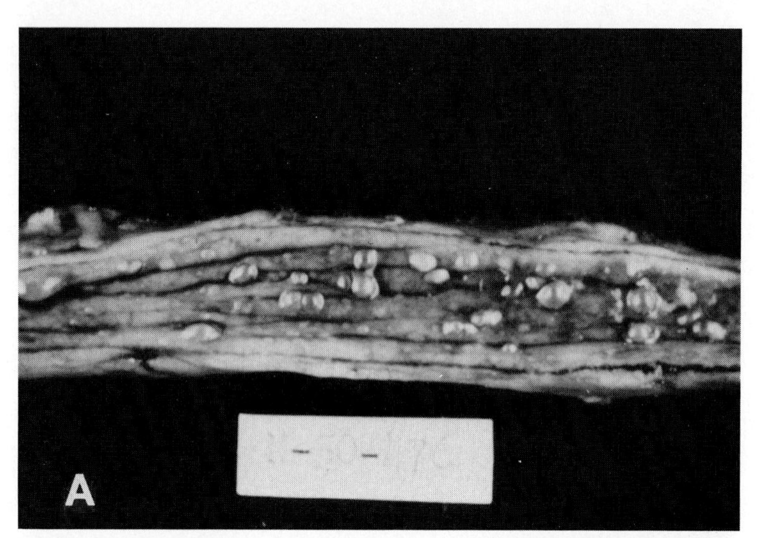

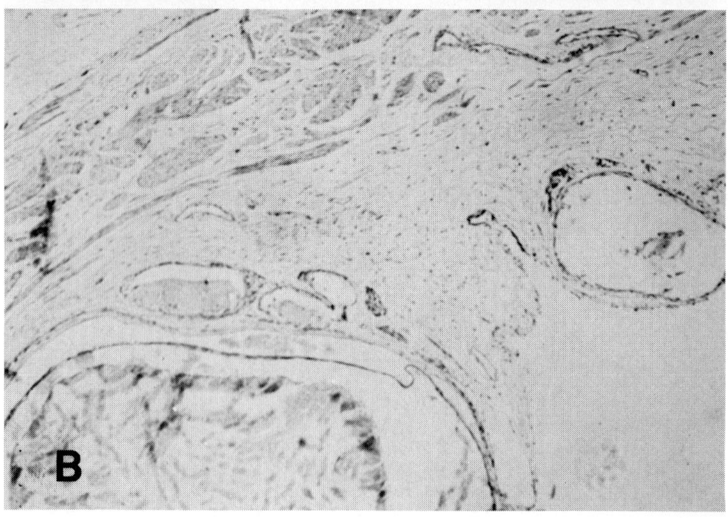

FIG. 15-11
URETERITIS CYSTICA. **(A)** Numerous cysts along ureteral mucosa give "cobblestone" appearance. (Gross) **(B)** These cysts within the edematous lamina propria vary in size and content of mucoid material. Cysts have a single layer of flattened epithelium or a stratified epithelium. Note the lack of significant inflammatory infiltrate. (× 40)

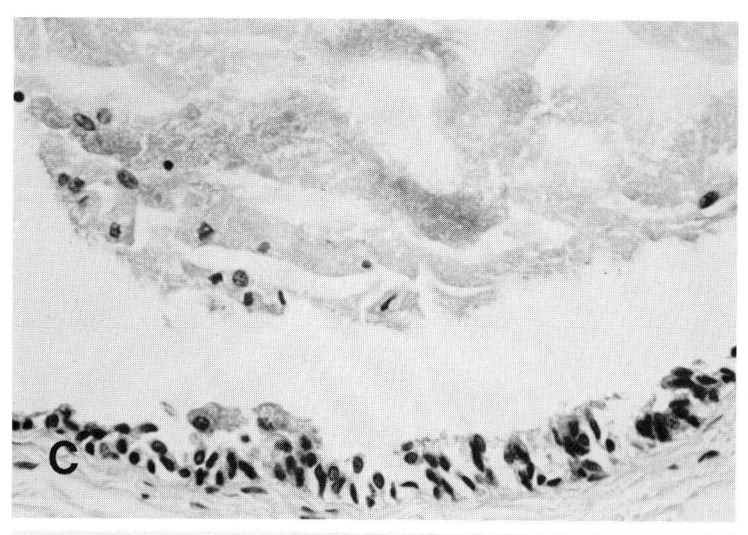

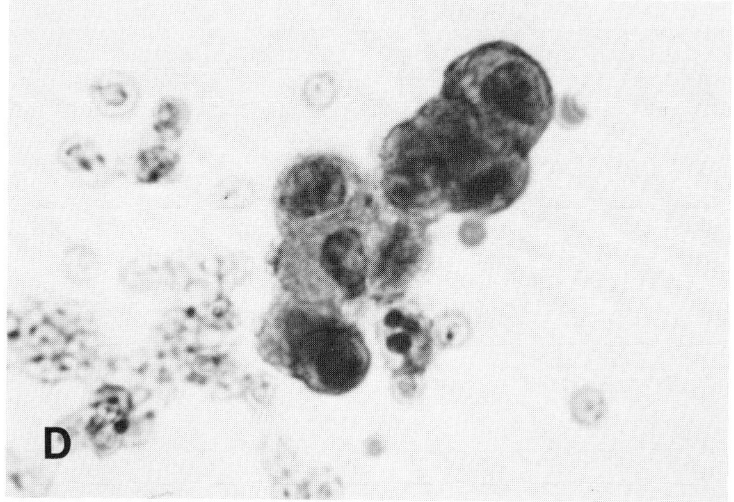

FIG. 15-11C, D *(continued)*
(C) Cysts with stratified epithelium. Superficial cells remain focally attached, but the majority have been exfoliated into the mucus. Occasional epithelial cells are elongated; however, a luminal layer of cuboidal, columnar, or mucus-secreting goblet cells, typical of ureteritis (or cystitis) glandularis, is not present. (× 250)
(D) Acinar epithelial fragment in urine sediment. Note nuclei with prominent nucleoli and finely vacuolated cytoplasm. The linear arrangement suggests a glandular origin. (× 1000)

16
UROTHELIAL ATYPIA

CYTODIAGNOSTIC APPROACH

In general, urothelial atypias in urine sediment are related to degenerative, regenerative, and proliferative lesions. These lesions may be induced by radiotherapy (Figs. 16-1–16-4), chemotherapy (Figs. 16-5 and 16-6), and lithiasis (Fig. 16-7). In addition, they may reflect an altered maturation with or without dyskaryosis of urothelium or indicate a low-grade malignancy (Figs. 16-8–16-12).

CYTODIAGNOSTIC APPROACH

Urothelial atypia can present in the urine sediment as exfoliation of individual cells and fragments. It is frequently accompanied by hematuria and variable inflammation.

An atypical urothelial cell in urine sediment has nuclear enlargement with an altered nuclear-cytoplasmic ratio and hyperchromasia. These changes may be degenerative in origin. When transiently present in urine they commonly reflect lithiasis (Fig. 16-7). Following radiation or chemotherapy, they may persist for longer periods of time in followup urine sediment examinations. In dyskaryotic cells there is a high nuclear-cytoplasmic ratio, and hyperchromatic nuclei have preserved chromatin texture. The persistence of these cells in followup urine sediment examinations requires clinical evaluation to exclude a malignant lesion (Figs. 16-10 and 16-11).

The presence of urothelial fragments in spontaneously voided urines, regardless of specific characteristics, is a significant finding. A major cause of a false-positive cytodiagnosis of transitional cell carcinoma is the misinterpretation of urothelial fragments occurring with instrumentation and lithiasis (Fig. 16-7). In this clinical setting, the numerous fragments must be more carefully evaluated to exclude the presence of fragments exfoliated from pre-existing or unassociated malignant lesions.

Urothelial fragments in the urine sediment should be evaluated for number of cell layers or thickness, cytoplasmic maturation, cell arrangement, and hyperchromasia (Table 16-1). A differential diagnosis can be given on the basis of the major characteristic of the urothelial fragments. These characteristics, when considered in combination and in the presence of exfoliated, individual atypical cells, may allow a more specific diagnosis. For example, atypical hyperplasia may present in the urine sediment as exfoliation of thick urothelial fragments containing tightly packed cells with hyperchromatic (dyskaryotic) nuclei and altered cytoplasmic maturation (Figs. 16-8 and 16-9). Exfoliation of individual atypical urothelial cells is usually a minor component. Similar fragments can exfoliate from papillary transitional cell carcinoma, grade II.

TABLE 16-1
MAJOR CHARACTERISTICS OF UROTHELIAL FRAGMENTS IN URINE SEDIMENT AND THEIR DIFFERENTIAL DIAGNOSIS

CHARACTERISTICS OF UROTHELIAL FRAGMENTS	CLINICAL CONDITIONS
MIXTURE OF DEEP AND SUPERFICIAL CELLS	Instrumentation (catheterization) Papillary transitional cell carcinoma (grades I–II)
ABNORMAL THICKNESS (> 7 CELL LAYERS)	Urothelial hyperplasia Papillary transitional cell carcinoma (grades I–II)
LACK OF CYTOPLASMIC MATURATION	Instrumentation (deep urothelial cells)
HYPERCHROMATIC CELLS	Degenerative: Radiation, chemotherapy, lithiasis Dyskaryotic Atypical hyperplasia Papillary transitional cell carcinoma (grades II–III)
SYNCYTIAL ARRANGEMENT	Urothelial carcinoma *in situ* Papillary transitional cell carcinoma (grades II–III)

In contrast to atypical hyperplasia, urothelial atypia may cause exfoliation of variably dyskaryotic individual cells without fragments (Figs. 16-10 and 16-11). It may be difficult to differentiate severely dyskaryotic cells from cells derived from carcinoma *in situ.* Urothelial carcinoma *in situ* frequently exfoliates syncytial fragments and large numbers of markedly dyskaryotic cells (Fig. 16-12).

BIBLIOGRAPHY

Koss LG: Diagnostic Cytology and Its Histopathologic Bases, 3rd ed. Philadelphia, JB Lippincott, 1979

Koss LG: Tumors of the Urinary Bladder. Atlas of Tumor Pathology, 2nd series, Fascicle 11. Washington, D.C., Armed Forces Institute of Pathology, 1975

Tweedale DN: Urinary Cytology. Boston, Little, Brown, 1977

Voogt HJ, Rathert P, Beyer-Boon ME: Urinary Cytology. Phase Contrast Microscopy and Analysis of Stained Smears. New York, Springer-Verlag, 1977

FIG. 16-1
ACUTE RADIATION CYSTITIS. **(A)** Histiocytic exudate with urothelial fragments in urine sediment. (× 250) **(B)** Degenerating deep urothelial cells with slight nuclear enlargement, chromatin clearing, and finely vacuolated cytoplasm. Background contains histiocytes and cellular debris. (× 1000) Urine sediment obtained within 24 hr after pelvic irradiation.

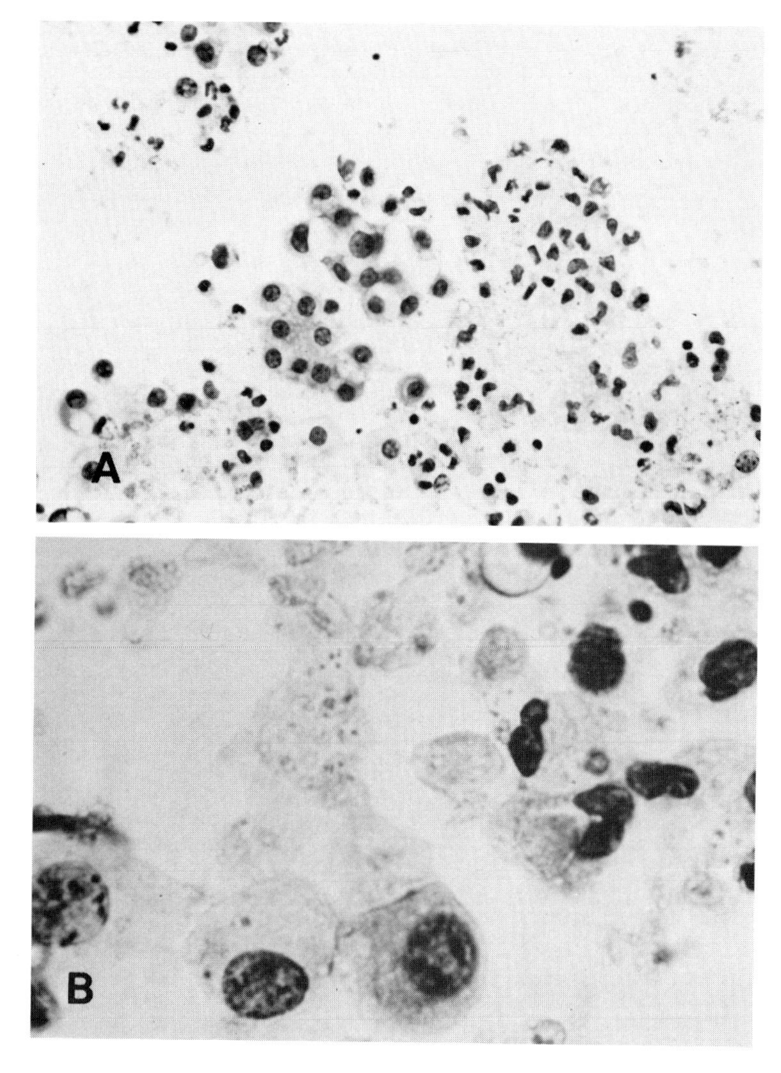

FIG. 16-2
RADIATION-INDUCED EPITHELIAL CHANGES. **(A)** Numerous binucleated urothelial cells with degenerated hyperchromatic nuclei and abundant foamy cytoplasm. (× 400)

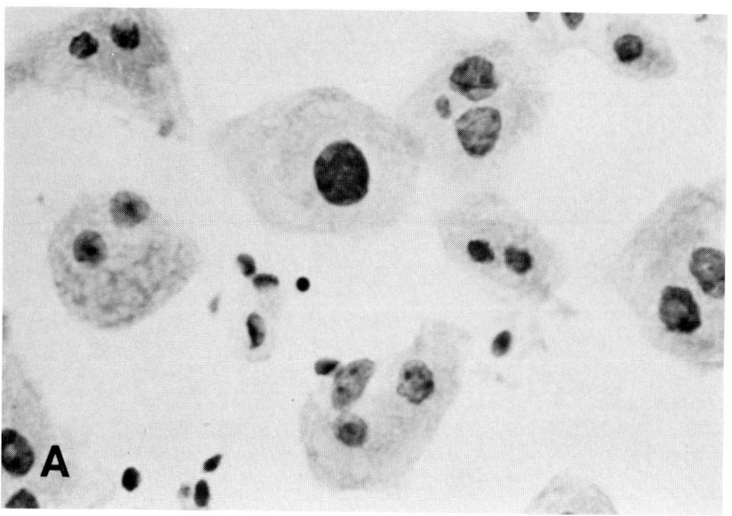

(continued)

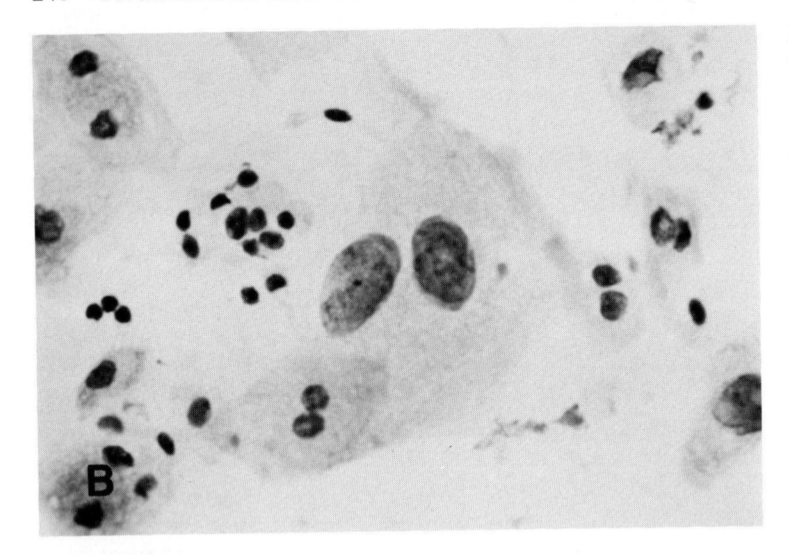

FIG. 16-2B *(continued)*
(B) Gigantic binucleated urothelial cell with nuclear enlargement, smudged chromatin pattern, and disintegrating foamy cytoplasm with frayed borders. (× 1000) Urine specimen obtained from patient following inguinal node irradiation for Hodgkin's disease.

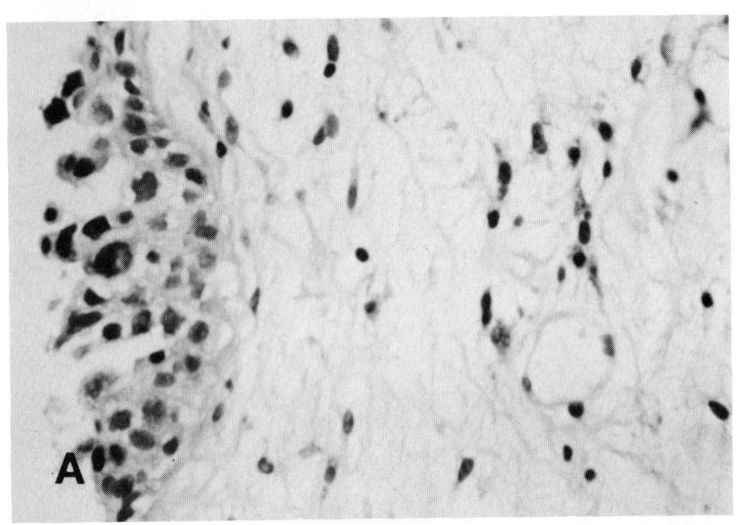

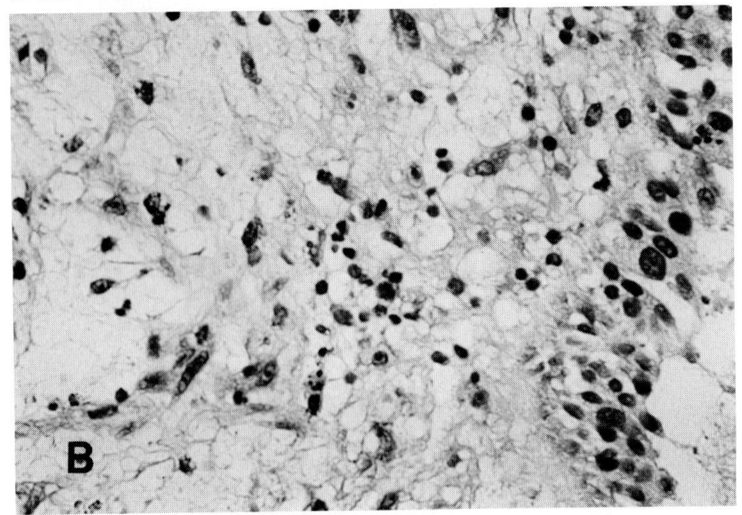

FIG. 16-3
RADIATION CYSTITIS. **(A)** The urothelium has diminished intercellular cohesiveness with variable cellular enlargement as well as nuclear enlargement and hyperchromatism or pyknosis. (× 250) **(B)** The edematous lamina propria contains reactive fibroblasts and hemosiderin-laden macrophages. (× 250)

(continued)

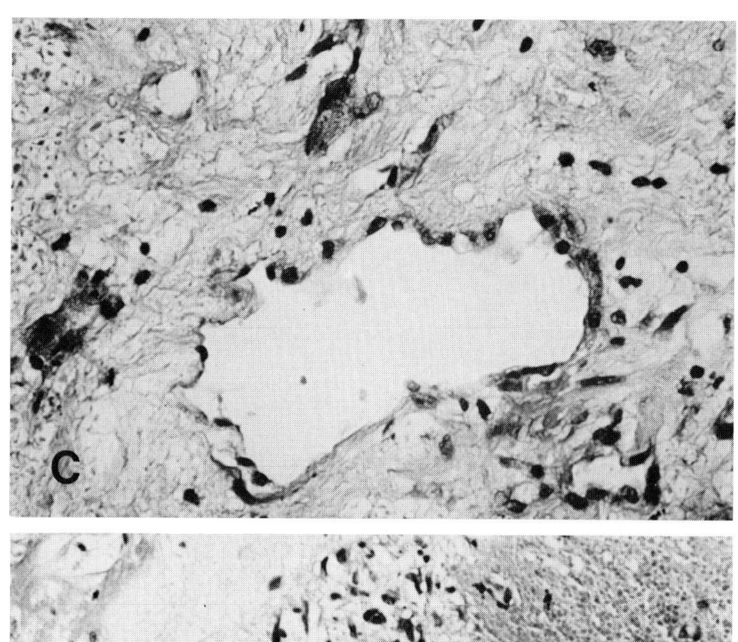

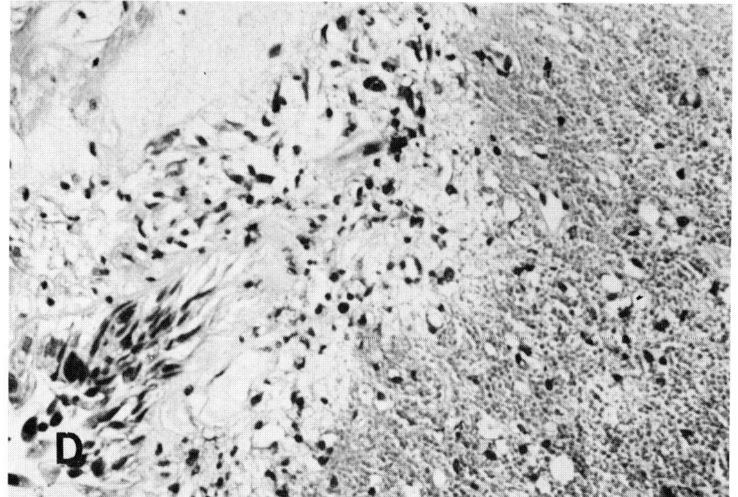

FIG. 16-3C-D *(continued)*
(C) A dilated capillary has swollen reactive endothelium. Note "bizarre" giant fibroblasts. (× 250) **(D)** Recent and old hemorrhage in lamina propria with acute and chronic inflammatory infiltrate. (× 160) Bladder biopsy 7 wk following radiation therapy.

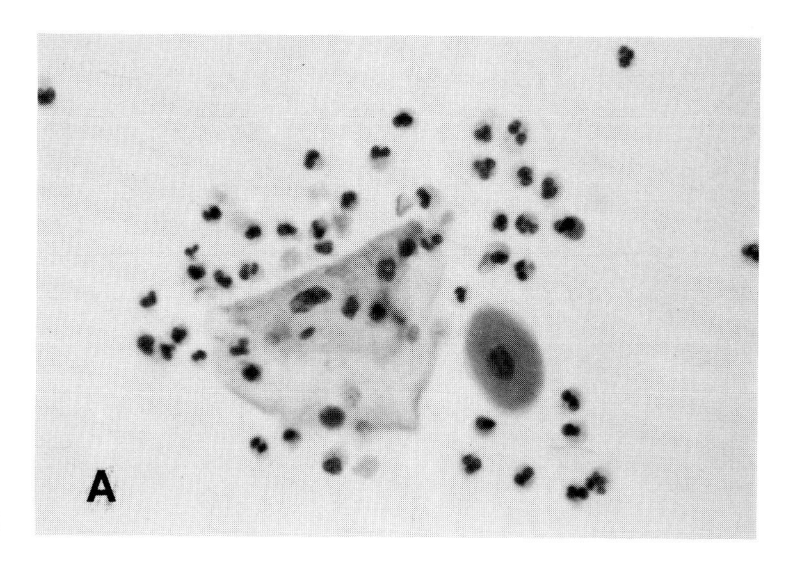

FIG. 16-4
CHRONIC RADIATION CYSTITIS. **(A)** Inflammation and urothelial cells. (× 400)

(continued)

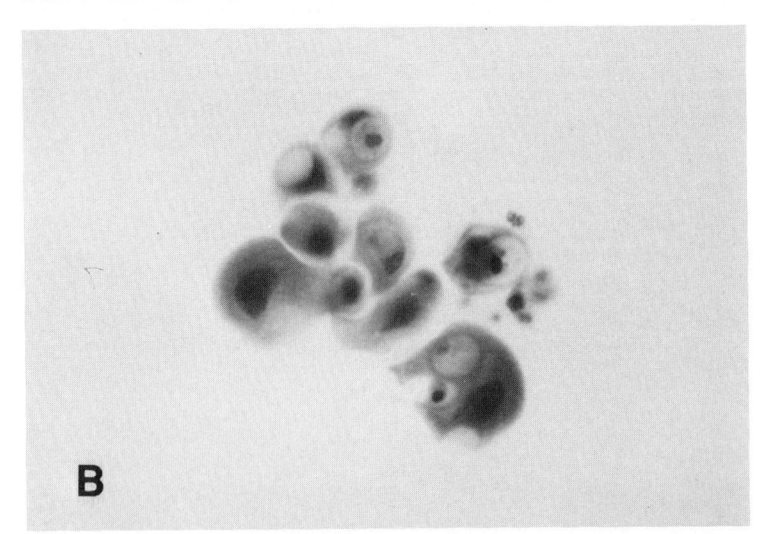

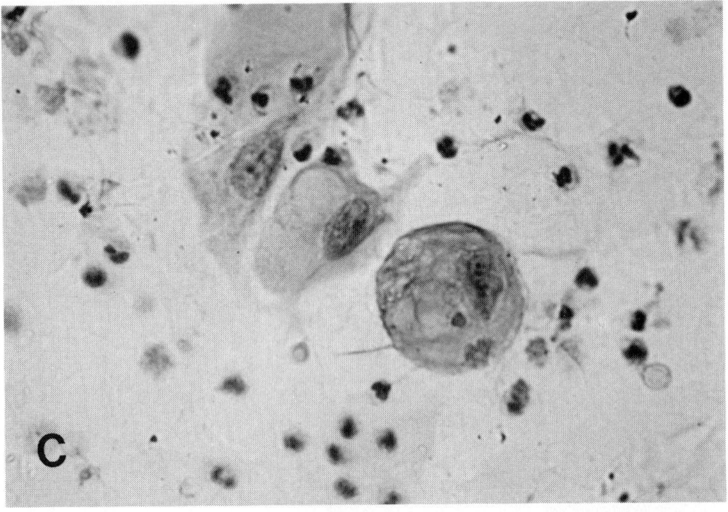

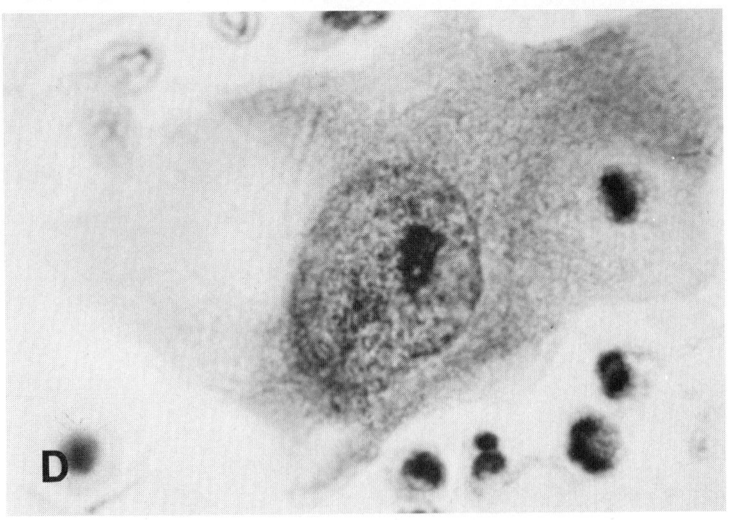

FIG. 16-4B-D *(continued)*
(B) Loose cluster of urothelial cells with nuclear enlargement, slight hyperchromasia, and polychromatic cytoplasm. (× 400) **(C)** Urothelial cells with nuclear enlargement and abundant cytoplasm with prominent vacuolization. (× 400) **(D)** Gigantic superficial urothelial cell with nuclear enlargement and a macronucleolus. (× 1000) Urine specimen obtained approximately 2 yr postradiation therapy.

FIG. 16-5
CHEMOTHERAPEUTIC CELLULAR CHANGES. **(A)** Epithelial cells with nuclear enlargement, hyperchromasia, and chromatin clearing and clumping. Eccentric nuclei and finely granular cytoplasm suggest possible renal origin. (× 1000) **(B)** Atypical epithelial cell with marked nuclear enlargement and hyperchromasia. (× 1000) Urine obtained following chemotherapy and radiotherapy for metastatic malignant melanoma.

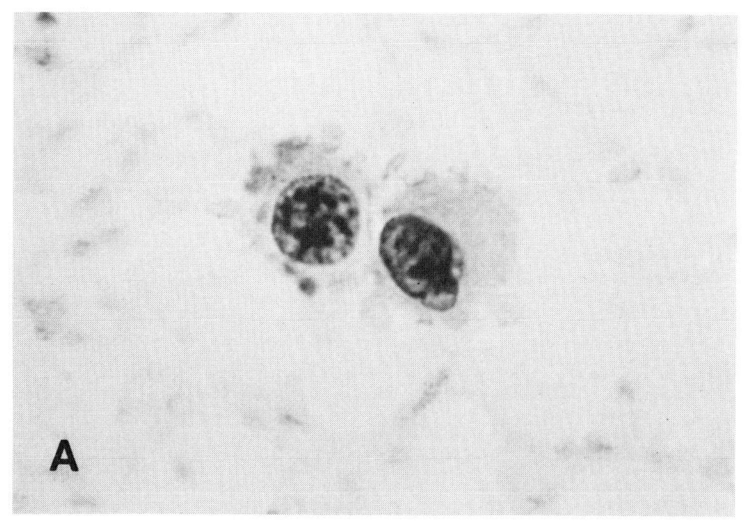

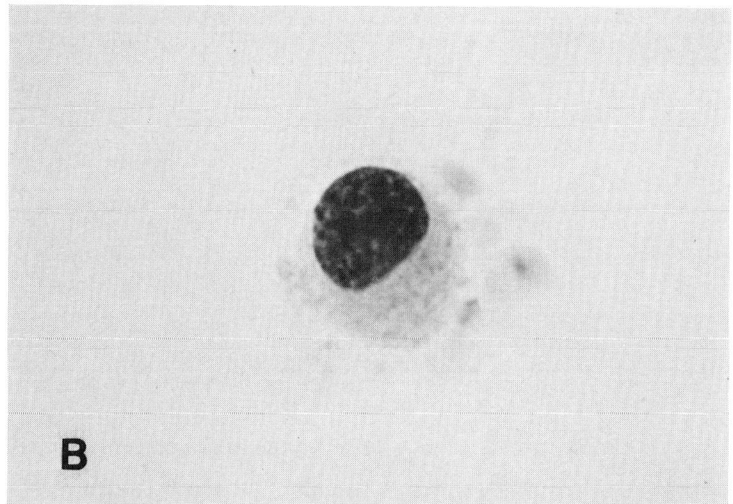

FIG) 16-6
CHEMOTHERAPEUTIC CELLULAR CHANGES. **(A)** Atypical epithelial fragment, urothelial cells, and mild inflammation. (× 400)

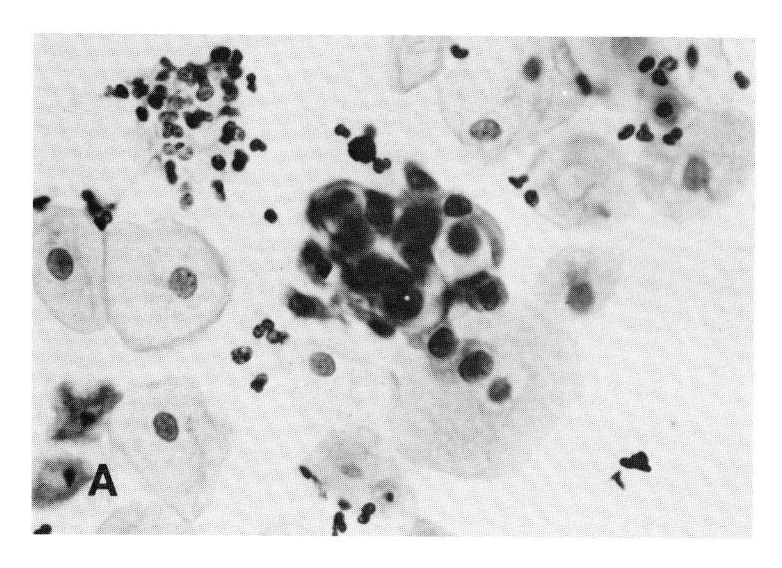

(continued)

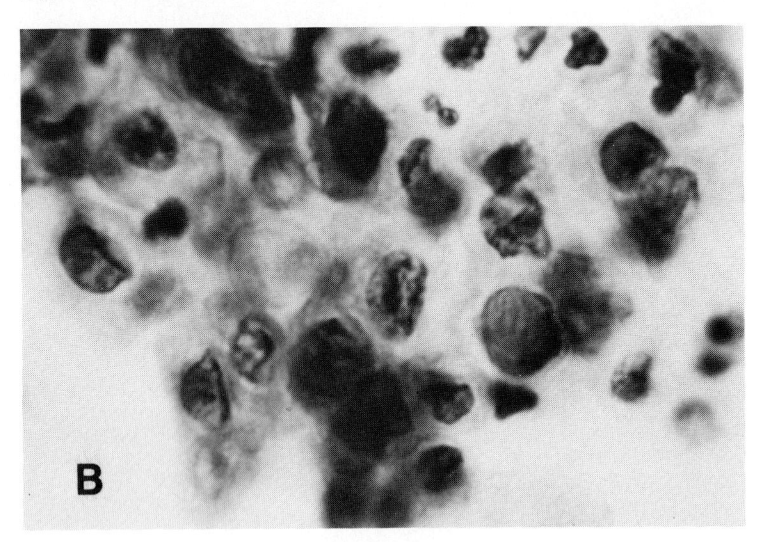

FIG. 16-6B *(continued)*
(B) Sheet of urothelial cells with enlarged, irregular hyperchromatic nuclei and dense cytoplasm. (× 1000)

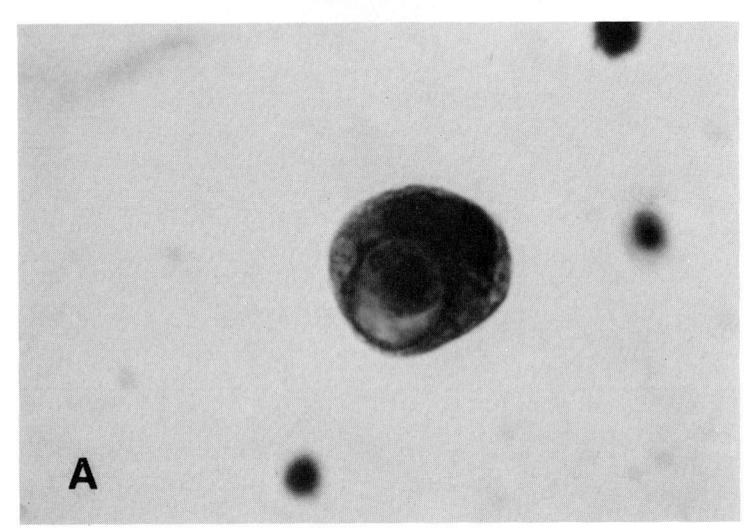

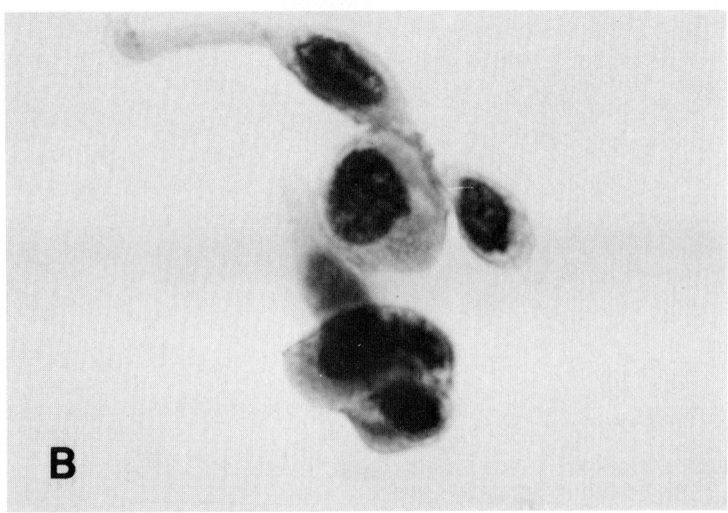

FIG. 16-7
CELLULAR CHANGES ASSOCIATED WITH LITHIASIS. **(A)** Atypical urothelial cell with enlarged degenerated nucleus and dense cytoplasm. Note degenerative cytoplasmic vacuole containing a neutrophil. (× 1000) **(B)** Atypical urothelial fragment. Note enlarged hyperchromatic nuclei, smudged chromatin pattern, and abundant dense cytoplasm with pale outer rim. Urine specimen obtained from patient during passage of renal stones. (× 1000)

FIG. 16-8
ATYPICAL HYPERPLASIA IN URINE SEDIMENT. (**A**) Atypical hyperplastic epithelial fragment. Note densely packed nuclei. (× 400) (**B**) The cellular features of nuclear crowding, the high nuclear: cytoplasmic ratio, and nuclear moulding are consistent with atypical hyperplasia. × 400) (**C**) The fragment contains a tight cluster of epithelial cells with nuclear enlargement, hypochromasia, occasional micronucleoli, and scant cytoplasm. (× 1000)

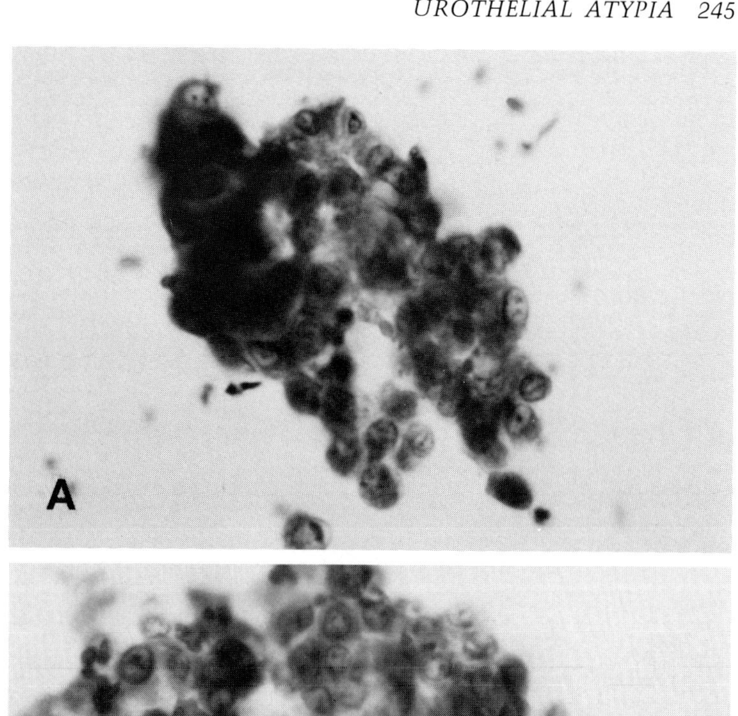

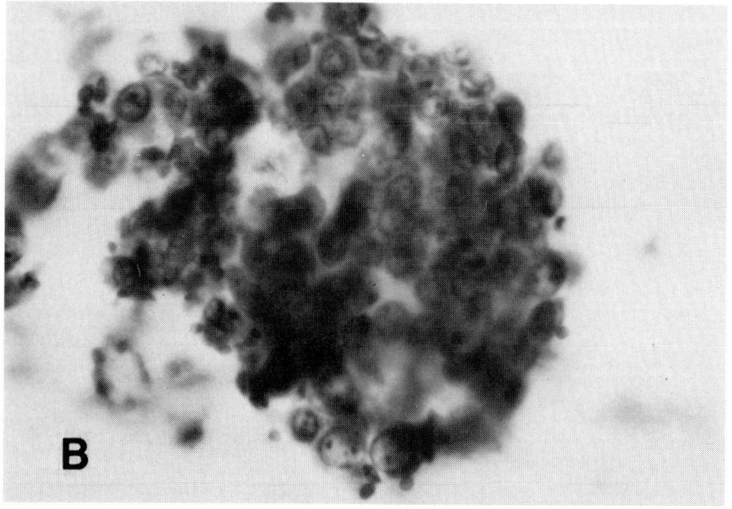

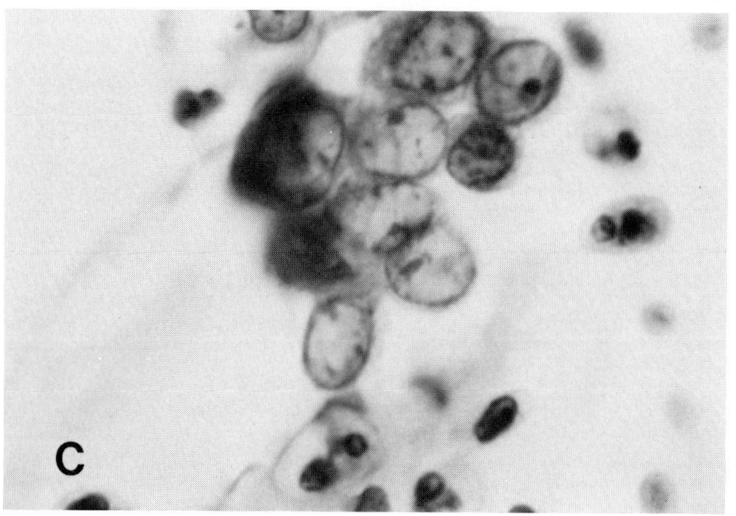

(continued)

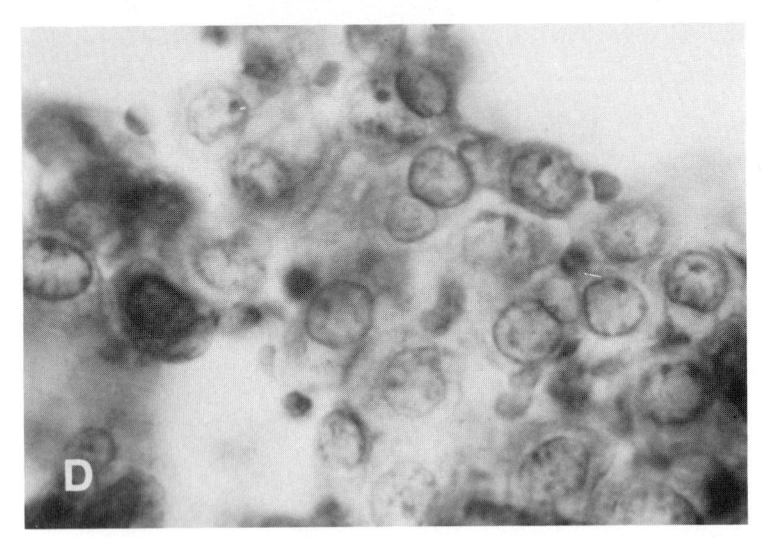

FIG. 16-8D *(continued)*
(D) A large urothelial fragment contains cells with nuclear crowding, moulding, enlargement, normochromicity, and scant cytoplasm. Note the lack of cytoplasmic maturation. (× 1000) This cellular pattern is consistent with atypical hyperplasia, but low-grade transitional cell carcinoma cannot be excluded.

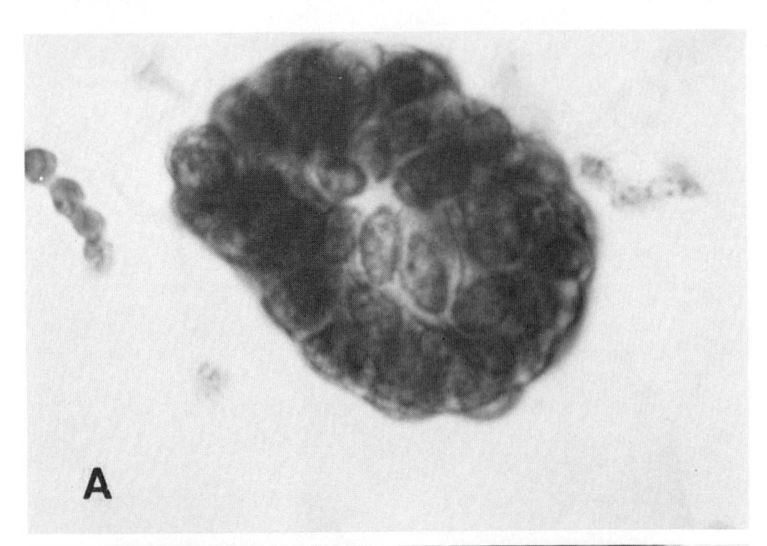

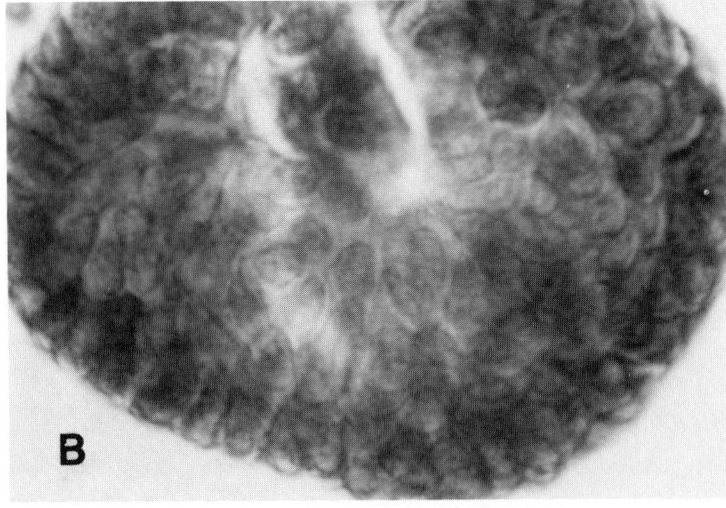

FIG. 16-9
ATYPICAL HYPERPLASTIC UROTHELIAL FRAGMENTS IN URINE SEDIMENT FOLLOWING CATHETERIZATION. **(A)** Papillary fragment with slight nuclear hyperchromasia and lack of cellular maturation. (× 400) **(B)** Large fragment with tightly packed nuclei and scant cytoplasm. (× 1000)

(continued)

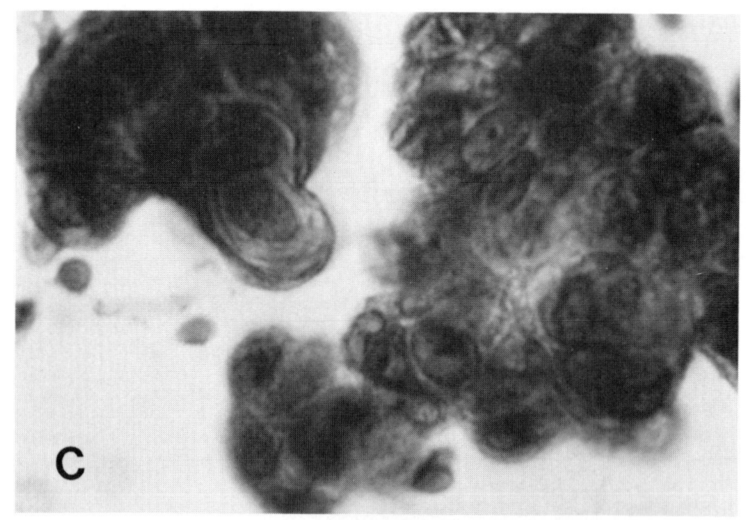

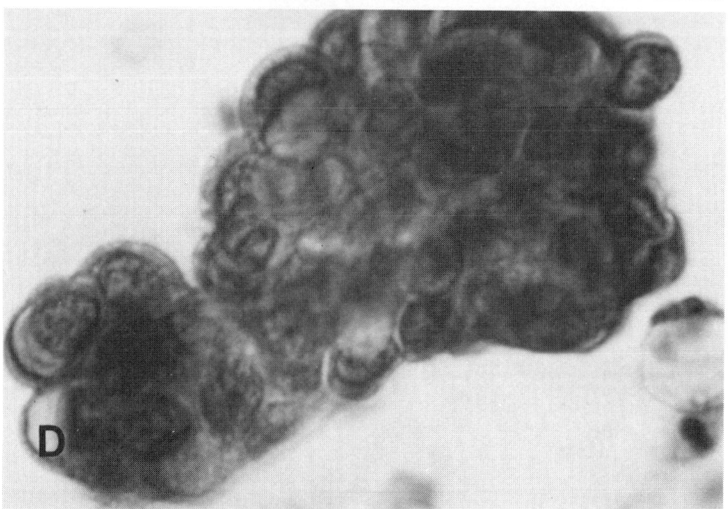

FIG. 16-9C-D *(continued)*
(C) Two atypical fragments with characteristic tightly packed nuclei and lack of cytoplasmic maturation. (× 1000)
(D) Note surface of urothelial fragment contains rounded, superficial cells with pale outer rims. (× 1000) Patient was diagnosed as having atypical hyperplasia, but low-grade transitional cell carcinoma could not be excluded.

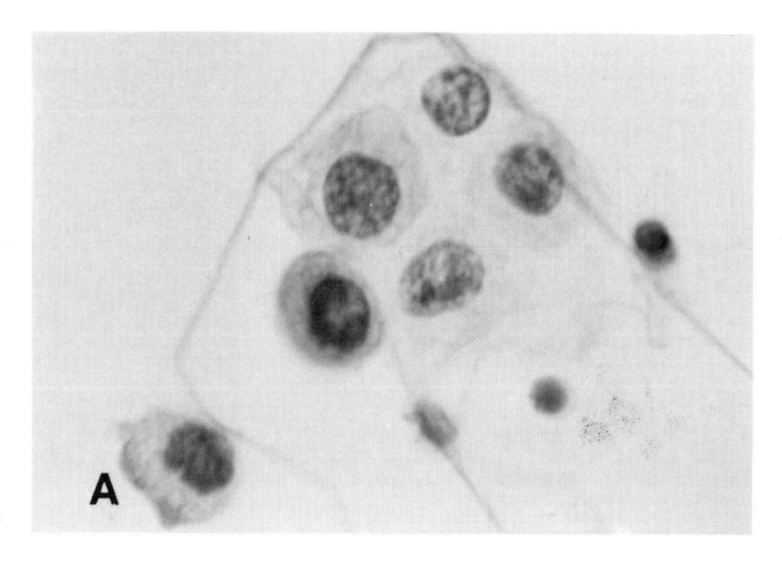

FIG. 16-10
UROTHELIAL ATYPIA (DYSKARYOSIS).
(A) Cluster of atypical urothelial cells. Note mild nuclear enlargement, hyperchromasia, and anisonucleosis. (× 1000)

(continued)

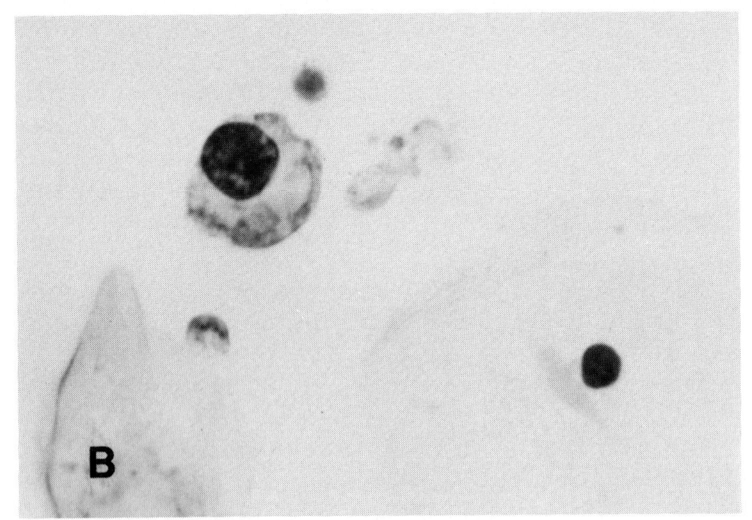

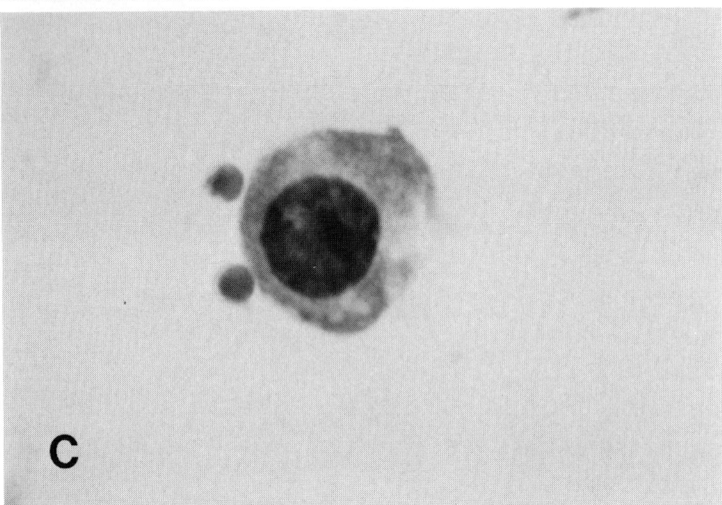

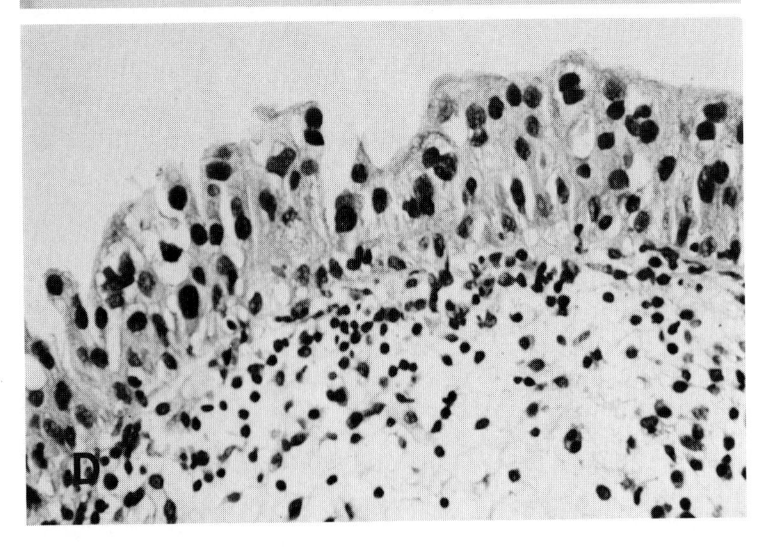

FIG. 16-10B-D *(continued)*
(B) Atypical cell with enlarged hyperchromatic nucleus and degenerated vacuolated cytoplasm. (× 1000) **(C)** Atypical cell with nuclear enlargement, hyperchromasia, chromocenter, and granular cytoplasm. Cellular features are consistent with urothelial atypia or low-grade transitional cell carcinoma. (× 1000) **(D)** Moderate urothelial atypia. This nonhyperplastic urothelium has disturbed maturation and obvious cytologic abnormalities. Cell crowding is nonuniform, and there is variable nuclear enlargement and hyperchromasia. (× 250)

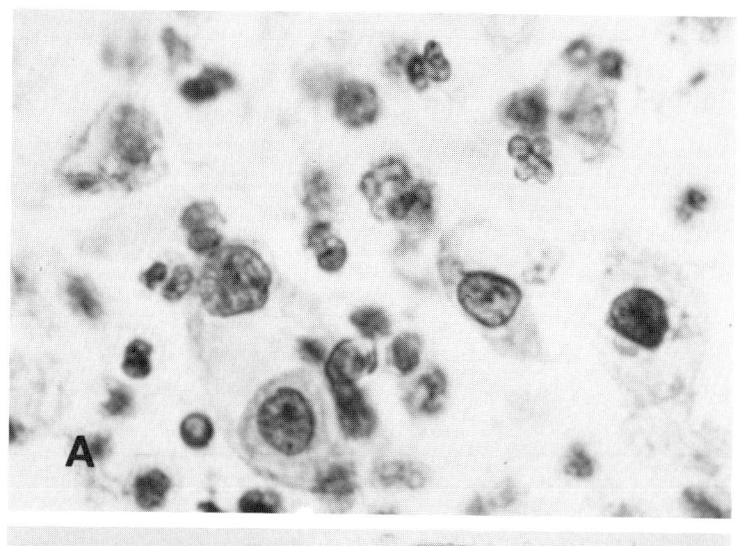

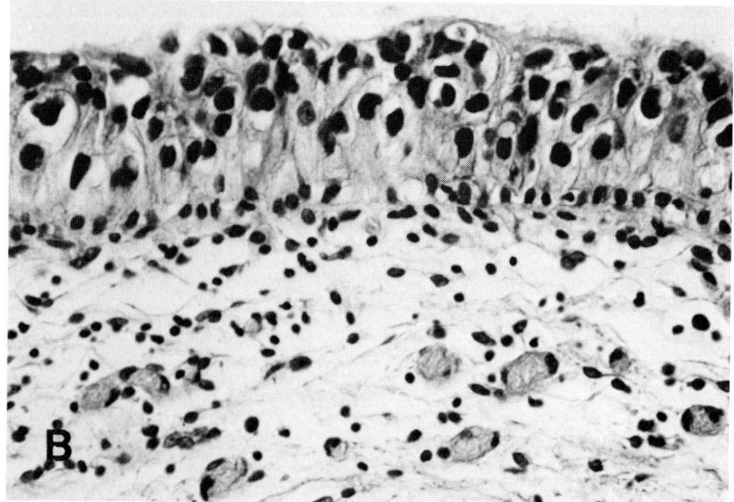

FIG. 16-11
UROTHELIAL ATYPIA (DYSKARYOSIS). **(A)** Highly atypical urothelial cells with nuclear enlargement, hyperchromasia, increased nuclear: cytoplasmic ratio, and chromocenters. (× 1000) **(B)** Marked urothelial atypia bordering on carcinoma *in situ.* Superficial layers have uniform nuclear enlargement and hyperchromasia with variable loss of polarity and crowding. Basal cells are irregularly preserved. (× 250) This patient also had a grade II papillary transitional cell carcinoma and an invasive grade III transitional cell carcinoma in separate foci.

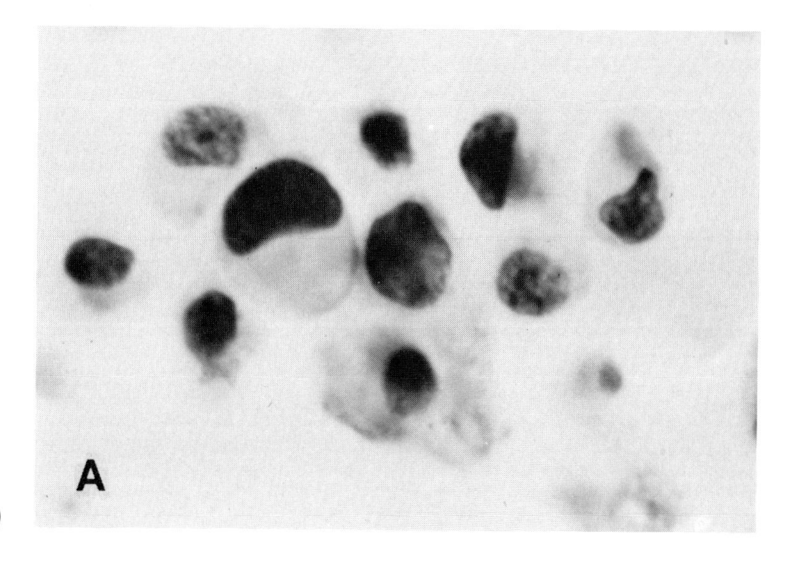

FIG. 16-12
UROTHELIAL CARCINOMA *IN SITU.* **(A)** Highly atypical small urothelial cells with increased nuclear: cytoplasmic ratio and marked hyperchromasia. (× 1000)

(continued)

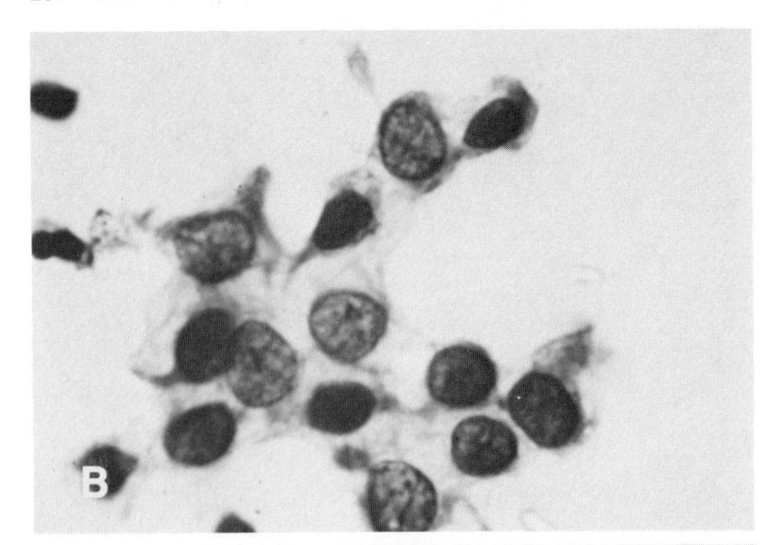

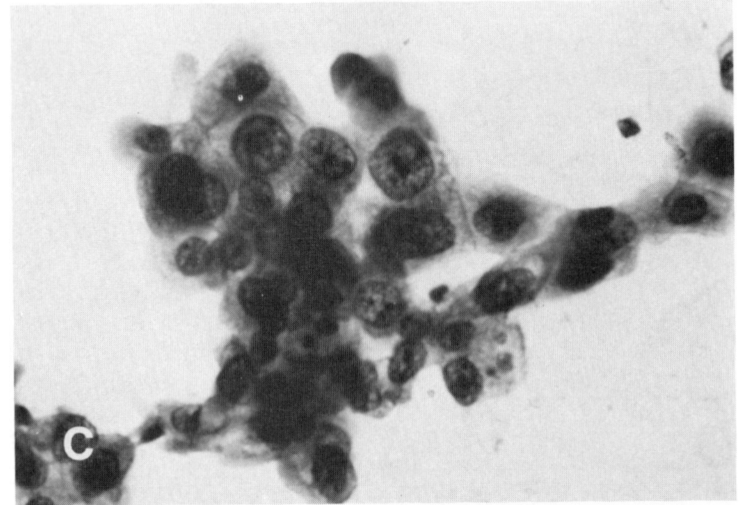

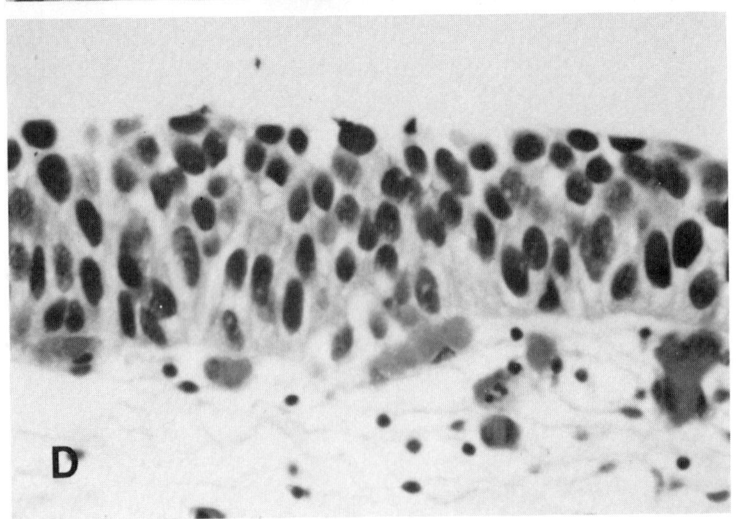

FIG. 16-12B-D *(continued)*
(B) Cells have a syncytial arrangement or lack distinct cytoplasmic borders. (× 1000) **(C)** A characteristic feature of the exfoliation of numerous atypical cells. Note marked anisonucleosis. (× 1000) **(D)** Urothelial cells are crowded and show no evidence of maturation. At all levels, nuclear: cytoplasmic ratios are abnormal, and there is uniform nuclear enlargement and hyperchromasia. Occasional nucleoli can be identified. Loss of intercellular cohesiveness is minimal. (× 400)

17
NEOPLASIA

PRIMARY NEOPLASIA
CARCINOMA *IN SITU*
PAPILLARY TRANSITIONAL CELL CARCINOMA
ADENOCARCINOMA
SQUAMOUS CELL CARCINOMA
METASTATIC NEOPLASIA

PRIMARY NEOPLASIA

CARCINOMA *IN SITU*

Carcinoma *in situ* (CIS) is a flat urothelial malignancy with a potential for invasion. It may present as a multifocal or diffuse lesion involving both ureter and urethra or it may coincide with or follow the appearance of a papillary transitional cell carcinoma. Cytohistologic criteria for diagnosing urothelial CIS are given in Chapters 6 and 16.

PAPILLARY TRANSITIONAL CELL CARCINOMA

Histologically, papillary transitional cell carcinoma is graded I to III (World Health Organization). In the urine sediment examination, exfoliated cells and fragments from papillary transitional cell carcinoma are generally described as being derived from low-grade (I–II) or high-grade (II–III) lesions.

Cells exfoliated from grade I carcinomas cannot be identified. Exfoliated fragments should be recognized as atypical, but are not diagnostic (Figs. 17-1–17-3). The characteristics and differential diagnosis of these atypical fragments are given in Chapter 16. It may be difficult to differentiate these atypical urothelial fragments from renal epithelial fragments, particularly in the setting of renal parenchymal ischemia (Fig. 17-1).

A definitive cytodiagnosis can be made from exfoliated urothelial cells of a grade II or a grade III lesion. In grade II lesions, the cytologic abnormalities evident histologically correlate with exfoliation of urothelium as individual cells and fragments having high nuclear:cytoplasmic ratio and hyperchromasia (Figs. 17-4 and 17-5). The presence of elongated cells, histologically characteristic of grade I to II lesions, may be observed in urine sediment (Fig. 17-4).

The loss of intercellular cohesiveness, characteristic of grade III carcinomas, results in marked exfoliation of cells and fragments. These malignant cells are recognized in urine sediment as hyperchromatic cells with prominent nucleoli, and anisonucleosis is often a prominent feature (Figs. 17-6–17-9). Fragments often contain cells in syncytial arrangement (Figs. 17-8 and 17-9).

Urine sediment diagnosis of malignant urothelial cells without a cystoscopically confirmed bladder lesion suggests the possibility of CIS or a pelviureteral tumor (Figs. 17-3 and 17-6). With persistence of these cells, multiple bladder biopsies and pelviureteral brushings or washings should be suggested for localization.

ADENOCARCINOMA

Primary adenocarcinoma of the lower urinary tract is not commonly encountered (Fig. 17-10). Malignant cells derived from an adenocarcinoma can be diagnosed in the urine sediment, but localization requires clinical and radiographic evaluation with biopsy confirmation. In a male, the presence in urine sediment of malignant glandular fragments strongly suggests bladder or urethral involvement by prostatic carcinoma (Fig. 17-11). One should not rely upon this occurrence for the diagnosis of prostatic carcinoma; prostatic fine-needle aspiration is the method of choice for cytodiagnosis.

SQUAMOUS CELL CARCINOMA

Primary squamous cell carcinoma of the urinary tract is encountered uncommonly in this country (Fig. 17-12). Malignant squamous cells in the urine sediment of a female usually result from widespread pelvic extension of a cervical carcinoma (Figs. 17-13 and 17-14).

METASTATIC NEOPLASIA

In patients with widespread metastases, the urinary system may be involved. Frequently, the type of malignancy can be diagnosed, that is, squamous cell carcinoma, adenocarcinoma (Figs. 17-15–17-17), or malignant melanoma. Lower urinary tract involvement by disseminated lymphoma is more difficult to diagnose and requires the presence of immature lymphoreticular cells with nuclear irregularities and nucleolar atypia (Fig. 17-18).

BIBLIOGRAPHY

Koss LG: Diagnostic Cytology and Its Histopathologic Bases. 3rd ed. Philadelphia, JB Lippincott, 1979

Koss LG: Tumors of the Urinary Bladder. Atlas of Tumor Pathology, 2nd series, Fascicle 11. Washington, D.C., Armed Forces Institute of Pathology, 1975

Tweedale DN: Urinary Cytology. Boston, Little, Brown, 1977

Voogt HJ, Rathert P, Beyer-Boon ME: Urinary Cytology. Phase Contrast Microscopy and Analysis of Stained Smears. New York, Springer-Verlag, 1977

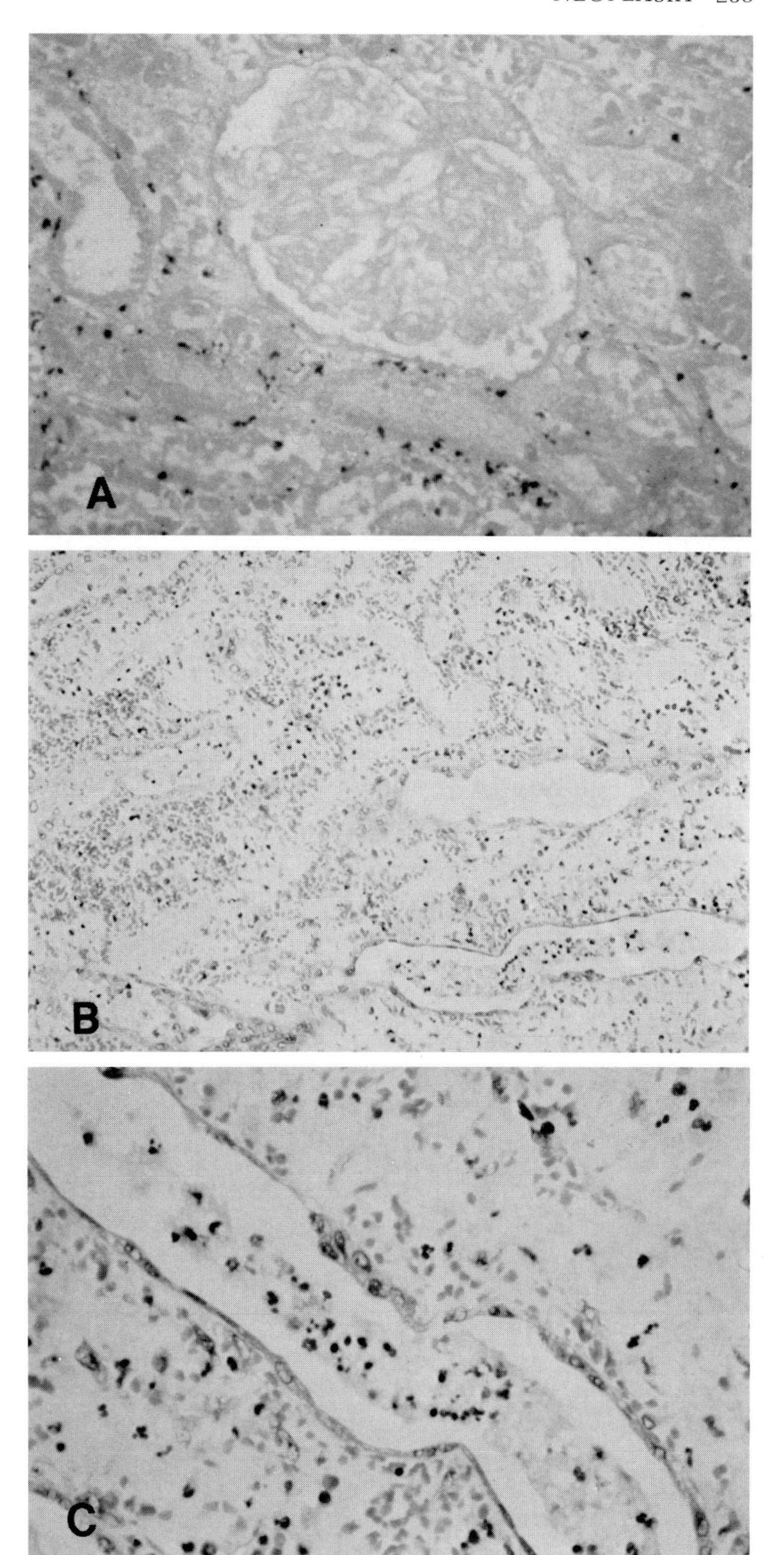

FIG. 17-1
PAPILLARY TRANSITIONAL CELL CARCINOMA, GRADE I TO II, IN RENAL TRANSPLANT PATIENT WITH INFARCTED ALLOGRAFT. **(A)** Renal allograft with infarcted cortex. (× 150) **(B)** Medullary portion of renal allograft. There is marked interstitial hemorrhage and edema with a scant chronic inflammatory infiltrate. Tubules have an attenuated or more actively regenerating epithelium and contain exfoliated cells and casts. (× 100) **(C)** Leukocytic cast. (× 250)

(continued)

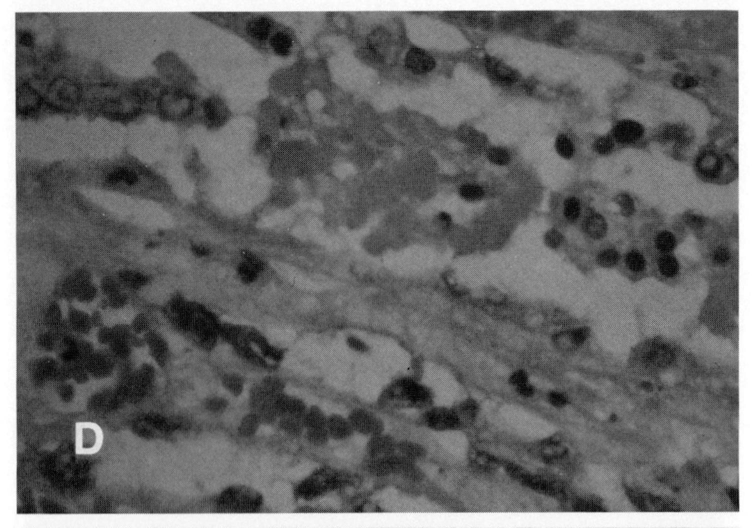

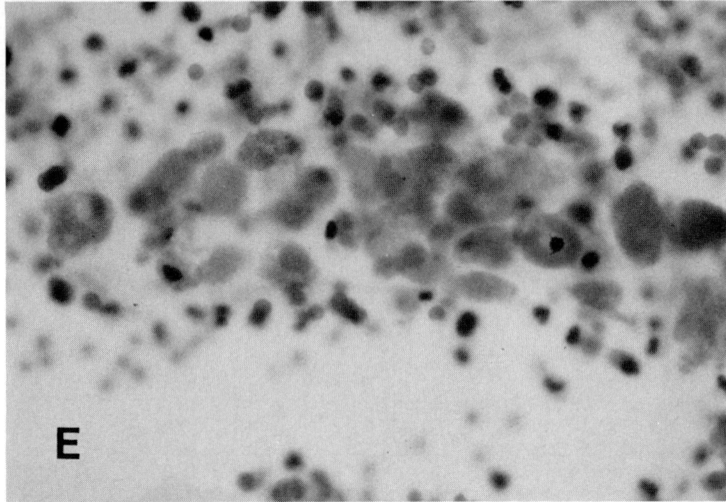

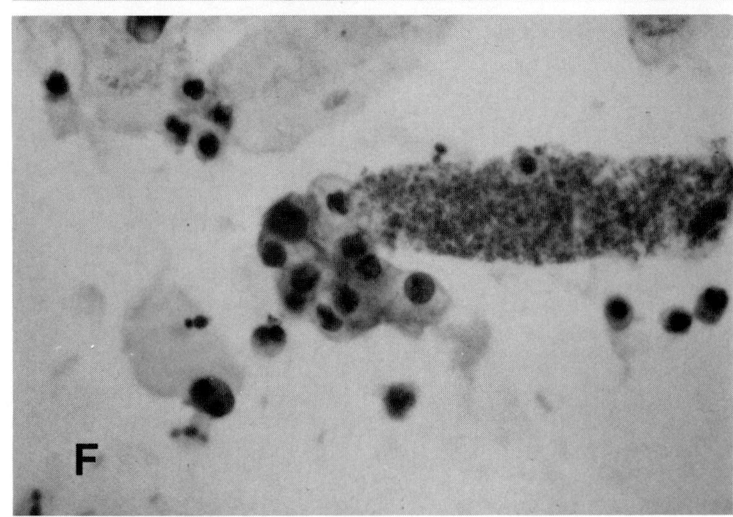

FIG. 17-1D-F *(continued)*
(D) Exfoliated regenerating epithelial cells, degenerating epithelial cells with pyknotic nuclei, and an aggregate of ghost cells within a collecting duct. (× 400) **(E)** Necrotic renal tubular epithelial cast. Note hematuria and inflammation in background. (× 250) **(F)** Renal epithelial fragment attached to a granular cast. The fragment contains cells with round nuclei and abundant finely granular cytoplasm. (× 400)

(continued)

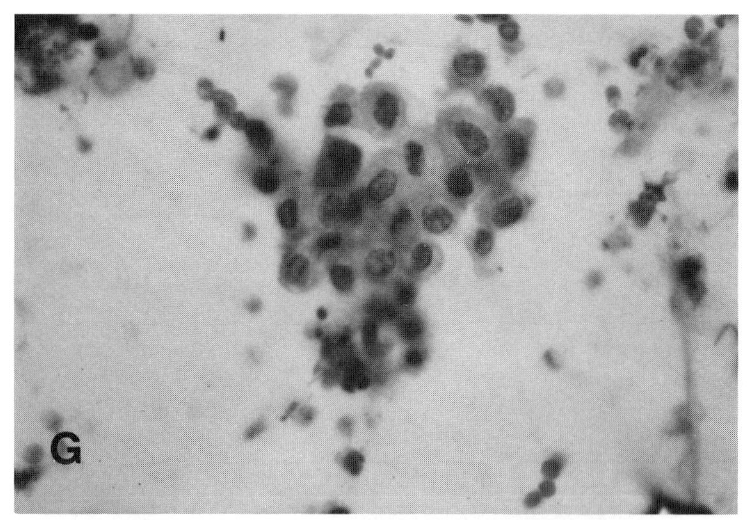

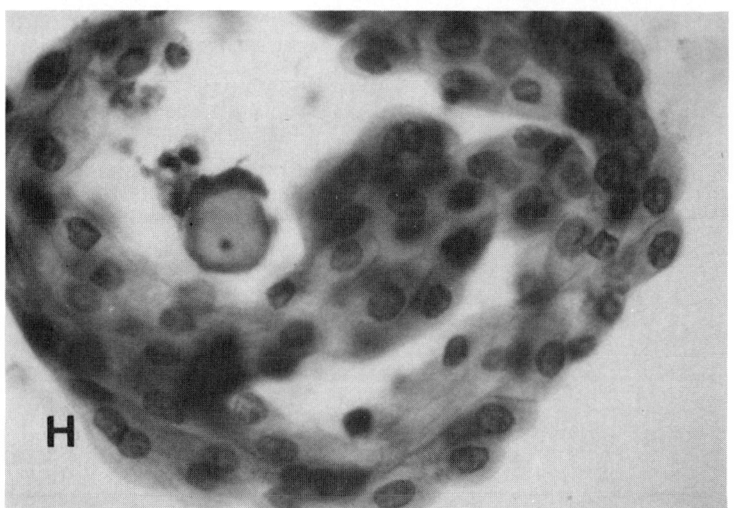

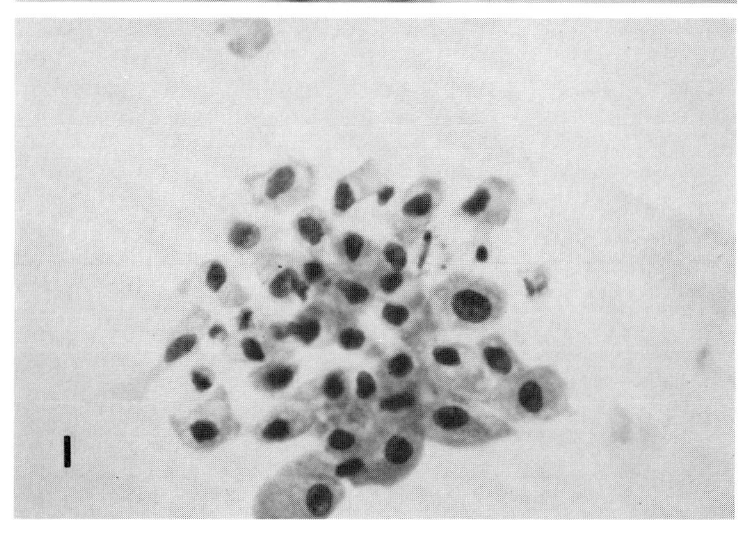

FIG. 17-1G-I *(continued)*
(G) Exfoliated renal epithelial fragment with honeycomb arrangement of cells. Note centrally located round to oval nuclei and finely granular cytoplasm with distinct borders. (× 400) **(H)** Large sheet of benign urothelial cells. These cells contain round to oval centrally placed nuclei and abundant dense cytoplasm. Large sheets of epithelium are commonly seen after instrumentation and following infarction of renal papillae. In this case, they probably represent exfoliation of the urothelium lining the calyceal system. (× 400) **(I)** Atypical urothelial fragment. Note large superficial and smaller deep urothelial cells. (× 400)

(continued)

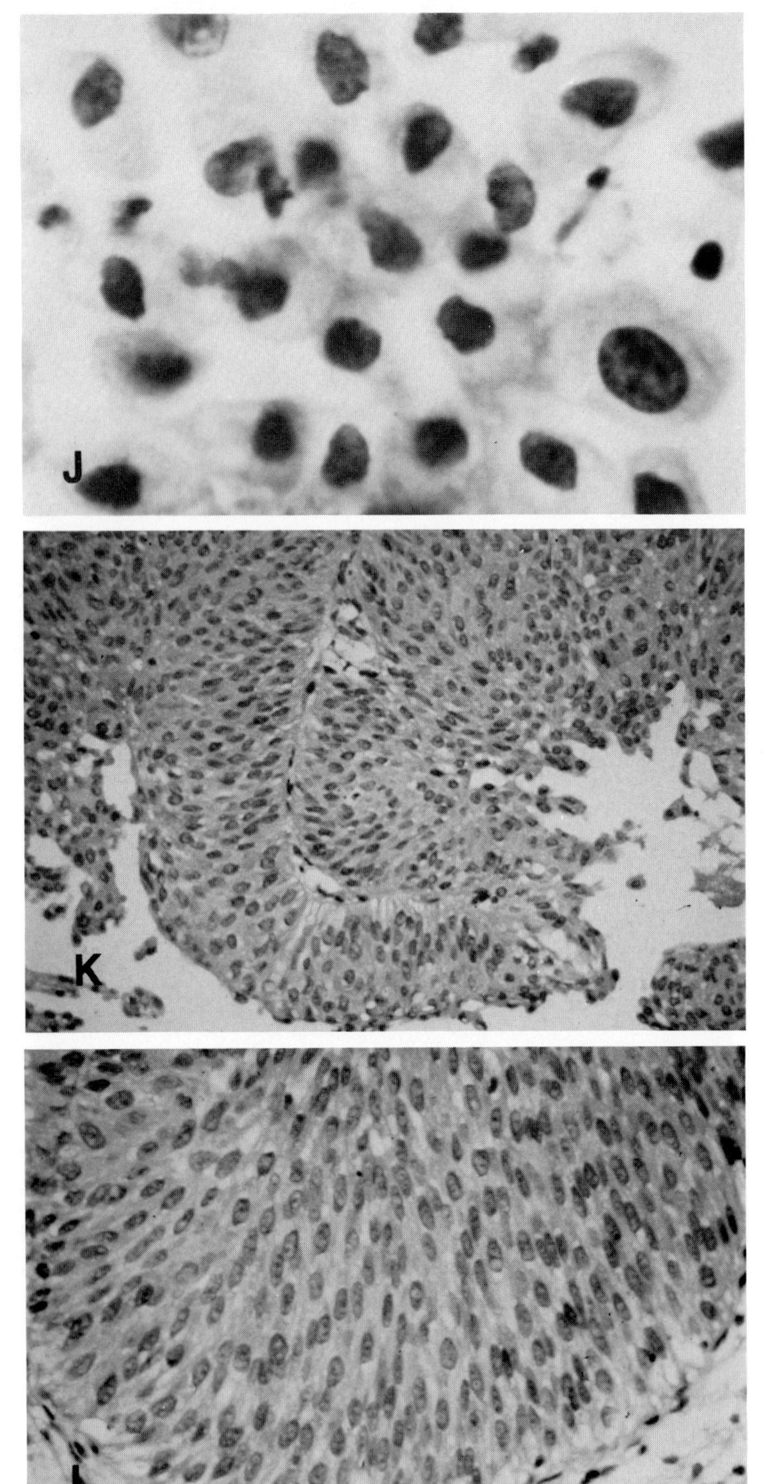

FIG. 17-1J-L *(continued)*
(J) Note nuclear enlargement and hyperchromasia consistent with a low-grade transitional cell carcinoma. (× 1000) **(K)** Papillary transitional cell carcinoma of bladder, grade I to II. Papillae are covered by hyperplastic urothelium. Superficial cells are focally lost. (× 150) **(L)** Note moderate nuclear enlargement and slight hyperchromasia. (× 250)

FIG. 17-2
LOW-GRADE TRANSITIONAL CELL CARCINOMA IN URINE SEDIMENT. (**A**) Epithelial fragment and benign urothelial cells. (× 400) (**B**) Urothelial fragment with anisonucleosis, mild hyperchromasia, and abundant dense vacuolated cytoplasm (× 1000) (**C**) Note nuclear enlargement and slight hyperchromasia (× 1000)

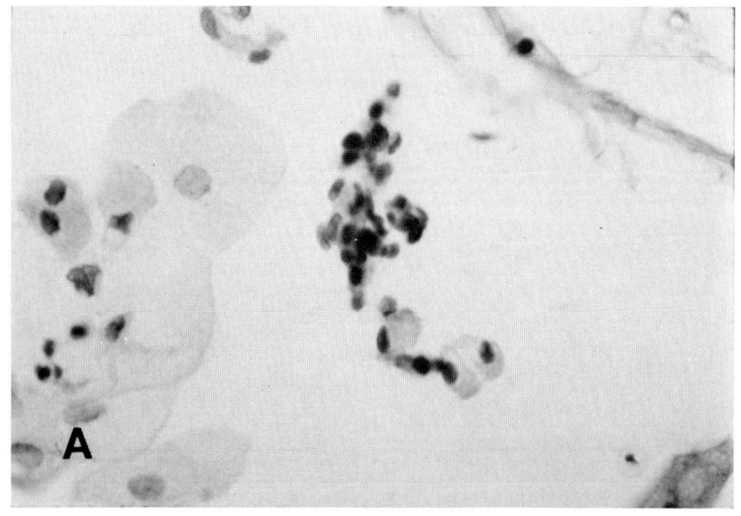

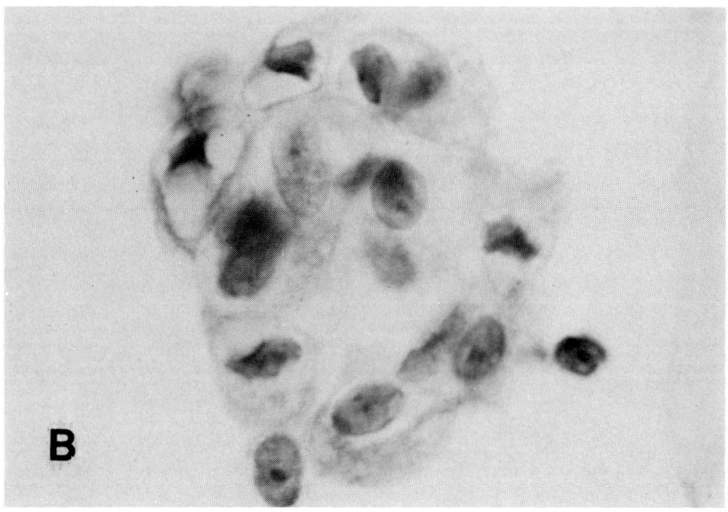

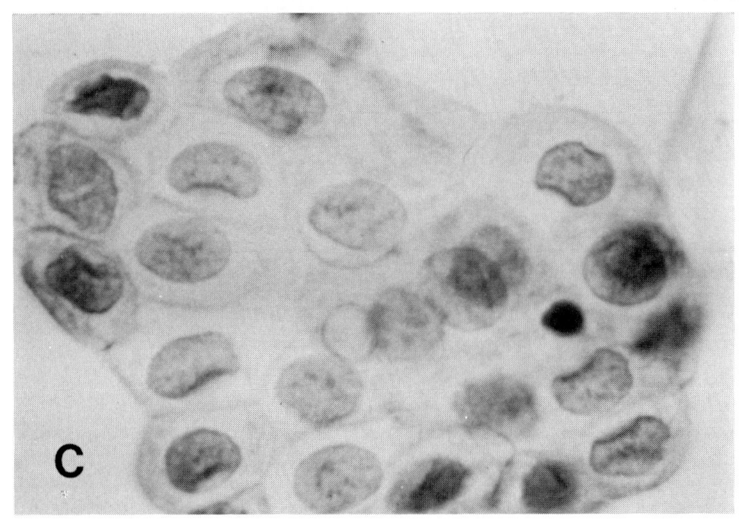

(continued)

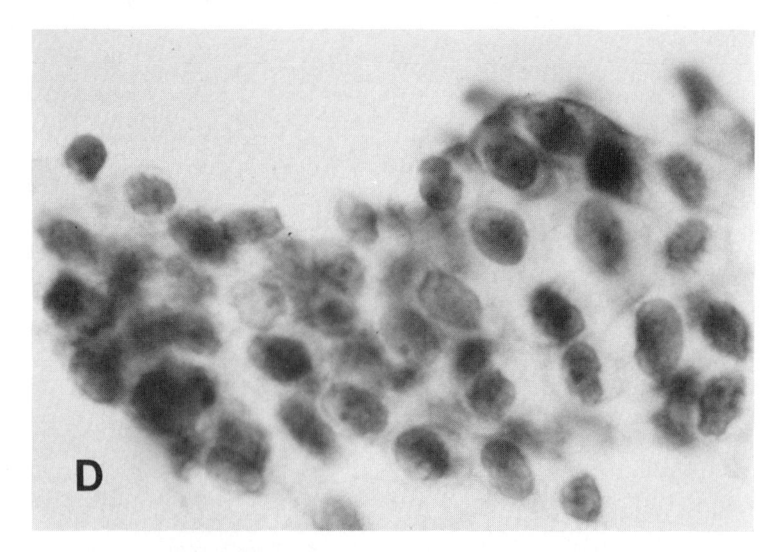

FIG. 17-2D *(continued)*
(D) Atypical urothelial fragment containing cells with slightly hyperchromatic nuclei and lack of cytoplasmic maturation. (× 1000) Urine specimen obtained from patient with biopsy-proven grade I transitional cell carcinoma of the bladder.

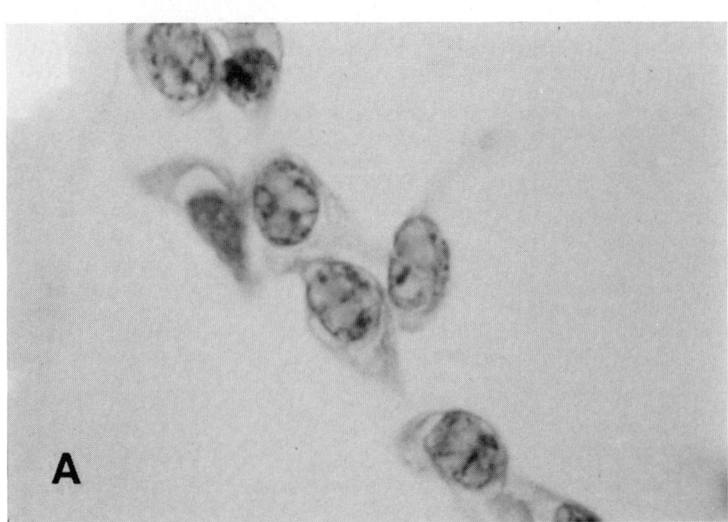

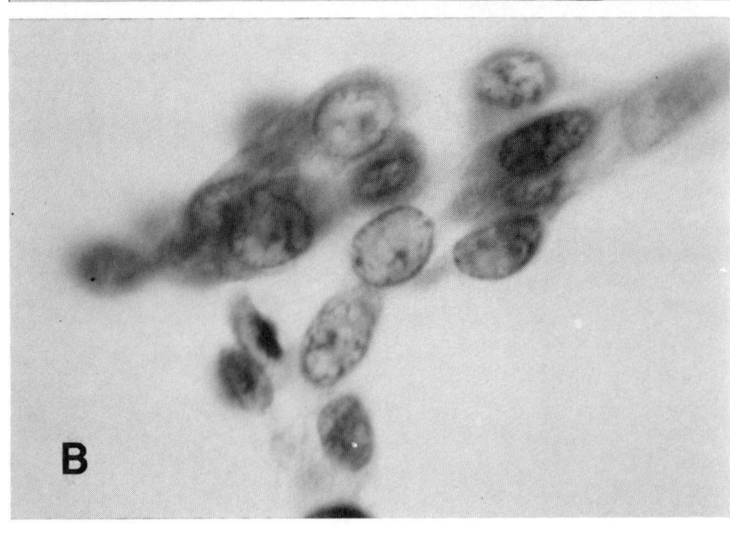

FIG. 17-3
PAPILLARY TRANSITIONAL CELL CARCINOMA OF RENAL PELVIS, GRADE I TO II. **(A)** Exfoliated urothelial fragment containing loosely cohesive cells with nuclear enlargement, nuclear membrane indentations, chromatin clumping and clearing, and dense cytoplasm. (× 1000) **(B)** Note anisonucleosis and chromocenters. (× 1000) Cellular features are consistent with urothelial atypia or low-grade transitional cell carcinoma.

(continued)

FIG. 17-3C-E *(continued)*
(C) Imprint of tumor. Note crowding and moulding of normochromatic nuclei and lack of cytoplasmic maturation. (× 400)
(D) A portion of the renal pelvis is filled with a papillary tumor. Note lack of invasion of peripelvic fat and renal parenchyma. (Gross) **(E)** Delicate papillary fronds are covered by a hyperplastic urothelium. Note nuclear crowding and hyperchromasia in comparison to normal urothelium (lower right) (× 100)

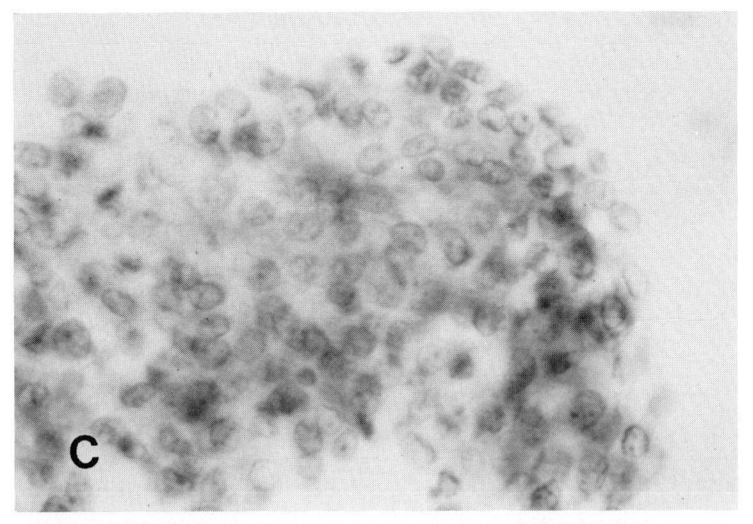

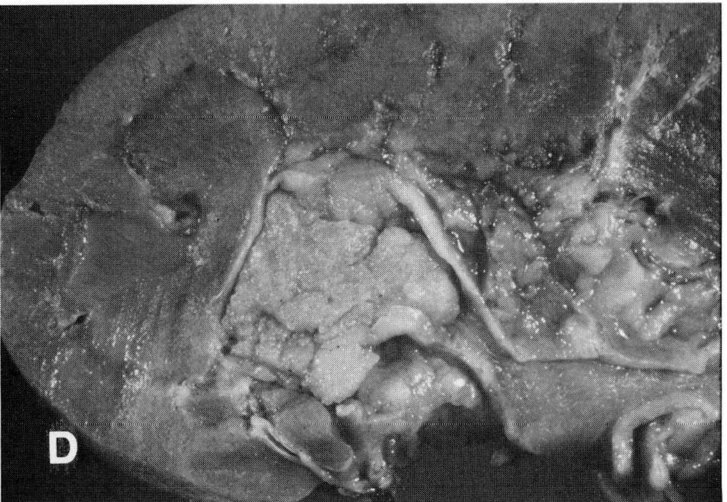

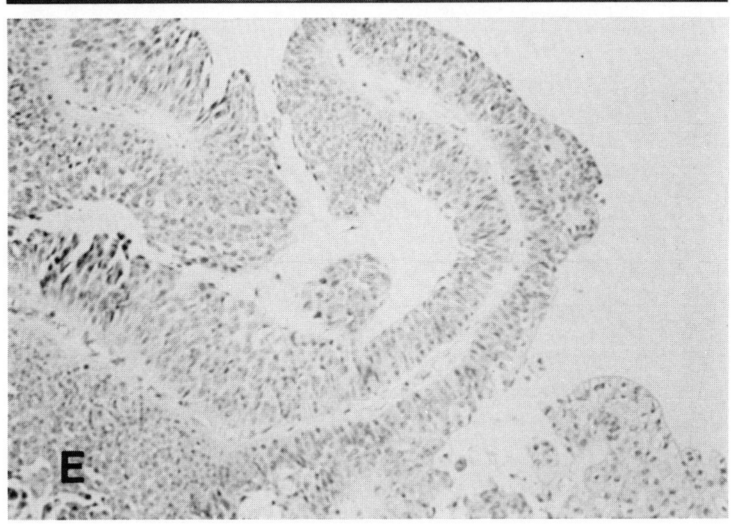

(continued)

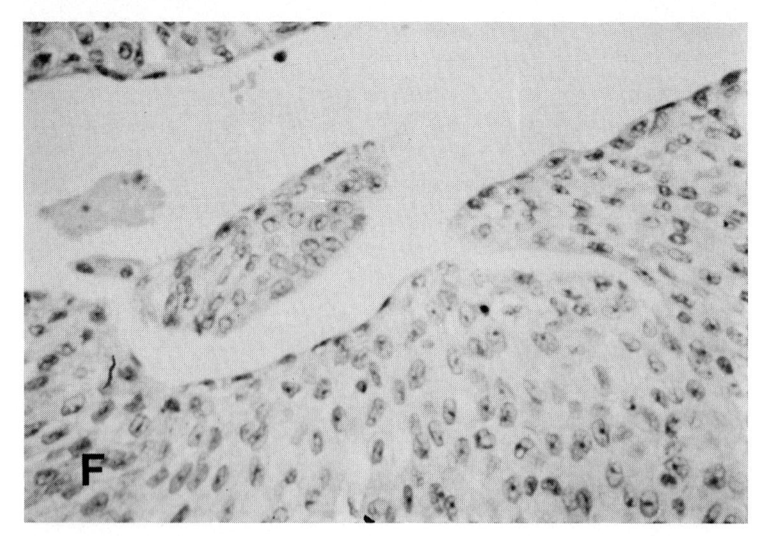

FIG. 17-3F *(continued)*
(F) The hyperplastic urothelium has uniform nuclear enlargement with mild hyperchromasia and micronucleoli. Flattened superficial cells are present. Similar cytologic features are evident in the detached urothelial fragment. (× 400)

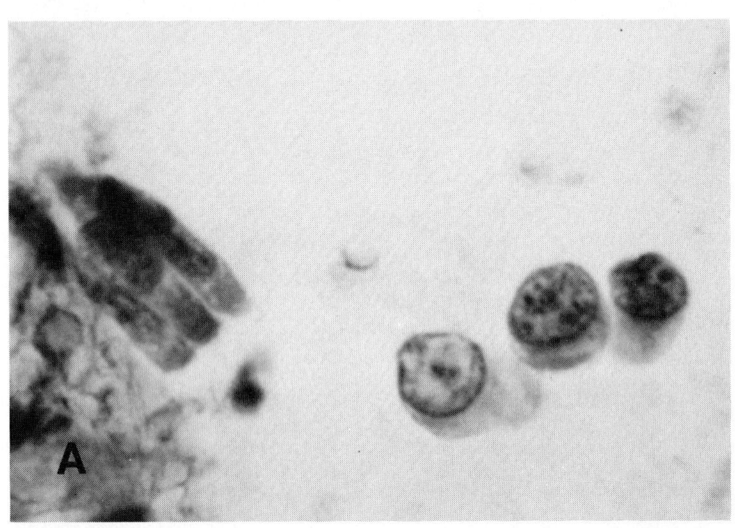

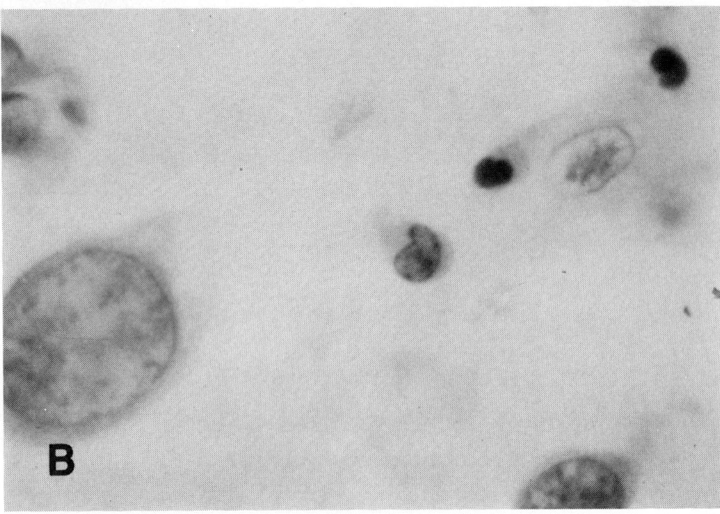

FIG. 17-4
PAPILLARY TRANSITIONAL CELL CARCINOMA OF BLADDER, GRADE I TO II. **(A)** Malignant urothelial cells. Small cells have enlarged hyperchromatic nuclei with chromatin clearing. Note elongated cells. (× 1000) **(B)** Large and small malignant urothelial cells have high nuclear: cytoplasmic ratio, hyperchromasia, and scant cytoplasm. Note irregular eosinophilic intranuclear inclusion in degenerating cell, suggesting viral infection. (× 1000)

(continued)

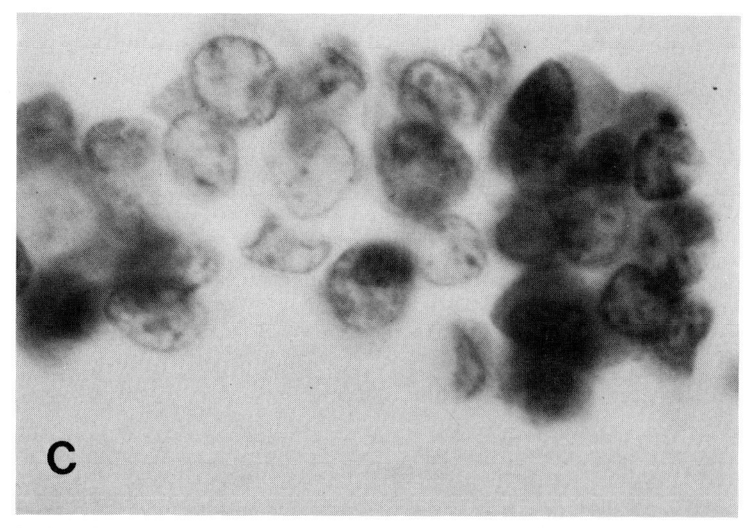

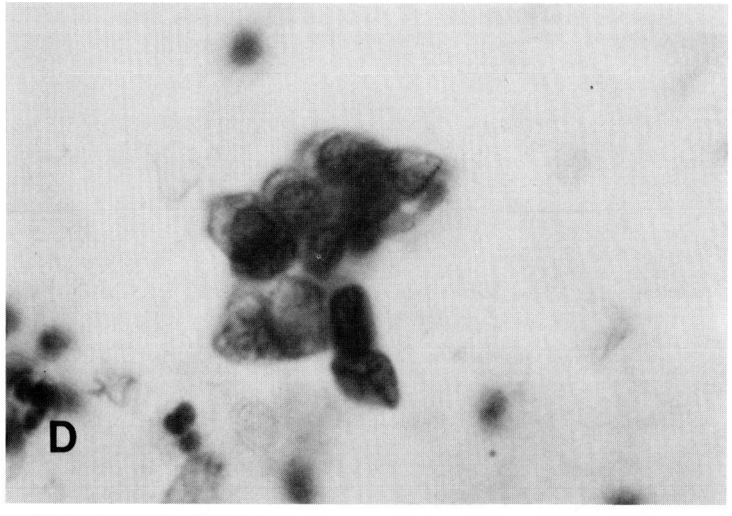

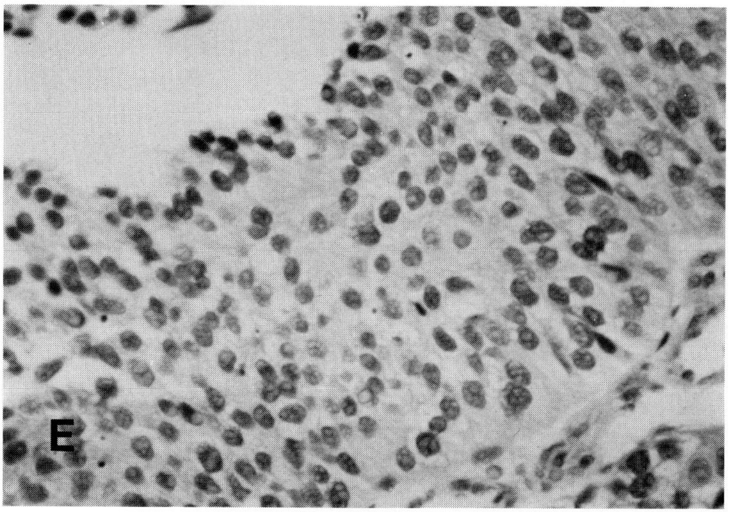

FIG. 17-4C-E *(continued)*
(C) Atypical urothelial fragment. Nuclear crowding, anisonucleosis and scant cytoplasm are features consistent with low-grade transitional carcinoma. (× 1000) **(D)** Fragment of malignant urothelium. The lack of prominent nucleoli characterizes low-grade transitional cell carcinoma. (× 1000) **(E)** The majority of this papillary tumor was lined by a hyperplastic urothelium showing moderate nuclear enlargement and mild hyperchromasia. (× 250)

(continued)

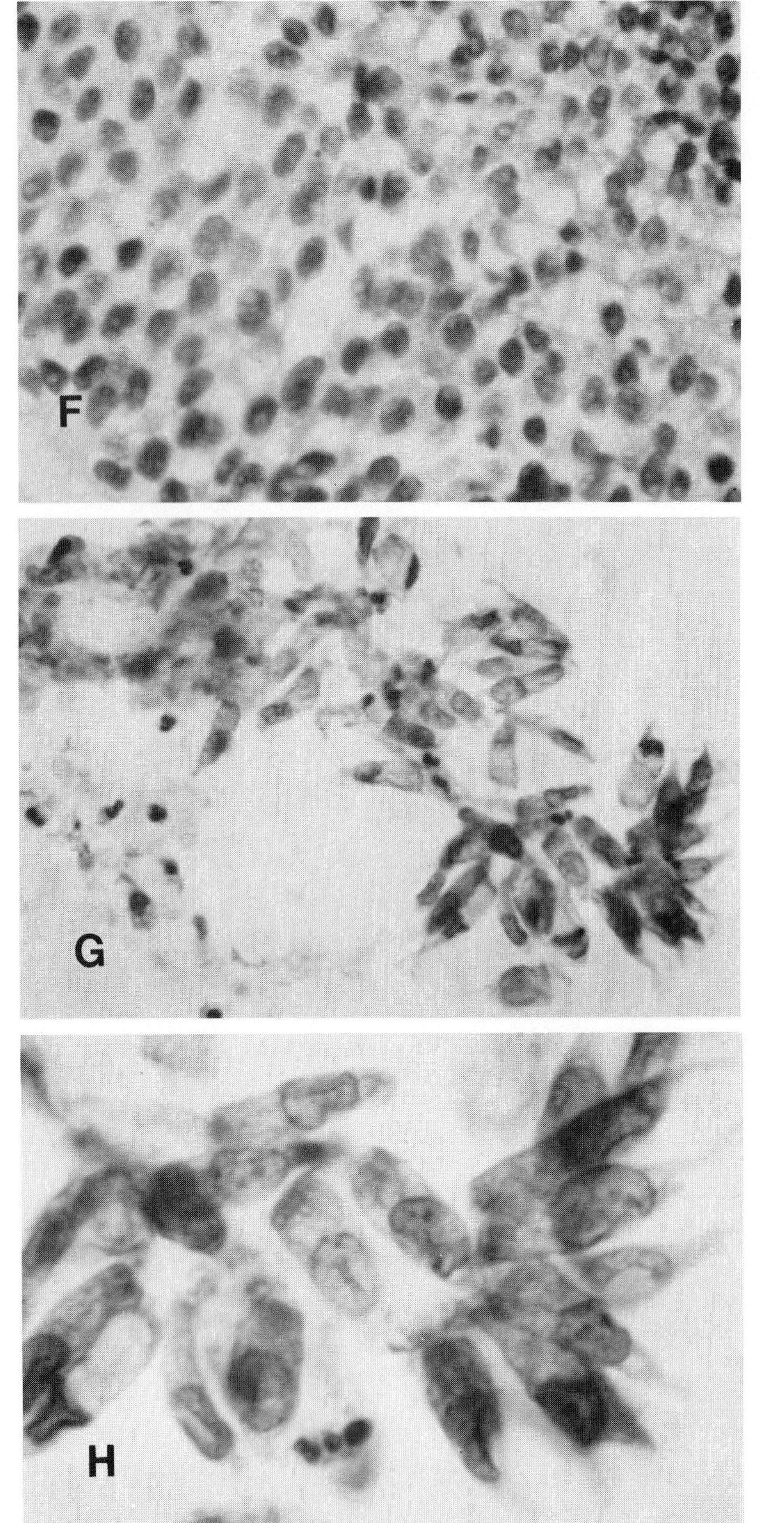

FIG. 17-4F-H *(continued)*
(F) Note lack of nucleoli. (× 400) **(G)** Numerous elongated or columnar epithelial cells with centrally placed round to oval nuclei. (× 400) **(H)** Note blunted and tapered ends. The cytoplasm is finely granular and contains occasional vacuoles. (× 1000)

(continued)

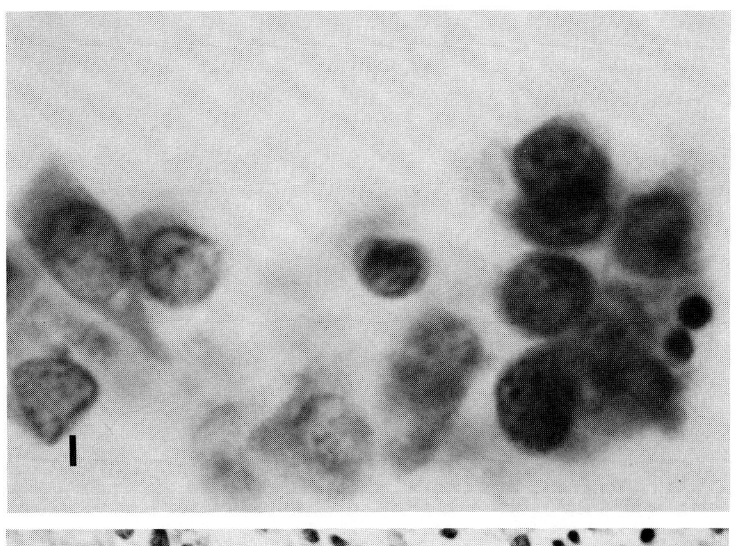

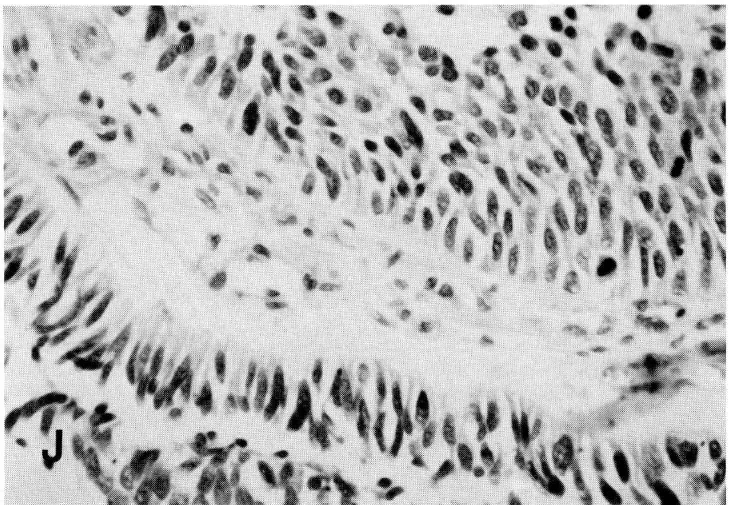

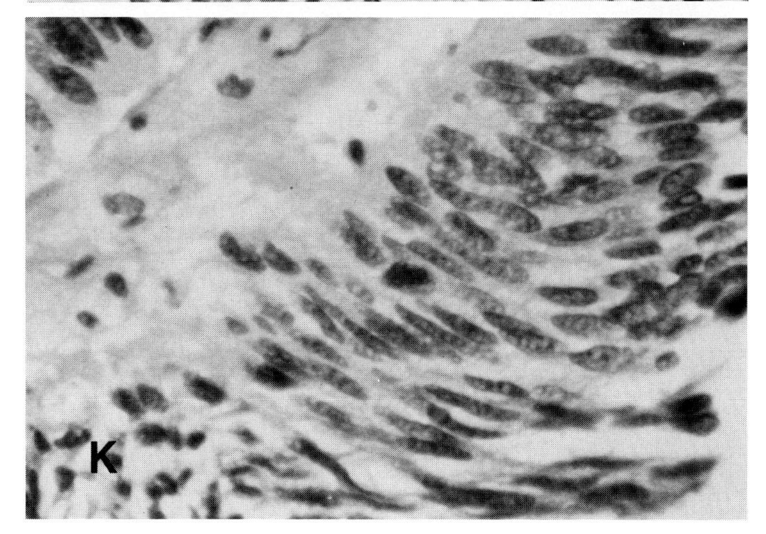

FIG. 17-4I-K *(continued)*
(I) Malignant urothelial fragment with cells having a high nuclear: cytoplasmic ratio and nuclear hyperchromasia. Note columnar differentiation at one end of frgament. Elongated or columnar cells may be seen in grade II transitional cell carcinoma. (× 1000) **(J)** In multiple foci elongation of cells and their nuclei produce a spindle-cell appearance. (× 250) **(K)** Cytologic characteristics remain otherwise unchanged. (× 400)

(continued)

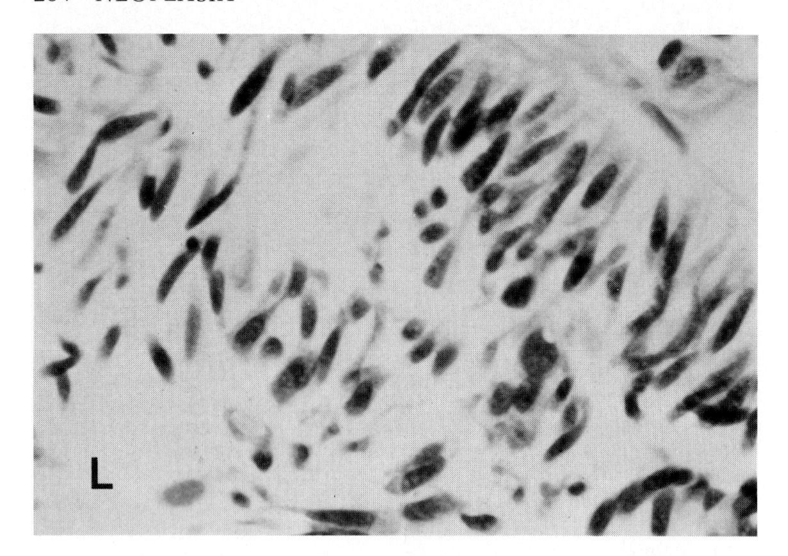

FIG. 17-4L *(continued)*
(L) Exfoliation from spindle cell foci yields elongated cells with two tapered ends or one blunted end corresponding to the site of basal attachment. (× 400)

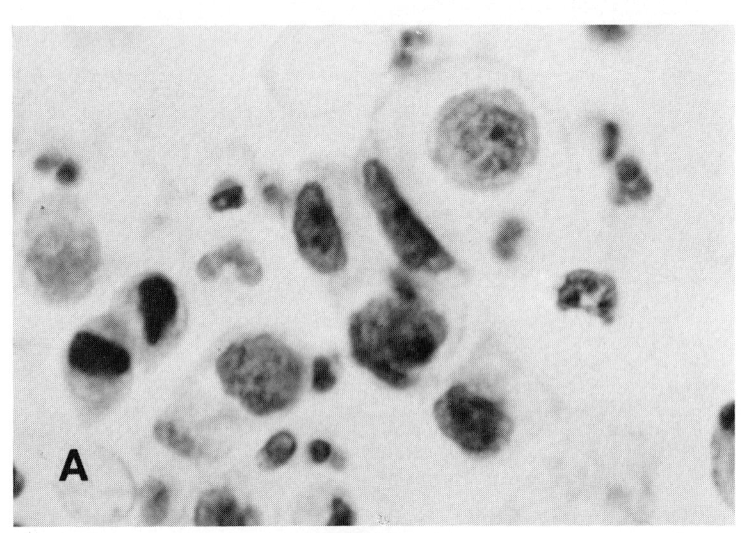

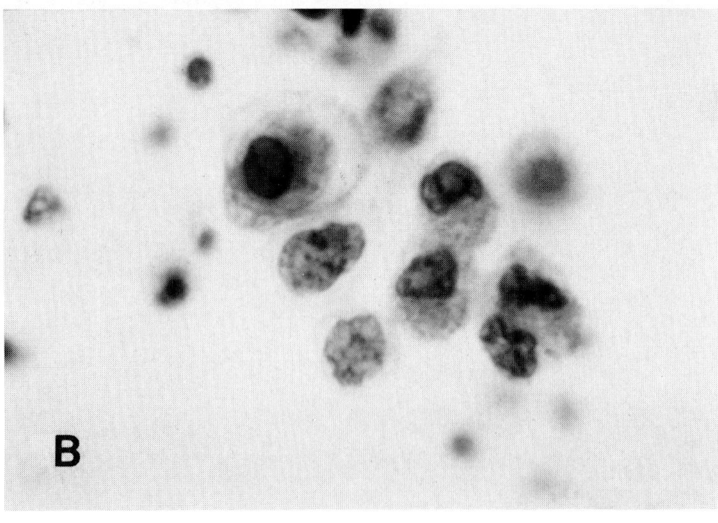

FIG. 17-5
PAPILLARY TRANSITIONAL CELL CARCINOMA, OF BLADDER, GRADE II.
(A) Malignant cells derived from transitional cell carcinoma. Note variable nuclear enlargement and hyperchromasia. (× 1000) **(B)** Small cells with high nuclear: cytoplasmic ratio and hyperchromasia. (× 1000)

(continued)

FIG. 17-5C-E *(continued)*
(C) Absence of prominent nucleoli are features of grade II transitional cell carcinoma. (× 1000) **(D)** This papillary tumor has a hyperplastic urothelium with diminished intercellular cohesiveness, resulting in marked exfoliation of individual cells and fragments. Note moderate nuclear enlargement and hyperchromasia. (× 250) **(E)** Multiple biopsies showed moderate to marked urothelial atypia with multifocal carcinoma *in situ.* (× 250)

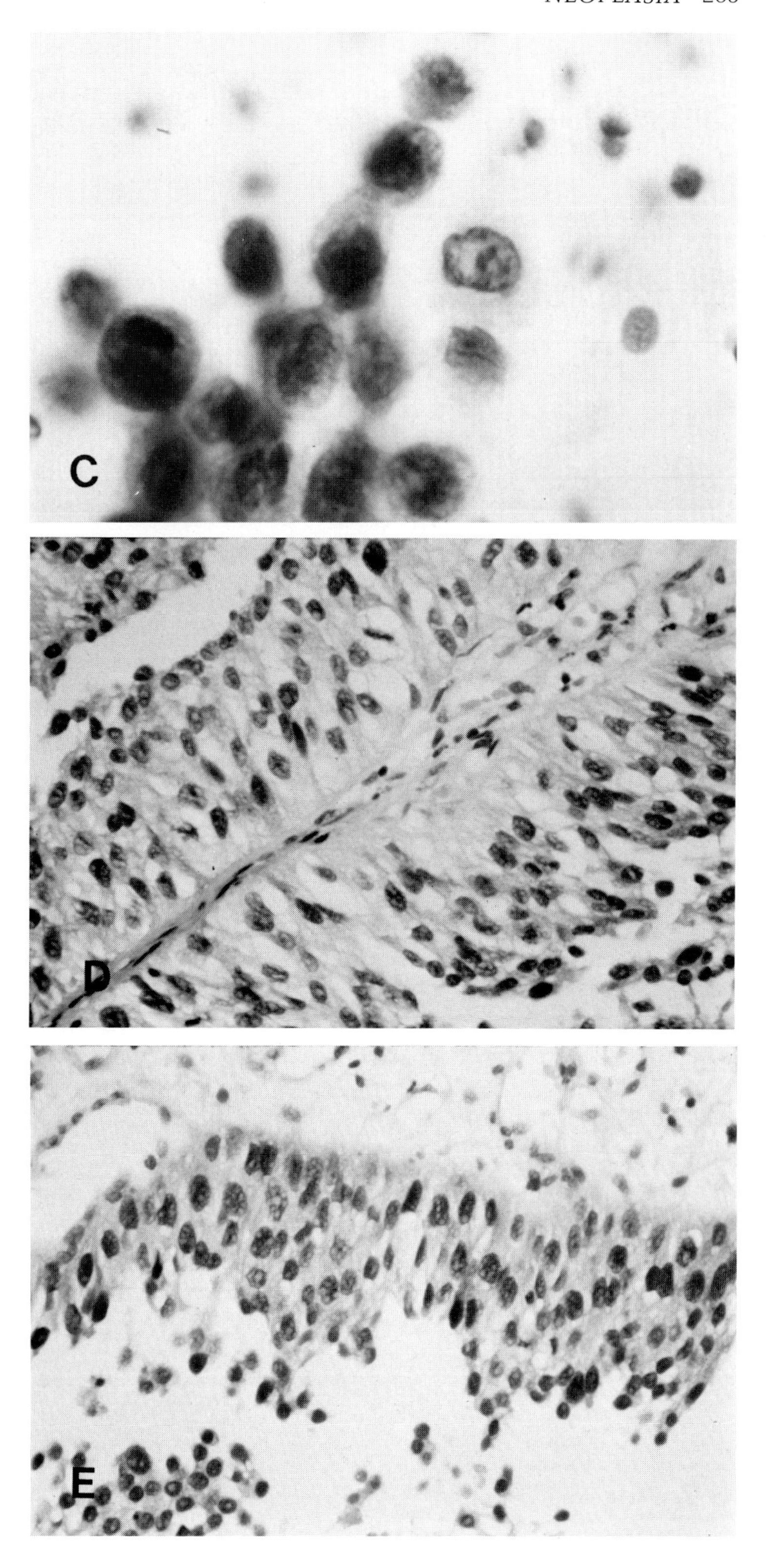

(continued)

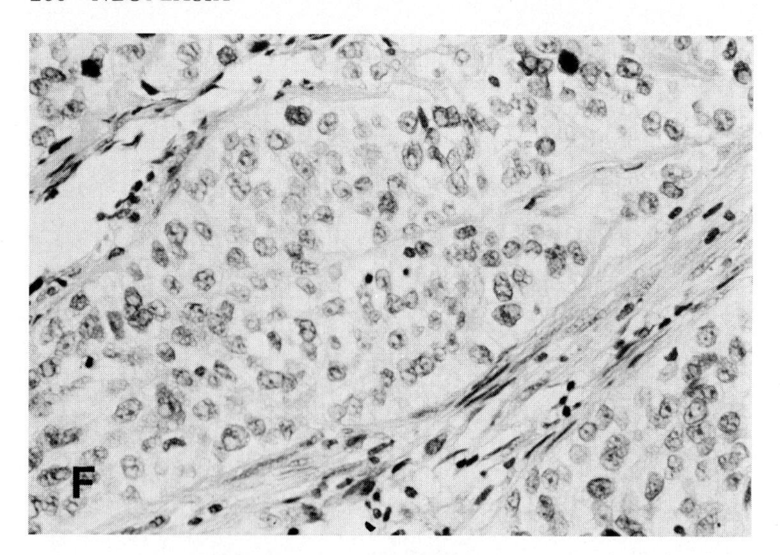

FIG. 17-5F *(continued)*
(F) Invasive grade II transitional cell carcinoma in bladder wall. Note prominence of nucleoli in tumor cells that infiltrate smooth muscle. (× 250)

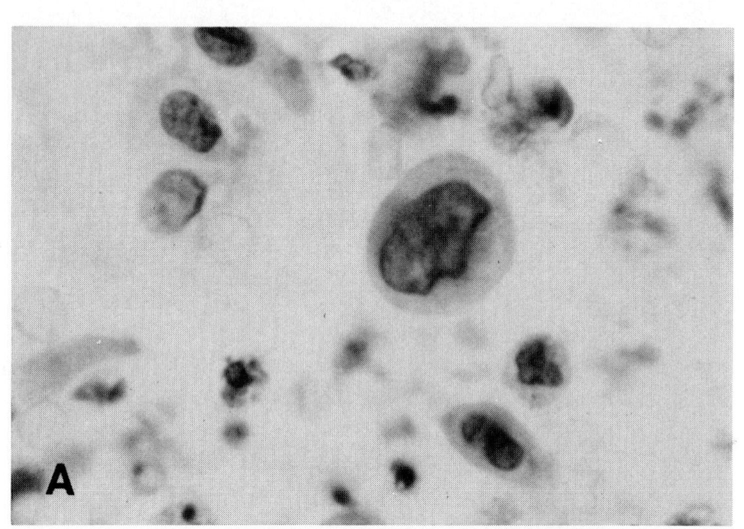

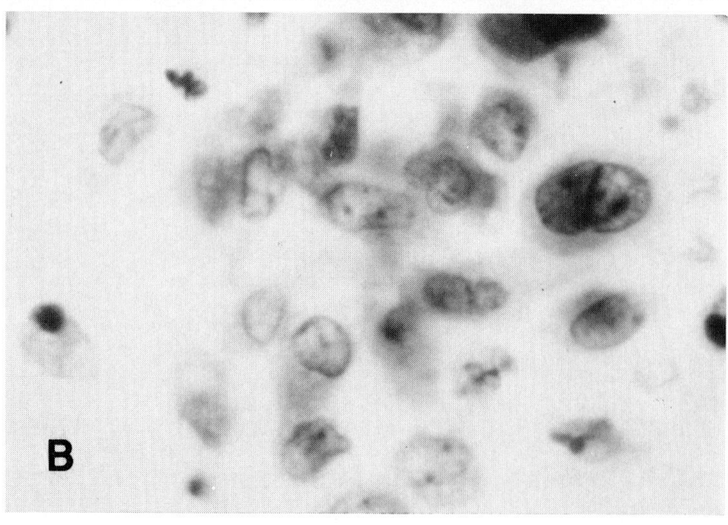

FIG. 17-6
PAPILLARY TRANSITIONAL CELL CARCINOMA OF RENAL PELVIS, GRADE II TO III. **(A)** Malignant urothelial cells in urine sediment. Note hyperchromatic nuclei and increased nuclear: cytoplasmic ratio. (× 1000) **(B)** Loose cluster of malignant urothelial cells with anisonucleosis, hyperchromasia, chromocenters, nuclear membrane irregularities, and indistinct cytoplasm characteristic of transitional cell carcinoma. (× 1000)

(continued)

FIG. 17-6C-E *(continued)*
(C) Atypical urothelial fragment obtained by renal brushing. Note nuclear enlargement, hyperchromasia, and dense cytoplasm characteristic of low-grade transitional cell carcinoma. (× 1000) **(D)** Extension of this papillary tumor into collecting ducts is a prominent feature. (× 25) **(E)** Carcinoma within collecting ducts replaces normal columnar epithelium. Note elongated appearance of cells with hyperchromatic nuclei. (× 160)

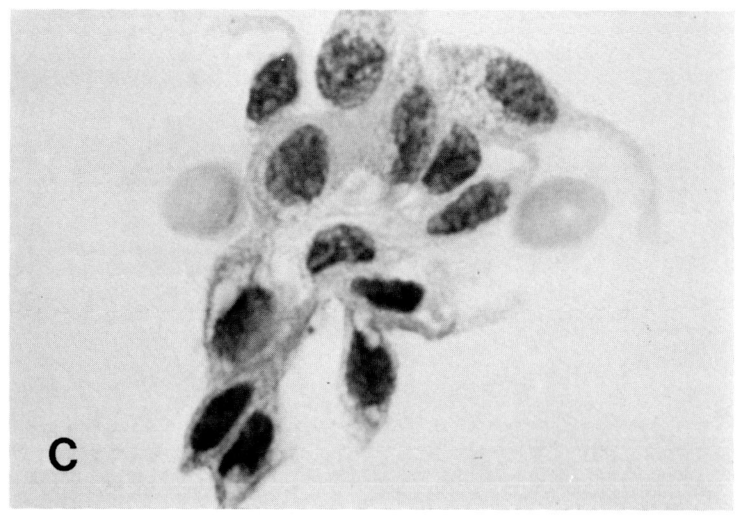

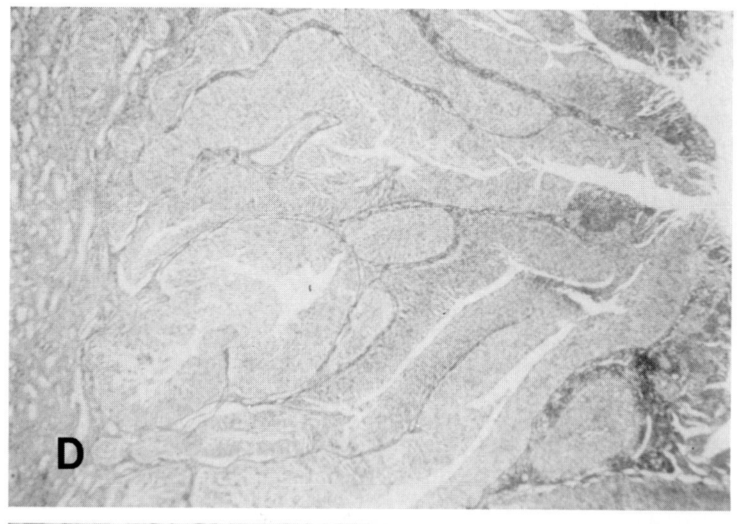

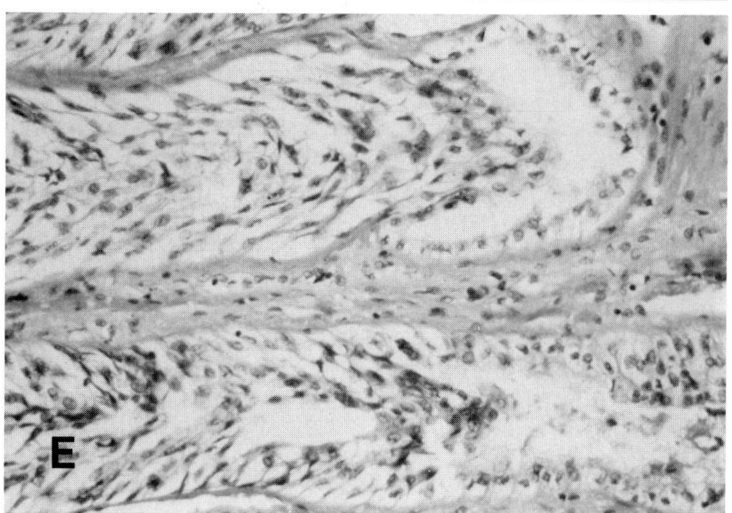

(continued)

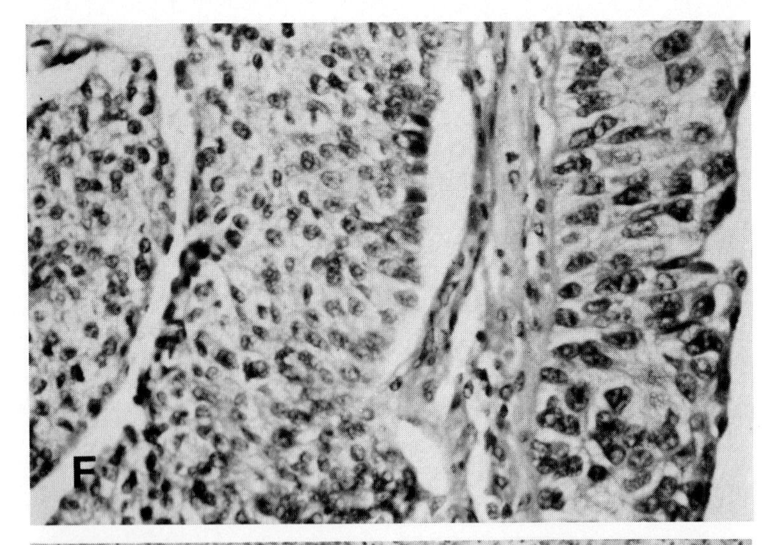

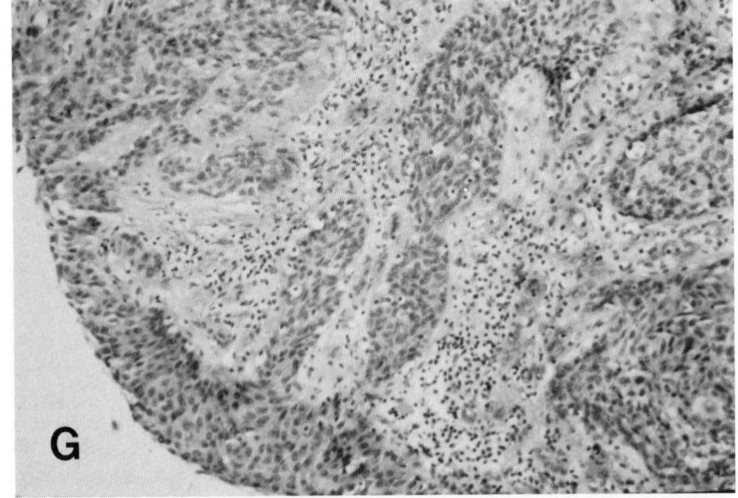

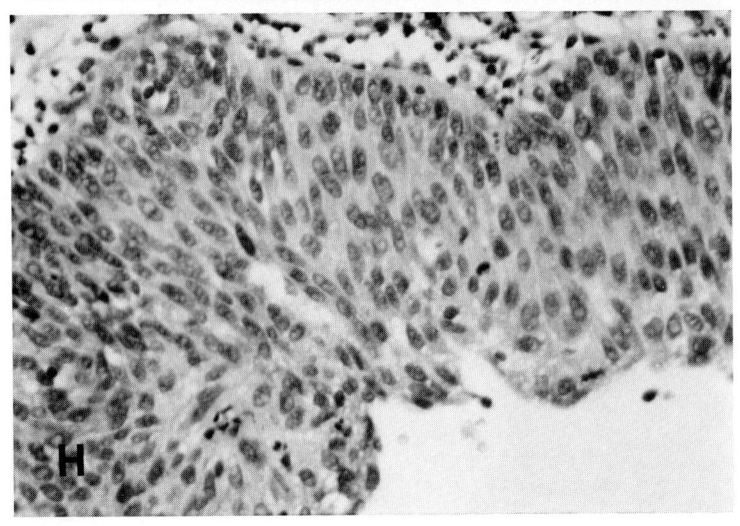

FIG. 17-6F-H *(continued)*
(F) Invasive grade II transitional cell carcinoma (left) has uniform, moderate nuclear enlargement and hyperchromasia with indistinct micronucleoli. Malignant urothelium on surface of papilla (right) has the cytologic characteristics of a grade III carcinoma. Note nuclear pleomorphism, marked enlargement and hyperchromasia and macronucleoli. ((× 250) **(G)** Invasive transitional cell carcinoma from area of (flat) carcinoma *in situ.* (× 100) **(H)** Nonpapillary carcinoma *in situ* involved large portions of the pelvic urothelium. Note full-thickness changes, with hyperplasia, loss of polarity and maturation, and crowding of enlarged, hyperchromatic nuclei. (× 250)

FIG. 17-7
INVASIVE PAPILLARY TRANSITIONAL CELL CARCINOMA OF BLADDER, GRADE III. **(A)** Fragments of malignant urothelium. Note inflammatory background. (Toluidine blue × 400) **(B)** Note nuclear enlargement, irregular indented nuclear membranes, hyperchromasia, and prominent nucleoli, which are features of high-grade transitional cell carcinoma. (Toulidine blue × 1000) **(C)** Fragment containing malignant urothelial cells with hyperchromatic nuclei and faint nucleoli. Note neutrophil in degenerative cytoplasmic vacuole. (× 1000)

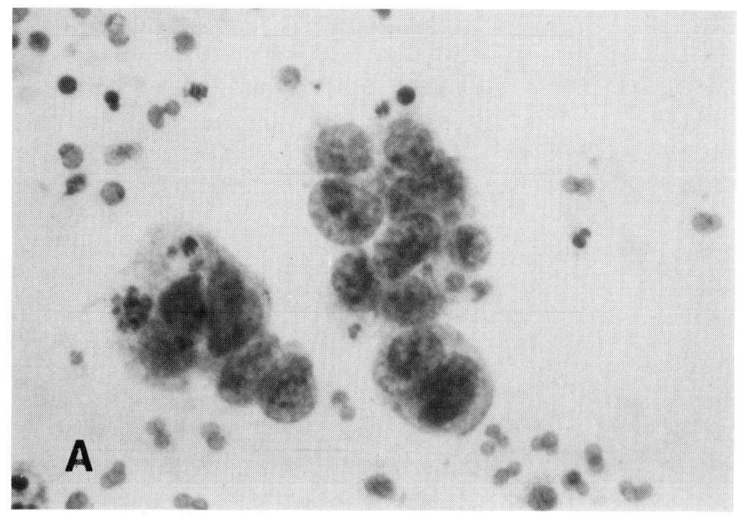

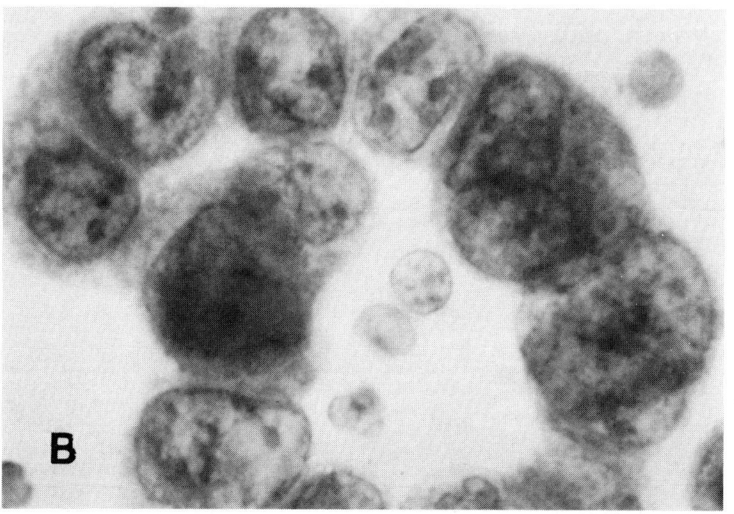

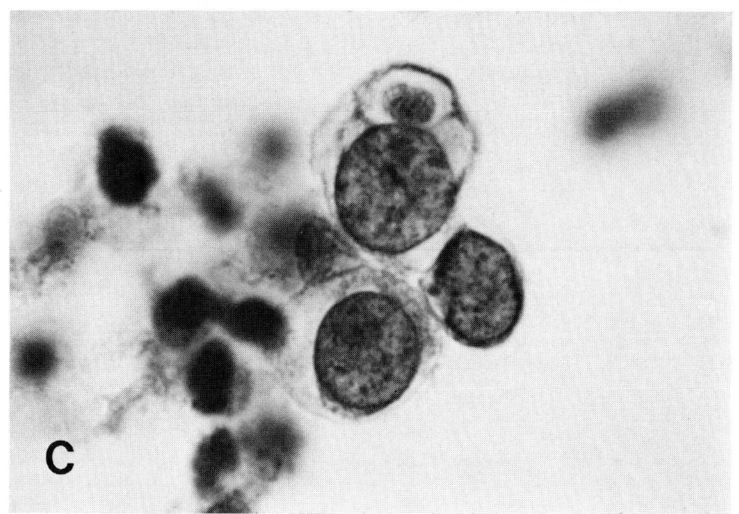

(continued)

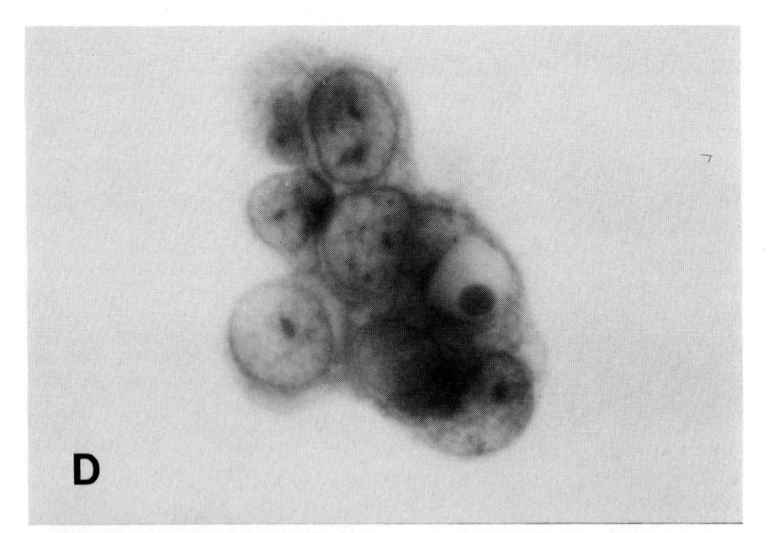

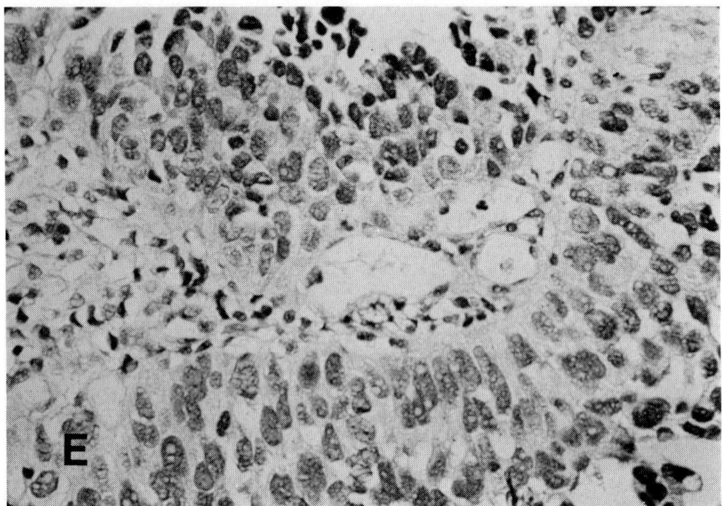

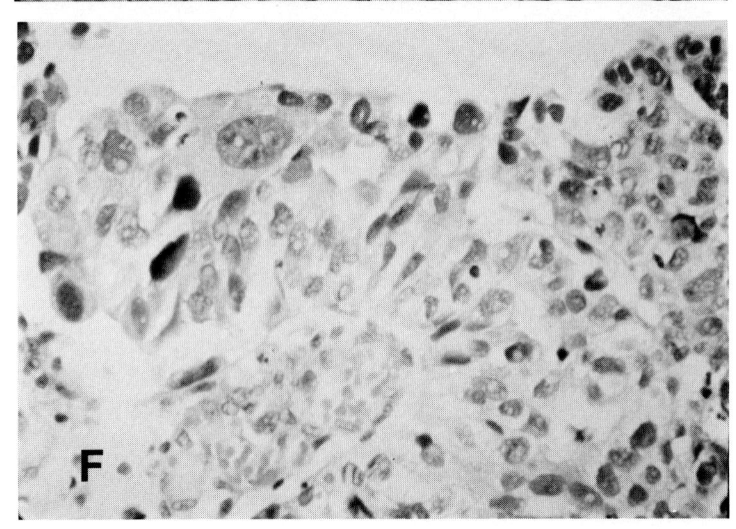

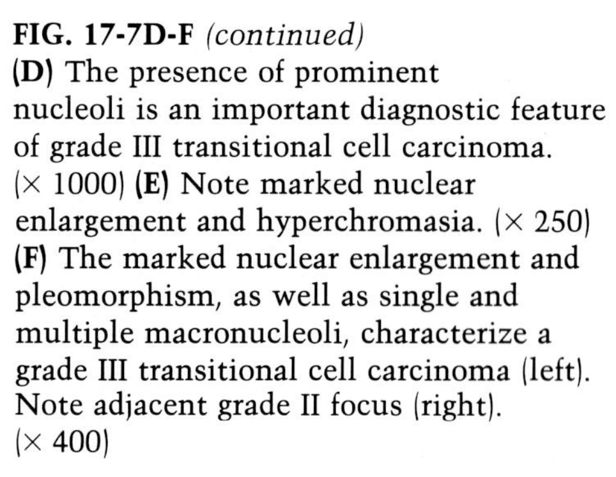

FIG. 17-7D-F *(continued)*
(D) The presence of prominent nucleoli is an important diagnostic feature of grade III transitional cell carcinoma. (× 1000) **(E)** Note marked nuclear enlargement and hyperchromasia. (× 250) **(F)** The marked nuclear enlargement and pleomorphism, as well as single and multiple macronucleoli, characterize a grade III transitional cell carcinoma (left). Note adjacent grade II focus (right). (× 400)

(continued)

FIG. 17-7G-H *(continued)*
(G) Malignant glandular focus. Note cribiform pattern with pseudostratified columnar epithelium lining gland spaces. A malignant glandular or squamous component is not uncommon in high-grade transitional cell carcinoma. (× 250) **(H)** Focus of invasion with a microacinus in the lamina propria. (× 400)

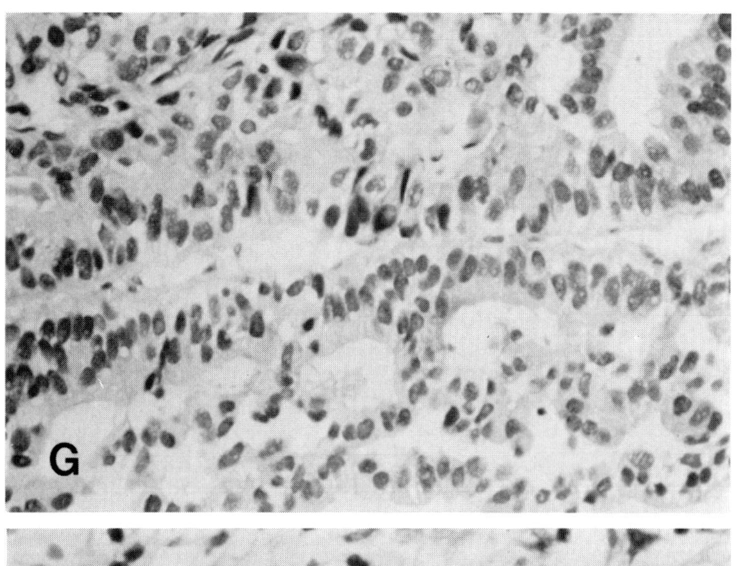

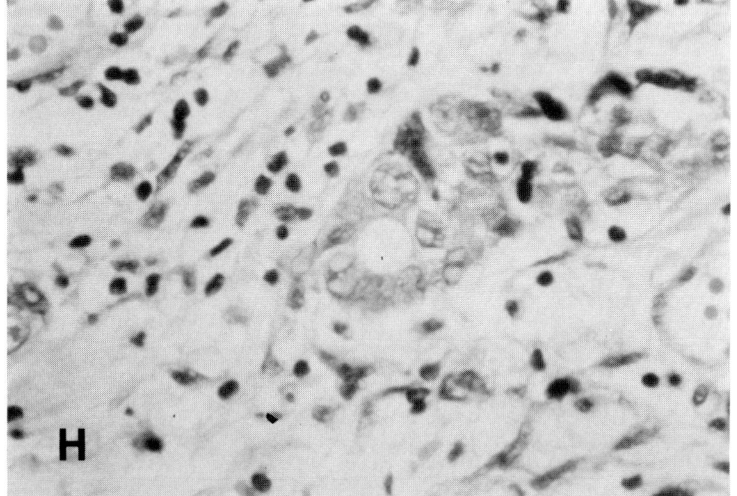

FIG. 17-8
TRANSITIONAL CELL CARCINOMA, GRADE III, IN URINE SEDIMENT. **(A)** Round atypical epithelial fragment with tightly packed nuclei and loss of cytoplasmic maturation. Note scattered atypical cells with increased nuclear: cytoplasmic ratio in background. (× 400)

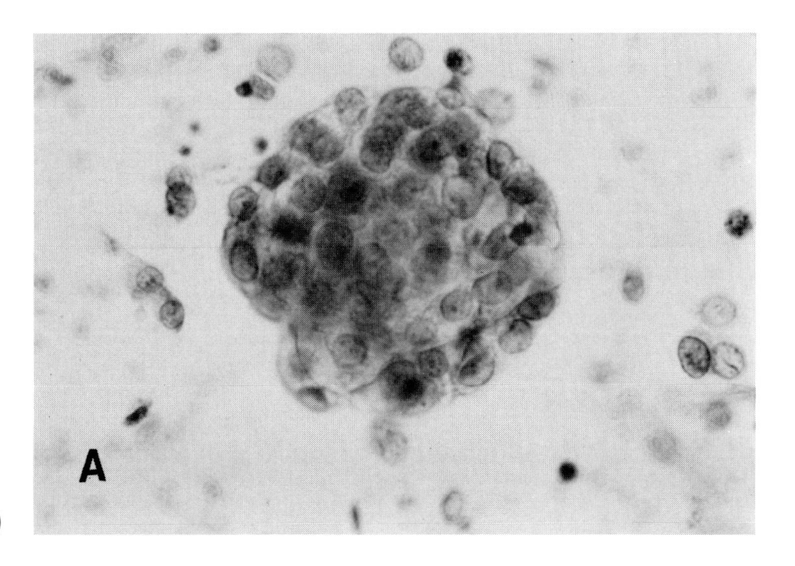

(continued)

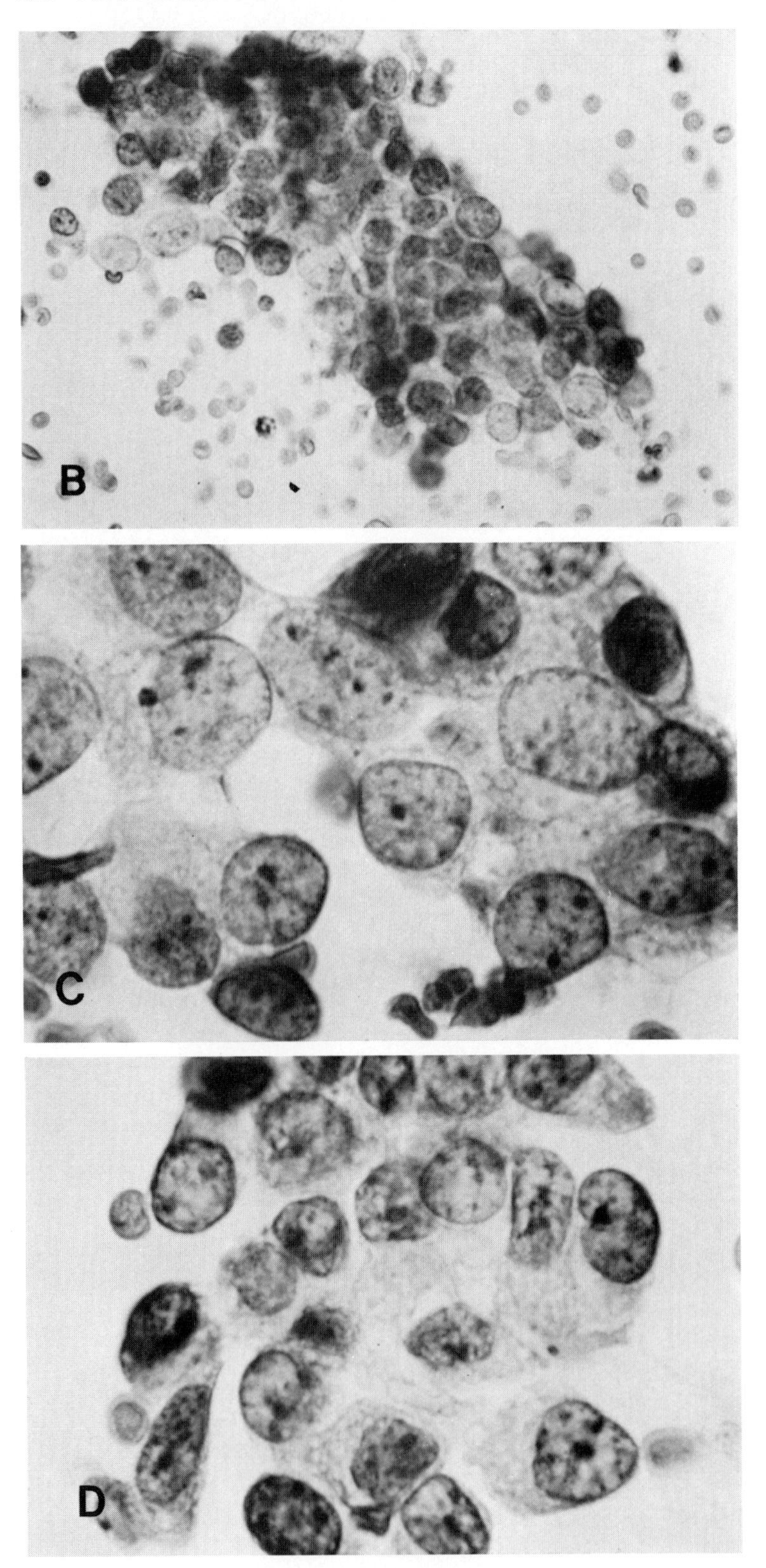

FIG. 17-8B-D *(continued)*
(B) Sheet of malignant cells with anisonucleosis and nuclear moulding. (× 400) **(C)** Syncytial arrangement of malignant cells with nuclear pleomorphism, hyperchromasia, and micronucleoli. (× 1000) **(D)** Note prominent nucleoli characteristic of high-grade transitional cell carcinoma. (× 1000)

FIG. 17-9
TRANSITIONAL CELL CARCINOMA, GRADE III, IN URINE SEDIMENT. **(A)** Malignant cell with increased nuclear: cytoplasmic ratio, prominent nucleolus, and scant dense cytoplasm. (× 1000) **(B)** Several pleomorphic malignant urothelial cells with prominent nucleoli. (× 1000) **(C)** Fragment of malignant epithelium. Note cells with hyperchromatic, irregular, and moulded nuclei with prominent nucleoli and indistinct cytoplasm. (× 1000)

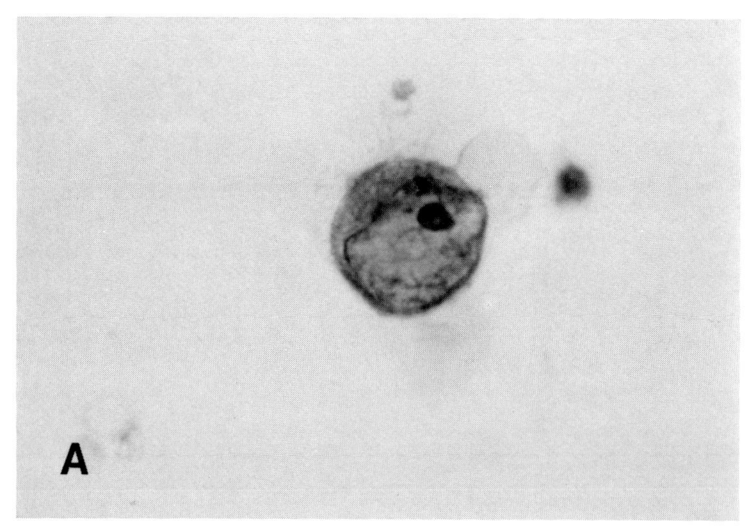

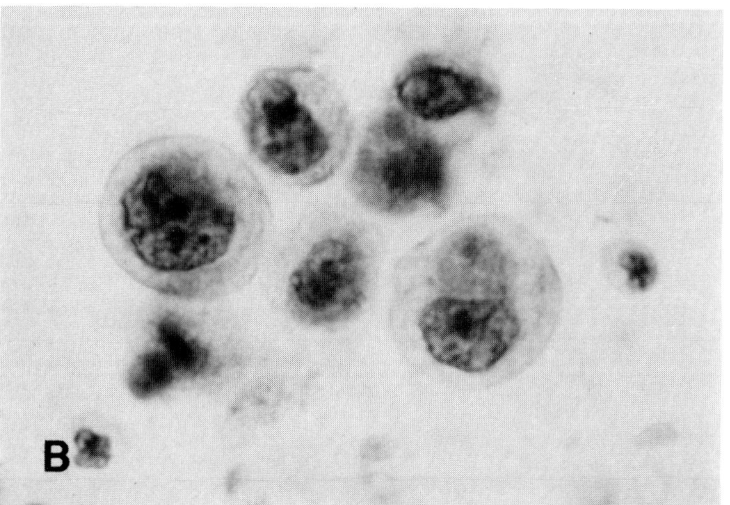

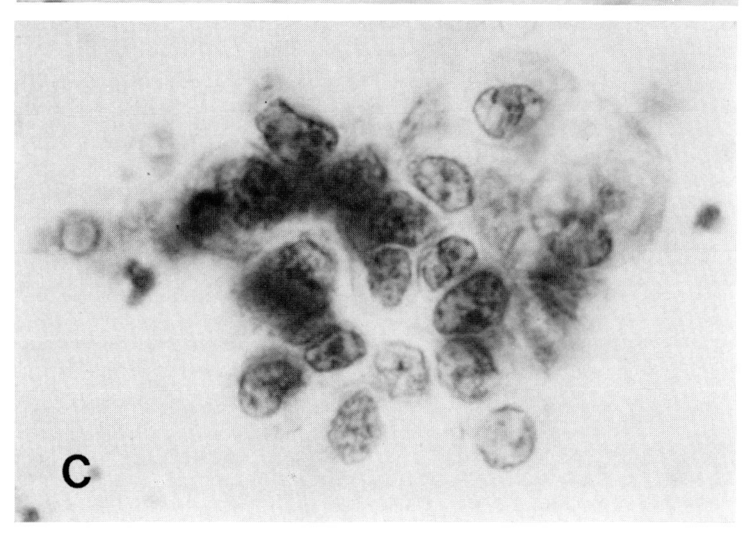

(continued)

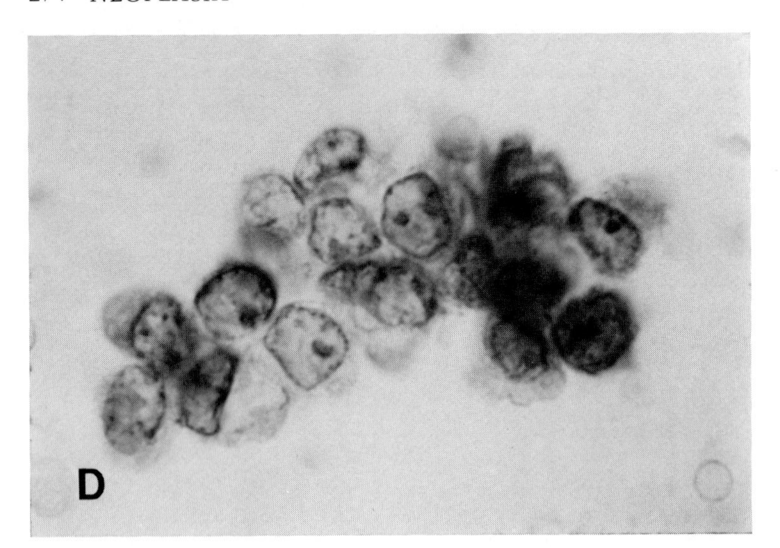

FIG. 17-9D *(continued)*
(D) Syncytial arrangement of malignant cells and prominent nucleoli are characteristic of high-grade transitional cell carcinoma. (× 1000)

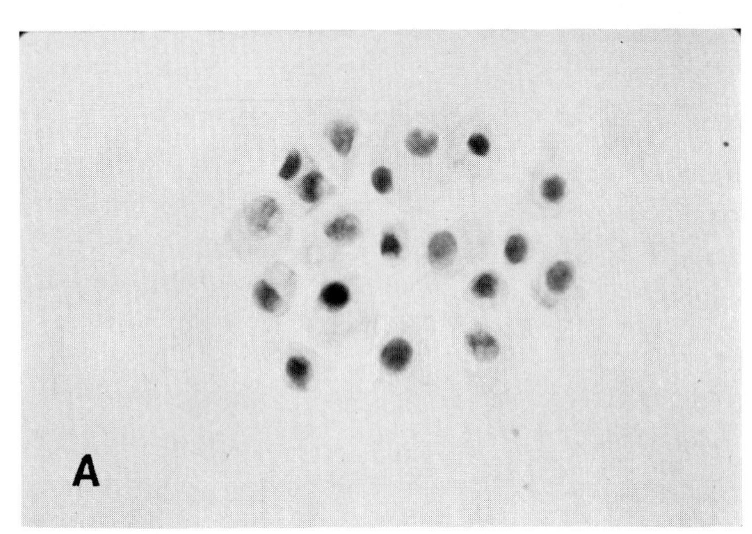

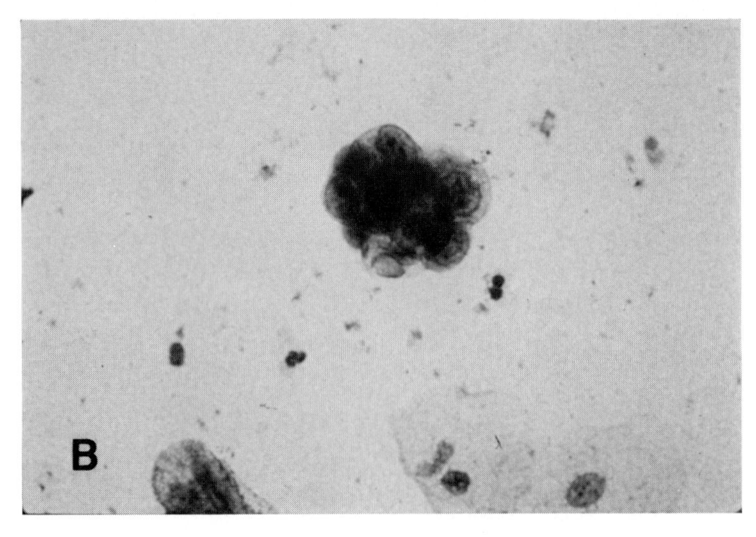

FIG. 17-10
PAPILLARY ADENOCARCINOMA OF RENAL PELVIS. **(A)** Loose cluster of atypical epithelial cells in urine sediment. Note polygonal shape of cells and slightly enlarged hyperchromatic nuclei. (× 400) **(B)** Three-dimensional atypical epithelial fragment. Note mild nuclear pleomorphism and hyperchromasia. (Membrane filter × 400)

(continued)

FIG. 17-10C-E *(continued)*
(C) Three-dimensional glandular fragment. Note nuclear enlargement and hyperchromasia. Cytoplasmic mucin was not identified. (× 400) **(D)** The papillary tumor (arrow) within the renal pelvis does not invade renal parenchyma. Calyceal deformities, retraction of papillae, and overlying cortical scars indicate chronic pyelonephritis. (Gross) **(E)** The tumor has a prominent papillary and glandular configuration. (× 40)

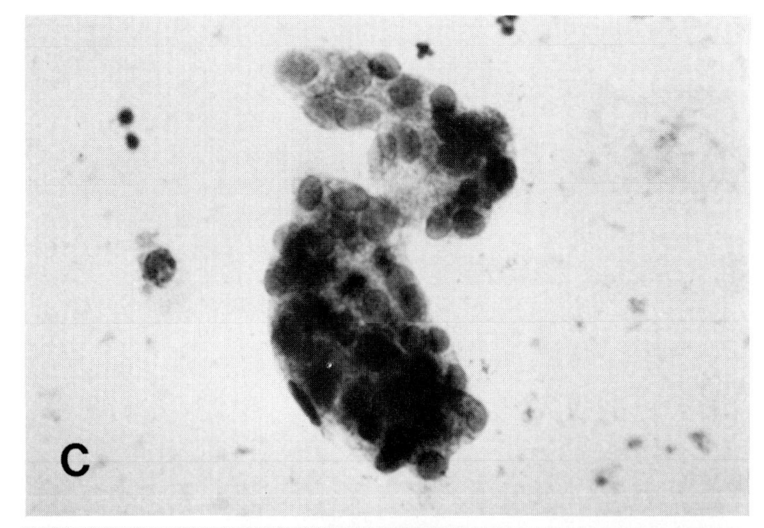

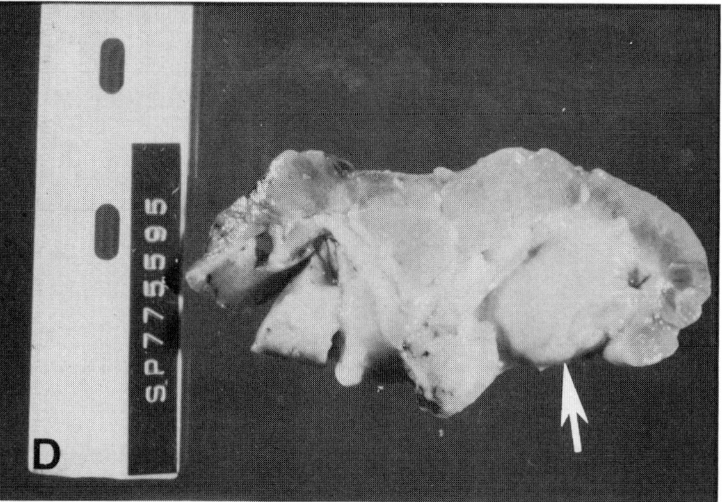

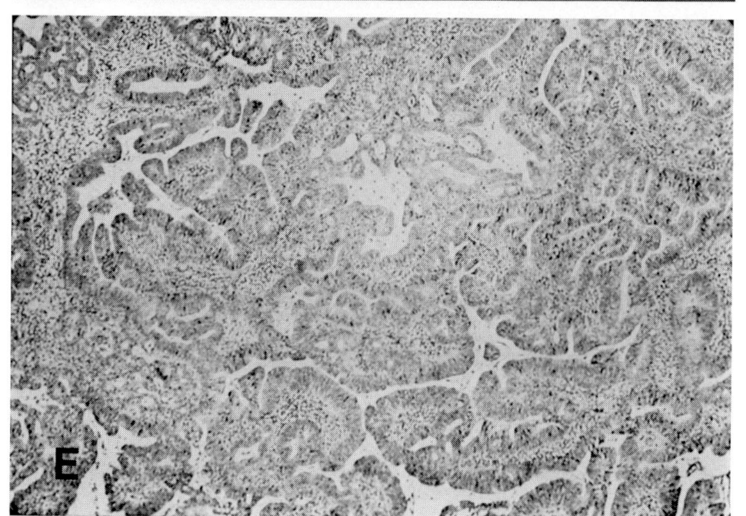

(continued)

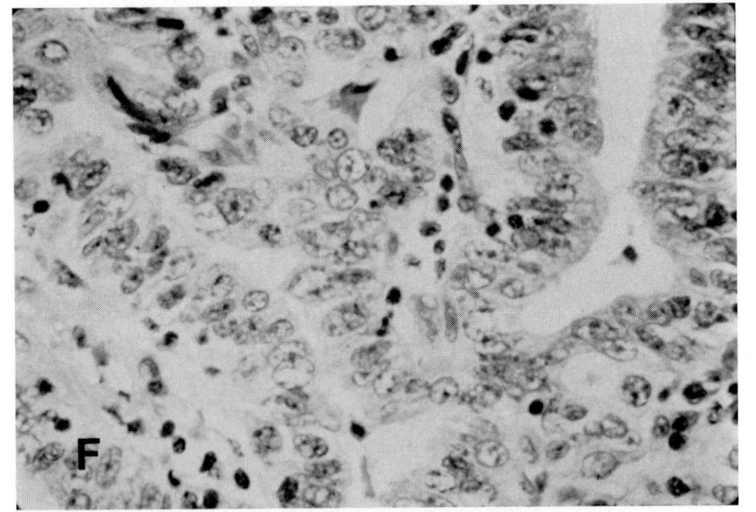

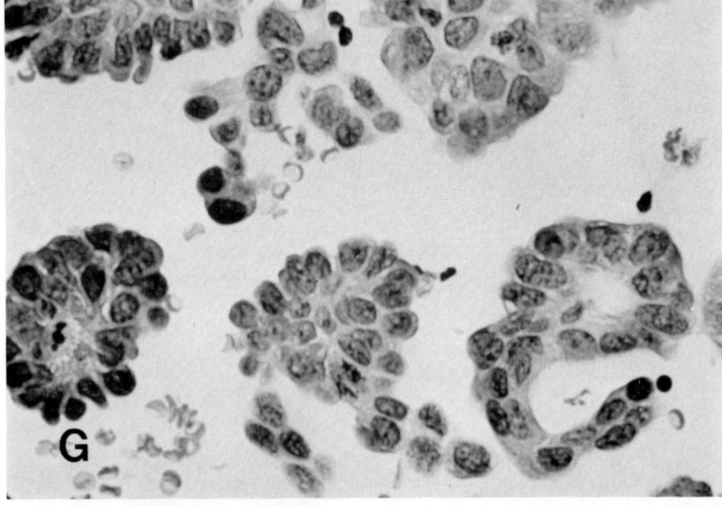

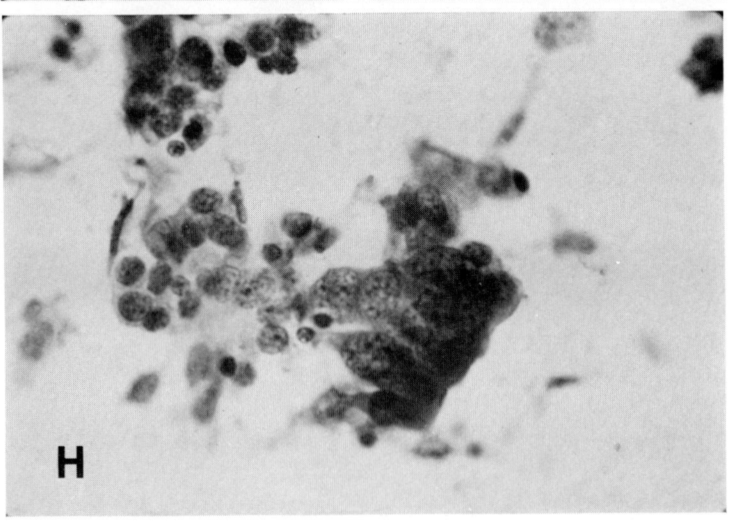

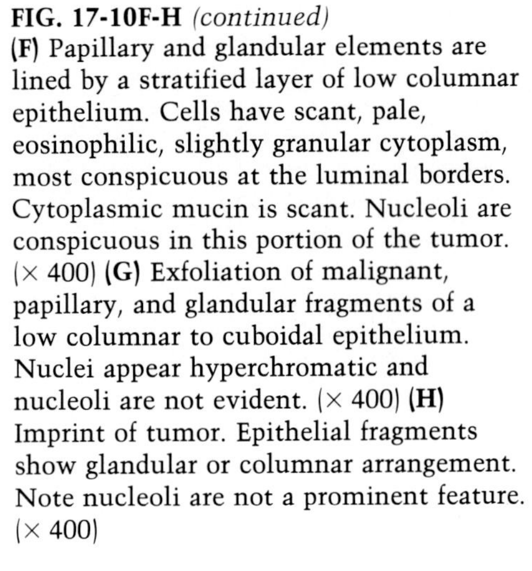

FIG. 17-10F-H *(continued)*
(F) Papillary and glandular elements are lined by a stratified layer of low columnar epithelium. Cells have scant, pale, eosinophilic, slightly granular cytoplasm, most conspicuous at the luminal borders. Cytoplasmic mucin is scant. Nucleoli are conspicuous in this portion of the tumor. (× 400) **(G)** Exfoliation of malignant, papillary, and glandular fragments of a low columnar to cuboidal epithelium. Nuclei appear hyperchromatic and nucleoli are not evident. (× 400) **(H)** Imprint of tumor. Epithelial fragments show glandular or columnar arrangement. Note nucleoli are not a prominent feature. (× 400)

FIG. 17-11
PROSTATIC CARCINOMA. **(A)** Acinar cluster of malignant cells. Nuclear crowding, enlargement, mild hyperchromasia, micronucleoli, and indistinct cytoplasm characterize this prostatic adenocarcinoma. (× 400) Prostatic fine-needle aspiration. **(B)** Malignant cells derived from adenocarcinoma. Note the syncytial arrangement or lack of cytoplasmic borders found in this acinar structure. (× 400) Prostatic fine-needle aspiration. **(C)** Adenocarcinoma in urine sediment. Note nuclear enlargement, irregularity, hyperchromasia, prominent nucleoli, and abundant finely granular cytoplasm with indistinct cell borders. (× 1000)

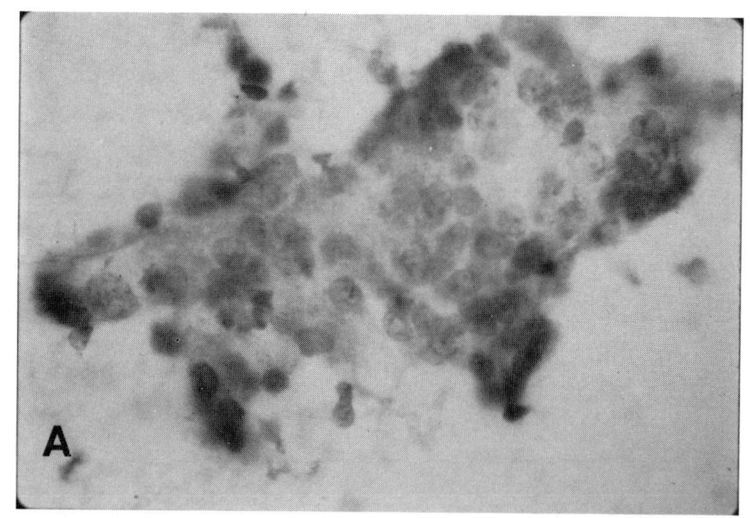

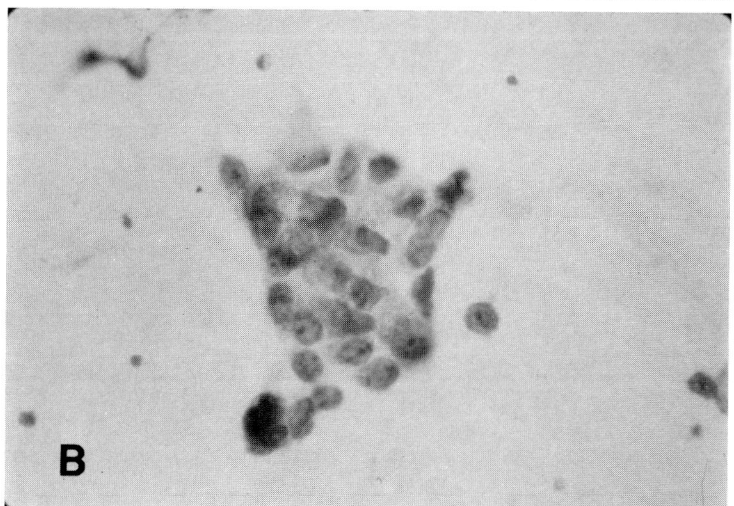

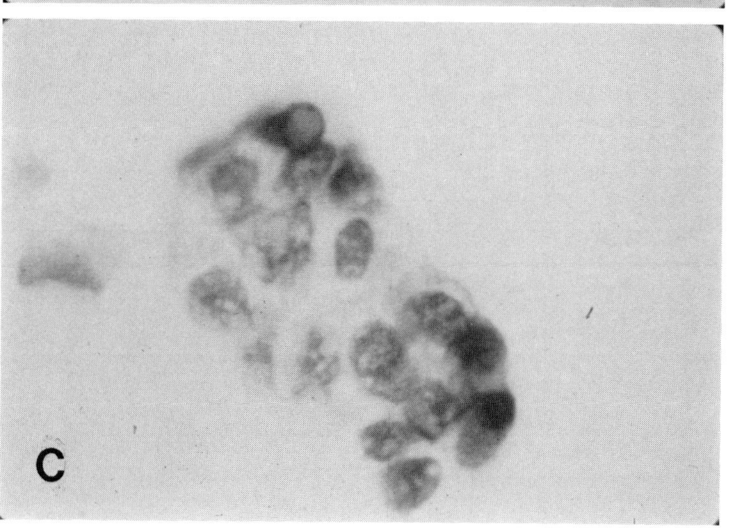

(continued)

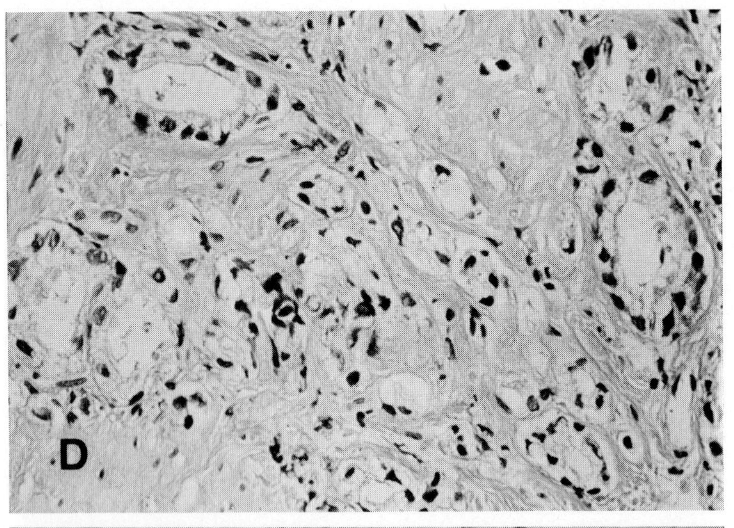

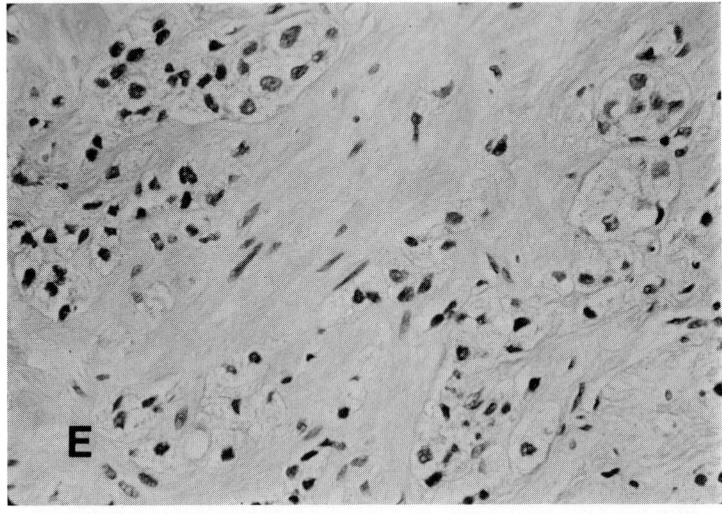

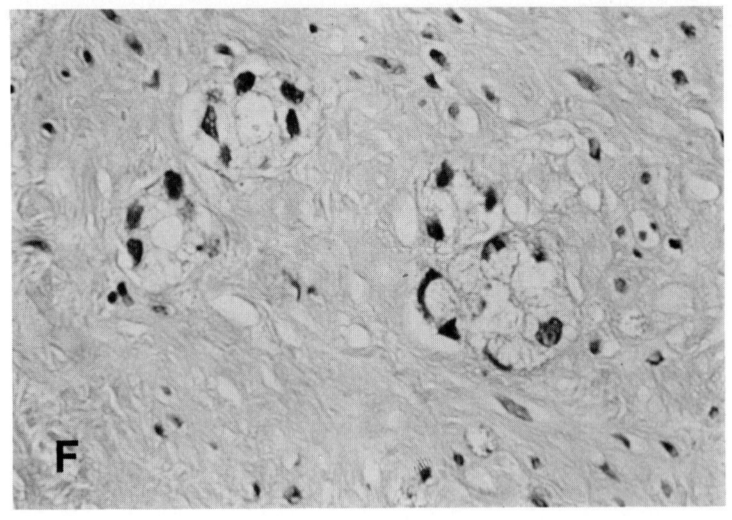

FIG. 17-11D-F *(continued)*
(D) Needle biopsy of prostate. Focus of well-differentiated prostatic adenocarcinoma. Small acini that haphazardly infiltrate the stroma are focally back to back and are lined by a single layer of cells with enlarged, hyperchromatic nuclei. (× 250) **(E)** Focus of prostatic carcinoma lacking glandular differentiation. Tumor infiltrates as nests and cords of cells. Note prominent nucleoli. (× 250) **(F)** Microacini in prostatic stroma. Note hyperchromatic nuclei and abundant vacuolated cytoplasm with indistinct cell borders. (× 400)

FIG. 17-12
SQUAMOUS CELL CARCINOMA OF BLADDER. **(A)** A large bulky tumor infiltrates the bladder wall. There is no mucosal papillary component. (Gross) **(B)** Nests of keratinizing squamous cell carcinoma (grade II) infiltrate the lamina propria. Overlying the tumor is a thickened metaplastic squamous epithelium. (× 100)

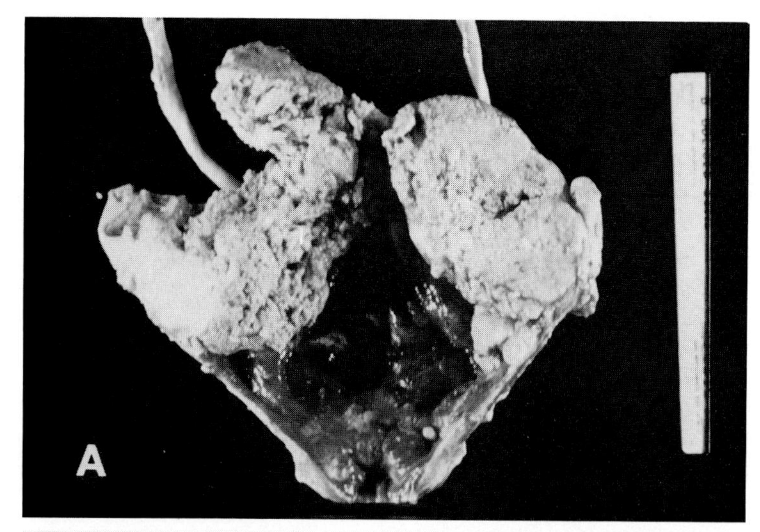

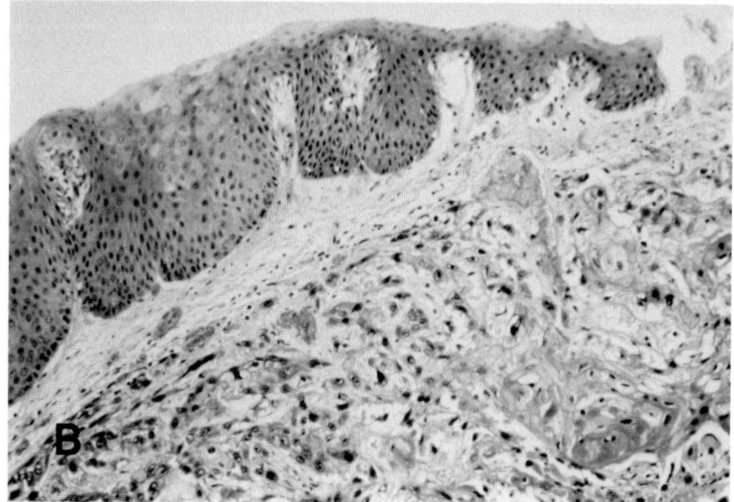

FIG. 17-13
METASTATIC SQUAMOUS CELL CARCINOMA FROM UTERINE CERVIX IN URINE SEDIMENT. **(A)** Note the highly atypical keratinized cell and necrotic cellular debris in background. (× 400) **(B)** Degenerated malignant squamous cells with pyknotic nuclei and orangeophilic cytoplasm. (× 1000) **(C)** Malignant spindle cell with necrotic background. Note elongated pyknotic nucleus. (× 1000) **(D)** Tadpole malignant squamous cell with abundant wavy orangeophilic cytoplasm. (× 1000) **(E)** Cluster of malignant squamous cells. Note distinct cell borders. (× 1000) **(F)** Better-preserved malignant squamous cells with nuclear enlargement, marked hyperchromasia, chromatin clearing, and abundant wavy orangeophilic cytoplasm. (× 1000).

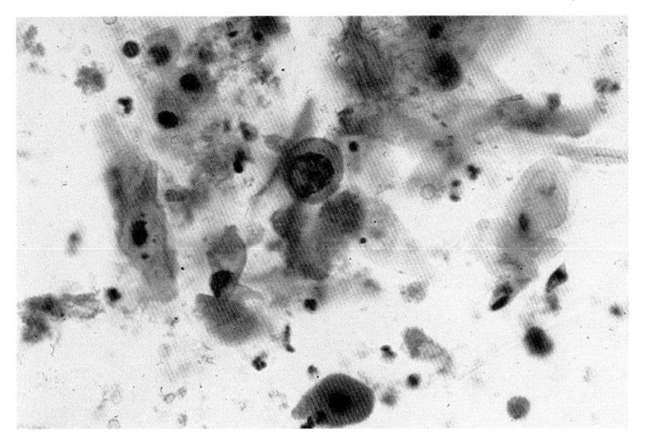

FIG. 17-13A

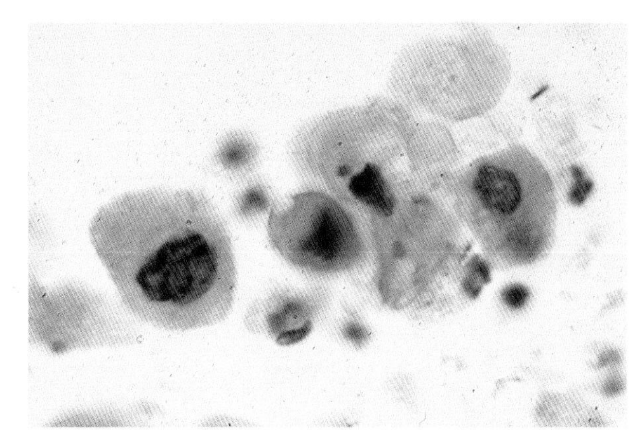

FIG. 17-13B

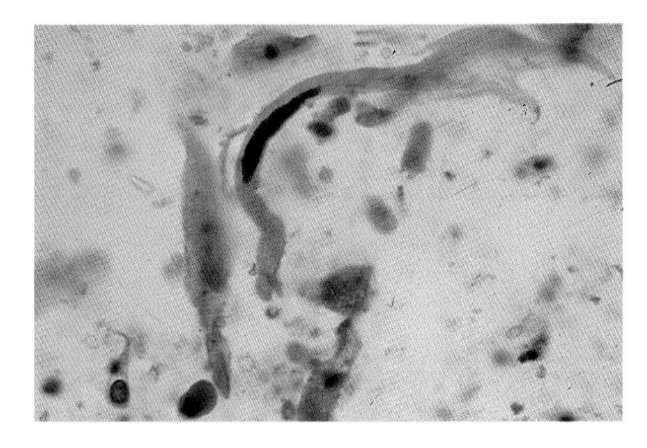

FIG. 17-13C

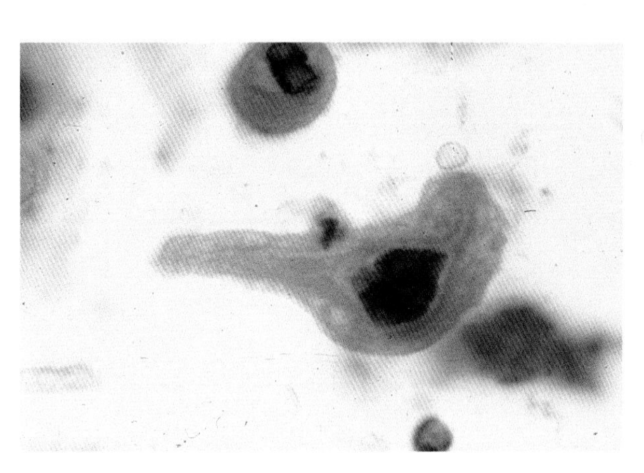

FIG. 17-13D

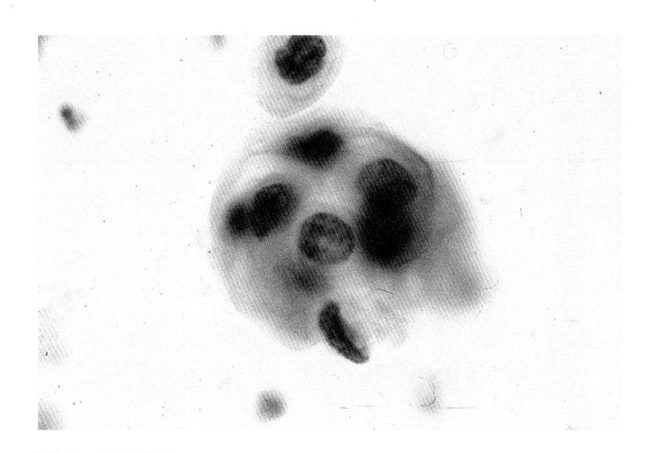

FIG. 17-13E

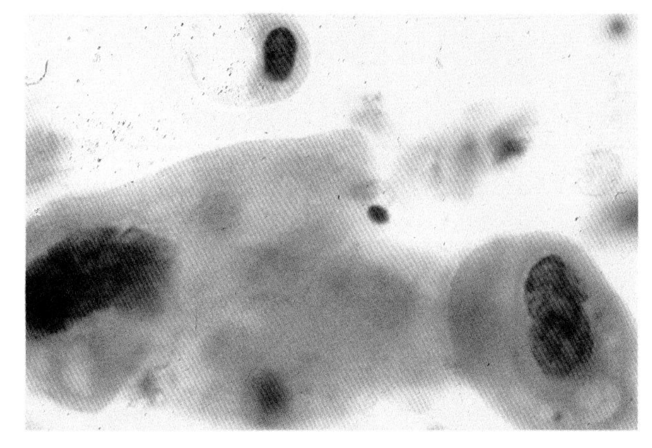

FIG. 17-13F

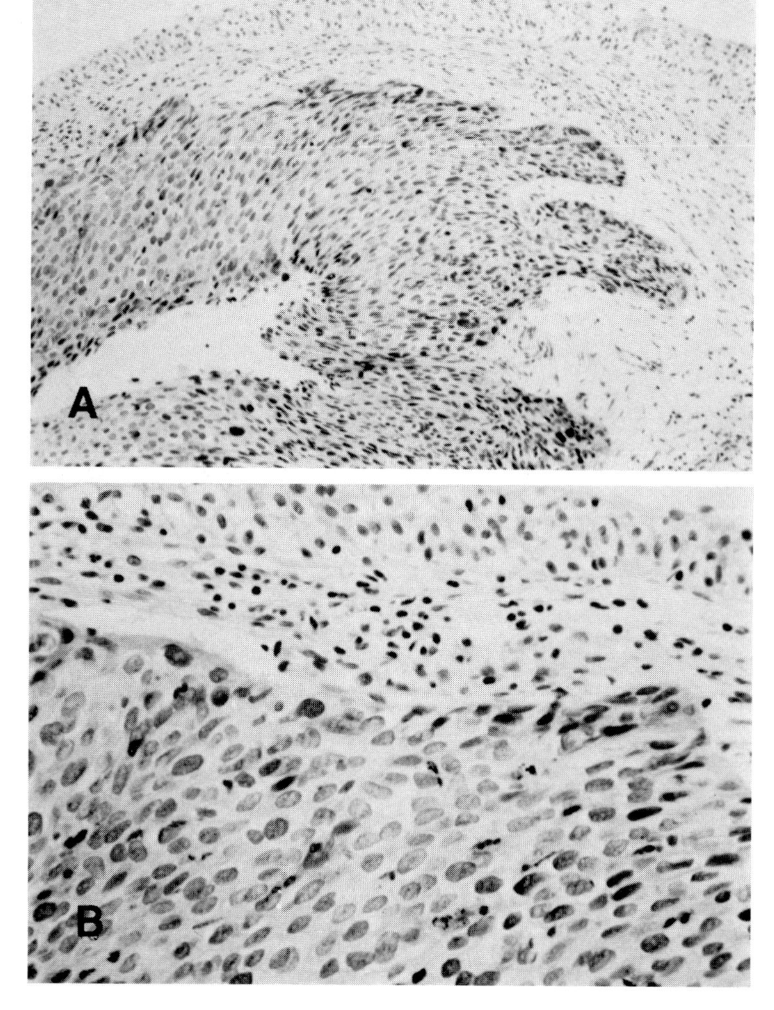

FIG. 17-14
METASTATIC SQUAMOUS CELL CARCINOMA FROM UTERINE CERVIX. **(A)** A focus of large cell nonkeratinizing squamous cell carcinoma in the lamina propria of the bladder. (× 100) **(B)** There is no evidence of keratinization; however, intercellular bridges are present. Note nuclear pyknosis and rare micronucleoli. (× 250)

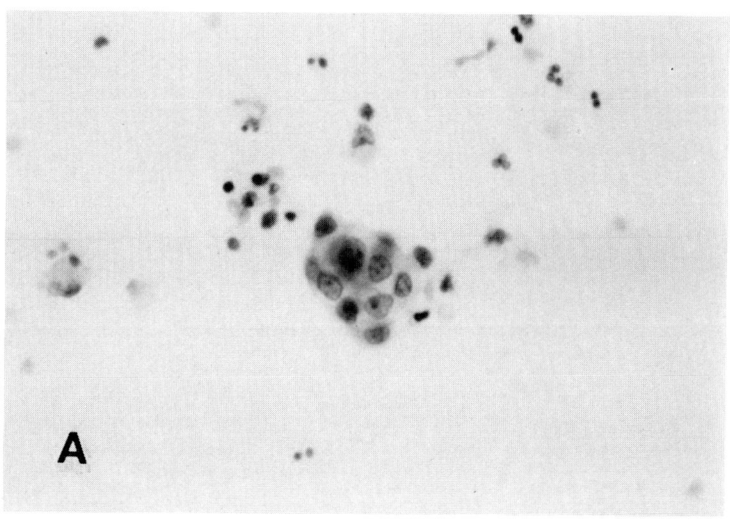

FIG. 17-15
METASTATIC ADENOCARCINOMA OF BREAST ORIGIN. Urine sediment with atypical epithelial fragment and scattered inflammatory cells. **(A)** (× 400)

(continued)

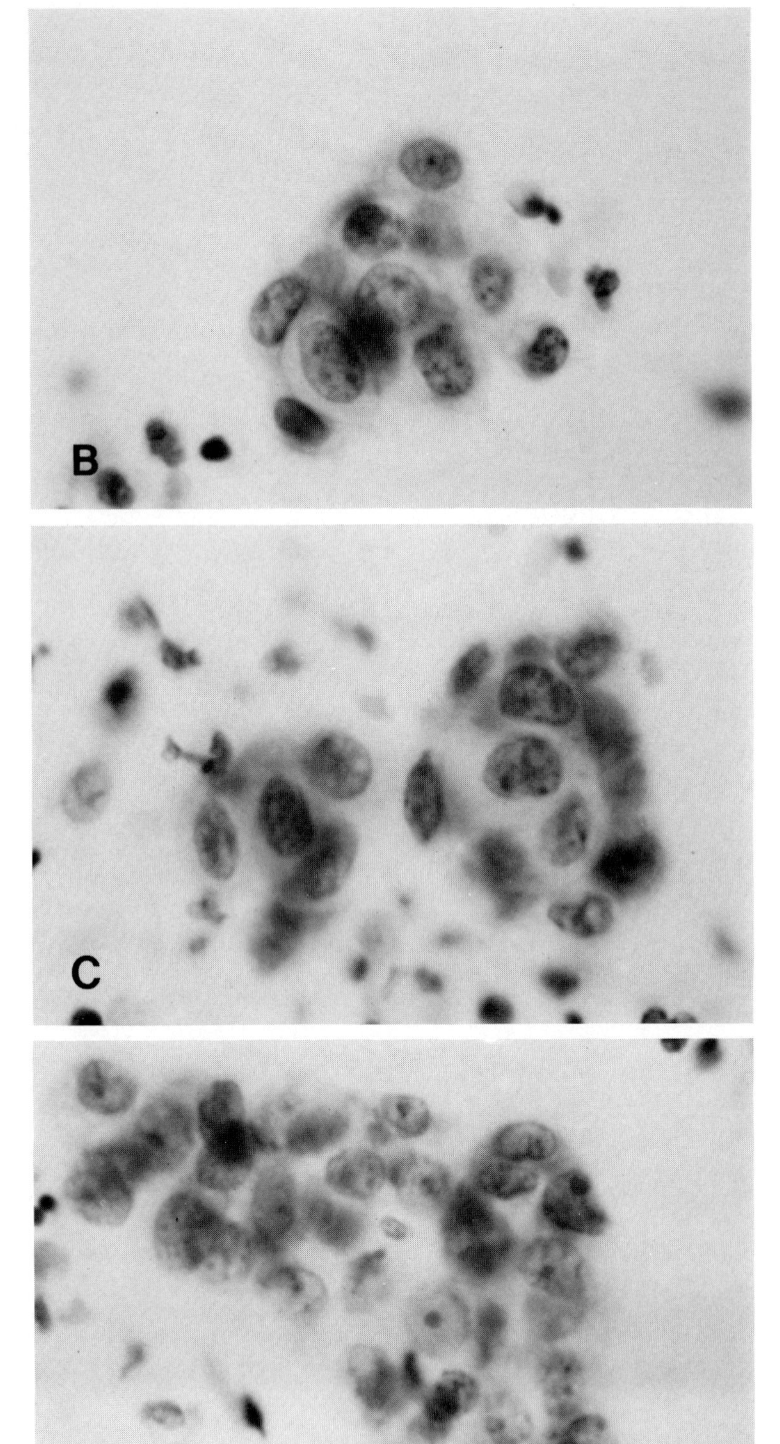

FIG. 17-15B-D *(continued)*
(B) Note acinar configuration of cells with mild nuclear enlargement, hyperchromasia, prominent nucleoli, and finely granular cytoplasm. (× 1000) Acinar fragments in urine sediment. **(C)** (× 1000). **(D)** Note prominent nucleoli. (× 1000)

(continued)

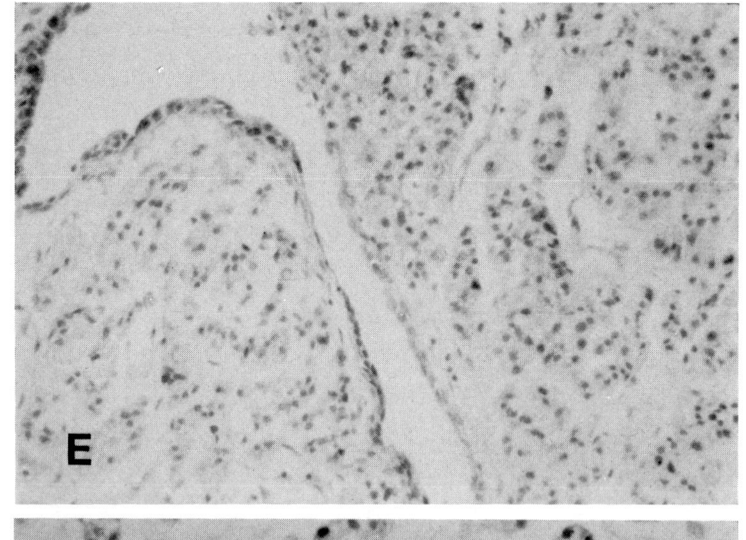

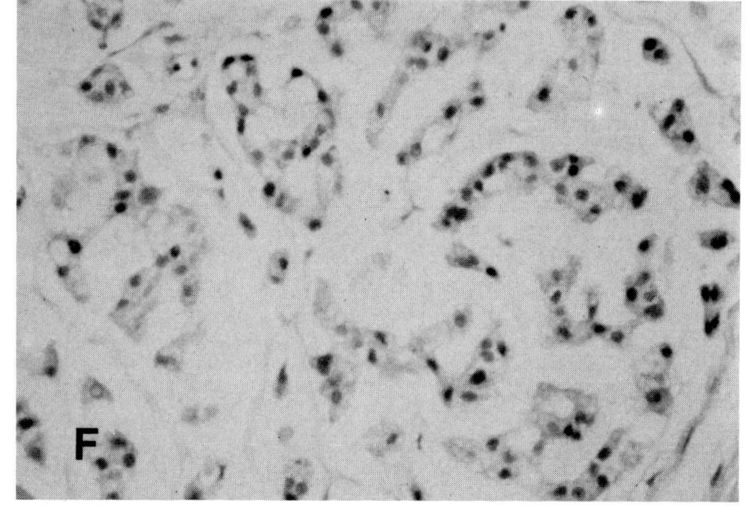

FIG. 17-15E-F *(continued)*
(E) The lamina propria of the ureter is infiltrated by tumor growing in nests and cords. (× 160) **(F)** Abundant intracytoplasmic mucin gives impression of a cribriform pattern. Extracellular mucin is present. (× 250)

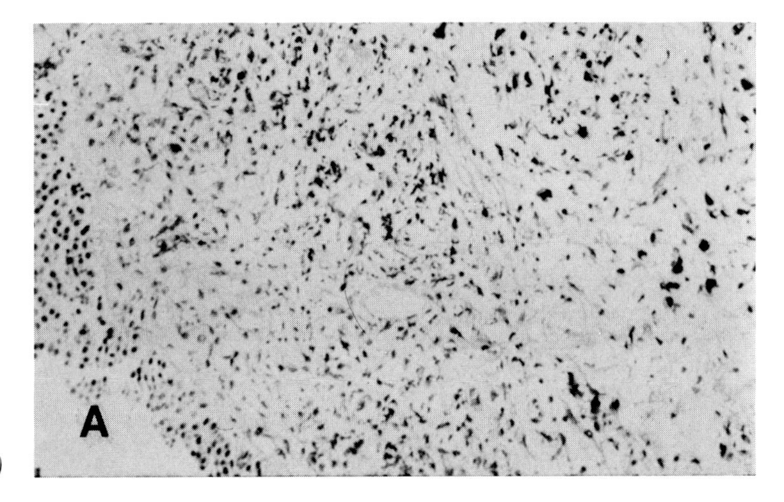

FIG. 17-16
METASTATIC SIGNET CELL CARCINOMA OF COLONIC ORIGIN.
(A) The lamina propria contains a distinct infiltrate that, at low magnification, appears inflammatory. Note individually scattered cells with large, hyperchromatic nuclei. (× 100)

(continued)

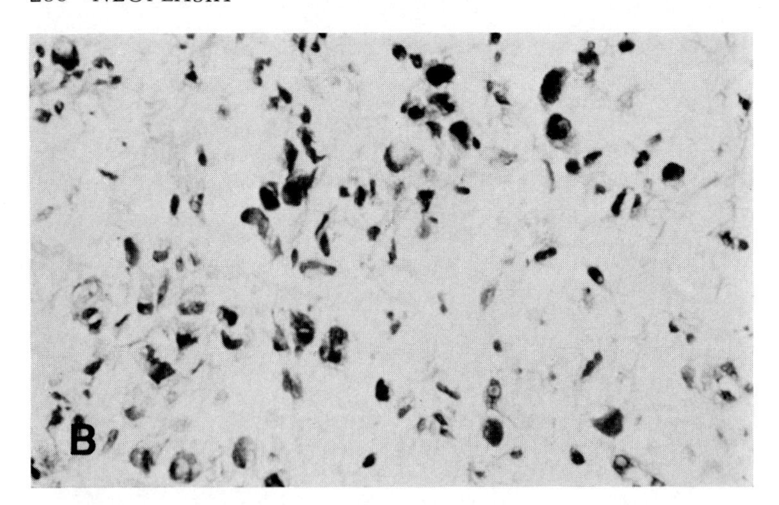

FIG. 17-16B *(continued)*
(B) Cells with large, hyperchromatic nuclei and abnormal nuclear: cytoplasmic ratio are prominent. Occasional nuclei are displaced to the cytoplasmic borders by intracellular mucin, characteristic of signet cell carcinoma. (× 250)

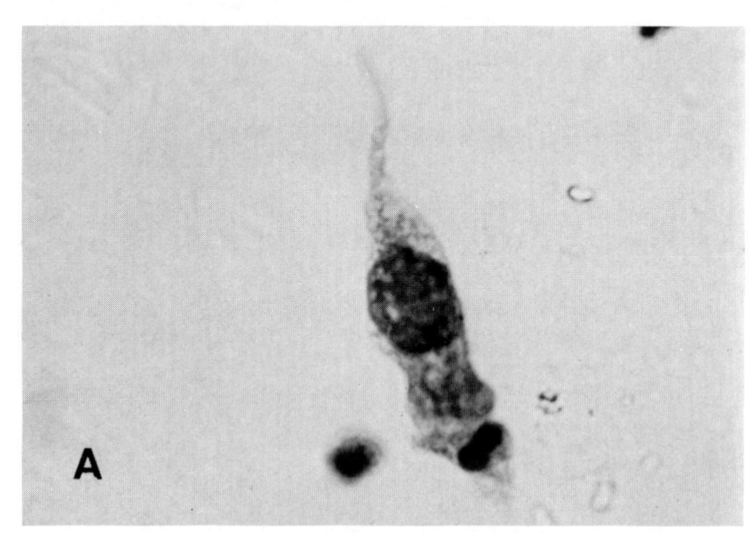

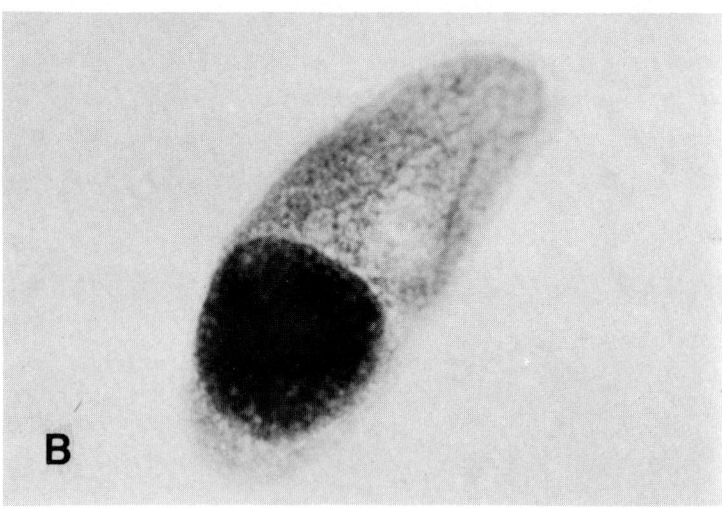

FIG. 17-17
METASTATIC COLONIC ADENOCARCINOMA IN URINE SEDIMENT. **(A)** Columnar-shaped cell with nuclear enlargement, hyperchromasia, faint nucleolus, and finely vacuolated cytoplasm. (× 1000) **(B)** Note eccentric, hyperchromatic nucleus containing nucleoli and abundant foamy cytoplasm. (× 1000)

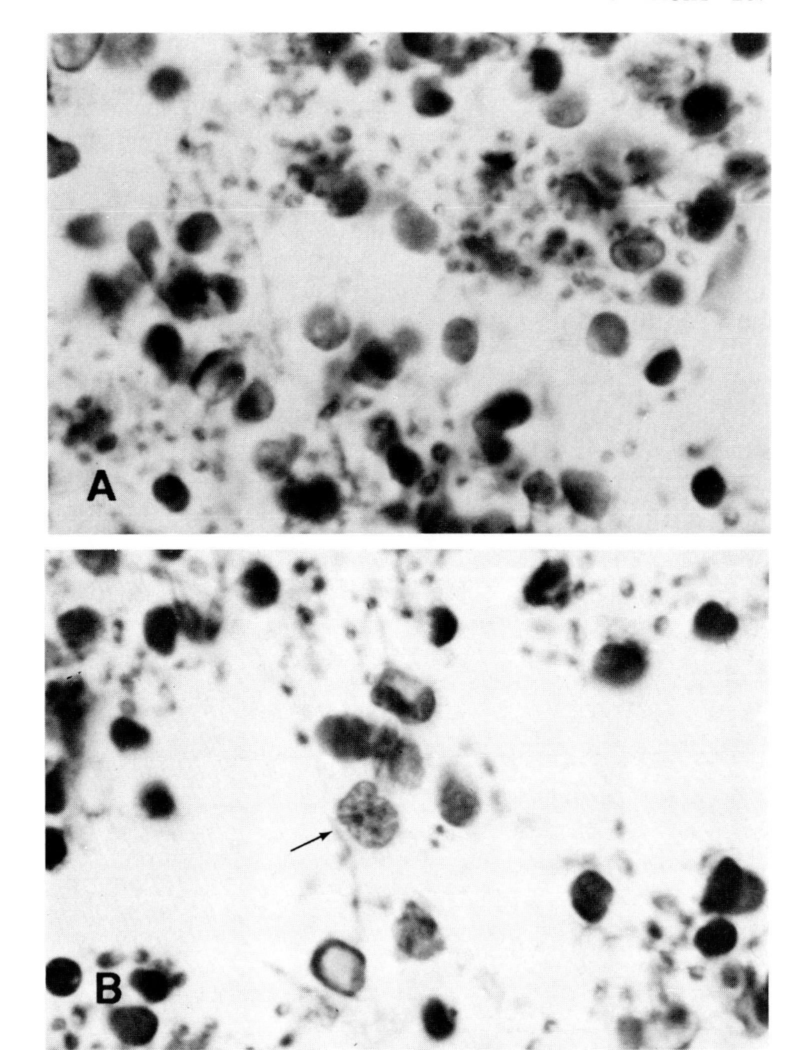

FIG. 17-18
ATYPICAL LYMPHOCYTES IN URINE SEDIMENT. **(A)** Numerous lymphocytes and fungi. (× 1000) **(B)** Note immature lymphocytes with nuclear membrane indentation (arrow) and micronucleoli. Urine obtained from patient suspected of having ureteral involvement by lymphocytic lymphoma. (× 1000)

INDEX

Page numbers for photomicrographs are in italics; tables are indicated by "t" following the page number.